Astrophysics and Space Science Library

Volume 472

The Astrophysics and Space Science Library is a series of high-level monographs and edited volumes covering a broad range of subjects in Astrophysics, Astronomy, Cosmology, and Space Science. The authors are distinguished specialists with international reputations in their fields of expertise. Each title is carefully supervised and aims to provide an in-depth understanding by offering detailed background and the results of state-of-the-art research. The subjects are placed in the broader context of related disciplines such as Engineering, Computer Science, Environmental Science, and Nuclear and Particle Physics. The ASSL series offers a reliable resource for scientific professional researchers and advanced graduate students.

Eleni Vardoulaki • Marta Dembska •
Alexander Drabent • Matthias Hoeft
Editors

Data-Intensive Radio Astronomy

Bringing Astrophysics to the Exabyte Era

Springer

Editors
Eleni Vardoulaki
Thüringer Landessternwarte
Tautenburg, Germany

IAASARS
National Observatory of Athens
Athens, Greece

Alexander Drabent
Thüringer Landessternwarte Tautenburg
Tautenburg, Germany

Marta Dembska
Institute of Data Science
German Aerospace Center (DLR)
Jena, Germany

Matthias Hoeft
Thüringer Landessternwarte Tautenburg
Tautenburg, Germany

ISSN 0067-0057 ISSN 2214-7985 (electronic)
Astrophysics and Space Science Library
ISBN 978-3-031-58467-1 ISBN 978-3-031-58468-8 (eBook)
https://doi.org/10.1007/978-3-031-58468-8

Cover illustration: Credit: Ziyue Ma (马梓越), Institute of International and Comparative Education, Faculty of Education, East China Normal University, Shanghai, China

This Springer imprint is published by the registered company Springer Nature Switzerland AG
The registered company address is: Gewerbestrasse 11, 6330 Cham, Switzerland

Preface

Radio astronomy is irreversibly moving towards the exabyte era. In the advent of all-sky radio observations, having efficient tools and methods to manage the large-data volume generated is imperative. A huge progress has already been made, through innovation and expertise in research, and large-data management has been the topic of discussion within the community for decades, with improved interferometers (unprecedented sensitivity and resolution, number of baselines, broad-band receivers). However, the challenges don't disappear and only grow. Thus the community is actively searching for efficient solutions which will take astrophysics to the exabyte era. Current radio telescopes (e.g. LOFAR, MeerKAT, ASKAP) have identified the difficulties of obtaining hundreds of Gbits per second of data and processing them through pipelines in order to produce science ready products, as well as providing efficient methods to the community to (re)use these data, and having efficient data storage for archiving.

Our vision for this special volume on Data-Intensive Radio Astronomy is based on a data-centric approach, addressing the data lifecycle to create a coherent flow of the content. Bringing together different communities, e.g. astrophysics, high-energy physics, data science, and computer science, our aim is to provide knowledge in the topic of current tools and ideas for the future development of data management. Diverse input and various approaches are meant to tackle the challenges that current and future instruments, including radio telescopes, are facing.

This special volume represents the vast amount of expertise that needs to come together to write a book about data-intensive radio astronomy. The book is separated in four parts that describe the data cycle, with a general introduction (Chap. 1) on radio astronomy and why we observe the radio sky, as well as the current and upcoming era of exabyte radio astronomy. Part I on 'Data Creation, Storage and Archives' (Chaps. 2–4) describes how the data are acquired using radio telescopes, and the need for long-term data archives and high-performance infrastructure to store data before they can be pre-processed. Part II on 'Data Processing' (Chaps. 5–7) narrates the challenges of processing large amounts of data efficiently and how different radio facilities around the world face these challenges, employing modern co-design and software architectures. Part III on

'Post-processing and Data Analysis' (Chaps. 8–11) describes the broad science that can be done using the data in hand. The data need to be processed depending on the scientific question, in order to create science-ready products. It further highlights how machine learning and artificial intelligence algorithms aid in the analysis of large data volumes, alongside modern techniques of visualisation using virtual reality and planetaria. Part IV on 'Data Access and Reuse' (Chaps. 12–14) explains the effort of radio astronomy and other disciplines to make data and science-ready products available for humans and machines, following the FAIR principles of Findability, Accessibility, Interoperability, and Reusability. It highlights the work done for the Virtual Observatory, as well as the need for proper metadata and data management. Finally, the epilogue (Chap. 15) gives a quick summary and outlook for the future of data-intensive radio astronomy, also keeping in mind the efforts of the community for sustainability in this data-intensive radio era.

Jena, Germany
December 2023

Eleni Vardoulaki
Marta Dembska
Alexander Drabent
Matthias Hoeft

Acknowledgements

Eleni Vardoulaki acknowledges support by the Carl Zeiss Stiftung with the project code KODAR. Alexander Drabent acknowledges support by PUNCH4NFDI and the BMBF Verbundforschung under the grant 05A20STA. Beatriz Mingo acknowledges support from the Science and Technology Facilities Council (STFC), under grant ST/T000295/1. Hongming Tang gratefully acknowledges the support from the Shuimu Tsinghua Scholar Program of Tsinghua University, the fellowship of China Postdoctoral Science Foundation 2022M721875, and long-lasting support from the JBCA machine learning group, DoA Tsinghua TAGLAB and machine learning group. Estela Suarez thanks the partners and funding agencies supporting the DEEP Projects, which have received funding from the European Commission's FP7, H2020, and EuroHPC Programmes, under Grant Agreements n° 287530, 610476, 754304, and 955606. The EuroHPC Joint Undertaking (JU) receives support from the European Union's Horizon 2020 research and innovation programme and Germany, France, Spain, Greece, Belgium, Sweden, Switzerland. Lucia Marchetti and Mattia Vaccari acknowledge financial support from the Inter-University Institute for Data Intensive Astronomy (IDIA), a partnership of the University of Cape Town, the University of Pretoria and the University of the Western Cape, and from the South African Department of Science and Innovation's National Research Foundation under the ISARP RADIOSKY2020 and RADIOMAP+ Joint Research Schemes (DSI-NRF Grant Numbers 113121 and 150551) and the SRUG HIPPO Projects (DSI-NRF Grant Numbers 121291 and SRUG22031677). Gregor Kasieczka acknowledges support by the Deutsche Forschungsgemeinschaft (DFG, German Research Foundation) under Germany's Excellence Strategy—EXC 2121 "Quantum Universe"—390833306. Manolis Marazakis, Estela Suarez, and Antony Chazapis thankfully acknowledge funding from the European Union's Horizon 2020/EuroHPC research and innovation programme under grant agreement 955606 (DEEP-SEA). National contributions from the involved state members match the EuroHPC funding. Caroline Heneka acknowledges support by the Volkswagen Foundation. Cristina Manzano and Arpad Miskolczi acknowledge support by the

Ruhr-Universität Bochum and Thüringer Landessternwarte Tautenburg for their hardware contribution to the Jülich LOFAR Long Term Archive. The National Radio Astronomy Observatory is a facility of the National Science Foundation operated under cooperative agreement by Associated Universities, Inc.

Contents

Contributors

Vishnu Balakrishnan Max-Planck-Institut für Radioastronomie, Bonn, Germany

Ian Bird Laboratoire d'Annecy de Physique des Particules (LAPP) - CNRS, Annecy-le-Vieux, France

François Bonnarel CDS, Université de Strasbourg, CNRS, Observatoire astronomique de Strasbourg - UMR7550, Strasbourg, France

Michelle M. Boyce Department of Physics and Astronomy, University of Manitoba, Winnipeg, MB, Canada

Sireesha Chamarthi Institute of Data Science, German Aerospace Center (DLR), Jena, Germany

Antony Chazapis Institute of Computer Science, Foundation for Research and Technology - Hellas (FORTH), Heraklion, Crete, Greece

Jordan D. Collier The Inter-University Institute for Data Intensive Astronomy (IDIA), Department of Astronomy, University of Cape Town, Rondebosch, Cape Town, South Africa

School of Science, Western Sydney University, Penrith, NSW, Australia

CSIRO, Space and Astronomy, Australia Telescope National Facility, Bentley, WA, Australia

Angus Comrie The Inter-University Institute for Data Intensive Astronomy (IDIA), Department of Astronomy, University of Cape Town, Rondebosch, Cape Town, South Africa

Marta Dembska Institute of Data Science, German Aerospace Center (DLR), Jena, Germany

Vandana Desai Infrared Processing & Analysis Center, California Institute of Technology, Pasadena, CA, USA

Ger van Diepen ASTRON, Dwingeloo, The Netherlands

Alexander Drabent Thüringer Landessternwarte, Tautenburg, Germany

Danielle Fenech SKA Observatory, Jodrell Bank, Lower Withington, Macclesfield, Cheshire, UK

Astrophysics Group, Cavendish Laboratory, Cambridge, UK

Hyoyin Gan Department of Physics and Astronomy, The University of British Columbia, Vancouver, BC, Canada

Jakob Gawlikowski Institute of Data Science, German Aerospace Center (DLR), Jena, Germany

Claudio Gheller Institute of Radioastronomy, INAF, Bologna, Italy

Yjan Gordon Department of Physics, University of Wisconsin-Madison, Madison, WI, USA

Yan G. Grange ASTRON, Dwingeloo, The Netherlands

Martin J. Hardcastle Department of Physics, Astronomy and Mathematics, University of Hertfordshire, Hatfield, UK

Caroline Heneka Institut für Theoretische Physik, Universität Heidelberg, Heidelberg, Germany

Hermann Heßling Hochschule für Technik und Wirtschaft (HTW) Berlin, Berlin, Germany

Matthias Hoeft Thüringer Landessternwarte, Tautenburg, Germany

Andrew M. Hopkins School of Mathematical and Physical Sciences, Macquarie University, Sydney, NSW, Australia

Aidan Hotan CSIRO Space and Astronomy, Bentley, WA, Australia

Minh T. Huynh CSIRO Space and Astronomy, Bentley, WA, Australia

International Centre for Radio Astronomy Research (ICRAR), The University of Western Australia, Crawley, WA, Australia

Marco Iacobelli ASTRON, Dwingeloo, The Netherlands

Michael Janssen Department of Astrophysics, Institute for Mathematics, Astrophysics and Particle Physics (IMAPP), Radboud University, GL Nijmegen, The Netherlands

Thomas H. Jarrett Department of Astronomy, University of Cape Town, Rondebosch, Cape Town, South Africa

Michael Johnson Max-Planck-Institut für Radioastronomie, Bonn, Germany

Institute of Data Science, German Aerospace Center (DLR), Jena, Germany

The University of Manchester, Manchester, UK

Gregor Kasieczka Institut für Experimentalphysik, Universität Hamburg, Hamburg, Germany

Kechil Kirkham The Inter-University Institute for Data Intensive Astronomy (IDIA), Department of Astronomy, University of Cape Town, Rondebosch, Cape Town, South Africa

Hans-Rainer Klöckner Max-Planck-Institut für Radioastronomie, Bonn, Germany

Michael Kramer Max-Planck-Institut für Radioastronomie, Bonn, Germany

Mark Lacy National Radio Astronomy Observatory, Charlottesville, VA, USA

Daizhong Liu Max-Planck-Institut für extraterrestrische Physik, Garching, Germany

Robert Main Max-Planck-Institut für Radioastronomie, Bonn, Germany

Cristina Manzano Jülich Supercomputing Centre (JSC), Forschungszentrum Jülich, Jülich, Germany

Manolis Marazakis Institute of Computer Science, Foundation for Research and Technology - Hellas (FORTH), Heraklion, Crete, Greece

Lucia Marchetti Department of Astronomy, University of Cape Town, Cape Town, South Africa

Institute of Radioastronomy, INAF, Bologna, Italy

The Inter-University Institute for Data Intensive Astronomy (IDIA), Department of Astronomy, University of Cape Town, Cape Town, South Africa

Joshua Marvil National Radio Astronomy Observatory, Socorro, NM, USA

Beatriz Mingo School of Physical Sciences, The Open University, Milton Keynes, UK

Arpad Miskolczi Jülich Supercomputing Centre (JSC), Forschungszentrum Jülich, Jülich, Germany

Vanessa A. Moss CSIRO Space and Astronomy, Epping, NSW, Australia

Julia Niebling Institute of Data Science, German Aerospace Center (DLR), Jena, Germany

Christopher O'Dea Department of Physics & Astronomy, University of Manitoba, Winnipeg, MB, Canada

Simon O'Toole Australian Astronomical Optics, Macquarie University, Sydney, NSW, Australia

André Offringa ASTRON, Dwingeloo, The Netherlands

Jack Radcliffe Department of Physics, University of Pretoria, Hatfield, Pretoria, South Africa

Jodrell Bank Centre for Astrophysics, University of Manchester, Manchester, UK

David A. Rafferty Hamburger Sternwarte, Universität Hamburg, Hamburg, Germany

Lawrence Rudnick Minnesota Institute for Astrophysics, University of Minnesota, Minneapolis, MN, USA

Stefano Alberto Russo INAF (Italian National Institute for Astrophysics) – Astronomical Observatory of Trieste Trieste, Italy

Jesús Salgado SKA Observatory, Jodrell Bank, Lower Withington, Macclesfield, Cheshire, UK

Sirko Schindler Institute of Data Science, German Aerospace Center (DLR), Jena, Germany

Dominik Schwarz Universität Bielefeld, Bielefeld, Germany

Kilian Schwarz Deutsches Elektronen-Synchrotron DESY, Hamburg, Germany

Biny Sebastian Department of Physics & Astronomy, University of Manitoba, Winnipeg, MB, Canada

Gary Segal School of Mathematics and Physics, University of Queensland, Brisbane, QLD, Australia

CSIRO Space and Astronomy, Epping, NSW, Australia

Alexander Sivitilli Department of Astronomy, University of Cape Town, Cape Town, South Africa

The Inter-University Institute for Data Intensive Astronomy (IDIA), Department of Astronomy, University of Cape Town, Cape Town, South Africa

Tim Sprenger Max-Planck-Institut für Radioastronomie, Bonn, Germany

Estela Suarez Jülich Supercomputing Centre (JSC), Forschungszentrum Jülich, Jülich, Germany

Computer Science Department, University of Bonn, Bonn, Germany

Giuliano Taffoni INAF - Astronomical Observatory of Trieste, Trieste, Italy

Hongming Tang Department of Astronomy, Tsinghua University, Beijing, China

Mattia Vaccari Department of Astronomy, University of Cape Town, Rondebosch, Cape Town, South Africa

Institute of Radioastronomy, INAF, Bologna, Italy

Thijs van der Hulst Kapteyn Astronomical Institute, University of Groningen, Groningen, The Netherlands

Cameron L. van Eck Research School of Astronomy & Astrophysics, The Australian National University, Canberra, ACT, Australia

Eleni Vardoulaki Thüringer Landessternwarte, Tautenburg, Germany

IAASARS, National Observatory of Athens, Athens, Greece

Prajwal Voraganti Padmanabh Max Planck Institute for Gravitational Physics (Albert Einstein Institute), Hannover, Germany

Stefan Wagner Zentrum für Astronomie der Universität Heidelberg, Heidelberg, Germany

Matthew Whiting CSIRO Space and Astronomy, Epping, NSW, Australia

Andreas Wicenec International Centre for Radio Astronomy Research, Crawley, WA, Australia

Wendy Williams SKA Observatory, Jodrell Bank, Lower Withington, Macclesfield, Cheshire, UK

Chapter 1
Introduction

Eleni Vardoulaki, Daizhong Liu, Marco Iacobelli, Jordan D. Collier, Michael Janssen, Matthew Whiting, Minh T. Huynh, Aidan Hotan, Vanessa A. Moss, Jesús Salgado, Danielle Fenech, Hongming Tang, and Dominik Schwarz

E. Vardoulaki (✉)
Thüringer Landessternwarte, Tautenburg, Germany

IAASARS, National Observatory of Athens, Athens, Greece
e-mail: elenivard@gmail.com

D. Liu
Max-Planck-Institut für extraterrestrische Physik, Garching, Germany

M. Iacobelli
ASTRON, Dwingeloo, The Netherlands
e-mail: iacobelli@astron.nl

J. D. Collier
The Inter-University Institute for Data Intensive Astronomy (IDIA), Department of Astronomy, University of Cape Town, Rondebosch, Cape Town, South Africa

School of Science, Western Sydney University, Penrith, NSW, Australia

CSIRO, Space and Astronomy, Australia Telescope National Facility, Bentley, WA, Australia
e-mail: jordan@idia.ac.za

M. Janssen
Department of Astrophysics, Institute for Mathematics, Astrophysics and Particle Physics (IMAPP), Radboud University, GL Nijmegen, The Netherlands
e-mail: m.janssen@astro.ru.nl

M. Whiting · V. A. Moss
CSIRO Space and Astronomy, Epping, NSW, Australia
e-mail: Matthew.Whiting@csiro.au; Vanessa.Moss@csiro.au

M. T. Huynh
CSIRO Space and Astronomy, Bentley, WA, Australia

International Centre for Radio Astronomy Research (ICRAR), The University of Western Australia, Crawley, WA, Australia
e-mail: Minh.Huynh@csiro.au

A. Hotan
CSIRO Space and Astronomy, Bentley, WA, Australia
e-mail: Aidan.Hotan@csiro.au

J. Salgado
SKA Observatory, Jodrell Bank, Lower Withington, Macclesfield, Cheshire, UK
e-mail: jesus.salgado@skao.int

E. Vardoulaki et al. (eds.), *Data-Intensive Radio Astronomy*, Astrophysics and Space Science Library 472, https://doi.org/10.1007/978-3-031-58468-8_1

From the moment we are born, we collect data, process them, analyse them, and store them. We do so in order to make sense of the things we see and experience and to understand the vastness of the world that surrounds us. Throughout the ages, humans have been asking questions about what is this megastructure that we are part of, our Universe. Driven by our inherent need for discovering the unexpected, we have been developing instruments, known as telescopes, and tools, known as algorithms, to help us probe into the unknown. By collecting and analysing these data, we aim to provide answers about how our Universe was created, how it is structured, how it evolved and how we came into existence. Thus with the development of astronomy and astrophysics, humans developed the so-called data lifecycle as a guidance to answer those difficult questions.

1.1 Why Radio Astronomy

The human eye is sensitive to a limited part of the electromagnetic spectrum. The corresponding wavelengths of visible light is 380–700 nanometres (nm). Across the Universe, light is emitted, scattered and absorbed over a much larger range of wavelengths (or frequencies) and thus astronomical observations from the ground should not be limited to visible light. The Universe is accessible from the ground mainly via two bands of the electromagnetic spectrum: the visible or optical band (and some parts of the infrared band when observed from high altitudes) and the radio band. Technically the radio band includes frequencies ranging from a few Hz to 3 THz, corresponding to wavelengths of 10^5 km to a fraction of a mm, but only frequencies between 10 MHz and about a THz can travel through the Earth's ionosphere and atmosphere to the ground. Observations outside the radio and visible bands require satellite missions which have extremely long lead times, are expensive and have significant environmental impact (see Chap. 15.1). Although sustainable astronomy must aim at optimally harvesting the information accessible via ground based astronomical facilities, space observations is the only way to access the X-ray, γ-ray, and parts of the infrared regions of the electromagnetic spectrum.

D. Fenech
SKA Observatory, Jodrell Bank, Lower Withington, Macclesfield, Cheshire, UK

Astrophysics Group, Cavendish Laboratory, Cambridge, UK
e-mail: danielle.fenech@skao.int

H. Tang
Department of Astronomy, Tsinghua University, Beijing, China
e-mail: hongmingt@mail.tsinghua.edu.cn

D. Schwarz
Universität Bielefeld, Bielefeld, Germany
e-mail: dschwarz@physik.uni-bielefeld.de

Observing the sky in the radio allows to obtain a wealth of information from astronomical objects that reveal their physical properties and mechanisms at play at the large wavelength of radio astronomy. Another advantage of observing at radio frequencies is that we can observe throughout the day and even if the sky is clouded. While optical light from astronomical objects is subject to daylight limitations, radio telescopes can observe during the day and the night, and almost at all weather conditions.

Making use of the radio part of the electromagnetic spectrum requires large collecting area in order to observe distant objects at high fidelity. We astronomers talk about sensitivity and angular resolution. The larger the collecting area, the better. But how large can you go before you reach the limits of material strength and structural integrity. A famous example is the 300-foot (90.44 m) Telescope at Green Bank in West Virginia, U.S.A., which collapsed in 1988 under its own weight due to a cracked steel connector.[1] A workaround is placing a telescope in a natural valley or sinkhole, like the famous Arecibo telescope,[2] a 304-m telescope which was even featured in the 007 film 'Golden Eye', and is unfortunately no longer operational (broken steel cables lead to breaking the telescope in 2020), or the Five-hundred-meter Aperture Spherical radio Telescope (FAST) in China (see Sect. 1.3). Still the most efficient way astronomers have developed in increasing the collecting area of a radio telescope is known as radio interferometry, in which many much smaller radio antennas are combined. A classic example is the Very Large Array (VLA), also known from the 1997 film 'Contact', the Low Frequency Array (LOFAR) and many more as described in Sect. 1.3.

A limitation in observing at radio wavelengths is the human parameter. Our devices, machines, home appliances, like microwaves, mobile phones, radio stations, windmills, even satellites (e.g. satellite trains), can cause radio frequency interference (RFI), contaminating or even destroying the observations. These need special care to be removed, but it is not always possible. This is why observatories try to establish protected zones (e.g. Green Bank Observatory[3] or LOFAR; Offringa et al. 2013).

One of the reasons we observe the radio sky is obtaining physical information about astronomical objects that populate our Universe. The type of astronomical objects that emit their energy at the radio part of the electromagnetic spectrum is very wide and diverse: from our Sun, the nearest star to the Earth and our source of energy, to stars, star-planet or planet-moon interactions, pulsars, masers, certain nebulae or supernova remnants, star-forming and active galaxies, fast radio bursts, the cosmic microwave background to other undiscovered phenomena. Despite the unique information we receive from the radio, multi-wavelength observations of astronomical objects are needed to put radio observations in an astrophysical context (see Part 3).

---

[1] https://public.nrao.edu/telescopes/300-foot-telescope/.

[2] https://www.naic.edu/.

[3] https://science.nrao.edu/facilities/gbt/interference-protection/ipg.

Astronomical objects emit thermal free-free and non-thermal synchrotron emission in the radio part of the electromagnetic spectrum (Condon 1992). Thermal free-free emission, also known as thermal bremsstrahlung, mainly dominates at 30–200 GHz part of the radio spectrum and depends on the temperature of the emitting object (also known as brightness temperature), which emits like a black body. Thermal emission can be observed due to ionised gas around star-forming regions or active galactic nuclei (AGN). Non-thermal synchrotron dominates the radio spectrum of star-forming galaxies, like our own, up to around 10–30 GHz and does not depend on the temperature of the emitting object. It is generated by relativistic electrons interacting with the magnetic field of the source or field. Similarly to thermal emission, it can originate from star-forming regions or the active nucleus. It can be difficult to distinguish from which process synchrotron emission is generated (see Part III). Above 200 GHz, thermal re-radiation from dust dominates the radio spectrum. It can be associated with emission from H_2 molecular clouds and star-forming regions. Apart from the continuum type of emission, we also observe spectral line emission from recombination and molecular lines, masers, and the 21 cm Hydrogen line.

A short summary of radio astronomy basic meanings is presented below. For more information about radio astronomy, we refer the reader to textbooks such as 'An Introduction to Radio Astronomy' (Burke and Graham-Smith 1997), 'Interferometry and Synthesis in Radio Astronomy' (Thompson et al. 2017), 'Tools of Radio Astronomy' (Wilson et al. 2013).

1.2 Radio Astronomy Basics in a Nutshell

Observations of astronomical objects depend on the instruments being used, which have their advantages but also limitations in area coverage, sensitivity, and angular resolution. Our measurements of the radio sky are a function of intensity and polarisation, angular resolution, and of course time. Astrophysical radio sources emit broad-band radio emission, partly or fully polarised. Grand example of sources being 100% polarised are the pulsars, and to characterise the radio emission from them we need a full set of Stokes parameters. In cases where the emission is not polarised, we only require the intensity to characterise a radio object. There are four Stokes parameters: *I*, *Q*, *U* and *V*. The *I* parameter denotes the intensity, proportional to the flux density of the radio source. The *Q* and *U* parameters denote the linear polarisation, and *V* the circular polarisation.

One of the important observable quantities is the beam solid angle, which is determined by the size of the telescope aperture. Large aperture provides a small beam, allowing for higher definition in the observations, and vice versa. Due to the fact radio sources have a finite size, their shape is characterised by the flux density, which is a function of observed frequency and beam solid angle, thus depends on the telescope used. In radio astronomy, flux densities are measured in Jansky (Jy). To obtain a 'true' observable flux-density one needs to remove from the observations the noise coming from the instruments or from the environment (RFI due to human-made machines). Radio astronomers have been developing techniques to minimise these effects and provide clean images of the radio sky to be analysed and provide answers to our astronomical questions.

1.3 Radio Observatories Around the World

Basic characteristics of major astronomical facilities around the world, which are presented in this special volume, are described below.

VLA

The Very Large Array (VLA) is an interferometer consisting of 28 dish antennas of 25-m diameter in the Plains of San Agustin desert (at an elevation of 2120 m of 6970 feet) in New Mexico, USA. It had a long history back to 1970s (construction 1973–1980). After about 30 years of operation, a major upgrade (Expanded Very Large Array, EVLA) was made in 2003–2012 and the interferometer was officially renamed as Karl G. Jansky Very Large Array (JVLA) in 2012. JVLA's working frequency ranges from 74 MHz to 50 GHz, covering 4, P, L, S, C, X, Ku, K, Ka and Q bands. Arranged in the iconic "Y" shape for a better uv coverage at its latitude of $+40°$, the antennae can move along the three arms of the "Y" shape to form various configurations/baselines. The most extended A configuration has a maximum baseline of 36.4 km, reaching $\sim 1''$ resolution at L band and $\sim$43 mas at Q band. The most compact D configuration has a maximum baseline of about 1 km, and $46''$ to $1.5''$ resolution from L to Q band. Depending on the sampler, the receiver bandwidth can be 2 or 8 GHz per polarisation.

ALMA

The Atacama Large Millimeter/submillimeter Array (ALMA; Wootten and Thompson 2009), is the largest submillemetre/millimetre interferometer located at the Chajnantor plateau in the Atacama Desert of northern Chile, which is the top world-class site for dry and thin atmosphere (latitude 5000 m or 16,000 ft) needed at these wavelengths. ALMA composites of 54 antennae with a 12 m diameter, and 12 antennae with a 7 m diameter. The 12-m array has a wide range of configurations with maximum baseline from $\sim$150-m up to 15-km. These compact and extended configurations cycle through the observing cycle of one year. The 7-m array, also named the Morita Array or Atacama Compact Array (ACA), has a fixed configuration with a maximum baseline of $\sim$45-m. The ALMA observing frequency ranges from 84 to 950 GHz, corresponding to eight bands from ALMA Band 3 to Band 10.

NOEMA

The NOrthern Extended Millimeter Array (NOEMA) of the Institut de radioastronomie millimétrique (IRAM) is the most advanced northern millimetre interferometer. NOEMA/PdBI is located at the Pic de Bure (elevation 2550 m) in the French

Alps. It was previously named Plateau de Bure Interferometer (PdBI) before 2014 with six 15-m antennae and old correlators (e.g., WideX with a 4 GHz bandwidth per polarization). Six new 15-m antennae were funded and built in the recent years and the new correlator PolyFiX has been commissioned, which processes a wide bandwidth of 16 GHz per polarisation. The full 12-antenna, 32 GHz dual-polarisation continuum bandwidth capability makes NOEMA an equally powerful continuum survey facility at mm wavelengths as ALMA in the other hemisphere.

FAST
The Five-hundred-meter Aperture Radio Telescope (FAST, Nan et al. 2011)is the largest filled-aperture single dish radio telescope built in a natural karst depression named Dawodang in Guizhou, southwest China. This 500-m diameter telescope has an illuminated aperture of 300 m and observes at frequencies from 70 MHz to 3 GHz (Peng et al. 2001; Li et al. 2013; Jiang et al. 2019).

EHT
The Event Horizon Telescope (EHT) is an ad-hoc VLBI array observing at 230 and 345 GHz. Currently, the array consists of phased ALMA (Wootten and Thompson 2009; Matthews et al. 2018; Goddi et al. 2019) and the Atacama Pathfinder Experiment (APEX; Güsten et al. 2006) in Chile, the Greenland Telescope (GLT; Inoue et al. 2014) in Greenland, the James Clerk Maxwell Telescope (telescope, JCMT) and phased Submillimeter Array (SMA; Ho et al. 2004) in Hawaii, the Kitt Peak 12 m Telescope (KP) and the Submillimeter Telescope (SMT; Baars et al. 1999) in Arizona, the Large Millimeter Telescope Alfonso Serrano (telescope, LMT) in Mexico, phased NOEMA (Guilloteau et al. 1992) in France, the Pico Veleta IRAM 30 m telescope (PV; Baars et al. 1987) in Spain, and the South Pole Telescope (SPT; Carlstrom et al. 2011) at the geographical South Pole, each recording high-bandwidth data with 64 Gbps locally. With this array, a maximum nominal resolution of 15 μas can be reached. The combined high angular resolution and high observing frequencies allows the imaging of the innermost region of AGN, which are unresolved and/or obscured by synchrotron self-absorption for other VLBI arrays. A detailed description of the EHT instrument is given in (Event Horizon Telescope Collaboration et al. 2019). A near-future upgrade of the EHT with additional small dishes and a 256 Gbps recording rate are being pursued through the next-generation EHT (ngEHT) project (Doeleman et al. 2019).

e-MERLIN
e-MERLIN (enhanced Multi-Element Radio Linked Interferometer Network) is a seven telescope array distributed across England and is operated by the University of Manchester from the Jodrell Bank Observatory. The change from MERLIN to e-MERLIN represented a major upgrade to the array including new wide-bandwidth receivers, a new correlator and new fibre optic links in 2008, enabling higher sensitivity and image fidelity observations.

e-MERLIN has a significant range of baseline lengths (min. and max. 217 km) offering high resolution observations at L (1.25–1.75 GHz), C (4.3–7.5 GHz) and

K-band (19–25 GHz), with resolutions of 0.2, 0.05, and 0.02" and sensitivities of 12, 10 and 120 μJybm^{-1} at L, C and K bands respectively.[4]

LOFAR

The LOw-Frequency ARray (LOFAR) is an interferometer operating at very low (10–250 MHz) radio frequencies (van Haarlem et al. 2013). Currently, the array consists of 54 stations distributed across Europe and each station having two sets of antennas to observe between 10–90 MHz (i.e. the Low Band Antennas, LBA) and 110–240 MHz (i.e the High Band Antennas, HBA). To enable a superb uv-coverage and sensitivity, three stations layout configurations are available providing different sizes for the array:

- 24 core station (CS) distributed in an area of about 2 $\times$ 3 km and providing baselines up to $\approx$4 km,
- 14 remote stations (RS) distributed in an area of about 120 $\times$ 60 km and providing baselines up to $\approx$120 km,
- 14 international stations (IS) scattered in 7 European countries and providing baselines up to $\approx$2000 km.

CS, RS and IS LBA antenna fields are similar, with the main difference being that for CS and RS only a subset of 48 dipoles (out of 96) are used. As a consequence, two main antenna sets (the LBA_Outer and the LBA_Sparse) are available trading between the size of primary beam and the suppression of side-lobes magnitude. In particular, the LBA_Sparse which offers larger field-of-view (FoV) and better suppression of station side-lobes is therefore recommended for survey observing that aims to maximise the area, at the cost of the computational resources needed for data processing.

The CS, RS, IS HBA antenna fields show major differences. To improve uv-coverage on short baselines, the CS set of tiles is arranged in two 24 tile antenna fields that can be operated independently (the so-called HBA_Dual antenna mode). The RS and IS have 48 and 96 tiles each, respectively, to provide good signal to noise ratios on medium and long baselines. While the IS always use the full set of antennas, the RS can limit the use up to the inner 24 tiles thus mimicking the layout of a CS. Such an (INNER) option combined to the CS antenna set (HBA_Dual) provides the best antenna mode for wide-field mapping: the HBA_Dual_inner mode.

MeerKAT

The MeerKAT radio telescope is South Africa's SKA-Mid pathfinder, operated by the South African Radio Astronomy Observatory (SARAO), and located in a remote region of the Northern Karoo, away from terrestrial interference. MeerKAT is an interferometer consisting of 64 fully-steerable dishes, each 13.5 m in diameter, with

[4] https://www.e-merlin.ac.uk/.

an offset-Gregorian design, containing an unblocked aperture for improved optics and system temperature.

A feed rotator or receiver index mounts L-band, UHF-band and S-band receivers. The L-band receivers operate between 856–1712 MHz, and due in particular to MeerKAT's highly sensitive low noise amplifiers (LNAs), achieves a system temperature divided by aperture efficiency of $T_{sys}/\eta \approx$ 20–25 K, and a System Equivalent Flux Density (SEFD) between ~390–490 Jy. The UHF-band receivers operate between 544–1088 MHz, and achieve $T_{sys}/\eta \approx$ 20–40 K and an SEFD between ~380–750 Jy. The S-band receivers, which recently became available for shared-risk early science, operate over an instantaneous bandwidth of 875 MHz, selected from five possible sub-bands between 1750–3500 MHz, and a preliminary analysis suggests they achieve an SEFD between ~440–630 Jy.

MeerKAT's 64 dishes include a dense core with a minimum baseline of 29 m, as well as many long baselines up to 7.7 km, meaning even short observations sample the (u,v) plane extremely densely. These specifications mean a typical 8 hour full synthesis observation achieves a highly-sensitive naturally-weighted noise level of ~2 μJy/beam, and between ~3.5–8 arcsec resolution, depending on weighting scheme.

MeerKAT's observing capabilities include a 4k 'wideband coarse' correlator mode, during which 4096 channels are recorded, as well as a 32k correlator mode, during which 32,768 channels are recorded, either across a wide ('wideband fine') or narrow band ('narrowband extended' or 'zoom band'), and integration times between 2, 4 and 8 seconds.

MeerKAT's specifications mean it is a significant survey instrument, and as such grants the majority of its time over the initial ~5 years of operations to Large Survey Projects (LSPs), each awarded over 1000 hours observing time. The LSPs can roughly be divided into time-domain, continuum and polarisation, and spectral-line surveys, although many projects cover more than one of these components. In between LSP observations, MeerKAT initially spends ~28% of its observing time on Open Time (OT) proposals, within an open call made every year or so. An additional ~5% of observing time is spent on Director's Discretionary Time (DDT) proposals.

ASKAP

The Australian Square Kilometre Array Pathfinder (ASKAP) is an SKA precursor radio telescope array located at Australia's SKA site, Inyarrimanha Ilgari Bundara, the CSIRO Murchison Radio-astronomy Observatory (MRO). ASKAP is designed specifically to maximise survey speed (Hotan et al. 2021) at frequencies between 700 and 1800 MHz. It consists of 36 dish antennas, each 12 m in diameter, equipped with Phased Array Feed (PAF) receivers designed by CSIRO Space & Astronomy. These receivers give the telescope a wide field of view (30 square degrees at 800 MHz) by sampling the radio signal with an array of 188 active elements in the focal plane of the dish. Digital signal processing hardware makes several weighted sums over these elements, forming 36 primary beams per antenna, which can be arranged to cover a wide field with nearly uniform sensitivity. However, this immediately increases the data rate of the telescope by a factor of the number of beams when compared to single-feed systems. After more than ten years of design, prototyping, construction, and commissioning, ASKAP became fully operational in 2019 and has conducted pilot surveys over the last few years. More than 2.5 PB of science data products have been produced and more than 80 scientific papers published. Full surveys with ASKAP began at the end of 2022.

SKAO
The SKA Observatory (SKAO; McMullin et al. 2020) is a next-generation radio astronomy-driven Big Data facility that will revolutionise our understanding of the Universe and the laws of fundamental physics. Enabled by cutting-edge technology, it promises to have a major impact on society, science and beyond. The SKA Organisation will operate two telescopes: one in South Africa's Karoo region (SKA-Mid) and one in Western Australia's Murchison Shire (SKA-Low). These telescopes are expected to have unprecedented sensitivity in radio astronomy, so they allow the project to address key subjects in astrophysics, fundamental physics, cosmology, and particle astrophysics among others.

1.4 The Era of Data-Intensive Radio Astronomy

Over the past decades, radio astronomy and the way we collect data has changed dramatically. From single dish telescopes to radio arrays connecting antennas in a single location, across continents (e.g. LOFAR) or around the world (e.g. VLBI), advances in the field have revolutionised how we build infrastructures to collect and store data, develop algorithms to process and analyse them, and how we make them accessible to the scientific community. The revolutions in the field include the expanded and broad-band observing capabilities as well as the wide field of view of the observations, an increased spatial and angular resolution, improved receivers and dedicated radio software to obtain and process the observations. Large radio-telescope facilities, pathfinders and precursors of the SKA have another thing in common: the increased data volumes obtained and ingested. Several issues arise when dealing with large volumes, in the same way these problems arise when you take one too many photos with your mobile phone: storage limitations. Current radio telescopes (e.g. LOFAR, MeerKAT, ASKAP) already generate TB of data, which will increase to TB per hour for the SKA. The result is that the raw data, what comes out of the receivers, cannot be stored. The radio astronomical community has adopted the solution where the raw data need to be averaged, processed and stored, before being purged. Some of these limitations could be eliminated with the use of artificial intelligent algorithms to process data fast. Another important issue to consider is data transport. Moving such large volumes is proving impractical, this is why the radio astronomical community is adopting practices of centralised long term storage facilities to store data which scientists can access now, but also in the future. Radio astronomers and dedicated experts are working on methodologies to advance the field of radio astronomy and bring it to the exabyte era.

1.5 The Data Lifecycle

This special volume is designed to follow the data lifecycle, starting with how we obtain the observations using radio facilities, creating the infrastructure to obtain, ingest and store the large amounts of data (Part I) before they are processed (Part II). Once the processing is over and radio images/data cubes are produced, data are analysed in post-processing (Part III) depending on the type of emission and the science case. To be able to be shared and re-used by the scientific community and other data users, FAIR principles (Findability, Accessibility, Interoperability, and Reusability; Part IV) are implemented to promote open, transparent, and collaborative data sharing practices, and to facilitate the reuse and re-purposing of data for a wide range of research purposes. The Epilogue provides an outlook for the future of radio astronomy in the data-intensive radio era of the SKA, and discuss about practices to enhance sustainability within the field.

References

J.W.M. Baars, B.G. Hooghoudt, P.G. Mezger, M.J. de Jonge, The IRAM 30-m millimeter radio telescope on Pico Veleta, Spain. Astron. Astrophys. **175**(1–2), 319–326 (1987)

J.W.M. Baars, R.N. Martin, J.G. Mangum, J.P. McMullin, W.L. Peters, The Heinrich Hertz telescope and the submillimeter telescope observatory. Publ. Astron. Soc. Pac. **111**(759), 627–646 (1999). https://doi.org/10.1086/316365

B.F. Burke, F. Graham-Smith, *An Introduction to Radio Astronomy* (Cambridge University Press, Cambridge, 1997)

J.E. Carlstrom, P.A.R. Ade, K.A. Aird, B.A. Benson, L.E. Bleem, S. Busetti, C.L. Chang, E. Chauvin, H.M. Cho, T.M. Crawford, A.T. Crites, M.A. Dobbs, N.W. Halverson, S. Heimsath, W.L. Holzapfel, J.D. Hrubes, M. Joy, R. Keisler, T.M. Lanting, A.T. Lee, E.M. Leitch, J. Leong, W. Lu, M. Lueker, D. Luong-Van, J.J. McMahon, J. Mehl, S.S. Meyer, J.J. Mohr, T.E. Montroy, S. Padin, T. Plagge, C. Pryke, J.E. Ruhl, K.K. Schaffer, D. Schwan, E. Shirokoff, H.G. Spieler, Z. Staniszewski, A.A. Stark, C. Tucker, K. Vanderlinde, J.D. Vieira, R. Williamson, The 10 Meter South pole telescope. Publ. Astron. Soc. Pac. **123**(903), 568 (2011). https://doi.org/10.1086/659879

J.J. Condon, Radio emission from normal galaxies. Ann. Rev. Astron. Astrophys. **30**, 575–611 (1992). https://doi.org/10.1146/annurev.aa.30.090192.003043

S. Doeleman, L. Blackburn, J. Dexter, J.L. Gomez, M.D. Johnson, D.C. Palumbo, J. Weintroub, J.R. Farah, V. Fish, L. Loinard, C. Lonsdale, G. Narayanan, N.A. Patel, D.W. Pesce, A. Raymond, R. Tilanus, M. Wielgus, K. Akiyama, G. Bower, A. Broderick, R. Deane, C.M. Fromm, C. Gammie, R. Gold, M. Janssen, T. Kawashima, T. Krichbaum, D.P. Marrone, L.D. Matthews, Y. Mizuno, L. Rezzolla, F. Roelofs, E. Ros, T.K. Savolainen, F. Yuan, G. Zhao, L. Blackburn, S. Doeleman, J. Dexter, J.L. Gomez, M.D. Johnson, D.C. Palumbo, J. Weintroub, J.R. Farah, V. Fish, L. Loinard, C. Lonsdale, G. Narayanan, N.A. Patel, D.W. Pesce, A. Raymond, R. Tilanus, M. Wielgus, K. Akiyama, G. Bower, A. Broderick, R. Deane, C.M.

Fromm, C. Gammie, R. Gold, M. Janssen, T. Kawashima, T. Krichbaum, D.P. Marrone, L.D. Matthews, Y. Mizuno, L. Rezzolla, F. Roelofs, E. Ros, T.K. Savolainen, F. Yuan, G. Zhao, Studying black holes on horizon scales with VLBI ground arrays. Bull. Am. Astron. Soc. **51**, 256 (2019)

Event Horizon Telescope Collaboration, K. Akiyama, A. Alberdi, W. Alef, K. Asada, R. Azulay, A.-K. Baczko, D. Ball, M. Baloković, J. Barrett, et al., First M87 event horizon telescope results. II. Array and instrumentation. Astrophys. J. Lett. **875**, L2 (2019). https://doi.org/10.3847/2041-8213/ab0c96

C. Goddi, I. Martí-Vidal, H. Messias, G.B. Crew, R. Herrero-Illana, V. Impellizzeri, H. Rottmann, J. Wagner, E. Fomalont, L.D. Matthews, D. Petry, N. Phillips, R. Tilanus, E. Villard, L. Blackburn, M. Janssen, M. Wielgus, Calibration of ALMA as a phased array. ALMA observations during the 2017 VLBI campaign. Publ. Astron. Soc. Pac. **131**(1001), 075003 (2019). https://doi.org/10.1088/1538-3873/ab136a

S. Guilloteau, J. Delannoy, D. Downes, A. Greve, M. Guelin, R. Lucas, D. Morris, S.J.E. Radford, J. Wink, J. Cernicharo, T. Forveille, S. Garcia-Burillo, R. Neri, J. Blondel, A. Perrigourad, D. Plathner, M. Torres, The IRAM interferometer on Plateau de Bure. Astron. Astrophys. **262**, 624 (1992)

R. Güsten, L.Å. Nyman, P. Schilke, K. Menten, C. Cesarsky, R. Booth, The Atacama Pathfinder EXperiment (APEX) - a new submillimeter facility for southern skies -. Astron. Astrophys. **454**(2), L13–L16, (2006). https://doi.org/10.1051/0004-6361:20065420

P.T.P. Ho, J.M. Moran, K.Y. Lo, The submillimeter array. Astrophys. J. Lett. **616**(1), L1–L6 (2004). https://doi.org/10.1086/423245

A.W. Hotan, J.D. Bunton, A.P. Chippendale, M. Whiting, J. Tuthill, V.A. Moss, D. McConnell, S.W. Amy, M.T. Huynh, J.R. Allison, C.S. Anderson, K.W. Bannister, E. Bastholm, R. Beresford, D.C.J. Bock, R. Bolton, J.M. Chapman, K. Chow, J.D. Collier, F.R. Cooray, T.J. Cornwell, P.J. Diamond, P.G. Edwards, I.J. Feain, T.M.O. Franzen, D. George, N. Gupta, G.A. Hampson, L. Harvey-Smith, D.B. Hayman, I. Heywood, C. Jacka, C.A. Jackson, S. Jackson, K. Jeganathan, S. Johnston, M. Kesteven, D. Kleiner, B.S. Koribalski, K. Lee-Waddell, E. Lenc, E.S. Lensson, S. Mackay, E.K. Mahony, N.M. McClure-Griffiths, R. McConigley, P. Mirtschin, A.K. Ng, R.P. Norris, S.E. Pearce, C. Phillips, M.A. Pilawa, W. Raja, J.E. Reynolds, P. Roberts, D.N. Roxby, E.M. Sadler, M. Shields, A.E.T. Schinckel, P. Serra, R.D. Shaw, T. Sweetnam, E.R. Troup, A. Tzioumis, M.A. Voronkov, T. Westmeier, Australian square kilometre array pathfinder: I. System description. Publ. Astron. Soc. Aust. **38**, e009 (2021). https://doi.org/10.1017/pasa.2021.1

M. Inoue, J.C. Algaba-Marcos, K. Asada, R. Blundell, W. Brisken, R. Burgos, C.C. Chang, M.T. Chen, S.S. Doeleman, V. Fish, P. Grimes, J. Han, H. Hirashita, P.T.P. Ho, S.N. Hsieh, T. Huang, H. Jiang, E. Keto, P.M. Koch, D.Y. Kubo, C.Y. Kuo, B. Liu, P. Martin-Cocher, S. Matsushita, Z. Meyer-Zhao, M. Nakamura, P. Napier, H. Nishioka, G. Nystrom, S. Paine, N. Patel, N. Pradel, H.Y. Pu, P.A. Raffin, H.Y. Shen, W. Snow, R. Srinivasan, T.S. Wei, Greenland telescope project: direct confirmation of black hole with sub-millimeter VLBI. Radio Sci. **49** (7), 564–571 (2014). https://doi.org/10.1002/2014RS005450

P. Jiang, Y. Yue, H. Gan, R. Yao, H. Li, G. Pan, J. Sun, D. Yu, H. Liu, N. Tang, L. Qian, J. Lu, J. Yan, B. Peng, S. Zhang, Q. Wang, Q. Li, D. Li, Commissioning progress of the FAST. Sci. China Phys. Mech. Astron. **62**(5), 959502 (2019). https://doi.org/10.1007/s11433-018-9376-1

D. Li, R. Nan, Z. Pan, The five-hundred-meter aperture spherical radio telescope project and its early science opportunities, in *Neutron Stars and Pulsars: Challenges and Opportunities After 80 Years*, ed. by J. van Leeuwen, vol. 291 (March 2013), pp. 325–330. https://doi.org/10.1017/S1743921312024015

L.D. Matthews, G.B. Crew, S.S. Doeleman, R. Lacasse, A.F. Saez, W. Alef, K. Akiyama, R. Amestica, J.M. Anderson, D.A. Barkats, A. Baudry, D. Broguière, R. Escoffier, V.L. Fish, J. Greenberg, M.H. Hecht, R. Hiriart, A. Hirota, M. Honma, P.T.P. Ho, C.M.V. Impellizzeri, M. Inoue, Y. Kohno, B. Lopez, I. Martí-Vidal, H. Messias, Z. Meyer-Zhao, M. Mora-Klein, N.M. Nagar, H. Nishioka, T. Oyama, V. Pankratius, J. Perez, N. Phillips, N. Pradel, H. Rottmann, A.L. Roy, C.A. Ruszczyk, B. Shillue, S. Suzuki, R. Treacy, The ALMA phasing system: a beamforming capability for ultra-high-resolution science at (sub)millimeter wavelengths. Publ. Astron. Soc. Pac. **130**(1), 015002 (2018). https://doi.org/10.1088/1538-3873/aa9c3d

J. McMullin, P. Diamond, A. McPherson, R. Laing, P. Dewdney, A. Casson, L. Stringhetti, N. Rees, T. Stevenson, M. Lilley, A. van Es, M.G. Labate, G. Swart, A. Schinckel, T. Cheetham, M. Caiazzo, M. Austin, L. Ball, S. Berry, R. Braun, A. Chrysostomou, The square kilometre array project, in *Society of Photo-Optical Instrumentation Engineers (SPIE) Conference Series*, vol. 11445. Society of Photo-Optical Instrumentation Engineers (SPIE) Conference Series, p. 1144512 (December 2020). https://doi.org/10.1117/12.2565117

R. Nan, D. Li, C. Jin, Q. Wang, L. Zhu, W. Zhu, H. Zhang, Y. Yue, L. Qian, The five-hundred aperture spherical radio telescope (fast) project. Int. J. Mod. Phys. D **20**(6), 989–1024 (2011). https://doi.org/10.1142/S0218271811019335

A.R. Offringa, A.G. de Bruyn, S. Zaroubi, G. van Diepen, O. Martinez-Ruby, P. Labropoulos, M.A. Brentjens, B. Ciardi, S. Daiboo, G. Harker, V. Jelić, S. Kazemi, L.V.E. Koopmans, G. Mellema, V.N. Pandey, R.F. Pizzo, J. Schaye, H. Vedantham, V. Veligatla, S.J. Wijnholds, S. Yatawatta, P. Zarka, A. Alexov, J. Anderson, A. Asgekar, M. Avruch, R. Beck, M. Bell, M.R. Bell, M. Bentum, G. Bernardi, P. Best, L. Birzan, A. Bonafede, F. Breitling, J.W. Broderick, M. Brüggen, H. Butcher, J. Conway, M. de Vos, R.J. Dettmar, J. Eisloeffel, H. Falcke, R. Fender, W. Frieswijk, M. Gerbers, J.M. Griessmeier, A.W. Gunst, T.E. Hassall, G. Heald, J. Hessels, M. Hoeft, A. Horneffer, A. Karastergiou, V. Kondratiev, Y. Koopman, M. Kuniyoshi, G. Kuper, P. Maat, G. Mann, J. McKean, H. Meulman, M. Mevius, J.D. Mol, R. Nijboer, J. Noordam, M. Norden, H. Paas, M. Pandey, R. Pizzo, A. Polatidis, D. Rafferty, S. Rawlings, W. Reich, H.J.A. Röttgering, A.P. Schoenmakers, J. Sluman, O. Smirnov, C. Sobey, B. Stappers, M. Steinmetz, J. Swinbank, M. Tagger, Y. Tang, C. Tasse, A. van Ardenne, W. van Cappellen, A.P. van Duin, M. van Haarlem, J. van Leeuwen, R.J. van Weeren, R. Vermeulen, C. Vocks, R.A.M.J. Wijers, M. Wise, O. Wucknitz, The LOFAR radio environment. Astron. Astrophys. **549**, A11 (2013). https://doi.org/10.1051/0004-6361/201220293

B. Peng, R. Nan, Y. Su, Y. Qiu, L. Zhu, W. Zhu, Five-hundred-meter Aperture Spherical Telescope project. Astrophys. Space Sci. **278**, 219–224 (2001). https://doi.org/10.1023/A:1013127316851

A. Richard Thompson, J.M. Moran, G.W. Swenson Jr., *Interferometry and Synthesis in Radio Astronomy*, 3rd edn. (2017). https://doi.org/10.1007/978-3-319-44431-4

M.P. van Haarlem, M.W. Wise, A.W. Gunst, G. Heald, J.P. McKean, J.W.T. Hessels, A.G. de Bruyn, R. Nijboer, J. Swinbank, R. Fallows, et al., LOFAR: The LOw-Frequency ARray. Astron. Astrophys. **556**, A2 (2013). https://doi.org/10.1051/0004-6361/201220873

T.L. Wilson, K. Rohlfs, S. Hüttemeister, *Tools of Radio Astronomy* (2013). https://doi.org/10.1007/978-3-642-39950-3

A. Wootten, A.R. Thompson, The Atacama large millimeter/submillimeter array. IEEE Proc. **97**, 1463–1471 (2009). https://doi.org/10.1109/JPROC.2009.2020572

Part I
Data Creation, Storage and Archives

Chapter 2
From the Data Generation to the Archive

Yan G. Grange, Hermann Heßling, Michael Kramer, Stefan Wagner, Jordan D. Collier, Minh T. Huynh, Andreas Wicenec, Ger van Diepen, André Offringa, and Matthias Hoeft

Y. G. Grange (✉) · G. van Diepen · A. Offringa
ASTRON, Dwingeloo, Netherlands
e-mail: grange@astron.nl; diepen@astron.nl; offringa@astron.nl

H. Heßling
Hochschule für Technik und Wirtschaft (HTW) Berlin, Berlin, Germany
e-mail: hessling@htw-berlin.de

M. Kramer
Max-Planck-Institut für Radioastronomie, Bonn, Germany
e-mail: mkramer@mpifr-bonn.mpg.de

S. Wagner
Zentrum für Astronomie der Universität Heidelberg, Heidelberg, Germany
e-mail: swagner@lsw.uni-heidelberg.de

J. D. Collier
The Inter-University Institute for Data Intensive Astronomy (IDIA), Department of Astronomy, University of Cape Town, Rondebosch, Cape Town, South Africa

School of Science, Western Sydney University, Penrith, NSW, Australia

CSIRO, Space and Astronomy, Australia Telescope National Facility, Bentley, WA, Australia
e-mail: jordan@idia.ac.za

M. T. Huynh
CSIRO Space and Astronomy, Bentley, WA, Australia

International Centre for Radio Astronomy Research (ICRAR), The University of Western Australia, Crawley, WA, Australia
e-mail: Minh.Huynh@csiro.au

A. Wicenec
International Centre for Radio Astronomy Research, Crawley, WA, Australia
e-mail: andreas.wicenec@icrar.org

M. Hoeft
Thüringer Landessternwarte, Tautenburg, Germany
e-mail: hoeft@tls-tautenburg.de

E. Vardoulaki et al. (eds.), *Data-Intensive Radio Astronomy*, Astrophysics and Space Science Library 472, https://doi.org/10.1007/978-3-031-58468-8_2

With the advent of the new infrastructures, radio astronomy is entering an era where handling, processing and analysing the vast amount of data becomes a tremendous challenge. In the preceding Chap. 1, several major new facilities have been introduced. Most of these are operated as interferometers, having a correlator as key element at which the signals of all antenna are combined and scientific raw data are generated. To illustrate the data challenge radio astronomy is facing, one may look at the throughput of those correlators. The LOFAR COBALT correlator is today processing data with a maximum rate of 80 Gbps (Broekema et al. 2018), producing several of Terabytes of raw data at every hour. For the Square Kilometre Array Observatory, this numbers are expected to be about three orders of magnitude larger.

To record as much precious information coming from the central processing, i.e. about distant galaxies or rare events, for subsequent scientific analysis, it is necessary to smartly reduce the stream of data and to have highly performant storage system.

In this chapter, we firstly address the fact that the data reduction generally leads to a loss of information and strategies to deal with this fact. Secondly, we describe the infrastructures for ingesting raw scientific data into large volume archives for selected telescopes.

2.1 Data Irreversibility

High-performance data processing is a central challenge not only for radio astronomy but for many areas of the modern society. The technical possibilities of data mining are increasing far faster than the possibilities of optimal data utilisation. Accordingly, new methods must be highly parallelised, self-learning and at the same time flexible. The diversity of challenges requires appropriate procedures to ensure the quality of algorithms and methods.

These requirements will coincide with those of future projects in commercial and civil use. There, too, so much data will be generated that only a fraction can be stored in the long term. In the future, archives will be much more dynamic than is usual today. The quality of archived data will have to be reassessed on a regular basis.

More and more sensors with ever higher resolution are contributing significantly to the rapid growth of the global stock of data. Soon, so much data will be produced in many areas—not only in science but also in industrial and social environments—that only a fraction of it can be stored in the long term.

Typical requirements of many fields are the fast and secure identification of events that require instantaneous decisions and reactions or events that are to be stored for follow-up measurements or for documentation purposes with high information content. Examples from basic research include continuous measurement processes that identify rare events and enable rapid responses or follow-up investigations. Examples from civil or commercial use are monitoring systems (process monitoring, dynamic pattern recognition, automatic quality control, traffic control systems, autonomous driving, etc.).

The largest civilian data producers are to be found in research and here in particular in particle physics and astronomy. Research projects such as the Large Hadron Collider (LHC) at CERN and, in the near future, the Square Kilometre Array Observatory (SKAO), which is being built in South Africa and Australia from thousands of antennas, produce larger data streams than currently flow on the global Internet.

At the SKAO, the data challenges are extremely high: only a small fraction of the huge raw data streams (zettabytes per year) can be stored in archives. Decisions about filtering, compression and what to archive are based on incomplete information and lead inevitably to irreversible information loss. This has an increasing impact on quality assurance, reproducibility and resource management. It is necessary to allow instantaneous decision processes during the data acquisition phase in order to identify rare events. To reduce data irreversibility as much as possible, feedbacks in realtime between archives and data taking devices are needed in order to optimise permanently decisions by adjusting the algorithms.

In high-energy physics, it has long been common practice to perform a rapid pre-analysis directly during data acquisition in order to extract the relevant data from the data streams and to store only a relevant fraction of it in archives. For the computing of the LHC, there are only a few IT centres worldwide where these archives are kept. During the workflow between data acquisition and transport to the archives, metadata are generated. Physicists perform their investigations using software frameworks that analyse the archived data on computer farms in the IT centres. Among other things, the metadata indicates what physical effects might be found in a dataset. Information of this kind makes it easier for physicists to focus on the parts of the data archives that are relevant to their investigations.

To support its scientific working groups in meeting the Big Data challenges, the Helmholtz Association launched the portfolio project "Large Scale Data Management and Analysis (LSDMA)" in 2012, in which seven university partners participated in addition to the four Helmholtz centres DESY, FZJ, GSI, and KIT (Jung et al. 2015). LSDMA was interdisciplinary from the beginning, and many scientists from different fields (computer science, physics, medicine, biology, life science, environment) successfully collaborated. To address the Big Data challenges, the Data Life Cycle Labs (DLCL) model was used. It allows the integration of data management, but leaves out the curation of data (Pouchard 2015).

The Curation Lifecycle Model of the Digital Curation Centre (DCC) describes aspects to be considered in the curation of data: Conception, data generation, evaluation and selection, transfer to archives, long-term preservation measures, storage, use, transformation, e.g., into other data formats (Higgins 2008). The "Big Data Life Cycle Model" extends the curation model of the DCC and is intended to support scientists, administrators, and librarians in planning their work processes. It takes into account the characteristic properties of Big Data: "Volume, Variety, Velocity, and Veracity." The verification of the quality of the data is of particular importance and is directly linked to the "Veracity" property. However, this model

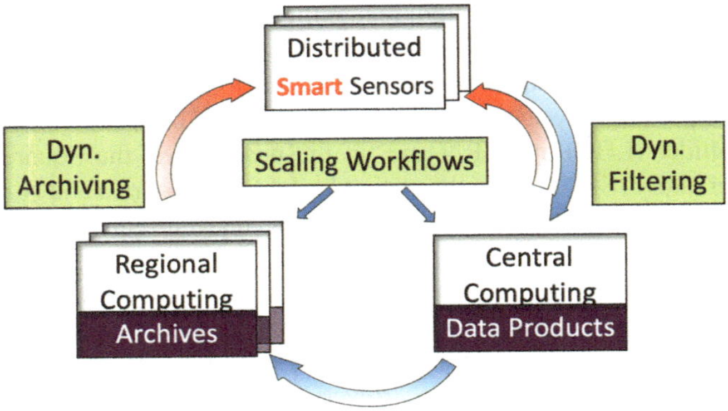

Fig. 2.1 Dynamic life cycle (Heßling et al. 2020)

does not sufficiently account for the dynamic nature of future research. It points out the need to archive processes that may be interconnected in the "Preserve" phase. It also points out that archiving processes is difficult because they may change over time. In terms of long-term preservation of data, this model focuses on questions such as:

- What type of storage is needed?
- When does data become obsolete?
- What legal issues need to be considered?
- Does it make sense to store derived data instead of raw data?

In particular, the question of whether raw data should be archived will hardly arise in the future: it will almost exclusively be possible to keep "derived data" in archives. This goes hand in hand with irreversible cuts in data processing chains and requires special consideration in future data lifecycle models, see Fig. 2.1. A crucial question is: How can the reproducibility of results be guaranteed under the condition that raw data is no longer available? How can the fundamental problem of data irreversibility be addressed?

In astronomy, the International Virtual Observatory Alliance (International Virtual Observatory Alliance, IVOA) defined in 2011 a "VOEvent", a recommendation for the use of metadata in observatories.[1] VOEvent is not sufficient, because essentially only information about astronomical observations can be stored and not how they were obtained. Further information, for example, which algorithms were used, is not provided.

[1] http://www.ivoa.net/documents/VOEvent/20110711/REC-VOEvent-2.0.pdf.

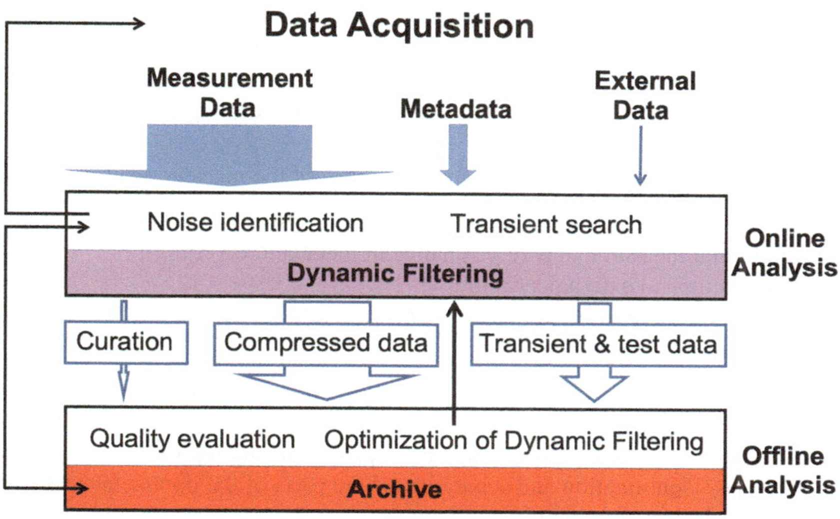

Fig. 2.2 Dynamic filtering

2.1.1 Dynamic Filtering

Many data-intensive investigations are based on very extensive measurement data that must be continuously recorded and analysed in a timely manner (Fig. 2.2). Online analyses of very large data streams are particularly necessary,

- to continuously check data acquisition,
- to identify disturbances and avoid costly failures,
- to instantaneously detect relevant (sought–after) variations,
- to archive relevant, non-variable information portions appropriately.

As data is continuously obtained, online analyses must process the full volume of data at the rate of data ingestion with all steps. As a rule, this is done with the help of ring buffers in which the data is analysed. The requirements of online analyses are challenging in this regard.

Errors caused by instrumental disturbances or interfering signals must be recognised in order not to influence the signal search.

- Variations not caused by instrumental disturbances or spurious signals must be identified in a complex, high-dimensional parameter space.
- Thresholds for identifying variations include complex correlations in non-contiguous parts of the parameter space.
- Threshold values may depend on the signal in the entire parameter space.
- Variable signals that were subcritical until the trigger threshold was reached must be captured in their overall time course. For this purpose, the entire data stream must pass through a ring buffer.

- Variable signals have no well-defined signature and are only defined as "non-constant". Efficient signal methods are necessarily self-learning.
- Variability includes change during data ingestion (i.e., while remaining in the ring buffer) and change from previous measurements. This requires an instantaneous archive matching.
- Identified variations influence the further course of the measurement process:
  - change in data taking (e.g., follow-up of the event),
  - additional measurements (e.g., change in metadata recording),
  - notifications to external systems,
  - change of archiving criteria (file rights, lossless storage of the data segments with variable signals).
- Control of archiving. The online analysis leads to an evaluation, which controls the archiving controls.
- Determination of the dynamic range and appropriate setting of the compression algorithms, identification and separation of the parts of the parameter space (not to be stored) that contain only noise signals.

So far, the online analysis uses criteria and filters that have been determined on the basis of offline analyses. Therefore, the decisions are static and thus sub-optimal. In particular, when there is a temporally limited but amplified background of spurious signals or noise from the sensors, there are currently limited concepts for dynamically adjusting the criteria, which must be generated by an interplay between sensor and archived metadata.

2.1.2 Dynamic Archiving

The offline analysis consists of the continuous "monitoring" of the quality criteria and the validation of the filtered data products with regard to their use for causal inference. Since the scientific question of the data analysis may change over time, it may be possible or even necessary to completely re-evaluate the archive data and implement corresponding new online filters or to adapt the entire online analysis.

A concrete example from radio astronomy is the new field of "Fast Radio Bursts" (FRBs), which clearly underlines the importance of transient astronomy. The first FRB was discovered in 2007 in archival data taken as early as July 2001. A strong radio pulse, several milliseconds long, was detected from the direction of the Magellanic Clouds, but was only discovered during the search for "radio pulsars" (galactic rotating neutron stars). The pulse was so strong that it drove the imaging sensors into saturation, and was disregarded as a presumed spurious signal. Realising the existence of such signals, astronomers planned a new high time resolution sky survey, the High Time Resolution Universe (HTRU) Survey, which began in 2009 (Keith et al. 2010). Technically, only off-line analyses were possible, which were eventually rewarded with the discovery of a cosmic population of FRBs (Thornton et al. 2013). The origin of the signals is still debated and can probably only be finally clarified by the simultaneous identification of signals in other frequency bands. Guided by the new findings and better technical capabilities, a new sky survey was planned and is still ongoing: the "SUrvey for Pulsars and Extragalactic Radio Bursts" (SUPERB). This survey uses ring buffers and online filters that sends automatic triggers to other telescopes whenever an FRB is detected. Here it quickly became clear that the parameter space covered by the online analysis was too small. This led not only to a complete re-evaluation of the archive data, but also to an adjustment of the online algorithms and thus to new FRB detections. Today, all searches for FRBs use the adapted strategies.

This described flow of online and offline analysis is typical, but benefits from the fact that (almost) all data could still be archived. In the future, dynamic curation of the data and adaptation of the online filters must take place on much shorter time scales and ideally with immediate feedback. This includes finding and classifying structural features in the multidimensional image space (location, time, frequency) as well as their statistical evaluation and error analysis. "Dynamic Archiving" focuses on developing generic concepts for workflows that can optimise this feedback between online and offline analysis (e.g., to simultaneously minimise the volume of data and the resulting error).

The boundaries between online and offline computing will become more and more blurred, helped by the fact that sensors will become "smart": they will be equipped with chip systems to allow machine learning methods to be applied as early as possible. Smart senors will thus become outsourced components of data centres.

2.1.3 Parallelisation

The analysis of large amounts of data requires the use of parallel computing. For this purpose, the software of the online and offline analysis must be executed on many nodes simultaneously. An important quality indicator for scaling parallel computing is the speedup, whereby the decisive question is: Can the duration for the online or offline analysis be halved by doubling the computing power?

Parallel computing is a core area of computer science and has been extensively studied. In recent years, however, there has been an increasing focus on how it relates to real-world applications. For years, many scientific disciplines have developed and optimised single-user solutions for analysing their data. At some point, it was then discovered that these solutions do not scale, i.e., they can no longer be used to comprehensively analyse the ever-increasing volumes of data.

An important strategy in parallel computing is to apply the Divide&Conquer method. A problem is split recursively into subproblems that are solved in parallel (divide step). Out of the partial solutions a solution of the whole problem is constructed (conquer step). An essential assumption is, also often not mentioned explicitly (mostly because it is fulfilled trivially), that the partial problems are of the same type as the problem itself.

While the Divide&Conquer method is well suited for sorting lists of numbers, it generally cannot be used directly for Big Data analysis where datasets are distributed across connected compute nodes. First, distributed workflows typically rely on the exchange of intermediate results, and it is well known that communication between nodes quickly becomes a bottleneck if the amount of data exchanged increases. Second, new objects are created when a large dataset is split into small pieces, i.e., the subproblems are no longer equivalent to the original problem, and a "black-box implementation" of the Divide&Conquer method can lead to strange anomaly effects, as shown by Heßling et al. (2021b).

2.1.4 Reproducibilty

Equal experiments should produce equal results. For this basic assumption of science it should be possible to reconstruct how decisions were taken during the online analysis phase. The data stored in the archives need to be compared with simulated data in order to determine the quality of data reduction decisions.

Simulations are essential for validation and understanding. In high-energy physics, about 50% of all data are obtained from simulations covering the entire collision process, including the interaction with the detector. In astronomy, simulations exist for selected areas. However, a complete simulation as in particle physics is not to be expected, because the universe is by far too complex. For selected regions in space and for subsystems of astronomical objects, however, a complete simulation is to be aimed at.

2.1.5 Dynamic Curation

Metadata describes what semantic information is available about data records. Especially in complex data structures, this "data about data" is of crucial importance. It should be as easy as possible to access metadata. This is made nearly impossible simply by the fundamental problem that metadata schemas tend to differ widely even between closely related scientific fields.

This is also true for observatories, which generally do not use uniform metadata schemas, but develop proprietary schemas for individual projects to describe the raw data from sensors. Information about the processes used to process this raw data is not stored in the metadata; moreover, the processing procedures do not scale in any way. At least standardized formats are available for exchanging data between observatories, e.g. FITS (Flexible Imaging Transport System) and MeasurementSet (MS).

There are a number of systems that have been developed for metadata management. For example, the EU project EUDAT has created the B2SHARE service. iRODS can be implemented on systems with distributed processor and memory resources, including grid architectures. Typically, these systems have either been optimized with the needs of specific communities in mind, making it difficult to use them in other disciplines, or they are so general that it is not easy to adapt them to a specific use case.

Storing more or less continuously the details of the Dynamic Filtering will have major impact: it has to be expected that its volume will become a significant part of the archived data. Traditional metadata systems are based on relational data base systems. This architecture will no longer be sufficient for two reasons. First, relational data base systems reach their limits when 1 TB or more of data has to be stored in individual tables. Second, a suitable database schema is quite complex: its creation not only requires a sufficient amount of time (of the order of a year), above all it is practically impossible to change or to extend it in a short time. This temporal flexibility is necessary, however, because it can be assumed that the algorithms of the Dynamic Filtering are constantly being extended, and it is not clear from the outset which variables of the workflows and control systems are to be stored.

The metadata solution MASi (Grunzte et al. 2018) uses a NoSQL database. The repository XTENS2 (Izzo et al. 2021) for biomedical data is based on a flexible JSON data model. The ATLAS Metadata Interface (AMI) is designed for connecting to metadata stored in heterogeneous databases (SQL, NoSQL) (Odier et al. 2019).

2.1.6 Smart Green Computing

Due to the rapidly increasing data challenges, previously rather peripheral areas in astronomy move into focus, such as theoretical computer science, machine learning on distributed data, scaling of workflows, large-scale simulation, and (numerical) mathematics.

In the future, computing centres should take special account of these new tasks and, in addition to the traditional basic services (high-performance computing, storage systems, networks, user support), set up departments dedicated to special topics—in coordination with the basic services (Heßling et al. 2022), see Fig. 2.3.

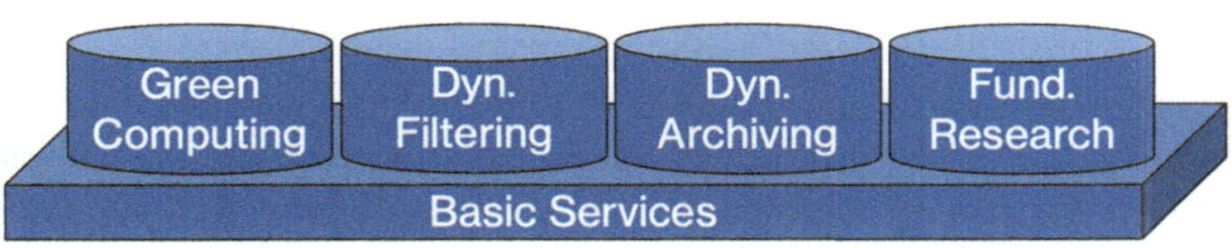

Fig. 2.3 Pillars of future computing centers

One of the main goals of "Dynamic Filtering" is to identify rare events already during data acquisition. For this purpose, it will be important to speed up the training significantly: It must be possible to quickly retrain sub-areas in a very high-dimensional parameter space of machine learning, where these sub-areas are correlated with characteristic properties, similar to the human brain, where different functions can be assigned to certain brain areas (e.g., speech center and vision center). This would allow to readjust control parameters of senors in real time. Here, the use of dedicated hardware could be helpful, such as the neuromorphic SpiNNacker chip system (Hoppner et al. 2022).

"Dynamic archiving" will considerably extend traditional offline computing. In addition to the already mentioned feedback to sensors in near real time, the fact that individual data objects exceed the petabyte limit has to be taken into account. For example, the thousands of telescopes and antennas of the SKAO will produce images of space with unprecedented resolution. To be able to analyze such "data monsters" sufficiently fast, the *memory wall* has to be overcome: processors wait "forever" for data from hard drives. The traditional processor-oriented computing must be replaced by a memory-based computing (Heßling et al. 2021b). Moreover, astronomy will have to take into account the fact that there will be significantly more simulation data (in high-energy physics, they account for about 50% of all data). Simulations are essential for understanding and verifying the information extraction decisions during dynamic filtering (Heßling et al. 2020).

Research on algorithms will be at the heart of future computing centers. Many areas face similar challenges. For example, data growth in biomedicine and genomic science is extremely high. The Internet of Things and Smart City will further intensify the data pressure. It can be assumed that the new methods to be developed for data reduction are sufficiently generic and suitable for various fields. This opens up new opportunities for green computing: in addition to the development of more energy-efficient hardware, the amount of data to be stored should be drastically reduced through the use of intelligent methods already during the data acquisition phase. The classic hardware-oriented Green Computing is being supplemented by Smart Green Computing (Heßling et al. 2021a),

2.2 Data Formats with Respect to Radio Astronomy

2.2.1 FITS Formats (in the Context of Big Data)

The Flexible Interchange Transport System (FITS) format (Wells and Greisen 1979) was originally created by a group of astronomers to overcome the compatibility issues between operating systems. The format was developed to be human and machine readable, self-documented, a "universally" readable format, and extensible. One of the features that have made FITS so popular is the edict to make FITS to remain forever backwards compatible (often referred to by the adage "Once FITS, always FITS"), making sure that any data written to FITS will always be accessible by a future generation of researchers.

Two formats have been developed to support storage of uv-data in FITS, being the FITS-IDI (storing the data in a binary table HDU; Greisen and Calabretta 2002) and the UVFITS (storing the data in a random group HDU; Greisen and Harten 1981) formats. In both formats the table contains rows for timestamp and baseline, and a multidimensional visibility array for that combination of timestamp and baseline. For timing data, the PSRFITS[2] format has been defined.

Even though the strong focus on backwards-compatibility makes the format very well-suited for community sharing of data, it also means that it is hard to implement optimisations for the computational challenges of big datasets, or to keep track of the provenance of data processing. The limitations of the format (see e.g., Thomas et al. 2015) have led to the adoption of other formats for radio telescope data, like MeasurementSet (van Diepen et al. 2018; Kemball and Wieringa 2000) and HDF5 (The HDF Group 2000–2010).

2.2.2 *Casacore MeasurementSet*

In 1992 the Astronomical Image Processing System (AIPS) was well developed and in common use. However, to meet growing needs it was decided in 1992 that the AIPS++ successor should be developed by an international team of experts. Some years later it was renamed to CASA (Common Astronomy Software Applications). In 2006 it was decided to split off Casacore, the core part of CASA, to separate the commonly applicable software from the software specifically developed for the ALMA and JVLA instruments. Casacore is now used at several radio telescopes around the world.

At the beginning of the AIPS++ project, Farris did a thorough analysis (van Diepen and Farris 1993) of the needs for the storage and processing of future radio-astronomical datasets, in particular visibility datasets. It showed that applications have quite different access patterns to the data, requiring flexible data access. For ad-hoc applications such as data inspection and plotting, flexible data selection is needed. It was concluded that the data system will play a key role for efficient and flexible data access to make the implementation of existing and new algorithms easily possible. The relational data model (Codd 1970) seemed quite well suited, but needed to be extended by the ability of storing large data arrays for a more efficient data storage format.

At that time only a few data storage systems existed. It was found that the Flexible Image Transport System (FITS; Wells and Greisen 1979) was quite basic and not well suited for the visibility data, certainly not for flexible data selection and access. At that time FITS was meant for transport and archival purposes using sequential

[2] https://www.atnf.csiro.au/research/pulsar/psrfits_definition/Psrfits.html.

access. The HDF group was in the early stages of the HDF5 development, the new version of their Hierarchical Data Format (HDF5; The HDF Group 2000–2010). The hierarchical nature of their system leads to a more rigid data ordering making it harder to get arbitrary data slices.

Therefore it was decided to design a data system, the Casacore Table Data System (CTDS; van Diepen 2015), meeting the requirements following from the analysis. The first development took place in 1993 and since then it has seen many improvements. It supports various important features:

1. Strong separation of logical and physical data model making it possible to choose the best storage representation for data.
2. The data can be stored in a tiled (chunked) way to make access in all directions equally fast.
3. A versatile SQL-based query language (TaQL; van Diepen 2010) can be used to easily select and manipulate the data in ad-hoc queries. It can be used in data inspection to reduce the size of a dataset by averaging, taking samples, etc.

CTDS is used as the basis for several datasets used in radio astronomy. The best known example is the MeasurementSet, but it is also used for the CASA calibration tables and images.

The initial version of the MeasurementSet could not represent all observation types. It was very soon replaced by version 2 (Kemball and Wieringa 2000) designed by Kemball and Wieringa. It consists of a main table containing the data (visibilities, flags, weights) and a number of subtables for the meta data such as antenna and frequency setup. It is very flexible as it can be extended with site specific columns and subtables. Figure 2.4 shows the schema of the MeasurementSet used by LOFAR.

MeasurementSet v2 has been in use for many years and still is. However, it has its shortcomings of which the most important is that the relational keys used to reference the subtables are based on the row number instead of an explicit id. With new instruments such as LOFAR and SKA its schema was too limited, especially for aperture telescopes with station fields consisting of many small antennae. For this reason MeasurementSet v3 (Dijkema et al. 2019) has been developed to address these issues. However, this version is not in common use yet.

At the start of the development of CTDS the amount of data was quite limited, but in the past years the amount of data has grown enormously. MeasurementSets can contain TeraBytes of visibility data that have to be stored and processed in an efficient way. It makes radio-astronomy science very data-intensive.

The separation of the logical and physical data model in CTDS proved to be very powerful. Dedicated storage managers have been developed to meet specific needs at various sites.

1. At the start of the LOFAR telescope, its data rate could only be handled by dumping the data directly to disk. Yet it was possible to handle the data as an normal MeasurementSet by developing the dedicated `LofarStMan` storage manager library which was automatically loaded by CTDS when needed.

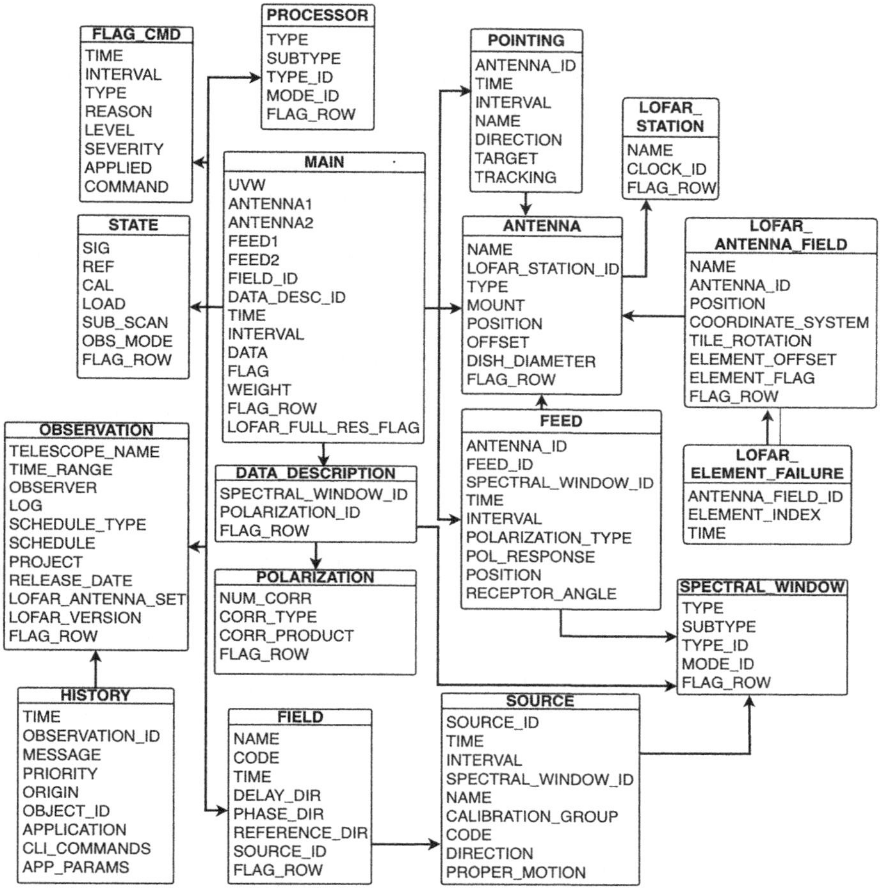

Fig. 2.4 The tables and their relations in the LOFAR MeasurementSets

2. Similarly, the `ASDMStMan` was developed by the CASA group to be able to treat ALMA datasets in its ASDM archive format as normal MeasurementSets avoiding an expensive copy of all data.
3. The `DyscoStMan` (Offringa 2016) was developed to compress the visibility data in a novel way. It is also implemented as a dedicated storage manager.
4. Unlike ADIOS (Liu et al. 2014), CTDS was not designed with parallel computing in mind. However, Wang and Tobar at ICRAR developed the Adios2StMan making it possible to use parallel IO for any CTDS table.

It is important to note that parallel computing can also be achieved without a parallel file system. The LOFAR data consists of many spectral bands, each 64-512 channels. Each spectral band is stored in a separate MS on a computing cluster such that each node typically holds 4 MSs. The flagging and calibration of each MS can be done independently whereafter the data can be averaged down considerably. A number of the resulting MSs (typically 10) have to be imaged jointly which requires some inter-node communication. If processing can be done that way, such a cluster scales linearly without the need of an expensive network.

The TaQL query language is a very important tool for MeasurementSet data inspection. It was used extensively in the Apertif pipeline for plotting purposes to select and integrate the visibility data. Its array functionality is a subset of numpy's; it also supports masks and units. The SQL-like GROUP BY clause can be used to operate on groups of data, for example per baseline. A few examples are given below to get an impression.

```
1.  SELECT mscal.ha() deg, means(DATA[FLAG], 0) from some.ms
2.  SELECT ANTENNA1,ANTENNA2,gmeans(DATA[FLAG])
          from some.ms GROUP BY ANTENNA,1ANTENNA2
3.  UPDATE some.img SET map = map - runningmedian(map,25,25)
```

1. This command averages in each row the unflagged data over all axes except 0. For a MeasurementSet it means averaging all frequencies. It also gives the hourangle in degrees using the `mscal` extension to TaQL.
2. This command forms a group for each ANTENNA1-2 combination, thus for each baseline. All data arrays are averaged element-wise, thus averaging is done over the time axis.
3. This command subtracts the background noise in an image using the median in a 51,51 box around each pixel. This is quite compute-intensive for large images.

2.2.3 ADIOS2/HDF5

The Hierarchical Data Format 5 (HDF5; The HDF Group 2000–2010) is being used in astronomy, including radio astronomy (see Sect. 2.2.2) already. The Adaptable Input Output System 2 (ADIOS2; Godoy et al. 2020) is much younger and less well-established in the astronomical community. ADIOS2 has been developed as part of the United States Department of Energy's Exascale Computing Project since 2016 and is a re-architecture of the original ADIOS code. Of the two, HDF5 arguably has a far broader community takeup and excellent tool support in many languages.[3] ADIOS on the other side had already been proven (Wang et al. 2016) to provide outstanding performance and scalability on typical supercomputing facilities and ADIOS2 continued this trend (Wang et al. 2020). There are distinct differences between the two, namely:

1. HDF5 is being used in a very broad range of sciences and industry.

[3] https://en.wikipedia.org/wiki/Hierarchical_Data_Format#Third-party_bindings.

2. HDF5's main paradigm, as the name suggests, provides hierarchical data structures. ADIOS2 essentially provides the layer below that. Hierarchies, or other high level structures, can be implemented on-top of ADIOS2.
3. HDF5 is mainly a file format with a supporting, highly optimised I/O library, while ADIOS2 also supports memory only and network transfers.
4. ADIOS2's architecture is based on parallel I/O, while in HDF5 this is an add-on.

ADIOS2 also supports the usage of the HDF5 Virtual Object Layer (VOL) and thus the two worlds can be merged.

2.2.4 Lossy Data Compression

Data is typically compressed to decrease the volume of the data, and reduce the amount of data transport necessary. The simplest method to do this is by averaging the data in the time and frequency directions. Particularly with wide-field instruments or high-resolution imaging (VLBI), care must be taken not to cause time or frequency smearing of sources in the target field, which may limit the amount of averaging possible. Averaging might also destroy information that might be relevant for selected science cases, such as spectral line studies that require high spectral resolution, or the search for transient objects that require high temporal resolution. In certain cases, a high data resolution might be required for calibration, such as to remove the sidelobes of strong sources over the target field.

To mitigate time and frequency smearing, it is possible to perform baseline-dependent averaging. With this approach, the time and frequency resolution limit for all baselines is no longer dictated by the longest baselines, but is rather scaled by the baseline distance between each pair of correlated antennas. This typically allows a further reduction of the data volume by factors of a few. Certain tasks, imaging in particular, may also speed up by this kind of averaging. The software support for this kind of averaged data is modest, and it is therefore not widely used in practice. Also, as for averaging in general, it might destroy information relevant for spectral or temporal studies.

The most widely used way to store a complex visibility, is in the form of two 32-bit single precision floating point values. Another method to reduce the data volume, is by storing every visibility in a compressed form. There is a little bit redundancy in each visibility (the exponent is for example will have a limited range), and this space can be reduced by using standard lossless compression methods from information theory. The amount of reduction in size reached with lossless methods is limited though, because visibilities are inherently noisy values, and random values are incompressible without loss. Lossy methods have therefore also been used, which typically involve linear quantisation. One could for example store the real and imaginary components of a visibility as 16 bit integer numbers, scaled in such a way that they cover the full range of expected values. When done correctly (i.e. with jittering), this leads to a bit of extra quantisation noise, which will normally be well under the system noise of the instrument. A step further is to apply the Dysco compression scheme (Offringa 2016). The Dysco scheme uses a normalisation scheme aimed at efficiently scaling radio data, together with a non-linear quantisation scheme and bit-packing of the result. The non-linear quantisation scheme is aimed to store Gaussian values with the lowest average least-squares error. This implies that more likely values (near zero) are more finely sampled compared to less likely values (away from zero). Compared to an approach with normal linear quantisation, Dysco can reach the same data quality with fewer bits per visibility. Using Dysco, complex visibilities can typically be compressed to two 6–10 bit numbers (hence compression by a factor of 5-3$\times$) with a noise addition that is well below the system noise.

2.3 Data Ingest into Archives—Data Management

The ingestion of data into the archive is the first step that is taken after an observation has been completed and first steps of post-processing, e.g., as described in Sect. 2.1, has been performed. The size and nature of the data, as well as the architecture of the archive, differs for each of the instruments. In this section, an overview of how data are ingested into the archive is given for several radio observatories.

2.3.1 LOFAR

The LOFAR (van Haarlem et al. 2013) Long-Term Archive (LTA; Renting and Holties 2011) is a distributed data archive. Data generated by the LOFAR telescope is pre-processed on a central processing cluster and distributed over multiple data centres for archiving. In 2023, three data centres contribute storage capacity to the LTA: SURF in the Netherlands, the Forschungszentrum Jülich in Germany, and the Poznan Supercomputing and Networking Center (PSNC) in Poland. Several petabytes of data are ingested into the archive each year. The data is registered in a catalogue based on the ASTRO-WISE system (see Valentijn and Belikov 2009), and made accessible through a web interface.[4]

The ingest workflow for LOFAR consists of several steps. The steps are illustrated in Fig. 2.5. First, metadata of the data products that is collected by the telescope management system is formatted in an XML document for exchange with the LTA systems. Then, the data products are transferred from the Central Processing system to the target LTA storage location as a data stream through a data transfer node. The destination for the data transfer is configured in the telescope management system, and typically set per science project. On each of the data

[4] https://lta.lofar.eu.

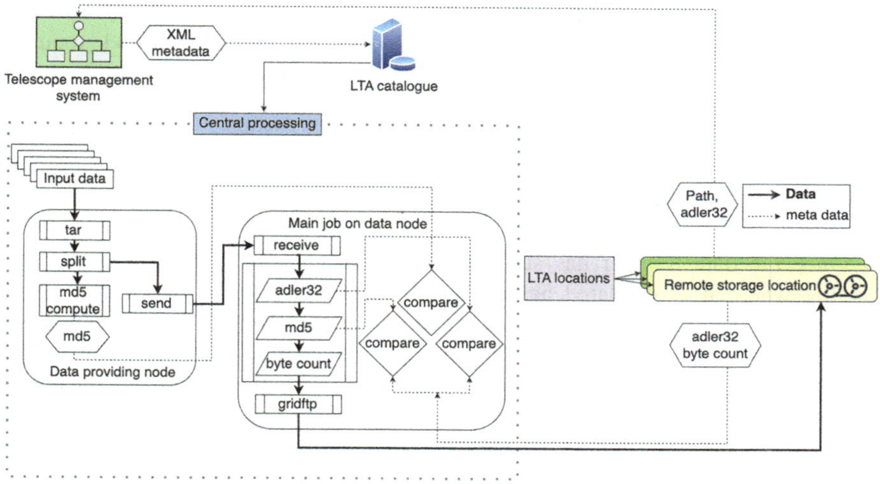

Fig. 2.5 Schematic of the LOFAR ingest flow. The data that has been processed on the Central Processing system of LOFAR (inside the dashed box) is read by several data providing nodes and gathered on a central data node from which it is sent to the remote storage system. To ensure data integrity over the system (i.e. that the data remains the same throughout the process), checksums and file sizes are compared at different places. Upon a correct copy, the telescope management system will provide the meta data, describing all properties of the observation and data products, to the catalogue

providing nodes on the central processing system, a script runs that reads the data from disk, packages multi-file data products in a single tar file, and duplicates the data stream where one stream is piped through an md5 checksum calculation algorithm while the other is forwarded to the data transfer node. At the end of the stream, the calculated checksum is also sent to the data transfer node. On the transfer node, the data stream is received and streamed through a custom-built process that computes byte count, adler32 (Deutsch and Gailly 1996) and md5 (Rivest 1992) checksums. The md5 checksum value computed by this tool is verified against the value computed on the data providing node. The reason that two different checksumming algorithms are used is that md5 is generally included in a standard Linux install, running on the data providing nodes, while the adler32 algorithm is used by default at the remote storage location.

At this stage, the data is stored using a temporary file path so that if anything goes wrong in the metadata registration of the file, new (retrial) transfers will not fail due the file already existing in the file system at the remote site.

The data stream received by the data transfer node is piped into a gridftp (Allcock et al. 2003) client which handles the transfer to a dCache[5] instance on the remote LTA site. The remote site will automatically initiate an Alder32

[5] https://www.dcache.org.

checksum computation, adding the checksum to the metadata of the file to the dCache file catalogue. Following successful transfer, the data transfer node requests the checksum and file size from the remote site and verifies those numbers against each other. In case of verification errors, the transfer will be automatically retried a set number of times. In case failures persist, the ingest will be marked as failed and require operator intervention for further handling. Once the transfer has successfully completed, the path on the destination storage is renamed to its definite value for archiving.

Once the data transfer has succeeded, the metadata is submitted to the LTA catalog where it is validated against a versioned schema and, when successful, persisted in the catalogue database, resulting in the data to be registered and therefore queryable. Data ingested into the LTA is available for direct retrieval from an online disk pool, typically for a week after the ingest. After this period, the data is only guaranteed to exist on tape and needs to be staged for download (see Sect. 3.2 for more information on the LTA hardware and architecture). Currently, no active data management takes place on archived data as long term availability of the data locations set at time of ingest is provided and there is no need for moving data between LTA sites.

2.3.2 *Apertif*

For Apertif (van Cappellen et al. 2022), the archive is called the Apertif Long-Term Archive (ALTA) . The ingest flow is shown in Fig. 2.6, and discussed below. The storage behind ALTA consists of 1.7 PB online storage, provided by four data servers located at ASTRON in Dwingeloo, and tape-based storage hosted by SURF in Amsterdam. Processed science-ready data products are stored on the online storage, and replicated to tape, while products with a lower level of processing are stored in the tape-backed archive only. The storage services are built on iRODS[6] which provides a rule-based interface for data management, as well as a single name space for users to access the data on the system. ALTA offers access through a REactJS[7] powered front-end in combination with a full REST[8] API access to the database. For more details on the ALTA see Holties et al. (2020). The ALTA data model has been designed such that the data in ALTA database entries can easily be mapped to VO compliant data structures as Apertif data releases are all published through VO services.

The custom part of the ingest procedure of data into ALTA is less complex than that of LOFAR, because the data logistics are taken care of by daemon services integrated with the iRODS backend. Ingest clients are provided on the science

---

[6] https://irods.org.

[7] https://reactjs.org/.

[8] https://www.django-rest-framework.org/.

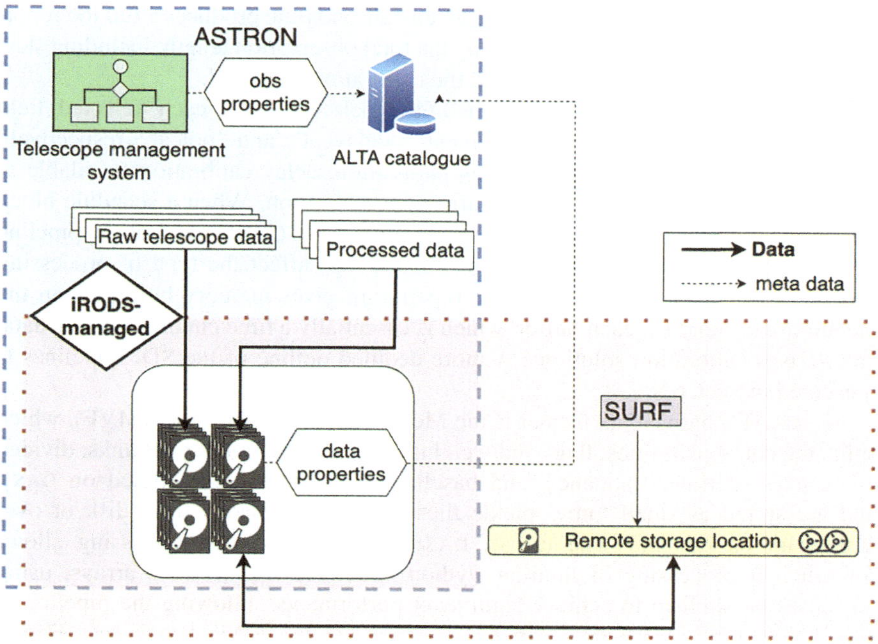

Fig. 2.6 Schematic of the Apertif ingest flow. Observational raw and processed data are both managed by iRods, which takes care of the data logistics. The main step in the ingest is extracting metadata from the data and putting it in the catalogue by the telescope management system

processing cluster. Files are transferred to a staging area in the ALTA iRODS environment where the main step in the ingest procedure is to extract the metadata from the files, and commit those to the observational database. On successful metadata extraction, files are moved to their (logical) archive location.

As is the case for the LTA, the tape data in ALTA needs to be migrated (replicated) to a disk pool before being user-retrievable. This is taken care of by submitting rules to the iRODS system where daemon processes take care of both staging the data as well as making online storage space available again after a configured time interval. Since it is not expected that users will regularly request data from tape, the staging process is not made directly available to users and this process requires users to submit helpdesk tickets to be picked up by Apertif Operations. More detials on the technical design of ALTA can be found in Sect. 3.4

2.3.3 *MeerKAT*

MeerKAT observers make use of the Observation Planning Tool (OPT), which manages the scheduling and simulation of a prospective observation. Running an observation through the simulator for a given date and time produces a full log report summarising the observation, specifying the total observation length, including slew time, time on spent on each target, and the calibration overhead.

During planning a MeerKAT schedule block via OPT, each selected field optionally includes tags 'bpcal', 'gaincal', 'delaycal', and 'polcal', respectively for bandpass calibration, complex gain calibration, delay calibration (available in combination via 'bfcal' tag), and polarisation calibration. When a schedule block ID (SBID) is executed, the Science Data Processor (SDP) calibration pipeline ('SDPcal') is run in real time, and the chosen tags affect the type of processing and reporting that is undertaken. The pipeline receives memory buffers from the 'accumulator', and for each buffer, which is essentially a time chunk, flags the data, and derives calibration solutions. A more detailed outline of the SDP pipelines is presented in Sect. 6.8.

MeerKAT's native data format is the MeerKAT Visibilty Format (MVF), which splits the data (visibilities, flags and weights) into a vast number of chunks, divided along axes of time, frequency, and baseline. These chunks are based on DASK and are stored as small '.npy' pickle files that can be loaded from disk or over a network on demand. The use of DASK arrays and graph processing allows for efficient processing of familiar Python structures like NUMPY arrays, using advanced parallelism to achieve significant performance, allowing the pipeline to run in real time. MVF data can be accessed via the KAT data access library (KATDAL).[9]

Once an observation is completed, the data, flags, and weights are stored along with the calibration solutions and metadata within a Ceph archive. This object-based storage cluster, accessible via the SARAO archive interface, is housed within the Centre for High Performance Computing (CHPC). This unique storage solution reduces data volumes compared to a scenario in which model and corrected data are stored separately, and allows different levels of solutions and flags to be applied to the data on the fly, giving the user much more control over their data. Figure 2.7 shows a flowchart summarising the real-time pipeline and its storage of the data products.

Once the data shows up on the SARAO archive, it can be accessed remotely via KATDAL, or can be converted to the standard MeasurementSet (MS) format during downloading and transferring the data. When downloading the data, the user is given the option for direct link or to push to a third party facility. A direct link provides a single URL for the user to download to any machine via WGET within 30 days. Pushing the MS data to a third party facility is available for limited facilities, such as CHPC and the Inter-University Institute for Data-Intensive Astronomy (IDIA), and makes use of an efficient transfer service using Globus/GridFTP, on a 10–100 GB/s network. During the initial download of the data, single time dumps of the data are downloaded and converted to MS, and written to Ceph storage, using dedicated transfer nodes as part of the SARAO archive transfer services.

[9] https://github.com/ska-sa/katdal.

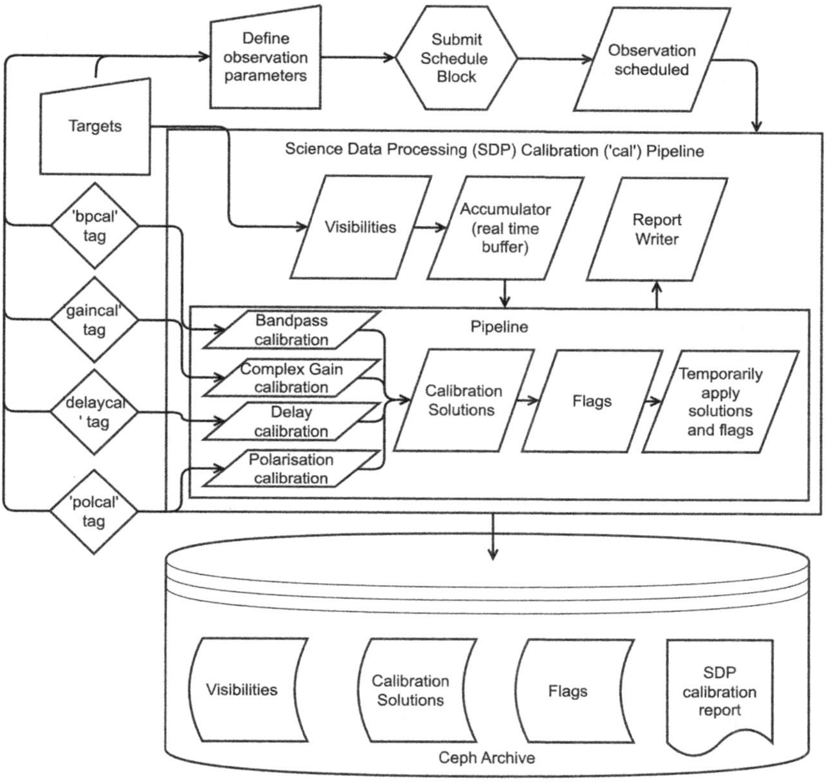

Fig. 2.7 Flow chart of MeerKAT observations and SDP calibration pipeline

During transferring MS data, the whole set of solutions and flags, stored under different levels of data and categories of flag, are available to the user for selection. Additional data selection and averaging parameters are also exposed to the user, including frequency channel, correlation/polarisation and data columns (i.e. including model data and corrected data columns), and time and frequency averaging.

2.3.4 ASKAP

ASKAP generates a vast amount of data. Signals from each telescope are correlated on-site at the MRO, the raw uv-visibilities are then transferred to the Pawsey Supercomputing Centre in Perth. The full uv-data rate amounts to about 70 to 80 PB per year. We do not attempt to store this huge amount of data. Instead the data are processed into science-ready data products (e.g. cubes, catalogues, spectra) by our custom software ASKAPsoft (see Sect. 6.9) and CSIRO ASKAP Science Data Archive (CASDA, see Sect. 3.3) makes these data available to the astronomical community around the world.

The ASKAPsoft data processing, CASDA deposit servers and CASDA storage are co-located at Pawsey. Data is processed and ingested into the archive on a 'scheduling block' basis. Once a scheduling block is ready for ingest into the archive, the ASKAPsoft pipeline generates an XML file which contains metadata that CASDA needs to ingest the data. The CASDA deposit module automatically recognises that a dataset is ready for ingest and initiates a deposit. The CASDA software additionally reads metadata from the individual FITS files to populate its database. The ASKAPsoft pipeline produces a MD5 checksum of each file and CASDA confirms after deposit that checksums match to ensure data integrity.

CASDA uses the new S3-based 60 PB object storage (Acacia) at Pawsey, The current size of data holdings is more than 2.5 PB, and this is expected to grow by 5 to 6 PB per year as ASKAP undertakes full survey operations. More detailed information on CASDA, including design philosophy, software stack, and details on services for users, can be found in Sect. 3.3.

2.3.5 *ALMA and MWA*

The Atacama Large Millimetre and Submillimetre Array (ALMA; Brown et al. 2004) archive has been designed as a distributed system from ground up. The concept of multiple regional centres, located in key geographical locations was a top-level requirement. The regional centres provide a number of critical functions for the observatory and its users, including locally accessible data products, processing capabilities as well as user and data reduction support. The ALMA data flow and archived volume is not that big (maximum <100 MiB/s), but the data requires a very complex set of meta data tables (ASDM[10]) to fully describe the intent, configuration and calibration of each observation and, further downstream, the science ready data products. In addition the telescope allows very flexible configuration options and observing strategies and the actual capabilities are constantly adjusted to the science demands. That also requires that the underlying data base schema needs to be adjusted accordingly.

The ALMA archive as a system consists of two, tightly integrated sub-systems, the ALMA Front-end Archive (AFA) and the ALMA Science Archive (ASA). The ASA provides the services commonly expected from an observatory data archive. In practice the AFA provides all required persistence, storage and database services for the whole observatory, including the storage and query services for configuration, monitoring and logging data as well as observation planning, scheduling and user and group management. The ASA can be seen as a specialised database view on

[10] http://aramis.obspm.fr/~alma/ssr/ASDM/asdm.html.

the AFA, collating the subset of information required by scientists to discover and access the data products they are interested in. Both the AFA and the ASA are replicated to the ALMA Regional Centres. In addition to the differentiation between AFA and ASA there is also a separation between the storage of bulk data and meta data. The bulk store is managed using a globally distributed object storage system called NGAS.[11] The meta data store is implemented as a high availability database system with one way replication across multiple sites.

The Murchison Widefield Array (MWA) is an instrument originating from a typical collaboration between research institutions (Tingay et al. 2013). Although there was a requirement to store and maintain the data longer term, the implementation initially lagged behind, although the data flow requirements are quite demanding (600 MiB/s).

References

W. Allcock, J. Bester, J. Bresnahan, S. Meder, P. Plaszczak, S. Tuecke, Gridftp: Protocol extensions to ftp for the grid (2003)

P.C. Broekema, J. Jan David Mol, R. Nijboer, A.S. van Amesfoort, M.A. Brentjens, G. Marcel Loose, W.F.A. Klijn, J.W. Romein, Cobalt: a GPU-based correlator and beamformer for LOFAR. Astron. Comput. **23**, 180 (2018). https://doi.org/10.1016/j.ascom.2018.04.006

R.L Brown, W. Wild, C. Cunningham, Alma – the atacama large millimeter array. Adv. Space Res. **34**(3), 555–559 (2004). ISSN 0273-1177. https://doi.org/10.1016/j.asr.2003.03.028. https://www.sciencedirect.com/science/article/pii/S0273117703011979. Astronomy at IR/Submm and the Microwave Background

E.F. Codd, A relational model of data for large shared data banks. Commun. ACM **13**, #6 (1970)

P. Deutsch, J.L. Gailly, Zlib compressed data format specification version 3.3 (1996). https://doi.org/10.17487/RFC1950

T.J. Dijkema, K. Golap, R. Hiriart, T. Nakazato, M. Pokorny, F. Pouzols, R. Raba, G. van Diepen, Measurementset definition version 3.0, casacore notes series 264. Technical report, Tech. Rep., Jan 2019. https://casacore.github.io/casacore-notes/264.html

W.F. Godoy, N. Podhorszki, R. Wang, C. Atkins, G. Eisenhauer, J. Gu, P. Davis, J. Choi, K. Germaschewski, K. Huck, A. Huebl, M. Kim, J. Kress, T. Kurc, Q. Liu, J. Logan, K. Mehta, G. Ostrouchov, M. Parashar, F. Poeschel, D. Pugmire, E. Suchyta, K. Takahashi, N. Thompson, S. Tsutsumi, L. Wan, M. Wolf, K. Wu, S. Klasky, Adios 2: the adaptable input output system. A framework for high-performance data management. SoftwareX **12**, 100561 (2020). ISSN 2352-7110. https://doi.org/10.1016/j.softx.2020.100561. https://www.sciencedirect.com/science/article/pii/S2352711019302560

E.W. Greisen, M.R. Calabretta, Representations of world coordinates in FITS. Astron. Astrophys. **395**, 1061–1075 (2002). https://doi.org/10.1051/0004-6361:20021326

E.W. Greisen, R.H. Harten, An extension of FITS for groups of small arrays of data. Astron. Astrophys. Suppl. **44**, 371 (1981)

R. Grunzte et al., The MASi repository service - comprehensive, metadata-driven and multi-community research data management. Future Gener. Comput. Syst. (2018). https://doi.org/10.1016/j.future.2017.12.023

[11] https://ngas.readthedocs.io.

H. Heßling, M. Kramer, S. Wagner, Data challenges at the Square Kilometre Array (SKA), in *EGI Conference 2020* (2020). https://indico.egi.eu/event/5000/contributions/14366/
H. Heßling, M. Kramer, S. Wagner, Next-generation green computing, in *EGI Conference 2021* (2021a). https://indico.egi.eu/event/5464/contributions/15661/
H. Heßling, M. Strutz, E.I. Buchholz, P. Hufnagl, On divide & conquer in image processing of data monster. Big Data Res. **25**, 100214 (2021b). https://doi.org/10.1016/j.bdr.2021.100214
H. Heßling, M. Kramer, S. Wagner, At the heart of future computing centers: research on algorithms, in *EGI Conference 2022* (2022). https://indico.egi.eu/event/5882/contributions/16760/
S. Higgins, The DCC curation lifecyle model. Int. J. Digit. Curation **3**, 134 (2008)
H. Holties, R. de Goei, N. Vermaas, ALTA, the APERTIF long term archive, in *Astronomical Data Analysis Software and Systems XXIX*, ed. by R. Pizzo, E.R. Deul, J.D. Mol, J. de Plaa, H. Verkouter. Astronomical Society of the Pacific Conference Series, vol. 527 (January 2020), p. 255
S. Höppner et al., The SpiNNaker 2 processing element architecture for hybrid digital neuromorphic computing (2022). https://arxiv.org/pdf/2103.08392.pdf
M. Izzo et al., XTENS - a JSON-based digital repository for biomedical data management, in *Third International Conference, IWBBIO 2015*, Granada, April 15–17, 2015, Proceedings, Part I & II (2021), pp. 123–130. https://www.researchgate.net/publication/276284556_XTENS_-_A_JSON-Based_Digital_Repository_for_Biomedical_Data_Management
C. Jung, J. Meyer, A. Streit, Helmholtz portfolio theme "Large-Scale Data Management and Analysis (LSDMA)". KIT Scientific Publishing (2015)
M.J. Keith et al., The high time resolution universe pulsar survey - I. System configuration and initial discoveries. Mon. Not. R. Astron. Soc. **409**, 619–627 (2010)
A. Kemball, M. Wieringa, Measurementset definition version 2.0, casacore notes series 229. Technical report, Tech. Rep. (Jan 2000). https://casacore.github.io/casacore-notes/229.html
Q. Liu et al. Hello adios: the challenges and lessons of developing leadership class i/o frameworks. Concurr. Comput. **26**(7), 1453–1473 (2014)
J. Odier et al., The ATLAS Metadata Interface (AMI) 2.0 metadata ecosystem: new design principles and features. EPJ Web Conf. **214**, 05046 (2019). https://doi.org/110.1051/epjconf/201921405046
A.R. Offringa, Compression of interferometric radio-astronomical data. Astron. Astrophys. **595**, A99 (2016). https://doi.org/10.1051/0004-6361/201629565
L. Pouchard, Revisiting the data lifecycle with big data curation. Int. J. Digit. Curation **10**, 176 (2015). https://doi.org/10.2218/ijdc.v10i2.342
G.A. Renting, H.A. Holties, LOFAR long term archive, in *Astronomical Data Analysis Software and Systems XX*, ed. by I.N. Evans, A. Accomazzi, D.J. Mink, A.H. Rots. Astronomical Society of the Pacific Conference Series, vol. 442 (July 2011), p. 49
R. Rivest, The md5 message-digest algorithm (1992). https://doi.org/10.17487/RFC1321
The HDF Group, Hierarchical data format version 5 (2000–2010). http://www.hdfgroup.org/HDF5
B. Thomas, T. Jenness, F. Economou, P. Greenfield, P. Hirst, D.S. Berry, E. Bray, N. Gray, D. Muna, J. Turner, M. de Val-Borro, J. Santander-Vela, D. Shupe, J. Good, G.B. Berriman, S. Kitaeff, J. Fay, O. Laurino, A. Alexov, W. Landry, J. Masters, A. Brazier, R. Schaaf, K. Edwards, R.O. Redman, T.R. Marsh, O. Streicher, P. Norris, S. Pascual, M. Davie, M. Droettboom, T. Robitaille, R. Campana, A. Hagen, P. Hartogh, D. Klaes, M.W. Craig, D. Homeier, Learning from fits: limitations in use in modern astronomical research. Astron. Comput. **12**, 133–145 (2015). ISSN 2213-1337. https://doi.org/10.1016/j.ascom.2015.01.009. https://www.sciencedirect.com/science/article/pii/S2213133715000104
D. Thornton et al., A population of fast radio bursts at cosmological distances. Science **341**, 53–56 (2013)
S.J. Tingay, R. Goeke, J.D. Bowman, D. Emrich, S.M. Ord, D.A. Mitchell, M.F. Morales, T. Booler, B. Crosse, R.B. Wayth, et al., The murchison widefield array: the square kilometre array precursor at low radio frequencies. Publ. Astron. Soc. Aust. **30**, e007 (2013). https://doi.org/10.1017/pasa.2012.007

E. Valentijn, A.N. Belikov, Lofar information system design. Memorie della Societa Astronomica Italiana **80**, 509 (2009)

W.A. van Cappellen, T.A. Oosterloo, M.A.W. Verheijen, E.A.K. Adams, B. Adebahr, R. Braun, K.M. Hess, H. Holties, J.M. van der Hulst, B. Hut, E. Kooistra, J. van Leeuwen, G.M. Loose, R. Morganti, V.A. Moss, E. Orrú, M. Ruiter, A.P. Schoenmakers, N.J. Vermaas, S.J. Wijnholds, A.S. van Amesfoort, M.J. Arts, J.J. Attema, L. Bakker, C.G. Bassa, J.E. Bast, P. Benthem, R. Beukema, R. Blaauw, W.J.G. de Blok, M. Bouwhuis, R.H. van den Brink, L. Connor, A.H.W.M. Coolen, S. Damstra, G.N.J. van Diepen, R. de Goei, H. Dénes, M. Drost, N. Ebbendorf, B.S. Frank, D.W. Gardenier, M. Gerbers, Y.G. Grange, T. Grit, A.W. Gunst, N. Gupta, M.V. Ivashina, G.I.G. Józsa, G.H. Janssen, A. Koster, G.H. Kruithof, S.J. Kuindersma, A. Kutkin, D.M. Lucero, Y. Maan, F.M. Maccagni, J. van der Marel, A. Mika, J. Morawietz, H. Mulder, E. Mulder, M.J. Norden, A.R. Offringa, L.C. Oostrum, R.E. Overeem, Z. Paragi, H.J. Pepping, E. Petroff, D.J. Pisano, A.G. Polatidis, P. Prasad, J.P.R. de Reijer, J.W. Romein, J. Schaap, G.W. Schoonderbeek, R. Schulz, D. van der Schuur, A. Sclocco, J.J. Sluman, R. Smits, B.W. Stappers, S.M. Straal, K.J.C. Stuurwold, J. Verstappen, D. Vohl, K.J. Wierenga, E.E.M. Woestenburg, A.W. Zanting, J. Ziemke, Apertif: phased array feeds for the Westerbork Synthesis Radio Telescope. System overview and performance characteristics. Astron. Astrophys. **658**, A146 (2022). https://doi.org/10.1051/0004-6361/202141739

G. van Diepen, Note 199 – table query language (2010). https://casacore.github.io/casacore-notes/199.html

G. van Diepen, Casacore table data system and its use in the measurementset. Astron. Comput. **12**, 174–180 (2015). https://doi.org/10.1016/j.ascom.2015.06.002

G. van Diepen, A. Farris, The aips++ table data system, aips++ project (1993)

G. van Diepen, T.J. Dijkema, A. Offringa, DPPP: default pre-processing pipeline. Astrophysics Source Code Library, record ascl:1804.003 (April 2018)

M.P. van Haarlem, M.W. Wise, A.W. Gunst, G. Heald, J.P. McKean, J.W.T. Hessels, A.G. de Bruyn, R. Nijboer, J. Swinbank, R. Fallows, et al., LOFAR: The LOw-Frequency ARray. Astron. Astrophys. **556**, A2 (2013). https://doi.org/10.1051/0004-6361/201220873

R. Wang, C. Harris, A. Wicenec, Adiosstman: parallelizing casacore table data system using adaptive io system. Astron. Comput. **16**, 146–154 (2016). ISSN 2213-1337. https://doi.org/10.1016/j.ascom.2016.05.003. https://www.sciencedirect.com/science/article/pii/S2213133716300452

R. Wang, R. Tobar, M. Dolensky, T. An, A. Wicenec, C. Wu, F. Dulwich, N. Podhorszki, V. Anantharaj, E. Suchyta, B. Lao, S. Klasky, Processing full-scale square kilometre array data on the summit supercomputer, in *SC20: International Conference for High Performance Computing, Networking, Storage and Analysis* (2020), pp. 1–12. https://doi.org/10.1109/SC41405.2020.00006

D.C. Wells, E.W. Greisen, FITS - a flexible image transport system, in *Image Processing in Astronomy*, ed. by G. Sedmak, M. Capaccioli, R.J. Allen (January 1979), p. 445

Chapter 3
Storage and Long Term Archives

Cristina Manzano, Kilian Schwarz, Yan G. Grange, Arpad Miskolczi, Minh T. Huynh, Andreas Wicenec, Jesús Salgado, and Vandana Desai

Our whole society is facing the risk of producing digital dark ages, where people in a few hundred years or so will not be able to access or even read vast amounts of the data we are collecting now. In our digital age there is nothing even vaguely comparable to the long-term preservation capability of information encoded with some symbols on physical media like stones or paper. The Rosetta stone did not

C. Manzano (✉) · A. Miskolczi
Jülich Supercomputing Centre (JSC), Forschungszentrum Jülich, Jülich, Germany
e-mail: c.manzano@fz-juelich.de; a.miskolczi@fz-juelich.de

K. Schwarz
Deutsches Elektronen-Synchrotron DESY, Hamburg, Germany
e-mail: kilian.schwarz@desy.de

Y. G. Grange
ASTRON, Dwingeloo, Netherlands
e-mail: grange@astron.nl

M. T. Huynh
CSIRO Space and Astronomy, Bentley, WA, Australia

International Centre for Radio Astronomy Research (ICRAR), The University of Western Australia, Crawley, WA, Australia
e-mail: Minh.Huynh@csiro.au

A. Wicenec
International Centre for Radio Astronomy Research, Crawley, WA, Australia
e-mail: andreas.wicenec@icrar.org

J. Salgado
SKA Observatory, Jodrell Bank, Lower Withington, Macclesfield, Cheshire, UK
e-mail: jesus.salgado@skao.int

V. Desai
Infrared Processing & Analysis Center, California Institute of Technology, Pasadena, CA, USA
e-mail: desai@ipac.caltech.edu

E. Vardoulaki et al. (eds.), *Data-Intensive Radio Astronomy*, Astrophysics and Space Science Library 472, https://doi.org/10.1007/978-3-031-58468-8_3

need any active preservation for thousands of years, but we have a very hard time even just reading tapes from a few decades ago.

In order to avoid this, groups like the Long-Term Data Preservation Task Force[1] within the EOSC Association (see Sect. 12.4) try to define the functions of repositories and data services necessary to create, store, curate, preserve, and engage with data (re)users. For instance, a Long Term Archive (LTA) should not just hold data, for some of the current LTA implementations would not live up to a proper long-term vision of retaining the data let alone access to it for various reasons, one of them being shortage of funding. An LTA should not simply be an extension of a project or even observatory archive but should preserve the data to lengthen their use over time and scope, including maintaining it even after a program has finished. The long-term digital preservation is an exercise that guarantees sustained access to digital objects, or in any case to the information comprised in them, indefinitely. For the organisations holding these archives this mean they should curate the digital objects with a long term FAIR (Wilkinson et al. 2016) enabling perspective on them and their designated community. FAIR stands for Findable, Accessible, Interoperable and Reusable and is further discussed in Sect. 14.1.

Not less important are open data policies. An LTA should be able to enable science allowing the possibility to reproduce, replicate and re-use scientific results, these depending like already mentioned on the long-term accessibility and assessibility of the underlying data. The users of these data are a certain collection of potential consumers who should be able to understand this particular set of information, but although this initial user community is firstly defined by the archive, this definition may change over time and new users should be allowed to re-use the data.

3.1 Storage Systems and Caching

In the next decade, astronomical observatories, such as the SKA, will be among the largest civilian data producers. This development is based on a continuous buildup of modern observatories, such as LOFAR or MeerKAT/ASKAP, which can be considered preliminary studies for the development of the SKA.

In order to cope with this huge amount of data, a heterogeneous structure is needed, a central storage infrastructure which provides storage capacity on different tiers, based on different software solutions or media. This strategy is used as well in the storage of HPC clusters and it plays an important role in a Modular Supercomputing Architecture (see Sect. 4.2).

The spectrum of different storage layers ranges from high-performance flash memory, spinning disks to tape capacity (for archives and backup). In this tiered solution new data are first ingested to the fast but more expensive media (spinning

[1] https://www.eosc.eu/advisory-groups/long-term-data-preservation.

disks, flash) and then migrated to tape libraries which are slower but inexpensive. The data which needs to be processed is first staged to cache (if not already there) and kept there as long as it is still needed. This approach offers both long term archiving on tape and efficient access to the international team of researchers for processing the data at a reasonable cost.

3.2 The LOFAR LTA in Jülich

The data collected at the LOFAR radio telescope is sent first to the central processor in Groningen, and then distributed later to one of the three sites which currently form the LOFAR Long Term Archive (LTA) via dedicated private networks: SURF in the Netherlands, Jülich Supercomputing Centre (JSC) in Germany and PSNC in Poland, see Sect. 2.3.1.

In this section, we describe the way the LTA infrastructure is set up at the JSC. However the LTA has been set up in such a way that the archive is accessible in the same way at any of the locations.

In the case of Jülich, the LOFAR LTA consists of a distributed heterogenous storage system based on the dCache (dCache 2022) software. dCache allows storing and retrieving huge amounts of data, distributed among a large number of different server nodes, under a single virtual file system tree with a variety of standard access methods.

The distributed storage system is made up of spinning disks for "online" or "hot" data and tapes for "offline" or "cold" data. The data ingested to the archive is first copied to the cache (disk pools) and asynchronously migrated to tape (Fig. 3.1) in a background process. The data which needs to be accessed is first staged to cache (if not already there) and kept there as long as it is still needed. The software interface ENDIT (Wadenstein et al. 2022) takes care of the communication between disk and tape. This approach offers both long term archiving on tape and efficient access to the international team of researchers for processing the data.

It is possible to do on-site processing of the large data volumes at the supercomputing facility of the JSC. For doing so, a gateway server gives access to the facility-wide file systems of the supercomputers, and provides grid tools for data access and data transfer.

LOFAR users who wants to access the data in the LTA go first to a web portal in ASTRON where they can select the data they are interested in (i.e. observations). From the archive view, users can choose to request that data be staged. This request is handled by a custom-built staging service which takes care of managing (e.g. retries, user communication, etc.) the staging requests for authorised users. This process will trigger a request in the site or sites (if the data is distributed among several of the LTAs) containing the data. In Jülich this request will be sent to the dCache management server, which will start a bring online process. This process is handled transparently to the user, who will only receive a notification once the data has been brought online.

Fig. 3.1 Tape library in the Jülich Supercomputing Centre. The data ingested to the LOFAR archive is kept on tape for long term archiving. Copyright: Forschungszentrum Jülich/ Ralf-Uwe Limbach. Used with permission

When the data that a user requested is brought online, also called "staged", the user has two options of getting the data. The first method is to use Grid Computing tools to download the data. For users that do not have access to these tools, the Jülich LTA offers regular HTTP downloads as the second method to retrieve data. For this purpose there is a server at the Jülich LTA that acts as an intermediary between the global users and the main dCache server. To able to download the data through this server, the user needs to authenticate with basic username/password credentials. These are checked through an LDAP server. If successful the download server will connect to the main dCache server and request the data. The server downloads

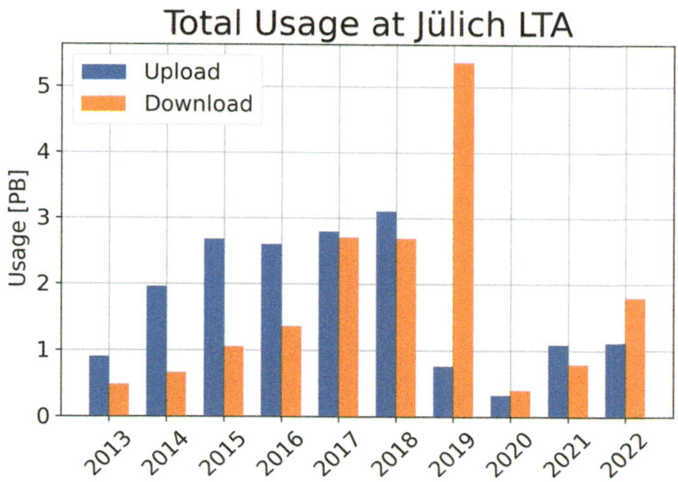

Fig. 3.2 Download and upload usage of the Jülich LTA per year

the data from dCache and streams it through regular TCP connections to the user somewhere on the globe.

This approach allows many users to download LOFAR data without needing to go through processes to acquire a certificate. This led to a high demand of data in 2019, as seen in Fig. 3.2 when the LoTSS data became public and many scientists started downloading the data for their specific research.

As of October 2022 the LTA at Jülich stores about 17.3 PB of LOFAR data. As shown in Fig. 3.3 the amount of stored data increased almost linearly since early 2013 with a temporary slow down from 2019 to 2021.

The obvious disadvantage using this approach is that every user that is concurrently downloading data shares the capabilities of the download server. In extreme cases this can lower the download speed significantly.

In the near future, the intention is to replace the dedicated download server by WebDAV services on the dCache nodes to which the authentication is done through Macaroons. Macaroons are a form of cookies containing a bearer token that can be used to perform user authorization. In addition the macaroons can be configured with multiple constraints, called caveats. For example, the macaroon can be configured to only apply to certain files or to work only during a limited time range. This macaroon can also be shared which can allow the downloading of files without exposing ones password or certificates.

Another advantage of the WebDAV protocol is its ability to redirect traffic. When enabled the download or upload request of a user can be redirected to the pools directly instead of using the bandwidth provided by dCache management node. At sites with multiple pools and a high traffic this can provide a more stable user experience and better utilize the available resources of the facility.

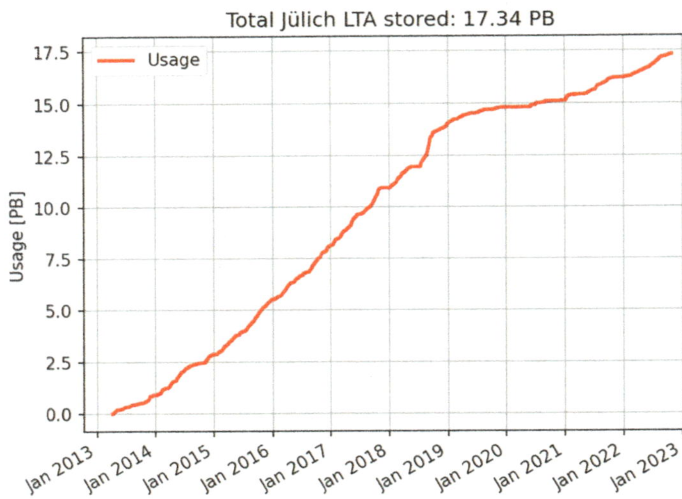

Fig. 3.3 Total usage at Jülich LTA from 2013 to October 2022

3.3 CASDA-ASKAP Archive

The CSIRO ASKAP Science Data Archive (CASDA) is a key part of the ASKAP system. Data from the observatory is processed by custom software ASKAPsoft (see Sect. 6.9) into science-ready data products, and CASDA then makes that data available to the astronomers across the world.

3.3.1 Design and Software Stack

The CASDA infrastructure and software application is deployed across two data centres, Pawsey in Perth and the CSIRO Canberra Data Centre in Canberra. CASDA shares the front end systems, e.g. web user interface, authentication and authorisation systems, with the CSIRO enterprise-wise Data Access Portal. The ASKAP data is stored and served from Pawsey. The operations that need to be close to the data are at Pawsey, e.g. deposit, data access, and Virtual Observatory services.

The CASDA modules are written in Java and use the Spring application framework. CASDA uses a PostgreSQL database running PGsphere to allow for quick queries and calculations with astronomical coordinate systems. CASDA requests and communications are done via RESTFUL web services. Job management and execution is done with SLURM.

The original CASDA infrastructure at Pawsey used a tape HSM system backed up by a CXFS filesystem. In this instance we managed storage to the tape system using the Next-Generation Archive System (NGAS) (Wu et al. 2013). Pawsey made available a 60 PB disk-based object storage system (named Acacia) in early 2022, and CASDA has fully migrated to this new disk-based system, and with this migration NGAS is no longer needed.

The philosophy behind CASDA is to make data easy to find and access, and lower the barrier for use and reuse of radio data. To this end we have implemented Virtual Observatory services throughout the system. Users also do not have to be radio interferometry experts as CASDA serves processed science-ready data. ASKAP Survey Science Teams provide basic data quality assessment and validation notes for ASKAP data. The data is team-only access before release, but is fully public after release.

3.3.2 Services and Data Types Supported

CASDA archives science ready data products produced by the ASKAPsoft pipeline (Sect. 6.9). The data formats are:

- Images and image cubes, in FITS format. These can be up to 3.5 TB in size.
- 1D spectra, in FITS format.
- Catalogues, in VOTable format.
- Visibilities, continuum resolution mostly, as CASDA measurement sets.

CASDA users query and access CASDA data products via the CSIRO Data Access Portal (DAP) user interface. The DAP also provides tools for validation and release of data products. The DAP generates Digital Object Identifiers (DOIs) for the various collections in CASDA.

The CASDA Observation Search portion of the DAP UI allows astronomers to search for ASKAP data via a simple web form. The DAP UI also has the CASDA Skymap Search service which allows astronomers to visually search for an object on the sky with an Aladin-lite interface. Clicking on a catalogued object then shows a summary page of the ASKAP measurements for that radio source and user can then download the associated ASKAP data.

We have very recently implemented a new service in the DAP UI to create cutouts around target objects from image datasets curated by ASKAP survey science teams. The initial dataset available through this cutout service is the Rapid ASKAP Continuum Survey Data Release 1 (AKA RACS-Low). This dataset is described in (Hale et al. 2021) and comprises uniform mosaicked tiles from the first RACS survey. In future, additional curated datasets from the various ASKAP Survey Science teams will be added to this cutout service. This CASDA Cutout Service is in addition to the cutout tool in the CASDA Observation Search, which a user can use to generate cutouts of specific images/cubes from a CASDA search result. This allows a user to download a region and channels of interest, instead of whole images/cubes which can be as large as 3.5 TB.

Virtual Observatory services make data easier to query and access. CASDA has a Table Access Protocol (TAP) service and ObsCore table for queries of catalogues and observational data products. We have also implemented Simple Cone Search (SCS) and Simple Image Access Protocol (SIAP) for access to catalogues and images/cubes. Server Side Operations for Data Access (SODA) is used to generate cutouts of images and cubes.

The VO protocols provide the backbone to some of the services (e.g. cutouts) provided through the UI, and allow for scripted access. We have contributed a CASDA module as part of the Python Astroquery package. This module provides an easy to use API for querying what data is available, accessing data products, and producing and downloading cutouts from the large images and cubes.

CASDA provides a lightweight URL endpoint for users to discover CASDA related events. This endpoint provides a list of VOEvent (as .xml) which describe when data is deposited, validated, released and updated in CASDA. It can be filtered by time (of the event in CASDA), type of event, project code and scheduling block ID. A user can use this endpoint to identify when a certain scheduling block has been deposited. Another use case is this endpoint could be queried by a science team to find out if new data has been deposited for their project, and they could automatically start a workflow based on that event.

3.3.3 Outlook

Over the last few years CASDA usage has grown as pilot survey data from the ASKAP Survey Science teams have been deposited and released in CASDA. With holdings now totally over 2.5 PB, CASDA usage has grown massively, averaging more than 500 web UI queries per month, and more than 3000 VO (non-HIPS) queries per month. We expect usage to grow even further with full ASKAP surveys having just begun (end of 2022). The CASDA data growth rate is expected to be 5 to 6 PB per year during full surveys.

The move to object storage at the Pawsey Supercomputing Centre allows us to meet the growth in data and storage, as it provides scalability. The metadata and 'buckets' associated with object storage allows for faster retrieval than traditional file or block-based systems. The disk-based system also results in better performance for users, as files can be served immediately instead of being queued to come off tape. We expect further improvements for users as we, and the astronomy community, learn to take full advantage of object storage in the future.

We plan to move to virtual machines at the Pawsey Supercomputing Centre and containerisation of the CASDA software application and modules. This will provide a more stable and reliable service, and an archive software suite which is easier to maintain and deploy.

We are also in the process of integrating the Cube Analysis and Rendering Tool for Astronomy (CARTA; Comrie et al. 2021) visualisation toolinto CASDA. CARTA is specifically designed for large images and cubes such as those produced from ASKAP and stored in CASDA. With this web-based application, users will be able to view and do basic analyses of ASKAP/CASDA images and cubes in a browser, reducing the need for large file downloads and improving astronomer workflows.

3.4 Apertif Long Term Archive

The Apertif Long-Term Archive (ALTA; Holties et al. 2020; van Cappellen et al. 2022) hosts the raw data generated by the instrument and the products generated by the standard processing pipelines developed by the Apertif science teams. It has been designed to hold millions of data products, adding up to 20 petabytes of data over the operational lifetime of the instrument.

The bulk of the data is stored on magnetic tapes (the "cold" storage). Those tapes are hosted at the Dutch national high-performance computing and e-Science support centre (SURF). For users to access the data that has been stored on the tapes, that data first needs to be copied to a local buffer of "online" storage (i.e. hard disk). To provide users with low-latency access to advanced data products, 1.8 PB of online storage is available that is hosted at ASTRON in Dwingeloo. Besides the storage, the architecture includes a PostgreSQL database server and a web server hosted at ASTRON, providing users with access to the data through the ALTA web interface. Scientific data releases are published on the ASTRON Virtual Observatory server which provides datalinks to the ALTA webserver for retrieving underlying data products. Internal and external network connections used for data transfers are based on 10 and 40 Gbit/s ethernet technology.

The data management within ALTA is based on the iRODS framework.[2] In iRODS, storage resources are organised in zones, which provide a logical namespace for data on physical storage that is managed by a given organisation. For ALTA, one zone has been created in iRODS. This zone spans the four data servers at ASTRON as well as the data server at SURF which acts as the buffer for the tape back end. Validation of data products during the ingest process, extraction of the metadata during the ingest process as well as the generation of inspection plots, and the migration between tape and disk storage are all services implemented in python and executed as scheduled services in the iRODS environment. The system allows for potential future implementation of user-oriented services (e.g. cutouts, image stitching).

Typically, advanced data products (e.g. FITS images or profiles from FRB detections in PSRFITS format) are kept on the online system and become part of the released data. Data collections that have been publicly released are made accessible through the Web Distributed Authoring and Versioning (WebDAV) protocol. However, not all data are made available for direct access. In particularly, low-level data products (e.g. raw visibilities, calibrated visibilities and corresponding calibration tables) are stored on tape only. This means that direct access to those products is

[2] https://irods.org.

not provided and the data needs to first be transferred to an online resource (this is known as "staging"). For ALTA, staging is performed on request by the operational team on a best effort basis.

The data model in ALTA builds on the same principles as the LOFAR data model, and is designed such that it can be mapped to data structures compliant with the Virtual Observatory (VO; also see Chap. 12). Provenance information is incorporated in the data model, highly inspired by W3C provenance Data Model (Moreau et al. 2013), adapted to fulfill the constraints of the Apertif data. The model is process-oriented and offers access to, and enabling the inspection of, the full chain of processing steps starting from the observations. The metadata are stored in the ALTA ProstgreSQL database.

The main access to ALTA for the Apertif science teams is through the web interface.[3] The system provides interactive access to users through a Graphical User Interface (GUI) as well as automated access through the Application Programming Interface (API). The GUI is provided by a ReactJS[4]-based front end, and the RESTful backend API that has been built using Django[5] and the Django REST framework.[6] The REST API exposes the full data model (including provenance information) and handles the interaction with the database and the data management component. For data that has not (yet) been released to the public domain, authentication and authorisation (AAI) policies are applied.

Although all data is hosted by the ALTA services describes above, data releases are published in the VO which provides the main intercase for users that are not a member of the Apertif science teams. The directly-accessible data products are directly linked and the metadata of the data on tape is exposed through VO tables. A link to the helpdesk where users can request for the data to be staged is provided where applicable. For more information on VO data collections at ASTRON in general, see Sect. 12.3.

3.5 ALMA

The ALMA archive architecture followed a different philosophy than most of the other archives described so far in the sense that it does not only offer the traditional archive and long-term archive functionality in a globally distributed regional centre setup. Those functions are based on the implementation of a persistence layer underpinning the whole data flow and integrates all persistence needs for almost all data and meta-data required by the observatory. This includes user data bases, hardware and environmental monitoring data, telescope configuration and observation planning data and all intermediate and final data products. The ALMA

[3] https://alta.astron.nl.

[4] https://reactjs.org.

[5] https://www.djangoproject.com/.

[6] https://www.django-rest-framework.org/.

archive subsystem is so fundamental, that it has to be started as one of the first systems during a cold-start procedure of the observatory. In order to achieve this, the central, distributed database has been designed and implemented using industry standard high-availability technologies, including automatic fail-over, heartbeats and synchronization between multiple replicated server instances. Replication to the three main regional centres in North America, Japan and Europe of both the databases and the science data holdings is also part of the ALMA archive system. The bulk science data is not ingested into the database, but kept on a separate, globally distributed and fully scalable storage server system NGAS,[7] but the required meta-data for that system is also part of the database service. Following this persistence layer philosophy, ALMA observational data is ingested into the archive directly from the correlator, there are only two other real-time subscribers accessing parts of that data stream in addition. All other subsystems are getting access to the observational and all other data through the archive. Currently the ALMA ingest data rates are maxing out at a moderate 64 MB/s, but an upgrade of the array and the correlator is planned to increase this by about an order of magnitude. The main limitation is not the archive system, but the network connecting the high site at 5000 m and the operations facility at 3000 m altitude, as well as the internal network switching infrastructure. As can be seen in Fig. 3.4 the system implements three different stores, the BulkStore, the XMLStore and the EventStore and provides access through the Observatory Data Model and the Science Data Model, which are the two main logical models used by ALMA. On-top of the ALMA Science Data Model sits the ALMA Science Archive, which provides the user facing access point for science queries and data requests.

The ALMA science data model is quite complex and consists of about 60 interrelated tables. Most of the ALMA data models, including the science data model are implemented as XML schemata. Consequently also the original ALMA archive implementation used a XML database, based on a traditional relational engine (Oracle).

3.6 Data Transfer Protocols

GridFTP allows sending data files over multiple TCP streams using multiple sockets across the same network path maximizing its use. GridFTP is an extension of the FTP protocol, supporting parallel and striped transfers. Also the protocol implements third-party transfer, which means directly transferring data between two end points from e.g. your local laptop. The authorisation and authentication is done by using X.509 certificates, which need to be obtained by the user through a trusted party, which originally was implemented by identifying users to a trusted employee.

---

[7] https://ngas.readthedocs.io.

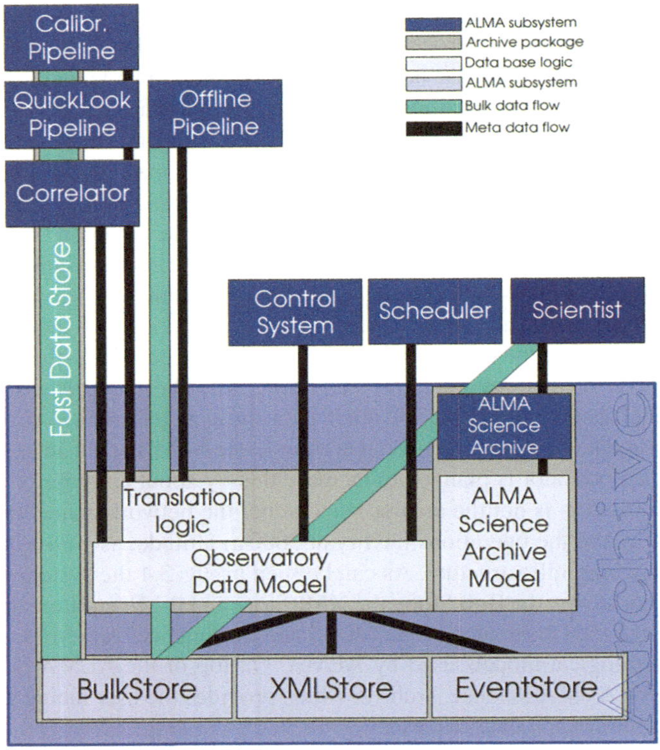

Fig. 3.4 ALMA Archive architecture showing the archive subsystem, represented by the light-blue box covering the lower half of the figure, providing persistent services to all other subsystems. Figure from original ALMA Archive Design document, 2002

This method was later replaced by federating the AAI for the certificate request to the users employer.

In recent times, the main GridFTP implementation has become closed-source and several instruments and collaborations have decided to adopt an alternative protocol for data transfers (e.g. Webdav using macaroons for AAI). Software like dCache (used in the LOFAR LTAs) runs GridFTP doors, which handle the communication with the client for ingesting and staging data.

In the case of LOFAR, ASTRON uses a Virtual Organization Membership Service (VOMS) server which keeps a list of users who registered with their personal grid certificates and their roles in the Virtual Organization (VO) LOFAR. The LTAs receive updates on that list and this way access to the data is granted.

3.7 Federated Storage Infrastructures in High Energy Physics

In the upcoming Runs 3 (2022–2025) and 4 (starting 2029) at the Large Hadron Collider (LHC[8] at CERN[9]) significantly more computing and storage resources will be required. With about 400 PB/year and 150 million HEPSPEC06[10]) these will be about one to two orders of magnitudes more than in the recent LHC Run 2 (2015–2018). Other experiments (Belle 2,[11] FAIR[12]) will face similar challenges. In order to be able to fulfill these demands relying alone on technological progress will not be sufficient. In addition to that the data management workflows of the experiments need to be far more efficient.

In the worldwide LHC computing Grid (WLCG[13]) the 15 largest computing centres provide about 50% of all resources while the hundred smallest computing centres together only provide about 20% of the resources. This inspired the idea to store the experiment data only on a few central data centres which would be interconnected with high speed bandwidth (1–10 Tb/s). This data cloud could be seen as a single virtual data centre or a data lake, which would be accessible from dedicated computing resources within WLCG but also from opportunistic resources as HPC centres and Clouds. Data processing could take place on distributed computing farms with reasonable bandwidth connection to the data lake and only local data caches on site. As a first step to such a central data cloud data federations could be created consisting of several computing centres with a single point of entry and caching possibilities for smaller centres. The concept of such a data cloud is visualised in Fig. 3.5.

Operations in such a scenario would in the long run be more cost efficient since data caches require significantly less expert personnel than maintaining a consistent storage infrastructure. In the end they may be operated completely autonomously. Another advantage is, that if more federated and commonly usable resources are available they can be more efficiently used which will help in closing the resource gap. Moreover this idea is supported by the fact that new protocols and tools have been developed which enable direct remote access via network on data which are stored on different computing centres without the necessity to create local copies before data processing could start. Also due to technological progress nowadays larger network bandwidths are available and many of the larger Grid centres are situated in a common network named LHCONE.[14]

---

[8] https://home.cern/science/accelerators/large-hadron-collider.

[9] https://home.web.cern.ch/.

[10] https://w3.hepix.org/benchmarking.html.

[11] https://www.belle2.org/.

[12] https://fair-center.eu/user/experiments.

[13] https://wlcg.web.cern.ch/.

[14] https://lhcone.web.cern.ch.

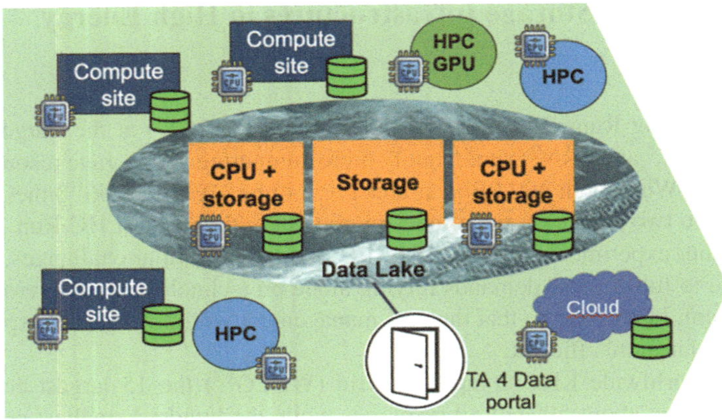

Fig. 3.5 A data cloud accessible from dedicated computing centres and opportunistic resources as well as via a data analysis portal

3.7.1 Data Lake Design

Principle design considerations for a data lake will be described in this paragraph. Access to the data lake should be as easy as possible. It should be supported that clients can access the data lake via terminal as well as by using a web browser using the https protocol. Files which are written to any data lake storage participant should be placed on one of them, transparently to the user. Files which are read from a different data lake storage participant than where they are actually placed should be transferred transparently. The storage namespace must be distributed amongst all storage participants. Each data lake storage participant should deliver the same (complete) view of the namespace to the users. A simple user interface for user interaction with a minimum set of required operations (as "ls", "read", "write", "delete") has to be defined, which needs to be implemented by all storage technologies used in the data lake. In addition to that an open and standardised interface between the data lake storage participants is needed which must include functions that enable data lake storage participants to perform placement and replication related tasks. For authentication and authorisation token-based access to the data lake needs to be enabled. A replication mechanism needs to be in place that continuously guarantees that replication requirements are met for each file or directory. Last but not least the responsibilities of software components should be clearly separated through well defined interfaces in order to allow relatively easy replacement of individual components via modular design. A data lake should support the full data life cycles of the experiments and user communities.

The development of data lake concepts needs to be done in close collaboration with the experiments and observatories as well as with the data lake concepts of large projects as WLCG/DOMA[15] and also in coordination with national and international initiatives such as EOSC[16] and ESCAPE[17] on European level or the National Research Data Infrastructure (NFDI[18]) on national level, e.g. in Germany.

The WLCG DOMA project consists of several activities in the area of Data Organisation, Management and Access, with a focus on the readiness for LHC Run 4, when the HL-LHC (High Luminosity LHC) era will start. DOMA acts as platform to exchange about corresponding research and development efforts happening in the LHC experiments and international projects in order to exploit synergies and to foster common approaches. Another important aspect of DOMA is the organisation of global data challenges that are evolving in terms of transfer volume and also in terms of employing more complex technologies likes software defined networks. While the main focus is on the LHC experiments the DOMA community links also to other data intense experiments like Belle 2, DUNE[19] and SKA, because there is large overlap regarding sites and common usage of research network infrastructure.

First community overarching data lake prototypes are being developed and setup in the context of European and national projects. Examples are the Data Infrastructure for Open Science (DIOS) of the ESCAPE project and German national projects such as FIDIUM in the context of ErUM-Data (Digitalisation in research of universe and matter) or PUNCH4NFDI[20] in the context of the NFDI (national research data infrastructure).

3.7.2 ESCAPE

The pilot data lake, which is being set up by work package 2 of the ESCAPE project, is using Rucio (Barisits et al. 2019) as data management system. This includes also work on replication rules and policies as data redundancy and location as well as mapping to storage quality of service. Rucio is an open-source software framework that provides scientific collaborations with the functionality to organise, manage, and access their data at scale. It provides a unified interface of the network and storage infrastructure. The data can be distributed across heterogeneous data centers at widely distributed locations. Rucio was originally developed to meet the requirements of the high-energy physics experiment ATLAS,[21] and now is

[15] https://twiki.cern.ch/twiki/bin/view/LCG/DomaActivities.

[16] https://research-and-innovation.ec.europa.eu/strategy/strategy-2020-2024/our-digital-future/open-science/european-open-science-cloud-eosc.

[17] https://projectescape.eu/.

[18] https://www.nfdi.de/.

[19] https://www.dunescience.org/.

[20] https://www.punch4nfdi.de/.

[21] https://atlas.cern/.

continuously extended to support next to ATLAS andCMS[22] at LHC also other diverse scientific communities. Drawing benefits from advances in Cloud and Big Data computations, it relies on a conceptual data model to ensure system stability. In order to reduce the operational overheads of the support staff, it employs also an automation framework. In the context of ESCAPE a Rucio extension is being developed which allows to interconnect the data lake infrastructure with analysis interfaces such as Jupyter notebooks. The data lake consists of a number of storage elements using different storage technologies, which includes EOS,[23] DPM, dCache (Mkrtchyan et al. 2021), StoRM, and XRootD.[24] All storage elements support the access protocols HTTP, xroot, and GridFTP (Allcock et al. 2005). As wide area file transfer service the CERN development FTS[25] comes into place. By integrating CRIC (Anisenkov et al. 2020) it was possible to store CRIC meta data as Rucio attributes. Authentication is planned to be token based and work is done to integrate this technology into the various layers of the data lake.

3.7.3 FIDIUM

In the German national project FIDIUM (Federated Digital Infrastructures for research in Universe and Matter), a joint project by the German communities of (Astro-) Particle and Hadron- and Nuclear Physics, more data lake related research but also the setup of testbeds is on the agenda. The research focusses on the setup of a real time monitoring system for data lakes, on technologies for data caching, on methods for data and workflow management in data lakes, as well as on efficient connection of users, computing centres and data sources and quality of service measurements.

A data lake monitoring system needs to be able to supervise the load and the amount of data stored on individual components of the data lake and also from what locations and how frequently datasets are accessed. As a first step this is planned to be developed based on the data storage related monitoring infrastructure provided by the XRootD framework.

The technologies for dynamic disk caching should enable efficient integration of non pledged opportunistic resources as HPC and Cloud systems into production environments in a transparent way. For this the XCache (Bauerdick 2014) system of the XRootD framework is considered an important ingredient. For being able to optimise the usage of dynamic disk caches in heterogeneous resources a simulation environment for complex systems including distributed caching and integration in existing workload and data management systems is being worked on. Also so called

[22] https://home.cern/science/experiments/cms.

[23] https://eos-web.web.cern.ch/eos-web/.

[24] https://www.academia.edu/88254944.

[25] https://fts.web.cern.ch/fts/.

on demand or ad hoc parallel file systems as burst caches in HPC systems are being investigated. Parallel file systems such as Gekko-FS (Brinkmann et al. 2020) , which can be created in user space, can provide added values here. A key question is here fast access to data to be analysed and an efficient automatic selection of data to be cached. Also secure writeback of the analysis results is an important topic. Local workload manager will be extended in such a way, that parallel on-demand file systems in HPC centres can be used as integrated cache systems.

Dynamic data lake workflows must be able to react on the real time monitoring information being developed in this project. Intelligent data management and an efficient resource usage can contribute to significantly reduced costs. One system which is being evaluated is the Rucio system which has already been mentioned before. In parallel to this hash based data placement and replication is being investigated. A possible implementation is based on consistent hashing and could help to reduce the dependency on classical file catalogues and data bases.

The quality of service (QoS) properties of a data lake can be shown following the CAP properties (consistency, availability, partitioning). Via intelligent QoS algorithms the limitations given by the CAP theorem should be reduced as much as possible. All existing data sources including data from observatories in astronomy or astro particle physics as well as data from high energy physics experiments must be included via open protocols into the data lake. Also existing computing and storage systems must be ingegratable in a transparent way, which includes also the globally distributed WLCG computing system of the LHC experiments at CERN. Therefore techniques are being developed to include dedicated experiment resources, but also opportunistic resources such as HPC and Cloud systems as well as dedicated analysis facilities into the data lake. Users from all these infrastructures must be able to access data to be analysed which are stored in the data lake, in an easy manner.

3.7.4 NFDI and PUNCH4NFDI

The German National Research Data Infrastructure (NFDI) is a digital distributed infrastructure which provides services and information to the scientific community in Germany and beyond mainly about the management of research data. The funding is about 90 million Euro per year from 2019 up to 2028. The non-profit association was founded to coordinate the activities and is based in the city of Karlsruhe. NFDI consortia are associations of various institutions within a research field, which work together in an interdisciplinary manner to implement the goal. PUNCH4NFDI is the NFDI consortium of particle, astro-, astroparticle, hadron and nuclear physics, representing about 9000 scientists with a Ph.D. in Germany. One central milestone of PUNCH4NFDI is to provide an integrated prototype package of dynamic digital research products, a science data platform, in combination with an integrated federated compute and storage environment coupled with single sign on (AAI, Authentication and Authorisation Infrastructure). This package of combined deliverables will allow to access open data and other open research products from a selected range of sources across PUNCH4NFDI and beyond using advanced workflows on federated compute and storage infrastructures. In Task Area 2, Data Management, PUNCH4NFDI is working, among others, on the topic of data lakes.

Towards the end of the 2020s, when the High-Luminosity LHC (HL-LHC) will start, the disk requirements are expected to reach Exabyte scales. At roughly the same time, the SKA will produce a similar data volume, and the experiments at the FAIR facility will require storage capacities of several 100 PB. Astronomy has a long tradition in surveying the sky, carefully documenting observations, and making the measurements available to the community and the general public. There are many examples in astronomy where projects have pioneered ways to store the large data collections obtained from systematic surveys and to provide open access to the data, as e.g. done by the SDSS.[26] A generation of new, large telescopes, e.g. LOFAR,[27] MeerKAT,[28] the Rubin Observatory,[29] and the SKA, require new concepts for data processing, storage, and access. LOFAR is already driving the transition by having its data stored in a long-term archive (LTA) distributed over several HPC centres. LOFAR data are processed on supercomputers at these sites. Many scientific goals can only be achieved if the most immense amounts of data, originating from experiments, observations or simulations, are managed and provided to the communities for scientific exploitation.

The landscape of storage technologies and data management tools used by the PUNCH community ranges from very simple setups to very advanced distributed installations. Smaller research groups often cannot afford to invest vast resources into their data infrastructure. The data are usually stored at the host laboratory of the experiment or the observatory. To access the data local accounts need to be negotiated. On the other end of the spectrum there is the Worldwide LHC Computing Grid WLCG.

The astronomical community has a long tradition of operating open data archives. In contrast to particle physics, however, these come without attached computing resources. An example is the CDS,[30] which hosts a large number of astrophysical catalogues that can be openly accessed. It is also available to smaller research groups for uploading and sharing their data. However, the most recent and next generations of astronomical instruments, like LOFAR, 4MOST,[31] the Rubin Observatory, or the SKA, reach data rates and volumes that do not allow to follow the same strategy. The LOFAR LTA is currently the largest astronomical data archive (order of 50 PB, mainly on tapes) worldwide. Individual measurement sets have sizes of tens of TB

[26] https://www.sdss.org/.

[27] https://www.astron.nl/telescopes/lofar/.

[28] https://www.sarao.ac.za/gallery/meerkat/.

[29] https://www.lsst.org/.

[30] https://vizier.cfa.harvard.edu/vizier//cats/cats.htx.

[31] https://www.4most.eu/cms/.

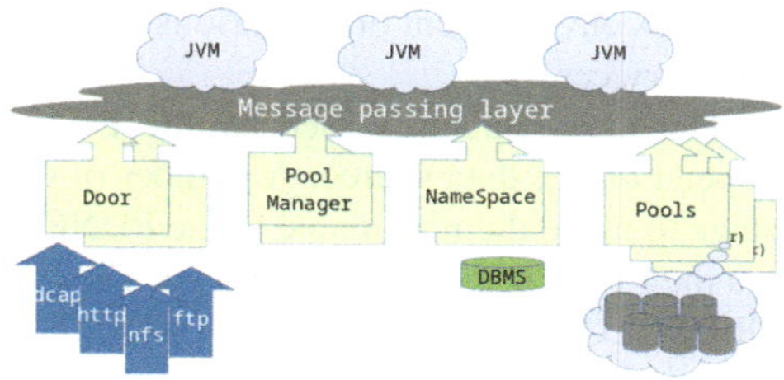

Fig. 3.6 A high level view of the setup of a dCache based storage element

and need compute clusters with high-memory nodes at the archive sites. This makes them hard to access, and it is expensive to move the raw data.

An infrastructure with standardised data access is needed to ease or enable analysis of distributed datasets. Hence archives and storage instances will be made interoperable leveraging established standard protocols. This functionality is a prerequisite to include the (dynamical) archives in the data flux of large experimental setups. The community, including members from the PUNCH4NFDI consortium, addresses the challenges for example in the ESCAPE project and the DOMA initiative by WLCG. Members of PUNCH4NFDI are actively involved in these international activities. The ongoing research is targeted at the needs of the international collaborations. Within PUNCH4NFDI, significant effort will be spent to employ these developments for the national needs and provide state-of-the-art solutions for local experiments at smaller accelerators that are hosted at universities.

Based on ongoing developments, prototypes will be provided to PUNCH4NFDI and beyond to allow the necessary integration work. This encompasses especially a prototype data lake which distributes over several centres which participate in the PUNCH4NFDI project and storage elements provided by those centres. The prototype data lake will make use of several storage technologies which are popular within the PUNCH communities, where especially dCache and XRootD need to be mentioned. A high level view of a dCache based storage setup can be seen in Fig. 3.6.

Moreover new developments as dynamic disk caches and standard access protocols are being integrated. That requires compatibility with existing and evolving systems from the international collaborations and the possibility to interconnect with infrastructure constructed within PUNCH4NFDI. The WLCG is presently converging to support two protocols for wide-area data transfer,[32] the standard protocol for file transfers over the internet, and XRootD, which is a protocol very specific to the High Energy Physics and Hadron and Nuclear Physics communities. The readiness for data transfers using these protocols has to a large extent already been established. AAI (Authentication and Authorisation Infrastructure) is of overarching interest for the consortium and beyond. A considerable amount of work is expected to setup the recommended AAI implementation and its validation in the

[32] https://web.archive.org/web/20120626092812/http://webdav.org/.

complex environment of the various communities, also in combination with the data lake services.

The data structures used by the PUNCH community are often complex and depend on the data source. A number of data sources will be addressed in an early stage to make them available to the PUNCH-SDP (PUNCH Science Data Platform) via the PUNCH data lake prototype. The identified pilot data sources are the DES,[33] GLOW-LOFAR[34] pulsar data, LoTSS,[35] data from one experiment at ELSA[36] at University of Bonn, data from experiments at MAMI[37]/MESA[38] at Johannes Gutenberg University of Mainz, and S-DALINAC[39] at Technical University of Darmstadt. Any approach to harmonise the access to those data (and metadata) can only be arranged on top of the existing systems and in collaboration with international partners.

A big focus will lay also on intelligent workflow techniques. Distributed data lakes should support replication of data objects and should enable client applications to find the nearest copy of a data object within a unified namespace. Data can be stored on arbitrary storage elements like tape, disk, or in memory, and grid sites can manage their data pools either as object storage, file systems, or databases. In particular the nuclear & particle physics community inside PUNCH4NFDI has developed open-source tools like dCache, XRootD, or StoRM[40] to connect data centres with data repositories, while frameworks like Rucio, Dynafed[41] or XRootD combine these pools and expose them under a unified namespace. The astronomical community faces similar challenges as the particle physics community. Data locations are often still statically managed and determined independent from workflows. In order to cope with the challenges of the research field, the storage capacity of individual data lakes, the capability to work across data lakes, and to move data between them have to increase by an order of magnitude. It is necessary to enhance the existing tools to enable them to efficiently use scale-out storage environments. This can be achieved by collecting information about the underlying storage infrastructures and the intended access patterns to the stored data, which then can be used to optimise data placement and data movements, while also steering workflows. Within PUNCH4NFDI one research focus with respect to data lakes is therefore to increase the capabilities of data lakes to intelligently place and

---

[33] https://www.darkenergysurvey.org/the-des-project/data-access/.

[34] https://www.glowconsortium.de.

[35] https://lofar-surveys.org/surveys.html.

[36] https://www-elsa.physik.uni-bonn.de/Beschleuniger/elsa.html.

[37] https://www.blogs.uni-mainz.de/fb08-nuclear-physics/accelerators-mami-mesa/the-mainz-microtron/

[38] https://www.mesa.uni-mainz.de/eng/.

[39] https://www.ikp.tu-darmstadt.de/forschung_kernphysik/experimentelle_geraete/\s_dalinac_details/index.en.jsp.

[40] http://italiangrid.github.io/storm/.

[41] https://lcgdm.web.cern.ch/dynafed-dynamic-federation-project.

move data by analysing monitoring information and predictions about future access patterns. The tasks to add intelligence to the management of data lakes will be closely aligned with international data management initiatives like the ESCAPE Data Infrastructure for Open Science (DIOS) project that is aiming to provide a scalable federated data infrastructure to manage Exabyte-scale data volumes and has been mentioned above. Several applicants of the PUNCH4NFDI proposal are actively driving the development of data lakes within ESCAPE. Intelligence can only be added to the PUNCH4NFDI data lake infrastructure by including monitoring information into the decision-making process of the applied tools. Here there is also a significant overlap with the work programme of the FIDIUM project mentioned above. One possibility is a coupling between underlying object storage solutions, like Ceph,[42] and grid tools like Rucio. Ceph provides a rich set of monitoring information that can be fed into the policy engine of Rucio. Rucio can then use this information to steer data placement within Ceph. These policies can then e.g. enforce the placement of hot data objects on an SSD tier and the placement of cold data on tape. Nevertheless, PUNCH4NFDI will use this example to also show the limitations of today's tool sets on different technological levels. First of all, there is still a mismatch between the capabilities of individual tools and the requirements to build distributed storage infrastructures that also enable fine-grained QoS. Data objects from huge PUNCH experiments are typically placed in a single storage pool, and IO rate limits to these storage pools should be based on dedicated data streams between hosting sites. Underlying technologies, like Ceph, often only allow rate limits for individual devices and have to be extended to also include individual jobs. Furthermore, it is necessary to develop interfaces between backend storage and frontend Grid tools. One task may therefore be to develop techniques to couple QoS between storage systems and Grid tools. Secondly, QoS only exists inside the current grid tool sets and in present data lake developments as a concept and first implementations are expected as results of ongoing projects in the context of WLCG DOMA and ESCAPE. Strategies how to use QoS within the data management of experiments are in a very early state. However, good concepts for data placement, data replication, and traffic steering would be beneficial to drive down the immense costs for storage that are coming along during HL-LHC and other data-intense endeavours. It is therefore desirable to implement QoS techniques for at least one major PUNCH4NFDI data archive. Finally, it is important to use forecasts about future access patterns to proactively migrate or replicate data or jobs. This information is (partly) available in workflow descriptions and can especially help workflows that rely on all input data being present before a computation can start. Computational problems can then either be divided to only cover what is present at a single location, jobs can be transferred, or data can be proactively prefetched. It is therefore important to include information from workflow descriptions to improve data management. Exabyte datasets also require new algorithms and protocols to scale concerning their management. An example is the design of distributed file

[42] https://ceph.com/en/.

catalogues. Each site participating in a data lake, for example, today needs to be aware of all participating storage elements and data caches to correctly locate objects and file copies. As a consequence, file catalogues do not scale well with the number of participants. An alternative is to use distributed hashing that refers clients to the correct metadata information even in case that this information is not locally available. Distributed hashing has shown to scale to (nearly) arbitrary sizes in peer-to-peer computing, while keeping the resource requirements of each participant extremely low. PUNCH4NFDI will investigate how such hashing strategies can be applied to data lakes and exemplary be combined with tools like Rucio or Dynafed. Also here an overlap with the FIDIUM project is given. New data lake technologies can also help to improve the design of workflows. An example is the introduction of an event interface, which can help to trigger workflow executions depending on the creation, modification, or deletion of files. Example implementations of event interfaces are the OpenIO object store, AWS Lambda,[43] and Google's cloud functions.[44] Event interfaces are currently mostly used in serverless computing, where individual object modifications trigger small computational tasks within a container. It is planned to investigate how these concepts can be transferred to PUNCH4NFDI to build dynamic workflows. The design of workflows on an adaptive storage substrate builds a direct link between the astrophysics and the nuclear & particle physics community. The current momentum to adopt workflow languages like the Common Workflow Language in the astrophysics community can be used to specify tool sets that are jointly developed by all PUNCH4NFDI partners and that can then also be exported to further NFDI communities and beyond.

3.8 Data Lakes in Astronomy

The analysis of using data lakes in the context of astronomical instrumentation requires analysis of the differences and overlaps between the data needs of astronomical archives and that of high-energy physics(HEP) experiments. A start for such an analysis has been done within work package 2 of the ESCAPE project, and can be found in Grange et al. (2022). It is clear that the overlap between the design of a distributed archive, with storage locations spread over external institutes as is the case in high-energy physics, maps very well to the model that particle physics experiments use for their data distribution.

The main difference between the data distribution in HEP and astronomy lies in the access model. Where in HEP access is generally granted to all the members of a collaboration, and members of the collaboration are generally actively managed. Because of this granularity in access control is not needed (if you have access to one file, you have access to all files). Also in HEP, access control has historically

[43] https://aws.amazon.com/lambda/?nc1=h_ls.

[44] https://cloud.google.com/functions.

been granted based on X.509 certificates (Boeyen et al. 2008), which is based on a model where users need to be authenticated with a trusted (by the collaboration) authority to obtain a certificate that can be used to authenticate to the system. In case of astronomical archives, the requirements on authorisation and authentication are very different. First of all, most instruments are openly accessible to any group of scientists who writes a proposal. Then for a set period of time, the data is only accessible to the team for which it was originally observed. After this time has passed the dataset is generally released to the public, ideally without any authentication at all. The implications for the infrastructure are that far more granularity is needed to provide access to files. This ideally extends to who is allowed to manage who has access to files in a project (this is generally the PI of the project). Also unauthenticated access to the data is currently not supported by the tooling that makes up the datalake.

Further, the group of users who need access to the archive will change over time (since it depends on who writes a proposal). This makes it much harder to define a fixed number of trusted parties to manage user identities, making that users find requesting an X.509 certificate cumbersome.

Recently, the WLCG formed a working group to review the methods of providing AAI in the field of HEP and it has worked on supporting and AAI flow that is based on JSON Web Tokens (JWT; for more details see Bockelman et al. 2020). Using tokens makes it possible to implement an authentication flow that is based on Federated AAI, which means that a service provider (in our case the archive) delegates user authentication to an external party (generally the users' home institute or employer), which (after consent from the user) provides the service provider with the information needed to make authentication decisions. This work flow makes it much easier to provide access to systems at all and will bring data lakes for astronomy archives one step closer.

In the current datalake architecture, which uses Rucio, the file catalogue supports use of metadata for files. By default the metadata keys consist of standard file properties (e.g. size, checksum) and some data keys that are accelerator physics specific. The system does allow for custom metadata to be provided in several ways (most notably using JSON and/or writing a plugin) but currently it is not yet possible to automatically apply rules based on those custom keys (e.g. "put all pulsar data at this position on the sky on a specific storage location"). Also rules are generally only based on qualities making it for example impossible to perform a cone query on the metadata (e.g. $01^h42^m30^s < \mathrm{RA} < 01^h45^m30^s$).

Within the SKA Regional Centres prototyping phase, the data lake as described in these sections, is being investigated as one of the candidates to manage the data flow between the instrument and the regional centres. Part of this prototyping involves further developing functionality for the system which is aimed at making the datalake more suited for the distributed astronomical archive use case. The two points mentioned in the previous section are for instance part of this ongoing effort.

3.9 SKA Regional Centres

The SKA Observatory's science data will result from processing and reduction in several steps. The data rate flow for the first phase of the SKA project construction can be summarised as follows:

- SKALow will produce around 2 Pbps that will be reduced by the correlator to around 7 Tbps that will be redirected to the Central Signal Processors
- SKAMid will produce around 20 Tb/s that will be redirected to the Central Signal Processors
- Both Central Signal Processors will produce a bandwidth of around 560 GB/s that will be redirected to the Science Data Processor (SDP) for processing
- The SDP will produce around 600–700 PB/year, including resilience overheads, that will be distributed to the different SKA Regional Centres (SRCs)

After this process, the amount of data that will be available to the scientific community, over 600 PB/year without considering the user collaborative data, will be of a new magnitude for the astronomical context (Chrysostomou et al. 2020). Providing this scientific data repository to the community represents a technical challenge for discovery, analysis, and exploitation tasks, which should include big data lakes exploitation techniques and a change of paradigm in the way the scientists will access the data.

These data will be distributed into several locations all around the planet called SKA Regional Centres (SRCs) that will provide access to the SKA data sharing the data resources.

The SRCNet is composed of SRC Net nodes, in different locations across the planet, that should share a set of basic services (including common data access libraries, provenance production, shared workflow systems, etc.), communication protocols between them and a certain level of federation on the services to permit an efficient use by the community of resources allocated into the different nodes across the planet and facilitate the science analysis. The resources of the SRCNet are provided by the different SRC partners through a pledging approach, i.e. by adding resources (in terms of hardware and development efforts) to a common pool. In terms of computational layer aspects, that implies a certain level of heterogeneity that needs to be provided, connecting nodes with different local cloud technologies, HPC technologies or local infrastructures. That implies that, in order to manage the execution of workflows in different SRC Net nodes, it is needed a definition of abstraction layers to harmonise the resources. This layer will define the signature of the execution interface and the translation to the underlying execution layer.

The development and implementation of the abstraction layers and modules of the SRCs are also a common effort that implies the creation of development teams distributed across the planet and creating synchronised development cycles. In summary, the SRCNet should solve the problems of access and analysis problems of the scientific community to the SKA project data from a heterogeneous big data lake sparse across many countries.

In general, big Data Lakes present difficulties for analysis as they usually contain unstructured data, latency accessing remote data, lack of data lineage and lack of integration with the computing resources. The SRCNet is being designed to integrate the computational layer with the data layer so the different nodes are able to process the data that has been allocated to this node. Following the data mesh approach, data parser libraries and executions were done in a closer location to the data so the latency produced by a preliminary data movement to the execution node is reduced as much as possible and the data travelling across the network is on the interface invocation parameters and the result of the remote parsing operation. Also, this integration of data, parsing libraries and computing resources required for the parsing creates the concept of domain data, which is a way to simplify access by, e.g. dividing the data lake into specific data formats or dividing the data by science domain.

This approach has been studied in certain contexts (e.g. manipulation of data cubes through a remote operations server, IVOA SODA) but the SRCNet will have a broader approach by the definition of client libraries that will allow remote operations on the data, hiding the complexity of the access to the users by underlying connections.

References

W. Allcock, J. Bresnahan, R. Kettimuthu, M. Link, The globus striped GridFTP framework and server, in *ACM/IEEE SC Conference, SC'05*, vol. 54, 2005. https://doi.org/10.1109/SC.2005.72. ISBN: 1-59593-061-2

A. Anisenkov, J. Andreeva, A. Di Girolamo, P. Paparrigopoulos, B. Vasilev, CRIC: Computing Resource Information Catalogue as a unified topology system for a large scale, heterogeneous and dynamic computing infrastructure. EPJ Web Conf. **245**(03032), 7 (2020). https://doi.org/10.1051/epjconf/202024503032

M. Barisits, T. Beermann, F. Berghaus et al., Rucio: scientific data management. Comput. Softw. Big Sci. **3**(11) (2019)

L.A.T Bauerdick et al., XRootd, disk-based, caching proxy for optimization of data access, data placement and data replication. J. Phys. Conf. Ser. **513**, 042044 (2014). https://doi.org/10.1088/1742-6596/513/4/042044

B. Bockelman, A. Ceccanti, I. Collier, L. Cornwall, T. Dack, J. Guenther, M. Lassnig, M. Litmaath, P. Millar, M. Sallé, H. Short, J. Teheran, R. Wartel, WLCG authorisation from X.509 to tokens, in *European Physical Journal Web of Conferences*, vol. 245. European Physical Journal Web of Conferences, November 2020, p. 03001. https://doi.org/10.1051/epjconf/202024503001

S. Boeyen, S. Santesson, T. Polk, R. Housley, S. Farrell, D. Cooper, Internet X.509 public key infrastructure certificate and certificate revocation list (CRL) profile. RFC 5280, May 2008. https://www.rfc-editor.org/info/rfc5280

A. Brinkmann, K. Mohror, W. Yu, P. Carns, T. Cortes, S.A. Klasky, A. Miranda, F.-J. Pfreundt, R.B. Ross, M.-A. Vef, Ad hoc file systems for high-performance computing. J. Comput. Sci. Technol. **35**, 4–26 (2020). https://doi.org/10.1007/s11390-020-9801-1

A. Chrysostomou, C. Taljaard, R. Bolton, L. Ball, S. Breen, A. van Zyl, Operating the square kilometre array: the world's most data intensive telescope, in *Observatory Operations: Strategies, Processes, and Systems VIII*, vol. 11449. Society of Photo-Optical Instrumentation

Engineers (SPIE) Conference Series, December 2020, p. 114490X. https://doi.org/10.1117/12.2562120
A. Comrie, K.-S. Wang, S.-C. Hsu, A. Moraghan, P. Harris, Q. Pang, A. Pińska, C.-C. Chiang, T.-H. Chang, Y.-H. Hwang, H. Jan, M.-Y. Lin, R. Simmonds, CARTA: The Cube Analysis and Rendering Tool for Astronomy. Zenodo, June 2021
dCache, dcache website (2022). https://www.dcache.org/
Y.G. Grange, V.N. Pandey, X. Espinal, R. Di Maria, A.P. Millar, Astronomical data organization, management and access in scientific data lakes (2022). https://arxiv.org/abs/2202.01828
C.L. Hale, D. McConnell, A.J.M. Thomson, E. Lenc, G.H. Heald, A.W. Hotan, J.K. Leung, V.A. Moss, T. Murphy, J. Pritchard, E.M. Sadler, A.J. Stewart, M.T. Whiting, The rapid ASKAP continuum survey paper II: first stokes I source catalogue data release. Publ. Astron. Soc. Aust. **38**, e058 (2021). https://doi.org/10.1017/pasa.2021.47
H. Holties, R. de Goei, N. Vermaas, ALTA, the APERTIF long term archive, in *Astronomical Data Analysis Software and Systems XXIX*, ed. by R. Pizzo, E.R. Deul, J.D. Mol, J. de Plaa, H. Verkouter, vol. 527. Astronomical Society of the Pacific Conference Series, January 2020, p. 255
T. Mkrtchyan, K. Chitrapu, V. Garonne, D. Litvintsev, S. Meyer, P. Millar, L. Morschel, A. Rossi, M. Sahakyan, dCache: inter-disciplinary storage system. EPJ Web Conf. **251**(02010) (2021). https://doi.org/10.1051/epjconf/202125102010
L. Moreau, P. Missier, K. Belhajjame, R. B'Far, J. Cheney, S. Coppens, S. Cresswell, Y. Gil, P. Groth, G. Klyne, T. Lebo, J. McCusker, S. Miles, J. Myers, S. Sahoo, C. Tilmes, Prov-dm: the prov data model (2013). http://www.w3.org/TR/prov-dm/. World Wide Web Consortium (W3C)
W.A. van Cappellen, T.A. Oosterloo, M.A.W. Verheijen, E.A.K. Adams, B. Adebahr, R. Braun, K.M. Hess, H. Holties, J.M. van der Hulst, B. Hut, E. Kooistra, J. van Leeuwen, G.M. Loose, R. Morganti, V.A. Moss, E. Orrú, M. Ruiter, A.P. Schoenmakers, N.J. Vermaas, S.J. Wijnholds, A.S. van Amesfoort, M.J. Arts, J.J. Attema, L. Bakker, C.G. Bassa, J.E. Bast, P. Benthem, R. Beukema, R. Blaauw, W.J.G. de Blok, M. Bouwhuis, R.H. van den Brink, L. Connor, A.H.W.M. Coolen, S. Damstra, G.N.J. van Diepen, R. de Goei, H. Dénes, M. Drost, N. Ebbendorf, B.S. Frank, D.W. Gardenier, M. Gerbers, Y.G. Grange, T. Grit, A.W. Gunst, N. Gupta, M.V. Ivashina, G.I.G. Józsa, G.H. Janssen, A. Koster, G.H. Kruithof, S.J. Kuindersma, A. Kutkin, D.M. Lucero, Y. Maan, F.M. Maccagni, J. van der Marel, A. Mika, J. Morawietz, H. Mulder, E. Mulder, M.J. Norden, A.R. Offringa, L.C. Oostrum, R.E. Overeem, Z. Paragi, H.J. Pepping, E. Petroff, D.J. Pisano, A.G. Polatidis, P. Prasad, J.P.R. de Reijer, J.W. Romein, J. Schaap, G.W. Schoonderbeek, R. Schulz, D. van der Schuur, A. Sclocco, J.J. Sluman, R. Smits, B.W. Stappers, S.M. Straal, K.J.C. Stuurwold, J. Verstappen, D. Vohl, K.J. Wierenga, E.E.M. Woestenburg, A.W. Zanting, J. Ziemke, Apertif: phased array feeds for the Westerbork Synthesis Radio Telescope. System overview and performance characteristics. Astron. Astrophys. **658**, A146 (2022). https://doi.org/10.1051/0004-6361/202141739
M. Wadenstein, L. Viklund, N. Edmundsson, Endit github repository (2022). https://github.com/neicnordic/endit
C. Wu, A. Wicenec, D. Pallot, A. Checcucci, Optimising NGAS for the MWA Archive. Exp. Astron. **36**(3), 679–694 (2013)
M.D. Wilkinson, M. Dumontier, I. Jan Aalbersberg, G. Appleton, M. Axton, A. Baak, N. Blomberg, J.-W. Boiten, L.B. da Silva Santos, P.E. Bourne, J. Bouwman, A.J. Brookes, T. Clark, M. Crosas, I. Dillo, O. Dumon, S. Edmunds, C.T. Evelo, R. Finkers, A. Gonzalez-Beltran, A.J.G. Gray, P. Groth, C. Goble, J.S. Grethe, J. Heringa, P.A.C 't Hoen, R. Hooft, T. Kuhn, R. Kok, J. Kok, S.J. Lusher, M.E. Martone, A. Mons, A.L. Packer, B. Persson, P. Rocca-Serra, M. Roos, R. van Schaik, S.-A. Sansone, E. Schultes, T. Sengstag, T. Slater, G. Strawn, M.A. Swertz, M. Thompson, J. van der Lei, E. van Mulligen, J. Velterop, A. Waagmeester, P. Wittenburg, K. Wolstencroft, J. Zhao, B. Mons, The FAIR guiding principles for scientific data management and stewardship. Sci. Data **3**(1) (2016). https://doi.org/10.1038/sdata.2016.18

Chapter 4
Computing Infrastructure

Stefano Alberto Russo, Estela Suarez, Antony Chazapis, Manolis Marazakis, and Giuliano Taffoni

At the scale of radio astronomy experiments for the next generation facilities discussed in this book, the computing infrastructure plays a crucial role, in particular with respect to High Performance Computing (HPC) systems. This chapter discusses the most recent advances in this space: in Sect. 4.1 we address a new type of computing facilities, based on combining both CPUs and accelerators as GPUs and FPGAs into heterogeneous HPC systems; in Sect. 4.1 we introduce a new architecture for enabling a more dynamical use of such heterogeneous hardware resources; and in Sect. 4.2 we take a deep dive in software containerisation, a key technology for effectively scaling complex codes and fully exploiting computing resources while greatly improving scientific reproducibility. We then conclude by providing an overview on how some recent yet widely adopted cloud-native technologies, as last-generation container orchestrators, can converge together with HPC-specific ones towards higher-level, unified frameworks.

S. A. Russo (✉)
INAF (Italian National Institute for Astrophysics) – Astronomical Observatory of Trieste, Trieste, Italy
e-mail: stefano.russo@inaf.it

E. Suarez
Jülich Supercomputing Centre (JSC), Forschungszentrum Jülich, Jülich, Germany

Computer Science Department, University of Bonn, Bonn, Germany
e-mail: e.suarez@fz-juelich.de

A. Chazapis · M. Marazakis
Institute of Computer Science, Foundation for Research and Technology - Hellas (FORTH), Heraklion, Crete, Greece
e-mail: chazapis@ics.forth.gr; maraz@ics.forth.gr

G. Taffoni
INAF - Astronomical Observatory of Trieste, Trieste, Italy
e-mail: giuliano.taffoni@inaf.it

E. Vardoulaki et al. (eds.), *Data-Intensive Radio Astronomy*, Astrophysics and Space Science Library 472, https://doi.org/10.1007/978-3-031-58468-8_4

4.1 Heterogeneous HPC Platforms

Traditionally, HPC cluster computers are built by attaching a large number of high-end general purpose processors to each other via a high-speed interconnect fabric. Such homogeneous system configurations are well established, so that both the hardware and software environments are mature. HPC applications are also prepared to run on these systems, as they are parallelized via message-passing strategies or run independent ensembles which scale up extremely well on the machines. For this reason, for many users, CPU-clusters have been the preferred platforms.

However, from the operations and particularly from the cost-perspective building very large scale CPU-only systems is not affordable. High-end CPUs implement deep superscalar pipelines with advanced features such as out-of-order execution, prefetching techniques and branch prediction, to name just some. All these features come at the prize of a high power consumption, leading to a moderate Flop/Watt ratio. Alternative compute devices, so called *accelerators*, are designed to deliver very high compute power for specific kinds of operations, compromising on the pipeline features, while focusing on floating point operations and extreme parallelism.

The most widely used acceleration devices in HPC are graphic processing units (GPUs), originally designed to make very fast visualization of graphics (e.g. for the gaming industry), but very efficient in calculating the ubiquitous floating point vector and matrix operations performed in HPC and AI applications. Other examples of accelerators are many-core processors and field-programmable gate arrays (FPGAs). Simply put, the difference between general purpose CPUs and accelerators is that the former employ a low number (tenths) of advanced and complex processing units, while accelerators use hundreds to thousands very simple processing units.

The HPC community, depending on the specific applications, needs however both advanced pipeline features and very large performance at an affordable energy consumption. Therefore, the approach taken in the last decade consisted on combining both CPUs and accelerators into heterogeneous HPC systems. This adds a level of complexity with which the management software, system software, compilers, programming environment, and monitoring and analysis tools must now cope with (see Sect. 7.2). Despite the very strong advances in all these layers of the stack, the complexity of the underlying heterogeneous hardware cannot be fully hidden, so that also application developers need to deal with it. The challenges are enormous, and affect code portability, performance, application efficiency, long-term maintainability, etc.

Heterogeneity is also growing on the memory side: DDR, High Bandwidth Memory (HBM), Non-Volatile Memory (NVM), etc. are combined in HPC systems, building up deeper memory and storage hierarchies.

The increasing growth of applications' memory demands is leading system architects towards combining diverse memory technologies. As these new memory technologies mature, conventional memory management should be reconsidered to better meet applications memory requirements by efficiently utilising the properties of each memory technology. However, there is currently no single widely-validated approach to achieve this goal. New memory technologies are proliferating, each with its own set of design trade-offs, thus resulting in differences in key characteristics such as latency, bandwidth, cost. Moreover, new memory technology differ in properties such as persistence and the degree of asymmetry in read and write access latency. Due to difficulties in sustaining empirical trends such as Moore's Law, and Dennard Scaling, and an explosion in data volumes, computing systems are becoming more heterogeneous. It is becoming increasingly evident that advances are needed beyond the hardware/software interfaces developed originally for more homogeneous systems, such as hiding memory technology diversity in a transparently hierarchical memory system. On the one hand, departure from broadly-used abstractions runs the risk of limiting the usability of advances in memory device technologies. On the other hand, limiting software stacks to only existing interfaces while operating on inherently heterogeneous system platforms will severely limit achieved performance levels and therefore cost-effective utilisation of memory system innovations. Currently ongoing research projects, such DEEP-SEA,[1] are contributing with improvements of software interfaces towards explicit management of heterogeneous memory systems. On a broader scope, it is worthwhile to invest more in design space exploration for upcoming generations of memory systems, particularly their integration in high-performance computing systems and their hardware/software interfaces, based on a cross-layer rethinking of the memory system hierarchy in anticipation of system designs with increasing heterogeneity.

From the system architecture perspective the key question is how to efficiently organise the large diversity of hardware resources that are integrated in modern HPC systems and mitigate the porting effort on the application side. One possibility is to connect the different compute devices within the node, and multiplying this heterogeneous entity as many times as necessary to reach the required total compute performance. However, fully utilising the hardware capabilities of this so-called *monolithic* architectures is challenging. The reason is that a single application can hardly use all compute devices in such *fat* heterogeneous node, and when a node is shared between different users they end up fighting for some common resources, typically memory and network bandwidth. The alternative are *segregated* architectures, in which compute devices of different kinds are grouped into partitions ore modules (see Sect. 4.2).

[1] DEEP projects: Programming Environment for European Exascale Systems. URL: https://www.deep-projects.eu.

4.2 Modular Supercomputing Architecture

Striving for a more dynamical use of heterogeneous hardware resources, the Modular Supercomputing Architecture (MSA; Suarez et al. 2019) has been developed within the EU-funded DEEP projects.[2] In the MSA compute devices of different kinds are grouped in so-called *modules*, which are cluster computers of potentially large size (Fig. 4.1). The configuration of each module is internally kept as homogeneous as possible, but each module present different node components. Put otherwise: ideally all nodes within a module contain the same kind of compute device (same CPU, or accelerator), but the node-design of two modules is different, so that the requirements of different kinds of applications can be fulfilled. A typical example of a modular system is a *cluster-booster* system, in which a module constituted only by CPUs (the *cluster*) is connected to another module made only by one kind of accelerator such as GPUs (the *booster*). Examples of modular supercomputers are the JURECA (Thörnig and von St. Vieth 2021) and JUWELS (Alvarez 2021) systems in Germany, and the MeluXina[3] system in Luxemburg. In the future, modular supercomputers will also integrate disruptive technologies. Current

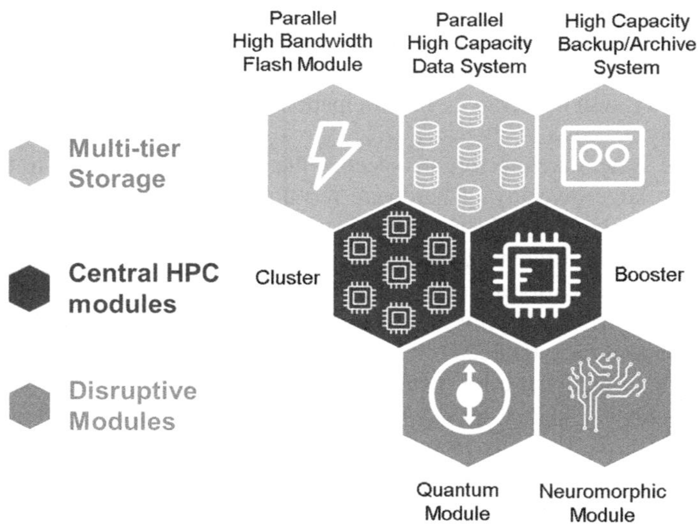

Fig. 4.1 Schematic view of the Modular Supercomputing Architecture. Each module is a cluster of potentially large size

[2] https://www.deep-projects.eu.

[3] https://docs.lxp.lu.

research focuses in the integration of a quantum module, for example within the EU-project HPCQS.[4]

A group of differently configured compute modules becomes a modular supercomputer when they are interconnected by a high-speed network and when a common software stack allows to allocate hardware resources and run applications across the module boundaries (see Sect. 7.2). In such system, each application can select any number of nodes on each side of the machine, depending on its specific requirements. For example, application parts that are serial or low-scalable and need high single-thread performance run on the cluster module, while the booster is used by highly parallelised codes. Application workflows can run across various modules, using the best suited hardware in each step of the workflow.

The specific configuration of each module depends on the kind of applications that it is intended for. A modular system for high-energy physics and radioastronomy could contain an FPGA-module to pre-process and merge the data coming from the different instruments, a CPU-based module to filter the relevant subsets of data, a GPU-based module to create images and visualise them, etc. Astron's applications to correlate and create images for the LOFAR radiotelescope have been ported and tested on the modular DEEP-EST prototype (Romein and Veenboer 2021). While the imager application ran on the data analytics module, a system provided with GPUs, FPGAs and large amounts of RAM and non-volatile memory, the correlator could run either on the booster or data analytics module, since its memory footprint is lower and both modules contain GPUs.

4.3 Software Containerisation

Software containers are lightweight, standalone, executable packages of software that include everything needed to run an application: code, runtime, system tools, libraries and settings. They were conceived as a solution to the problem of how to reliably move software from a computing system to another.

Containers are important as they allow to avoid the common pitfalls that arise in such task, as for example entering the so called "dependency hell", but also to greatly improve reproducibility. In the context of data-intensive radio astronomy this is important not only for the scientific aspects, but also for the practical ones, as for example to prevent wasting CPU power if running a data processing task which does not behave as expected and thus generates unusable data. At the scale of radio astronomy experiments for the next generation facilities discussed in this book, this is a topic of paramount importance for effectively scaling the analysis tasks.

The purpose of this section is not to provide a detailed elaboration on how containerisation is implemented, but rather to frame and describe its philosophy and the various technologies.

[4] https://www.hpcqs.eu/.

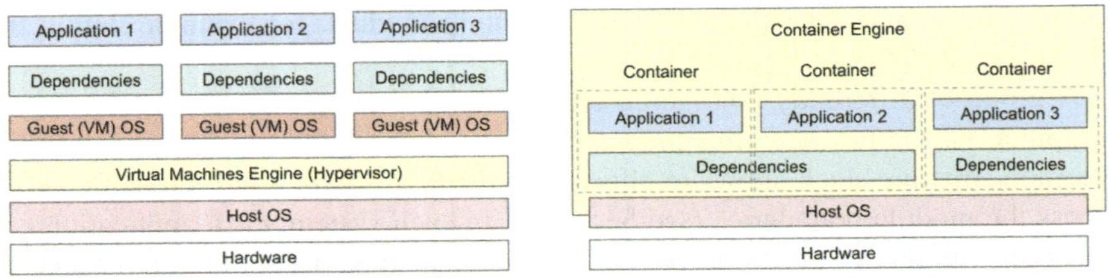

Fig. 4.2 Comparison of virtual machine (left) and container (right) stacks at OS-level

4.3.1 General Concepts

The main concept behind containerisation is relatively simple: instead of virtualising an entire computing system (as classic virtual machines do), the kernel is shared between the host and the containers, where only the file system and process space are virtualised, implementing what is commonly referred to as a "chroot jail" sandbox.

This approach, which is schematised in Fig. 4.2, allows for a number of benefits including no memory pre-allocation, no container boot time, and pure bare-metal performance as there is no hardware virtualisation in place. Moreover, depending on the container engine, this can also provide a layered, incremental file system where the same layer can be shared with multiple containers. This feature, together with the aforementioned ones, completes the set of the so called "killer features" which lead containers to gain popularity and overtake virtual machines. If for example we build three containers on top of a base Ubuntu root file system (which is, in other words, a vanilla Ubuntu installation), the same base layer will be shared between all the three containers and stored on the host file system only once, thus greatly optimising disk space. But there is more: since container images can be also pushed and downloaded from container registries, as it will be introduced in Sect. 4.3.4, if downloading a container based on the same Ubuntu root file system or the same set of dependencies used by other containers are already present on the host system, the base layers are not required to be downloaded again. Instead, the local copies can be used to reconstruct the container image. The same logic applies when pushing containers to registries. This mechanism, which is under the hood implemented using an hashing algorithms for the contents of each layer, made it very fast and efficient to both build and share containers. Docker (both as a technology and as a company, as it will be explained in Sect. 4.3.1.1) is behind one of the first implementations of this concept, which was then released as an open-source project.

On the cons side, it has to be noted that since containers share the kernel with the host operating system, they are not as much isolated as classic virtual machines. In particular, they are prone of privilege escalation, so that if an attacker can get a root access inside a container, it can potentially exploit kernel bugs to get a privileged (root) access to the host system.

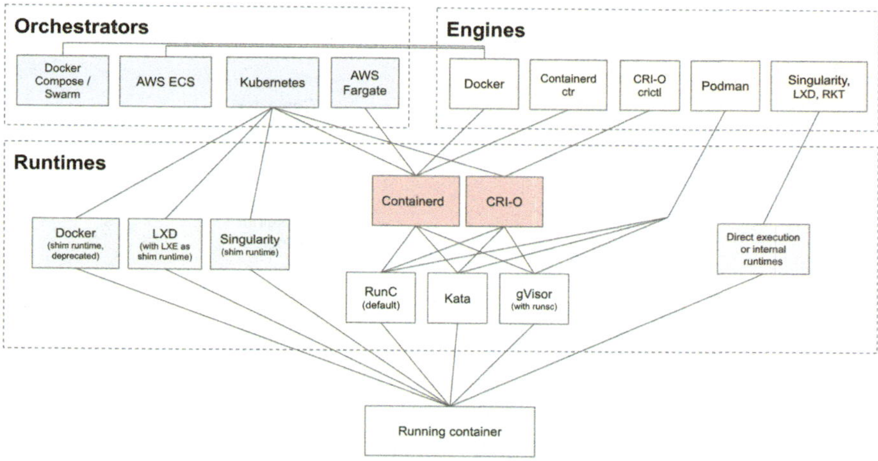

Fig. 4.3 Schematic representation of the relationships between container engines, runtimes and orchestrators

However, the pros of using containers are quite relevant, and as of today they are seen as a great compromise between usability, performance and security. There are also some projects as Kata and gVisor which are working in the direction of re-introducing in the container world a certain degree of kernel or hardware virtualisation, but this of course comes with a price in terms of memory pre-allocation and/or performance decrease.

Before digging any deeper in the containerisation ecosystem, a set of definitions is required, as its complexity grow significantly over the last years. In the transition from standalone, monolithic projects as LXC and Docker to the Open Container Initiative (OCI) standards, the terminology got indeed quite convoluted, and the same component can behave as two different ones, depending on how it is framed (e.g. an engine can became a runtime of another engine). The diagram in Fig. 4.3 tries to summarise the situation as of today, and most importantly to clarify the relationships between the various moving parts.

Not all container orchestrators, engines and runtimes are included in this diagram nor discussed here, and most notably Nabla, Shifter, Nomad and Marathon are not covered. We indeed wanted to keep the discussion light, and to focus on high level concepts in order to then generalise.

- A *container engine* is a piece of software that accepts user requests, including command line options, build and pulls images, and from the end user's perspective, runs the container.
- A *container runtime* is a software component which is in charge of managing the container lifecycle: configuring its environment, running it, stopping it, and so on. They can be thought as what's inside the engine (i.e. valves and pistons). Runtimes can be further sub-divided in two types: high level and low level. high

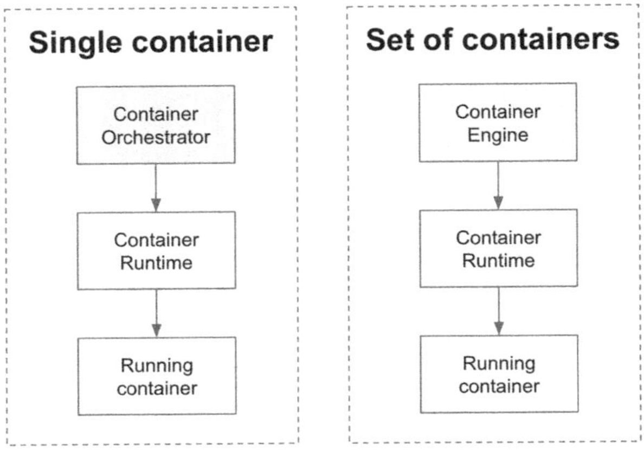

Fig. 4.4 Running single containers vs. set of containers

level container runtimes, or container runtime interfaces (CRI) for Kubernetes. Following the engine analogy, they can be thought as the valves which in turn feeds the pistons. low level container runtimes, or CRI runtimes for Kubernetes. You can think about them as the pistons which do the heavy lifting. Also note that some engines can behave as runtimes and can be thus used from within other engines, or orchestrators.

- Lastly, a *container orchestrator* is a software in charge of managing set of containers across different computing resources, handling network and storage configurations which are under the hood delegated to the container runtimes (or in some nearly legacy cases, engines).

The general rule, schematised in Fig. 4.4 is thus as follows. If running single containers, the interaction is via a container engine, which will in turn interact with a container runtime, either monolithically built-in or as a module. This leaving out the added complexity of engines that can behave as runtimes and that can thus be used from other engines. If instead running set of containers, the interaction is via a container orchestrator. Which one to use will depend on the use case and deployment complexity. Moreover, this is true if we leave out a new trend of building orchestrators on top of other orchestrators (e.g. Portainer), in which case the orchestrator will interact with another orchestrator. Moreover, no one prevents you from using an orchestrator to run a set of containers with a single element, thus effectively running a single container, which is something you might want to do in particular in Cloud environments (e.g. in AWS ECS or Fargate).

4.3.1.1 Container Engines

Docker is the most common container engine and a nearly "de facto" standard. Being a monolithic project in the beginning, it has since then been refactored to support both the need of an open source ecosystem and architectural flexibility. At different stages, the Docker GitHub repository was renamed in Moby and the internal, built-in runtime was extracted as Containerd. Docker identified indeed for a long time many things: a container engine, a runtime, a registry, an image format, a project and also a company. It is a normal part of software projects to get refactored, however with Docker and Kubernetes this generated a bit of confusion. As of today, the Docker Engine is to be intended as an open source software for Linux, while Docker Desktop is to be intended as the freemium product of the Docker, Inc. company for Mac and Windows platforms. From Docker's product page: "Docker Desktop includes Docker Engine, Docker CLI client, Docker Build/BuildKit, Docker Compose, Docker Content Trust, Kubernetes, Docker Scan, and Credential Helper".

Podman is a daemonless container engine for developing, managing, and running OCI Containers on your Linux System. Containers can either be run as root or in rootless mode. Podman is a near drop-in replacement for the Docker engine which can run container in userspace. On shared systems, it is probably the best possible tradeoff between usability and security, as it allows to operate both as root and standard user. Moreover, unlike other userspace containerisation solutions, in Podman users can become root inside containers even if they are just standard users on the host system, which makes it extremely powerful.

Singularity (now Apptainer) should be thought perhaps more as a virtual environment on superpowers rather than as a container engine. It is a solution that operates entirely in user-space and that is designed to be easily pluggable into existent HPC systems. However, in order to achieve this goal, it prioritizes integration over isolation. More in detail, by default the `/home`, `/tmp`, `/proc`, `/sys`, and `/dev` folders are all shared with the host, environment variables are exported as they are set on host, and the PID namespace is not created from scratch. Network and sockets are also shared with the host, and cannot be decoupled. Moreover, Singularity maps the user outside the container as the same user inside it, meaning that every time a container is run the internal user UID (and name) can change, making it very hard to handle permissions. Two issues opened on the former Singularity project are quite self-explanatory: "Same container, different results[5]" and "Python3 script fails in singularity container on one machine, but works in same container on another[6]". In both cases the issue was due to lack of isolation between the container and the host.

[5] https://github.com/apptainer/singularity/issues/476.

[6] https://github.com/apptainer/singularity/issues/3484.

Containerd (with ctr), which will be introduced in the runtimes section, it is not intended to be directly used as an engine (being a runtime), but with the Containerd CLI (`ctr`) utility it can behave as such.[7] Contained is included in the list for completeness and to show how mutable definitions can be in the container ecosystem, as for CRI-O crictl below.

CRI-O (with crictl), which will be introduced in the runtimes section as well, is not intended to be directly used as an engine too. However, with the `crictl` command line utility it can behave as such, mainly for debugging purpose. To underline that CRI-O is not intended to be directly used from a command line (being a runtime), the official CRI-O code repository states that "any CLIs built as part of this project are only meant for testing this project and there will be no guarantees on the backward compatibility with it". In any case, there are tutorials[8] showcasing how it can be used in practice for demo purpose.

LXD (and RKD) is something tangential to a container engine, as it allow to manage both containers and virtual machines, offering "a unified user experience around full Linux systems running inside containers or virtual machines". LXD uses LXC as internal runtime. Rocket (RKD) was instead a command line utility for running containers on Linux directly using kernel-level calls, similarly as for LXC, and is as of today an ended project.

4.3.1.2 High-Level Container Runtimes (or CRI)

Containerd is an high-level container runtime originated from Docker, and extracted out from Docker itself for flexibility over the years. A default Docker engine installation will install Containerd as well. Containerd is also the default Kubernetes CRI runtime. Containerd uses runC as its default low-level runtime.

CRI-O is an "implementation of the Kubernetes CRI (Container Runtime Interface) to enable using OCI (Open Container Initiative) compatible runtimes". It basically tried to fill some gaps along the Kubernetes development and is now a direct competitor (if it makes sense to call it as such) to Containerd. CRI-O uses runC as its default low-level runtime as well.

4.3.1.3 Low-Level Container Runtimes

runC is an OCI-compatible container runtime. It implements the OCI specification and runs the container processes. RunC is called the *reference implementation* of OCI.

[7] https://github.com/projectatomic/containerd/blob/master/docs/cli.md.

[8] https://github.com/cri-o/cri-o/blob/main/tutorials/crictl.md.

gVisor is a runtime developed by Google which implements kernel virtualisation. In other words, each container has its own kernel, unlike other container runtimes where the kernel is usually shared between the host and the containers. It allows for more security than other runtimes while allowing to share host resources without pre-allocation.

Kata containers is a runtime implementing hardware virtualisation (aka: a virtual machine). The idea is to have a runtime which behave as running software containers but that under the hood spawn a new virtual machine and run the container inside it. It is very interesting in terms of security and hardware emulation for multi-architecture tests. On the cons side, it requires pre-allocation of resources, and in particular of the memory which is set by default to 2GB per container.

4.3.1.4 Container Orchestrators

Docker compose allows to define simple multi-service applications where all the containers run on the same node. It creates a dedicated network for the containers on the host from which they can all talk each others, and a `docker-compose.yml` file describes how to assemble them. It the simplest orchestrator, and very useful for local and simple deployments. Docker compose has support only for the Docker APIs, and Podman can work with it by emulating Docker.

Docker Swarm is similar to Docker compose but it can manage multi-node deployments, or on other words a cluster of Docker engines called a "swarm". As for Docker compose, Docker Swarm supports only the Docker APIs. An in-between solution, but still very useful for small teams where using managed orchestrators is not possible and configuring Kubernetes would require too much effort.

Kubernetes is the full-featured solution for container orchestration, supporting a variety of settings, network topologies and container runtimes. In 2021 it dropped support for Docker, which generated some panic over the internet. What it actually happened is that it dropped support for Dockershim in favour of directly using Containerd, and nothing changed for the users. Kubernetes adds the notion of "pod" to the container ecosystem, and can support multiple container runtimes by defining pods with different settings. Mastering Kubernetes is hard, and even an entry-level setup can take time. Kubernetes can be accessed both using a CLI and a set of REST APIs.

AWS ECS, which stands for Amazon Web Services's Elastic Compute Service, is Amazon's internal implementation of a Kubernetes-like solution. Amazon virtual machines require the Docker Engine to be installed in order to be managed using ECS. Alternatively, customers can directly use a virtual machine image pre-build by Amazon which is already configured for using it with ECS. The main point is that AWS ECS use the Docker engine, and not a container runtime.

AWS Fargate is likely one of these "definitive" solutions that will become the new normal for a large number of use cases (as it happened with RDS). Fargate allows executing containers in a serverless fashion, on the Amazon Web Services computing infrastructure, and to entirely forget about the underlying OS (and hardware, of course). Interestingly enough, by probably being a project younger than ECS, it could make the strategic move of stopping to rely on container engines in favour of adopting container runtimes. In particular, with Fargate platform version 1.4 in April 2020, they replaced the Docker Engine with Containerd as Fargate's container execution engine.

4.3.2 Building Containers

Containers are built by interacting with an engine, and starting from a sort of source code of the container. The result of the building phase is a container *image*, which can then be run by an engine or an orchestratror using a runtime.

The most common source format for containers is the Dockerfile, which is a file containing a series of statements to be executed in order to build it, but other formats are possible (as Singularity recipes or LDX config files). The Dockerfile can be built by a number of engines besides Docker, as for example Podman and Singularity. The name hardcode "Docker" for historical reasons. Here is an example skeleton of a Dockerfile:

```
FROM <base image>
RUN <a setup command>
COPY <source file on the host OS> <dest file in the container>
RUN <another setup command>
CMD <default container command>
```

The <base image> can refer to a local image or to an image stored on a registry, for example: FROM my_base_image. Depending on the engine building the Dockerfile, if the image "my_base_image" is not found locally, it is then looked up on a registry. Registries will be briefly discussed in Sect. 4.3.4, for now they can be thought just as repositories for container images. Docker, for example, will automatically look up images not available locally on Docker Hub, its own registry service, so that the statement FROM ubuntu:22.04 will automatically download and build a container on top of the base Ubuntu 22.04 image. To explicitly use a container image on a registry, this can be done by including the registry URL in the statement: FROM docker.io/library/ubuntu:22.04. It is also possible to build an image from scratch, using the statement FROM scratch.

The RUN statement runs a command in the container image being built, for example: RUN apt-get update && apt-get install -y apache2. This will update the Ubuntu package manager cache and install the Apache 2 web server.

The COPY statement copies a file (or a directory) from the host file system to the container internal one. For example, to add a custom default page to Apache: COPY index.html /var/www/html/index.html.

The `CMD` statement sets the default container command and its arguments. For example: `CMD ["/usr/sbin/apache2ctl", "-DFOREGROUND"]`. There is also a variant of the `CMD` statement, the `ENTRYPOINT`. The main difference is that the default command set with the `ENTRYPOINT` statement cannot be overwritten at runtime, with the command set via the `CMD` statement can.

To build the container, usually the command is `<engine> build <folder>` where `<folder>` is where the Dockerfile is stored. For example, to build a Dockerfile in the current directory with Docker: `docker build .` (note the dot after the build argument).

Technically, each build step runs in a temporary container, so that the build phase actually involves also running container, which is discussed in the next subsection. Each build step also generates a layer of the final container image file system, and it is associated with an hash. As already mentioned this is used for caching purpose, but it can also be used to recall a container image to run.

Since running containers referencing them by hashes is inconvenient, most of container engines also support *tagging* containers. This is an operation which is often contextual with the build phase (even if a container can be tagged or re-tagged afterwards), so the build command usually support also tagging it. With Docker, for example: `docker build . -t my_apache_image`.

4.3.3 Running Containers

We already saw the general concept of running containers in the introduction of this section. Let's now get a bit more hands-on and see how we can actually use an engine to run a container. Usually the command is in the form `<engine> run <container_image>`. For example, to run the Apache container used as an example in the previous subsection with the Docker engine: `docker run my_apache_image`. This will run the container but in non-interactive mode, so that you cannot even kill it from the current shell. To run a container in interactive mode with Docker, the -it switch is required: `docker run -it my_apache_image`.

This will run the container in interactive mode and allocate a tty (`-it` is actually a shortcut for both the `--interactive` and `--tty` options).

To execute a custom command at container startup instead of the default one, this can just be appended to the run command. For example, to get a shell access: `docker run -it my_apache_image /bin/bash`. This will override the command set with the `CMD` statement of the Dockerfile and open a Bash shell in the container.

A running container can be usually managed with a series of commands to perform list, delete and inspect operations. For example, `docker ps` will list all container running on the host, while `docker stop` will stop a given container.

A running container is ephemeral: every change made inside it is not persisted on container exit. This makes it very easy to reproduce runs, but does not allow

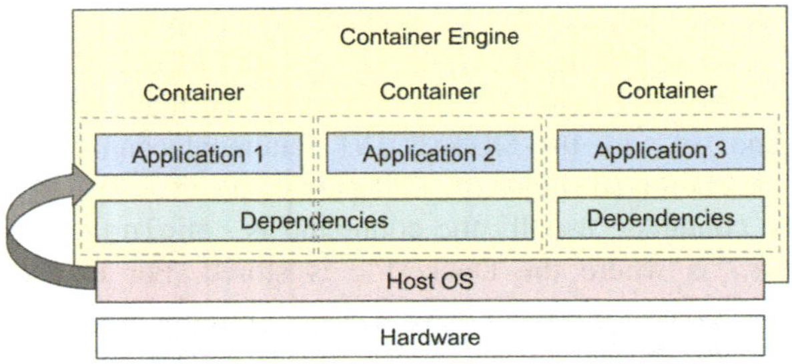

Fig. 4.5 Container volume bridge

to save any data. To persist data in a running container, we need *volumes*. These are virtual disks that are plugged into the container and managed by the container engine. Usually the option is `-v`, for example: `docker run -v my_www_data_volume:/var/www/ my_apache_image`. In this case an empty volume named `my_www_data_volume` is created, mounted on /var/www inside the container and stored on the host. Another option for persisting data is to use a volume which actually binds a host directory instead of creating a new virtual disk. This allows the running container to access portions of the underlying host file system, by creating a sort of bridge, as shown in Fig. 4.5. In this case the first value of the `-v` argument has to be the folder on the host file systems, e.g. `docker run -v /home/my_user/my_www_data_folder:/var/www/ my_apache_image`.

A running containers is (usually) isolated also in terms of network. It has its own IP address on a dedicated subnetwork shared with the container runtime. Because of this, the Apache web server running in the container of our example is accessible only if pointing a web browser to the container IP address.[9] However, the common scenario is usually to expose the service running in the container on the host, in order to make it accessible from the outside world. To achieve this goal, most container engines support a port mapping feature which allows to map a port on the container to a port on the host. For Docker, this is done using the `-p` argument, similarly to volumes. For example, to expose the port 80 of the container on the port 8080 on the host: `docker run -p8080:80 my_apache_image`.

Using an orchestrator to run set of containers recall similar concepts, usually using a configuration file. The following is a Docker Compose configuration file (`docker-compose.yml`) for a typical web application, including a database and a proxy service. The `database` and `webapp` services have some persistent data using volumes, and the `proxy` service exposes the 80 and 443 ports for HTTP and HTTPS protocols.

```
version: '3'
services:
```

[9] For technical reasons, accessing a container via its IP address is not possible if using Docker Desktop on Mac and Windows.

```
database:
  image: "database"
  environment:
    - SAFEMODE=False
  volumes:
    - ./data/postgres:/data

webapp:
  image: "webapp"
  volumes:
    - ./data/webapp/:/data

proxy:
  image: "proxy"
  ports:
    - "80:80"
    - "443:443"
```

To run this set of containers with Docker Compose, it is just required to issue the command `docker-compose up`, while to stop them, the command is `docker-compose down`. Both commands are required to be run in the same folder where the `docker-compose.yml` file is stored, and in container images must be already built and available either locally or on a registry. Once containers are run as part of the container set specified in the configuration file, they can be looked up by their name: to ping the `database` container from the `webapp` one, it is just required to type `ping database`.

Other orchestrators share similar concepts and configuration files, but a detailed overview is out of the scope of this book.

4.3.4 Sharing Containers

Containers come in various shapes, but in the end they are stored somewhere on disk. For simpler container engines, this can be just a folder or an image file (e.g. LXC and Singularity), while for more sophisticated ones, as Docker and Podman, this can be a "dedicated" portion of the host file system where the various layers are stored. In the fist case, sharing containers is as easy as just compressing a folder or copying the container image files. For the latter, things get a bit more complicated. To share a container as a file, an export phase is indeed required, where the engine put together all the various layers composing the container image and generates a compressed archive. For example, with Docker: `docker save my_container > my_container.tar`, and `docker load < my_container.tar`. A big pro of sharing containers as archive files is that they can be stored on "standard" file repositories, as for example shared POSIX file systems or archive services as Zenodo.

However, this approach is very limiting when the container offers a layered, incremental file system, since its main feature of reusing shared layers cannot be used. This is why container registries born, which allow not only to store containers, but also to fully support file system layers.

Registries are basically repositories where the stored entities are container images, referred either by hash or tag. Because of lifting the abstraction level from mere files to the images, it is entirely up to the registry (and the engine) how to exchange the actual "bits" composing the image, which can be therefore optimised. Not re-exchanging the layers with the same hash is perhaps the most important optimisation, and as mentioned earlier one of their main features. Registries are also closely bind to image formats, which will be better discussed in the next subsection.

4.3.5 Interoperability and the Open Container Initiative

As every new technology, containers born to solve a specific need, with respect to a well defined use-case. In the early days containers were built and run on the same host, mainly as a security measure or in-lieu of virtual environments to manage multiple sets of complex dependencies.

As years passed and containers got more traction and adoption, some limitations started to arise. First of all, containers where assumed to be built and run on the same OS and architecture. Since a container is no more than just a set of files, and shares the kernel with the host, it is by design not possible to run them on another OS and/or architecture than the one they were built on. But perhaps most importantly, there was not even a way to detect an OS or architecture mismatch.

In Docker, this was solved by adding a `platform` meta tag for each container, a string composed by the OS and the architecture, separated by a slash. For example, `linux/amd64`. In this way the engine could check against the container image if the container OS and architecture matched the host ones, and if not it could raise an error or resort to other solutions (as running a temporary VM emulating the required architecture, which is what Docker Desktop does as of today by default).

At this point, the multi-OS and multi-architecture support was introduced. This is somewhat in contrast to reproducibility, as what happened is that Docker made co-exist *multiple* container images under the same tag. That is, the official Python container on Docker Hub for the tag 3.9-alpine provides actually seven different container images, one for each of the supported OS/architecture combinations: linux/386, linux/amd64, linux/arm/v6, linux/arm/v7, linux/arm64/v8, linux/ppc64le and linux/s390x. Note that also darwin/arm64/v8 or windows/amd64 would have been perfectly valid additions to the list. Other container engines started to apply the same logic, which is today the standard approach as the Docker image format become central in the standardisation process, as it will be explained later in this subsection.

But the other, huge problem with container interoperability consisted in the impossibility of running a container with a different engine than the one used to build it: a Docker container could be run run only using Docker, and the same applied for all the other engines. This is because of the image format, and every engine having its own standard. The most common image format is the Docker Image Manifest, where an image is defined as "an ordered collection of root filesystem changes and the corresponding execution parameters for use within a container runtime". LXD currently supports instead two LXD-specific image formats: the first is a unified tarball, where a single tarball contains both the instance root and the needed metadata, while the second is a split model using two files, one containing the root and the other containing the metadata. The Singularity Image Format is a compressed squashfs filesystem that has a block organisational structure to include metadata and content for the container definition file, the header, partition, signatures (if present), and of course, the container binary itself.

To improve standardisation and unify all these formats, the Open Container Initiative (OCI) was born. Quoting its official website: "The Open Container Initiative is an open governance structure for the express purpose of creating open industry standards around container formats and runtimes. Established in June 2015 by Docker and other leaders in the container industry, the OCI currently contains three specifications: the Runtime Specification (runtime-spec), the Image Specification (image-spec) and the Distribution Specification (distribution-spec). The Runtime Specification outlines how to run a *filesystem bundle* that is unpacked on disk. At a high-level an OCI implementation would download an OCI Image then unpack that image into an OCI Runtime filesystem bundle. At this point the OCI Runtime Bundle would be run by an OCI Runtime."

Container images in OCI are defined as a "format for encoding a container as a filesystem bundle—a set of files organised in a certain way, and containing all the necessary data and metadata for any compliant runtime to perform all standard operations against it". Most interestingly, since this standardization is based on the Docker image manifest V2, Schema 2 format, it was already supported by most container registries.

While the efforts towards standardisation and interoperability are still ongoing, thanks to the Open Container Initiative a common ground was reached relatively fast, and is as of today working for most use-cases.

4.4 HPC-Cloud Convergence

Cloud and HPC computing environments are becoming increasingly similar in hardware specifications, both offering large-scale, distributed platforms, combining general–purpose and domain–specific processing units with access to high-speed, low-latency storage and networking. In practice, however, they still differ largely in the software stack and how it manages available resources. In the Cloud it is typical to use virtualization mechanisms to facilitate sharing, valuing colocation of workloads to improve infrastructure efficiency, whereas the HPC community prefers to rely on job queueing mechanisms to coordinate among tasks, favoring dedicated use of physical resources by each application.

As modern applications become more complicated, it is not uncommon for deployments to include parallel provisioning of backend services (such as databases or message queueing systems), as well as on-demand execution of data analytics pipelines and HPC codes. For such workloads, it is essential to either accommodate both resource allocation schemes on the same hardware infrastructure—exploiting resource sharing while avoiding interference as much as possible—or *bridge* Cloud and HPC via higher-level frameworks that provide the necessary abstractions. The latter approach is now possible as HPC installations increasingly support containers: reusable units of integrated, pre-packaged software that runs in an operating system sandbox with minimal performance overheads.

In Zervas et al. (2022), the authors explore the convergence of Cloud and HPC in a common, Kubernetes-based environment. To run HPC applications in Kubernetes[10] (the most prominent distributed container orchestration framework[11]), they introduce the concept of the *virtual cluster*, as a group of multiple container instances that function as a private HPC cluster for a user. Each node in a virtual cluster embeds all necessary libraries and utilities, as well as a full Slurm deployment; the user working inside a virtual cluster can only view and manage jobs submitted from within the same context. To coordinate resource allocations between tasks running within virtual clusters and other Kubernetes services, they extend the Slurm controller with a custom protocol that requests resources from the Kubernetes scheduler, effectively placing Kubernetes in charge of resource management for the whole cluster. Moreover, they use *Genisys*, a custom Kubernetes scheduler that distinguishes between "HPC" and "data center" services (typical Kubernetes deployments that run in other containers), in order to apply different container placement policies. In cases where HPC workloads do not consume all node-local resources, Genisys colocates data center services, while constantly satisfying their user-defined performance targets. For this, it uses a control loop, based on Skynet (Sfakianakis et al. 2021), that dynamically adjusts application resources (number of cores, memory size, I/O bandwidth, and network bandwidth) depending on their runtime performance. Evaluation results of this "embedded" HPC approach indicate that it is not only possible to colocate data center tasks with HPC jobs when remaining CPU cycles are available, but with appropriate scheduling it can be beneficial to overall performance, compared to unmanaged (no scheduling coordination) and partitioned setups.

Most other Cloud-HPC convergence approaches assume that some of the available nodes are already allocated to and configured as part of a large HPC cluster (probably also under a separate administration scheme), thus focus on interfacing mechanisms for submitting HPC jobs from the Cloud side. This scheme has benefits for both userbases: Cloud users can easily tap on the vast amount of high-performance computing resources available in HPC centers, while HPC users can

[10] https://kubernetes.io.

[11] https://k8s.vmware.com/state-of-kubernetes-2020/.

leverage the expressiveness and the interactivity of the Cloud software ecosystem, which places particular emphasis on portability and integration issues.

In López-Huguet et al. (2020), the authors use a utility called *hpc-connector* that acts as an HPC job proxy: Users submit respective Kubernetes jobs with the appropriate settings, and hpc-connector forwards them to the HPC cluster, monitors their execution, and collects their results. Docker and Singularity containers are used to address portability issues. Furthermore, the authors identify five requirements for a system that allows running the same workloads on Cloud and HPC. The main focus is on HPC job management with a Kubernetes-compatible interface; the HPC and cloud clusters are treated as two separate environments and users have to format their workloads to explicitly use the job-forwarding utility. A similar approach is presented in Zhou et al. (2020), where a Kubernetes installation is interfaced to a Torque-based HPC cluster, using a custom tool called *Torque-Operator*. Although this work offers the flexibility of running containerized Cloud and HPC jobs over the same frontend interface through the *WLM operator*,[12] it again uses a different language for describing jobs targeted for the HPC cluster.

StreamFlow (Colonnelli et al. 2021), on the other hand, focuses on separating the workflow definition from the peculiarities of each execution environment. The StreamFlow runtime extends the workflow language with declarative descriptions of execution sites and their relationship to workflow nodes, in order to transparently submit workflows to a mixture of heterogeneous Cloud and HPC setups, while automatically copying required data where needed. Groups of workflow steps may require specific environments to run, which are translated to runtime deployment dependencies. A proof-of-concept implementation uses the Common Workflow Language (CWL).

Rosetta (Russo et al. 2022) science platform is another example of tool developed towards the HPC-Cloud convergence. A science platform is an environment designed to offer users a smoother experience when interacting with remote computing and storage resources. While on one hand the science platform do make it easier to access and use remote computing resources, on the other, since they are mainly focused on web-based and integrated analysis environments built on top Jupyter Notebooks or similar software, most of the science platforms developed today rely on technologies originating from the Cloud IT industry (e.g. Kubernetes) and require deep integration at system-level, which is often hard to achieve in the framework of HPC clusters and data-intensive system.

Built on top of a novel architecture based on framing user tasks as microservices—independent and self-contained units—Rosetta allows to fully support custom software packages, libraries and environments, including remote desktops and GUI applications, besides standard web-based analysis environments as the Jupyter Notebooks. Its user tasks are implemented as software containers, which allow for safe, effective and reproducible code execution, and that in turn allows users to add and use their own software containers on the HPC platforms.

---

[12] https://github.com/sylabs/wlm-operator.

Rosetta is designed with real-world deployment scenarios in mind, and thus to easily integrate with existing computing and storage resources including HPC clusters and data-intensive systems, even when they do not natively support containerisation, in a way similar a user interacts with a Cloud environment, Rosetta is able to run cloud native workflows on an HPC cluster based on SLURM.

The platform architecture follows a standard approach where a set of services implement the various functionalities. These comprise a web application service for the main application logic and the web-based UI, a database service for storing internal data and a proxy service for securing the connections. The web application service functionalities can be further grouped in modules which are responsible for managing the software containers, interacting with the computing and storage resources, orchestrating the user tasks, handling the user authentication and so on.

In particular:

- *Software* functionalities allow to track the software containers available on the platform, their settings and container registries;[13]
- *Computing* functionalities allow to interact with both standalone and clustered computing resources, hosted either on premises (e.g. via Openstack) or on cloud systems (e.g. on Amazon AWS);
- *Storage* functionalities allow browsing and operating on local and shared file system (as Ext4, NFS, BeeGFS);
- *Task* functionalities allow submitting and stopping tasks as well as viewing their logs, by interacting with the computing resources workload management systems (WMSs) as Slurm and Kubernetes and/or their container engines (e.g. Docker, Singularity, Podman);
- *Account* functionalities provide user account and profile management features including user registration, login and logout, supporting both local and external authentication (e.g. OpenID Connect, Shibbolet).

Rosetta has been tested for a number of use cases, including the LOFAR data reduction pipelines, the SKA data challenges, post processing of large cosmological simulations output, quasar spectral analysis.

KNoC (Kubernetes Node on HPC Cluster; Maliaroudakis et al. 2022) is a virtual node for Kubernetes, that transparently manages the container lifecycle on a remote HPC cluster using Slurm and Singularity. This technique effectively allows users to employ existing cloud-native tools, such as the popular Argo Workflows[14] language to express complex data-processing pipelines for both Cloud and HPC. Argo Workflows is quickly gaining ground as the industry-standard workflow environment in the Cloud, providing a language and runtime to model and execute applications as directed acyclic graphs (DAGs). In Argo, every node of the

[13] A container registry is a place where container images are stored, which can be public or private, and deployed both on premises or in the Cloud. Many container registries can co-exist at the same time.

[14] https://github.com/argoproj.

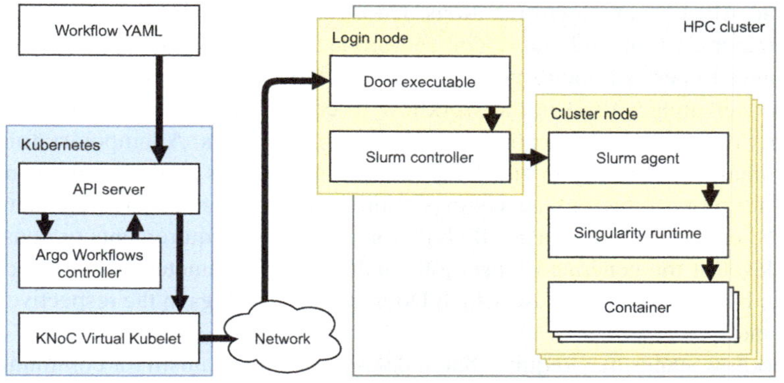

Fig. 4.6 Running a cloud-native workflow on an HPC cluster using KNoC

graph is a container. The Argo controller processes each workflow by submitting respective containers for execution, monitoring their status, and collecting their outputs; all presented via a user-friendly, interactive, web-based frontend. The frontend also allows organizing workflows using templates, as well as planning repeated execution with a crontab-like syntax.

An overview of the main components involved in KNoC's deployment environment is shown in Fig. 4.6. The main goal is to run the container-based workflow provided by the user in Argo format (top-left) on the HPC cluster with Singularity containers (bottom-right). The Argo Workflows engine receives the workflow YAML and creates the corresponding Pods through the Kubernetes API. Then Kubernetes selects "KNoC" as the execution node to run the respective containers. If other, physical nodes also exist, this can be accomplished using node-selection constraints in the workflow language.

The kubelet is the primary agent that runs on each node of a Kubernetes cluster. It is responsible for starting and stopping containers, reporting on their status, gathering their logs, etc. Each kubelet receives *PodSpecs*, which are Kubernetes objects that describe Pods and ensures that the containers comprising those Pods are running and healthy. Kubelets running on physical machines practically implement the interfacing between Kubernetes and the underlying container runtime; typically Docker or containerd. When KNoC receives a request to create a container as part of a Pod specification, it forwards the request to the remote HPC system supplied in the configuration. The assumption is that all interfacing with the remote environment can only happen by running commands through a secure shell (ssh) connection. KNoC is implemented as a Virtual Kubelet[15] plug-in.

To simplify the integration and add an abstraction layer at the remote end, KNoC automatically deploys an executable at the HPC side, called *Door*, which provides

[15] https://github.com/virtual-kubelet/virtual-kubelet.

a simple API for container operations. Door receives requests from KNoC related to container creation and tear down, and produces the necessary Singularity/Slurm commands to perform the respective actions. Singularity will automatically convert any Docker images used in the workflow to create Singularity-compatible image files that run with the same commands as their entrypoints. A temporary folder for each container at the cluster side keeps all runtime outputs and state, and is used by KNoC for monitoring the status of execution. To access special hardware features of the HPC environment (such as GPUs), or specify other requirements or constraints at the level of the generated Slurm job (such as MPI parameters), the user can use specific labels in the workflow, which Door includes as flags in the respective Slurm command.

Workflow stages may require shared datasets or a mechanism for communicating processed data from one stage to the next, which can be achieved by using an Argo *artifact repository*, as a common place to deposit files, or a *shared folder*, mounted across all containers at a known path. Argo supports many S3-compatible services for artifacts; the runtime will copy in specific files so they are available to running containers before startup and copy out results after stage completion. With shared folders, a preparatory workflow step may download required datasets in the execution environment, and subsequent steps use it for writing out intermediate files and results. In HPC, a user home folder is typically available at all cluster nodes and the default Singularity behavior is to mount it and make it available within running containers.

The landscape of tools that bridge the gap between Cloud and HPC is continuously growing. Applications, expressed as workflows, deal with increasingly large and diverse datasets, requiring more and more processing capacity, as well as the integration of a variety of tools from both domains. The general direction is towards embracing the heterogeneity of available platforms under a common higher-level workflow language or other abstraction and enable workloads to seamlessly exploit all available resources.

References

D. Alvarez, JUWELS cluster and booster: exascale pathfinder with modular supercomputing architecture at juelich supercomputing centre. J. Large-Scale Res. Facil. **7**, A183 (2021). ISSN 2364-091X. https://doi.org/10.17815/jlsrf-7-183. https://juser.fz-juelich.de/record/904929

I. Colonnelli, B. Cantalupo, I. Merelli, M. Aldinucci, Streamflow: cross-breeding cloud with hpc. IEEE Trans. Emerg. Top. Comput. **9**(04), 1723–1737 (2021). ISSN 2168-6750

S. López-Huguet, J. Damià Segrelles, M. Kasztelnik, M. Bubak, I. Blanquer, Seamlessly managing hpc workloads through kubernetes, in *High Performance Computing*, ed. by H. Jagode, H. Anzt, G. Juckeland, H. Ltaief (Springer International Publishing, Cham, 2020), pp. 310–320. ISBN 978-3-030-59851-8

E. Maliaroudakis, A. Chazapis, A. Kanterakis, M. Marazakis, A. Bilas, Interactive, cloud-native workflows on hpc using knoc, in *Proceedings of the 5th Workshop on Interactive High-Performance Computing*, InteractiveHPC 2022 (2022)

J. Romein B. Veenboer, *Radio Astronomy with the GPU Correlator, the GPU Imager and the FPGA Imager*, Schriften des Forschungszentrums Jülich IAS Series, vol. 48 (Forschungszentrum Jülich GmbH Zentralbibliothek, Verlag Jülich, Jülich, 2021), pp. 81–108. https://juser.fz-juelich.de/record/905868

S.A. Russo, S. Bertocco, C. Gheller, G. Taffoni, Rosetta: a container-centric science platform for resource-intensive, interactive data analysis. Astron. Comput. **41**, 100648 (2022). ISSN 2213-1337. https://doi.org/10.1016/j.ascom.2022.100648. https://www.sciencedirect.com/science/article/pii/S2213133722000634

Y. Sfakianakis, M. Marazakis, A. Bilas, Skynet: performance-driven resource management for dynamic workloads, in *2021 IEEE 14th International Conference on Cloud Computing (CLOUD)*. IEEE (2021)

E. Suarez, N. Eicker, T. Lippert, *Modular Supercomputing Architecture: From Idea to Production; 3rd*, vol. 3, (CRC Press, Boca Raton, 2019), pp. 223–251. ISBN 9781138487079. http://hdl.handle.net/2128/22212

P. Thörnig, B. von St. Vieth, JURECA: data centric and booster modules implementing the modular supercomputing architecture at Jülich supercomputing centre. J. Large-Scale Res. Facil. **7**, A182 (2021). ISSN 2364-091X. https://doi.org/10.17815/jlsrf-7-182. https://juser.fz-juelich.de/record/902211

G. Zervas, A. Chazapis, Y. Sfakianakis, C. Kozanitis, A. Bilas, Virtual clusters: isolated, containerized hpc environments in kubernetes, in *Proceedings of the 17th Workshop on Virtualization in High-Performance Cloud*, VHPC'22 (2022)

N. Zhou, Y. Georgiou, L. Zhong, H. Zhou, M. Pospieszny, Container orchestration on hpc systems, in *2020 IEEE 13th International Conference on Cloud Computing (CLOUD)* (2020), pp. 34–36

Part II
Data Processing

Chapter 5
Challenges of Big Radio Data Processing

André Offringa, Cameron L. van Eck, Danielle Fenech, Andreas Wicenec, and Michael Janssen

The use of radio telescopes has been the source of a number of significant advances in understanding across astronomy, pushing the boundaries of our knowledge by enhancing the capability of the instruments used. The main developments focus around improving the image quality and increasing the sensitivity to detect a greater number of objects or more of their associated emission. Some of the challenges facing radio astronomy today have always existed but the wish for increased performance requires higher accuracy in processing. There are associated challenges in how the processing can be done as this needs to take into account more subtle factors in correcting the data. Fundamentally this adds complexity to the major challenge of processing data on a significantly large scale. With the advent of

A. Offringa (✉)
ASTRON, Dwingeloo, The Netherlands
e-mail: offringa@astron.nl

C. L. van Eck
Research School of Astronomy & Astrophysics, The Australian National University, Canberra, ACT, Australia
e-mail: cameron.vaneck@anu.edu.au

D. Fenech
SKA Observatory, Jodrell Bank, Lower Withington, Macclesfield, Cheshire, UK

Astrophysics Group, Cavendish Laboratory, Cambridge, UK
e-mail: danielle.fenech@skao.int

A. Wicenec
International Centre for Radio Astronomy Research, Crawley, WA, Australia
e-mail: andreas.wicenec@icrar.org

M. Janssen
Department of Astrophysics, Institute for Mathematics, Astrophysics and Particle Physics (IMAPP), Radboud University, Nijmegen, The Netherlands
e-mail: m.janssen@astro.ru.nl

E. Vardoulaki et al. (eds.), *Data-Intensive Radio Astronomy*, Astrophysics and Space Science Library 472, https://doi.org/10.1007/978-3-031-58468-8_5

'bigger and better' radio facilities, the sheer volume of data produced requires a shift to high-performance processing at dedicated facilities. For both planned and in progress upgrades for several existing instruments, and the construction of the SKA, much of the performance improvements expected rely on addressing these challenges. This chapter will discuss some of the current major challenges encountered in radio astronomy data processing (post correlation).

There are a number of computational challenges in processing data from radio telescopes and interferometers from correlation through to imaging. Major advances in radio astronomy and particularly interferometric imaging are achieved through increased sensitivity, improved image fidelity and for some cases higher resolution. However the difficulty comes from how these are put into practice.

Image Quality
Both advances in sensitivity and image fidelity usually arise from one of two options (and in many cases both). The first is increasing the number of antennas to more completely fill the aperture (uv-plane) which improves the overall sensitivity but also to specific spatial scales that may otherwise be missing. This provides both a greater collecting area and a more complete aperture for synthesis observations. Both an increase in antennas n, or bandwidth ν would reduce the expected noise level. This can improve the processing and imaging as there is more complete information about the astronomical source. However, it also presents a challenge in terms of volume of data/data-rates both pre-correlation (from increased number of antennas) and post-correlation processing and imaging with full dynamic range. For n antennas present in an array, the number of independent baselines for which the data are correlated is

$$\frac{n\,(n+1)}{2} \tag{5.1}$$

meaning the data volume increases dramatically even with the addition of a small number of antennas.

Increased Bandwidth
The second major advancement in upping sensitivity has been the advent of improved receivers that are increasingly sensitive to broad frequency ranges, increasing the overall bandwidth available. This again improves both the sensitivity and uv-coverage but requires taking into account changes in the frequency response of the instrument, the atmosphere and the astronomical source itself. These effects have to be taken into account during calibration through e.g. frequency-dependent, flux density, bandpass corrections and complex gain corrections.

Radio emission fundamentally has a spectral index i.e. change in brightness as a function of frequency as a result of the emission mechanism (e.g. synchrotron or thermal in the continuum case). There is also a frequency dependence to the primary beam response causing a change in sensitivity across the field-of-view. The wider the frequency range the greater this issue can be, in the most extreme cases this can result in sources appearing outside of the main lobe and into the first null of the primary beam at certain frequencies. When imaging both this fundamental frequency dependence of the radio emission as well as the change in primary beam response must be taken into account to achieve high dynamic ranges and correct flux densities.

This can also become a compounded problem with standard astronomical calibration sources used to provide absolute amplitude scaling and solve for the bandpass response. These sources must be well-modelled across the whole band in order to account for their frequency response. Direction-dependent corrections and wide-field imaging in many cases become the de-facto requirement in order to prevent artefacts appearing from sources much further out in the primary beam even when scientifically focused on the centre of the field.

The data received from the antennas and subsequently correlated is typically channelised with the width of a channel limited by the required unaberrated field-of-view (Thompson et al. 2017). Increasing the bandwidth therefore also increases the total number of channels present and the total volume. The end result being the need for higher data throughput and storage.

There are limited frequencies in the radio range that are protected for the purposes of astronomy observing, so with the advent of wider bandwidths brings a greater prevalence of RFI in the data (for a more detailed discussion, see Sect. 5.2.1)

Time-Domain Processing
For a number of astronomical sources whose emission varies significantly in time, high-time resolution processing is required to understand and characterise the emission. This typically requires sampling at micro to millisecond integration times and as for other areas of processing the increase in instrument size fundamentally increases the data throughput and data volumes.

Whilst solving these problems is already required for radio astronomy data processing, the shift in scale necessitates improved algorithms, efficient processing and large dedicated computing facilities. The new and upgraded instruments are already producing significant volumes of data and the SKA is expected to produce data rates of the order of $\mathrm{Pb\,s^{-1}}$, high-performance algorithmic implementations are a necessity. Large storage capability in large-scale computing facilities located close to the physical instruments are required to enable efficient processing. There remain a number of unanswered questions as to how this is achieved—how do we scale the processing to be able to enable processing from full raw data at the Pb scale to produce science-ready images? What will the driving limitations be: data throughput, memory, storage and access? What algorithms are required to handle large data volumes efficiently?

5.1 Management and Workflows for Large-Scale Data and Reproducible Workflows

Scientific workflows, workflow management, reproducibility and execution are big topics in other science communities since many years (e.g., bioinformatics (The Galaxy Community 2022; Wratten et al. 2021), and geo sciences (Altintas et al. 2011)) and supported by large grants. In astronomy there had been a few attempts to introduce them as well with varying success and sustainability (ESO Reflex,[1] LOFAR, CyberSKA[2]). There are also a number of more generic workflow systems around, which are being used by some astronomy projects (e.g., Kepler,[3] Taverna,[4] Pegasus,[5] SPARK,[6] Nextflow,[7] Snakemake,[8] CWL [9]). However, currently there is no more general adoption of workflow systems. One of the reasons behind this 'reluctance' might be the fact that astronomy has a very long history of implementing a very much related concept, albeit with a different, maybe less flashy name: Pipelines! Pipelines fundamentally are workflows, but mostly implemented in a very programmatic way, sometimes going to the extreme of being a single compiled executable. Arguably C source-code, or any source code in any formal computing language for that matter, are perfectly valid descriptions of all the individual processing steps, their relationships and dependencies. Provided that the code compiles and runs with valid results, this is obviously a correct statement. However, when it comes to human consumption or just verification or visualisation of the steps and relationships involved, this approach fails to deliver. Depending on the quality of the implementation and documentation, it also has issues with maintainability, reproducibility, collaboration and also re-use of algorithmic modules from other sources. In addition to these pipelines the astronomical community has invested a lot of effort into developing and maintaining multiple very large, more or less self-contained data reduction software packages like IRAF, AIPS, AIPS++ (discontinued and replaced by CASA), CASA, MIDAS, and GILDAS and more recently ASKAPSoft (based on casacore), ALMA (based on CASA). Even commercial products like IDL[10] have been used extensively in some projects. This has resulted in a lock-in effect of individuals and whole

[1] https://www.eso.org/sci/software/esoreflex/.

[2] https://www.canarie.ca/software/platforms/cyberska/.

[3] https://kepler-project.org/.

[4] https://incubator.apache.org/projects/taverna.html.

[5] https://pegasus.isi.edu/.

[6] https://www.sparknit.com/.

[7] https://www.nextflow.io/.

[8] https://github.com/snakemake/snakemake.

[9] https://www.commonwl.org/.

[10] https://www.nv5geospatialsoftware.com/Products/IDL.

organisations and makes it very hard to break up silos and arguably makes in-depth multi-wavelength research much harder than it could/should be. Efforts in the International Virtual Observatory Alliance (IVOA[11]) are trying to mitigate these issues by defining interoperability standards, but these are mostly concentrating on the data, data access and provenance side, not so much on the processing component side. The excellent effort of astropy(Astropy Collaboration et al. 2022) to combine a lot of astronomy related software into a single, modular package should also be mentioned here.

Modern astronomy and astrophysical research is inherently multi-wavelength and now also multi-messenger (with gravitational waves). Just looking at radio, optical or X-ray data individually is not sufficient to get a good understanding of the physical processes we are observing in the universe. On the other side reducing data from a multitude of instruments scattered around the electromagnetic spectrum requires to get familiar with multiple, usually fairly complex, data reduction eco-systems before even attempting to bring the individual results together. Apart from the software/system issues described above, new dimensions of challenges are emerging. For instance, combining data on a larger scale from the future Vera Rubin optical telescope with data from the SKA is already planned, but will be extremely challenging because of the data volumes, rates and cadence coming from either side. That requires not just some Python software to work seamlessly, but it will require, at least in critical places, highly optimised code in order to be efficient and even feasible. For such projects we will need to go all the way from software running on single computers or maybe small clusters to the domain of Many Task Computing (MTC), which is a combination between High Performance Computing (HPC) and High Throughput Computing (HTC; Raicu 2009). This requires a concerted effort from experts in the various fields, but the question is: How can this concertation be instrumented and enabled? One possible answer comes with scientific workflow systems and the clear separation of concerns enabled by them, when implemented carefully. Essentially the system we would look for needs to make sure that people can work mostly independently on challenges in their area of expertise, while maintaining the overall workflow goals. Figure 5.1 depicts the various levels of concerns and their respective area of expertise.[12]

5.2 RFI Mitigation, Calibration, Imaging

Many science cases in radio astronomy require high-quality images of the sky. Making such images requires extensive processing of the raw visibilities. The large volume of visibilities created by the correlator need to go through several processing steps that finally produce a continuum image or spectral image cube of the sky.

---

[11] https://www.ivoa.net/.

[12] https://daliuge.readthedocs.io.

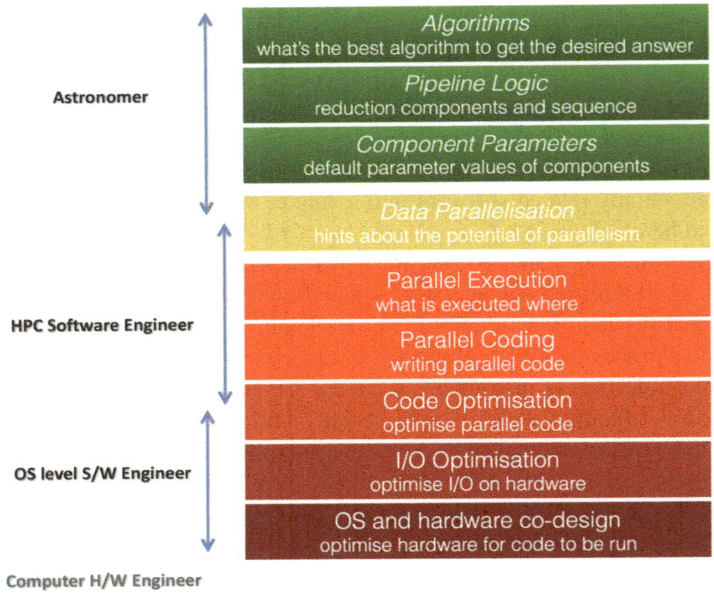

Fig. 5.1 Developing complex, parallel and efficient data reduction workflows is a task requiring a multitude of skills. There are only very few people, if any, who would be able to cover all of them. The figure depicts the various skills required in the coloured bars as well as the kind of people required to implement and optimise them on the left. A carefully implemented scientific workflow system separates these concerns and allows these people to essentially work independently and within their field of expertise to a large degree. One such system, DALiuGE, has been used to perform the largest radio astronomy simulation and reduction run to date, using 95% of the then fastest computer in the world: Summit (Wang et al. 2020)

Pipelines that produce these images typically include at least these fundamental algorithms: Radio-frequency interference (RFI) detection, calibration and imaging.

These algorithms are highly complex and require specific expertise in their design. At the same time, these steps must meet many requirements: they need to be fast enough to process large data volumes, yet they need to be flexible and modular enough to support multiple science cases, should work in a highly automated and stable way and should be user friendly enough to make it possible for astronomers to adapt the algorithms to their science case.

A scenario where astronomers need to adapt algorithms is for example the detection of RFI. For continuum science, triggering may happen on spectral and temporal structure, as both are a symptom of RFI. However, for spectral (line) imaging, it may be undesirable to detect and remove spectral structures — even if the consequence of this is that more RFI is left in the data. Similarly, in transient science cases it is often not desirable to remove temporal structures. The balance between undetected RFI and incorrectly classified RFI is dependent on the science case, and often requires astronomers to tweak their RFI detection strategies.

5.2.1 Radio-Frequency Interference Mitigation

In rare cases, RFI can be filtered or subtracted from the data stream. However, in most cases, such an approach is not feasible, and the presence of RFI is merely detected and flagged. During calibration and imaging, flagged visibilities will then be ignored, such that the RFI does not impact the image quality. Fortunately, in most scenarios RFI has a relatively small bandwidth or short time-scale, causing a loss of only a few percent of the full data.

RFI detection is one of the first steps in processing, and is ideally performed on high time and frequency resolution. After RFI detection, data may be averaged in time and frequency to reduce its volume. This however implies that the result of RFI detection becomes an irremovable part of the data, and if major detection errors are made, the resulting data becomes unusable. As a result, the automated RFI detection must be accurate and robust.

Most RFI detectors find visibilities that show an unlikely high amplitude or an abruptly change in time or frequency direction. To do this, both "classical" signal processing techniques have been employed as well as machine learning techniques—both with varying success. Typical signal processing techniques use operations such as thresholding, line detection, matched filters and morphological detection. These have the advantage of being computationally fast and provide simple ways of adaptation to different scenarios (e.g., disabling detection of spectral features). Machine learning techniques employ convolutional neural networks or auto-encoders. As of yet, they provide a detection accuracy that matches with signal processing techniques. In the future, these may be used to detect more challenging RFI features in the data, thereby outperforming the accuracy of signal processing techniques, but this is yet to be seen. They do have the downside that they are significantly slower (order of magnitude), require training and they are somewhat of a black box operation.

To date, most RFI detection is performed on the CPUs. Detection of different time-frequency data chunks or baselines of the same observation are independent, and can therefore be parallelized or distributed. For the same reason, RFI detection is highly scalable. Certain steps of the detection process might be easy to transfer to GPUs, such as the application of a high-pass filter on the time-frequency data. In typical processing, the computational performance of RFI detection is limited by data transfer, and not by the computations. Simple CPU processing therefore suffices.

Both approaches require processing of a set of visibilities that is contiguous in time and frequency, as this is where RFI is most easily recognized. Accurate detection requires roughly a few hundred timesteps as well as a few hundred of channels of the same baseline (pair of antennas) to be simultaneously in memory. This is somewhat challenging, because this requires a different ordering of the data compared to the natural ordering in which the data comes in. With the natural ordering, channels and baselines are stored together for each timestep. Later processing steps, such as calibration, also require data to be ordered in this natural ordering. RFI detection must therefore perform a temporary transpose of the high-resolution data. This requires careful buffering of the full data stream.

5.2.2 *Calibration*

To obtain science data products, the data from a radio interferometer needs to be calibrated. This consists of correcting for various phase and amplitude effects, and possibly polarisation leakage effects. This is often performed iteratively, where first-order calibration is used to obtain an initial model of the sky, followed by one or more iterations of self-calibration that make use of an increasingly more complete model. When the science products require a large field of view or very high dynamic range, it may be required to perform direction-dependent (DD) calibration, in which the solved effects vary spatially. Solving beam-modelling errors may require DD calibration. At lower frequencies, DD calibration is necessary to correct for the DD effect of the ionosphere.

DD calibration is computationally expensive. Various algorithms exist, each with their own trade-offs between features, computational efficiency and accuracy. The optimal algorithm of choice may dependent on the instrument being used and the exact science case being. Algorithms do share some properties though. First of all, they generally find solutions for small time intervals at a time, thereby making it possible to stream over the data once. However, to keep the degrees of freedom in a DD solver acceptable, it may be necessary to involve larger data spans in the calibration, to perform spectral, spatial or temporal regularisation or to model physical behaviour, such as to solve directly for the total electron content (TEC) of the ionosphere. This complicates a streaming approach, because in this case the required buffer that holds the data for a single solution interval and all directions can become very large. Solutions exist that distribute the data buffer by employing consensus optimisation with a head node and worker nodes; or by simply solution smoothing or fitting after calibration.

5.2.3 *Imaging*

The visibilities from an interferometer follow a Fourier relation to the sky. Because the Fourier plane is not sampled in full, reversing the Fourier transformation that the instrument measures, results in an image of the sky that is convolved by the so-called point spread function (PSF) of the instrument. Interferometric imaging consists of performing a Fourier transform followed by deconvolution of the resulting image, thereby obtaining a truthful image of the sky. Both of these steps are computationally challenging, and multiple algorithms exist to perform these steps, each with their own merits.

Performing a direction Fourier transform, where the input consists of N visibilities and the output consists of a gridded image of size $M \times M$, requires $O(NM^2)$ operations, and is excessively slow for data volumes produced by any modern observatory. This can be mitigated by using a fast Fourier transform (FFT), if the visibilities are first gridded onto a two-dimensional plane. This results in two new challenges: visibilities as they are measured do not fall perfectly onto a two-dimensional grid; and visibilities from different pairs of antennas that are not coplanar, observe Fourier modes of the sky in different projections.

The issue that visibilities may fall between grid cells, is solved by using convolutional gridding, where each visibilities is convolved by a small kernel. If the kernel is analytic in nature, the contribution of non-gridded visibilities to a gridded plane can be calculated fully accurately. Due to the convolutional theory, convolving the Fourier plane results in an image that is multiplied with a known shape, which can be corrected afterwards. While performing this convolutional gridding is expensive operation, it is typically many orders of magnitude faster than a direct Fourier transform.

The second mentioned issue, that different pairs of antennas observe in a different projection, can be solved by a method called w-correction. This correction basically converts the visibilities to the same plane of projection. Typical methods to perform this step, are w-stacking, w-projection, facetting or image-domain gridding—sometimes multiple approaches are combined to obtain better performance.

Computationally, gridding is a task that is easily made scalable. Unlike previously described processing steps such as RFI detection and calibration, a gridder typically doesn't require visibilities to arrive in any specific order. The only exception is that it may be necessary to group visibilities of the same frequency group, in particular when performing spectral imaging. Making gridding scalable is therefore trivial: visibilities can be split in time, frequency or even (almost) randomly, and can be gridded in parallel or even distributed. The resulting grids can be added together before or after Fourier transform, to obtain the final result.

The result of gridding and Fourier transforming is an image of the sky convolved with the PSF. Deconvolution algorithms such as CLEAN work in image space, and model the true (deconvolved) sky iteratively. An image-space deconvolution can not model all the (direction dependent) effects that may affect an observation. To remedy this, the full deconvolution process is split into minor iterations that operate in image space, and major iterations that, once a first order model is obtained, forward predict this model. This corrects for the simplifications that are made during the image-based deconvolution. Subsequently, the residual visibilities are again gridded. As a consequence, the gridding process needs to iterate multiple times over the data, typically about 10 times. A full streaming approach is therefore not possible.

To perform deconvolution, the visibility data from all timesteps, channels and baselines needs to have been gridded. Therefore, there is a synchronisation barrier between the gridding step and the deconvolution, where the final image (cube) needs to be transported and collected. While deconvolution needs all frequency information of the image cube together in one place for accurate deconvolution, it is possible to split the image in small sub-images that are deconvolved independently. Finding an appropriate division is somewhat challenging when the image contains very large scales (that may extend over multiple sub-images), but splitting algorithms have been designed that take advantage of the major-minor style deconvolution, together with careful dynamic splitting of the image in sub-images. These work well in practice, producing results that match an approach without splitting into subimages. Once images have been split into subimages, the subimages can be processed in parallel.

5.3 Very Long Baseline Interferometry

There are several aspects in which very long baseline interferometry (VLBI) data needs to be handled differently from single-dish and connected-element interferometer radio data. RFI does not correlate over the long distances between pairs of telescopes and can therefore be ignored in most cases, unless the interfering signals are so strong that they saturate the receiver of a telescope or when auto-correlation data are used for calibration steps (such as a scalar amplitude bandpass correction). The long distances between VLBI stations prohibit the use of a shared frequency standard. Instead, accurate local clocks (commonly masers but less stable crystals can also work) are used to ensure the stability of the receiving systems and enable a later overlap of the time-tagged baseband data at a dedicated correlator facility. GPS, while not being accurate enough on short timescales, is used as an absolute time reference for VLBI stations.

A model used at the correlator corrects for differences between the recorded time-tagged baseband telescope data and the absolute Universal Time as well as a delay model to facilitate the computation of cross-correlations. This model does however not correct for all data corruption effects, e.g., atmospheric phase turbulence and pathlength differences can lead to residual phase and delay errors in the visibilities that need to be corrected in post-processing. Through fringe-fitting steps, antenna-based[13] delay and delay-rate errors are found. Without good prior information about the magnitude of these residual errors, large search spaces have to be covered, where false fringes due to noise can easily occur (Rogers et al. 1995). Care must therefore be taken to get trustworthy source detections, especially when the correlated source flux density is not much higher than the thermal noise. Such situations can easily arise for weak sources, sources that are resolved out by milliarcsecond to microarcsecond VLBI resolutions, telescopes with small collecting areas, or small observing bandwidths when continuum sources are observed. For the correction of atmospheric phase turbulence, simple self-calibration methods, fringe-fitting on short segmented sub-scan timescales, or ad-hoc fitting algorithms can be used (see

[13] In geodesy experiments, baseline-based delays are the quantity of interest, which is not covered here.

e.g., Janssen et al. 2022, for a recent review). Very-low-frequency data may be additionally affected by dispersive delays from the ionosphere and space-based VLBI antennas may be subject to second-order delay-rate (acceleration) terms.

Another special aspect of VLBI data is the lack of unresolved calibrator sources. Typically, all observed sources are variable and resolved on many baselines in a VLBI observation. The flux density calibration can therefore not be performed by adjusting telescope gains from an observation of a source with a "known" flux density. Instead, telescope metadata is used for this calibration. The total noise power along the telescope's signal chain is measured as a system temperature on a regular basis during an observation. These system temperatures are then used to scale the correlation coefficients formed at the correlator to a physical scale of flux density units in Jy. The necessary gain conversion between noise temperature and Jy is realized by single-dish observations of primary calibrators (e.g., solar system objects) outside of the VLBI session.

Compared to modern connected-element interferometers, VLBI arrays are sparse and consists of a few tens of dishes at most. Thus, the baseband and visibility data volumes are comparatively small. In particular, computational efficiency and hardware scaling is much less of a problem for VLBI image reconstructions. On the other hand, telescope gain errors and sidelobes are more difficult to deal with in VLBI. Furthermore, the fewer VLBI visibility data points make the measurements more "precious" and potentially problematic data cannot be discarded (flagged) as liberally. Currently, the baseband recording data rates are in the order of a couple gigabit per second (Gbps) at a VLBI telescope, with the notable exception of the 64 Gbps recording of the Event Horizon Telescope. Even at small data rates, realtime streaming of the data over long distances to a correlator ("e-VLBI") is usually not feasible. For moderate data rates, the stored data can be transferred online to a correlator with a transfer rate that is smaller than the data recording rate. For this "e-shipping", the transfer of the data takes therefore more time than the observation itself. For the highest data rates or at remote and poorly connected telescope sites, the duration of online data transfer is not a sustainable option. Instead, the recording media hard drives are physically shipped via mail to the correlator. In this "offline" correlation mode, the recording data rates (i.e., available bandwidth sensitivity for continuum sources) are limited by the cost of hard drive storage media for the baseband data and the parallel processing and data ingest capacities of the correlator.

5.4 Polarisation Calibration and Imaging

Polarisation adds an additional dimension to radio data processing, both to the processing itself as well as the resulting data products: the Stokes dimension. As mentioned in Chap. 1, the four Stokes parameters, I, Q, U, and V, are needed to describe the full polarisation state of the signal being observed. Stokes I, which describes the total intensity of the signal, is enough for analyses that don't need the polarisation information; when the linear polarisation is desired Stokes Q and U

must be added, and when circular polarisation is desired Stokes V must be processed as well. While the processing is mostly common to all Stokes parameters up until imaging, the processing from imaging onwards must be repeated in parallel for each desired Stokes parameter (up to 4 times, if all 4 Stokes parameters are needed).

The total intensity calibration process, for most telescope designs, involves only correlations of matching antenna polarisations (e.g., XX and YY correlations, for linear feeds). The 'cross-handed' correlations, of the orthogonal polarisation antennas (e.g., RL and LR, for circular feeds), are only used when looking at the polarisation state of a signal. This calibration leaves one unknown parameter in the measurement equation: the relative phase between the two orthogonal polarisations. This parameter effectively controls mixing between the Stokes Q, U, and V parameters, and is calibrated by observing a source with a known non-zero polarisation state.

Additional calibration is needed for effects known as polarisation 'leakage', where signal from one Stokes parameter contributes to (or 'leaks into') another Stokes parameter. As radio sources appear most strongly in total intensity and more weakly in the other Stokes parameters, the dominant concern is usually leakage from Stokes I into the other Stokes parameters.[14] Leakage is generally divided into two contributions: 'on-axis' (direction independent) and 'off-axis' (direction dependent) terms. On-axis leakage is associated with feed misalignment and the electronic signal chain, and can be calibrated by observing a source of known (possibly zero) polarisation—any difference from the known polarisation is assumed to be caused by Stokes I leaking into polarisation. This leakage is typically a function of time, frequency, and antenna, and is calibrated as such; corrections are applied directly to the visibilities.

Off-axis leakage is associated with imperfections in the optical path, and as such is typically not a strong function of time (unless something changes the optical properties, such as changes to beam-forming in a phased-array feed) but is a function of sky position (relative to the antenna pointing), frequency, and antenna. The position dependence of the leakage is most often measured using holography, which uses a strong calibrator source to record the antenna response as a function of direction and frequency. The result is a primary beam model that predicts the leakage as a function of position and frequency; this model can then be applied in the image domain to remove the leakage. Holographic observations to characterise the primary beam response can occupy a significant amount of observational (and computational) time for telescopes where the beam may change significantly over time.

Full polarisation data products end up being quite large, for two reasons. As mentioned previously, the Stokes parameter acts effectively as an additional data axis, with 4 elements. But in addition to this, the polarisation (especially

[14] In extremely high precision experiments, such as Epoch of Reionization experiments (e.g., Asad et al. 2015), leakage of other Stokes parameters into I can be a cause for concern and additional calibration.

linear polarisation) typically has a strong, non-linear frequency dependence. Unlike Stokes I, which can be averaged over frequency or modelled in a way that the frequency dependence can be captured in a few images (i.e., Taylor-term imaging with multi-frequency synthesis), the other Stokes parameters must be imaged per-channel, making the resulting data four-dimensional: position-position-frequency-Stokes. The imaging does not necessarily need to be done for every single frequency channel; if exceptionally fine channels are observed, some averaging can occur without damaging the data significantly. Averaging the frequency-dependent polarised signal leads to bandwidth depolarisation, which degrades the signal and affects the scientific quality. All polarimetric imaging must carefully consider the ideal balance between keeping the per-image bandwidth narrow to minimize bandwidth depolarisation and averaging as many channels as possible to reduce computational needs. This balance is further complicated by the fact that having many narrower channels across the same total bandwidth reduces the signal-to-noise ratio per channel, which can negatively affect per-channel processes (such as image deconvolution) further down the processing. Even with these factors driving down the desired number of final images, most instruments will produce data cubes with tens to hundred of channels, making the imaging and storage of the resulting image products a computationally intensive process. Table 5.1 gives some data volume properties for selected modern polarimetry surveys.

Large polarisation survey projects require sophisticated software pipelines to manage the processing required to generate science-ready data products. Figure 5.2 shows a polarisation pipeline block diagram from the POSSUM survey, and gives a qualitative sense of the computational burden of each step (combining CPU and RAM use with disk I/O and data volume). The early steps in the pipeline are

Table 5.1 Data properties of selected current polarisation surveys

Survey	Pixel size (")	Sky area (sq. deg.)	# pixels	# channels	Data volume[a] (TB)
POSSUM[b,c]					
Observatory	2	20,627	$157 \cdot 10^9$	288	725
Processed	3.2	20,627	$26 \cdot 10^9$	288	120
VLASS[b,d]	0.6	33,885	$1.2 \cdot 10^{12}$	16	313
LoTSS[e]	4.5	20,627	$117 \cdot 10^9$	480	900
Apertif (Adams et al. 2022)	4	2300	$1.8 \cdot 10^9$	24	24
GALFACTS[f]	60	12,700	$46 \cdot 10^6$	4096	3

[a] Assuming 4 Stokes parameters and 32 bits per pixel; neglecting overheads from FITS headers or other metadata
[b] Survey data specifics not fully finalised; values current as of Dec 2022
[c] POlarization Sky Survey of the Universe's Magnetism (Gaensler et al. 2010). ASKAP will directly archive individual POSSUM observations; POSSUM will use a processing pipeline to combine observations and reduce data volume
[d] Very Large Array Sky Survey (Lacy et al. 2020). Data volume for a single VLASS epoch
[e] LOFAR Two-metre Sky Survey (Shimwell et al. 2017)
[f] The G-ALFA Continuum Transit Survey (Taylor and Salter 2010)

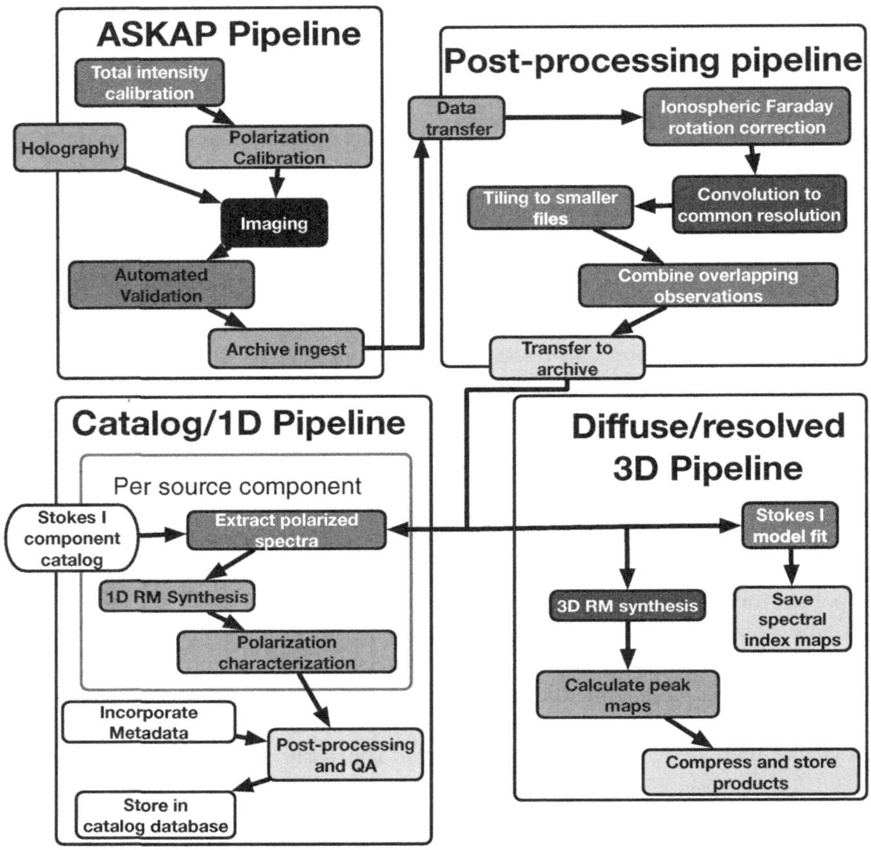

Fig. 5.2 Simplified block diagram of the POSSUM data pipelines. Each block is coloured by approximate computational cost (with white as least and black as most computationally heavy), as a qualitative combination of CPU use, RAM needed, disk I/O, and file size

weighted heavily by the large file size: the imaging step produces FITS cubes of 170 GB (per Stokes parameter). These are very difficult to work with; loading them into memory requires nodes with at least that much memory, and involves significant disk I/O. During pipeline development, the problem of working with such large files came to be fully appreciated, and a step of breaking the cubes into smaller tiles (approximately 4.6 GB each) was implemented; the tiles make it much easier to load and manipulate data in later steps of the pipeline. A major challenge in the development of future pipelines intended for even larger datasets will be to establish a framework with which intermediate data can be held in memory through several steps, minimising the number of required (and very slow) read-write cycles that move data from memory to/from disk. The rate limiting step is often the reading of data from disk; within the POSSUM catalogue pipeline extracting the source spectra can take hours to days for a single observation while the remainder of the pipeline can run on a laptop in less than half an hour. The other major challenge of polarisation pipeline development, albeit one not unique to polarisation, is making the pipeline efficiently parallelised. Elements with the observatory pipelines benefit from the attention of specialists in parallel and high-performance computing, making those parts of the pipeline very efficient; the later parts of the pipeline associated with the scientific analysis are generally written by specialists in the astronomical methods being used, and are consequently less optimised from a computing standpoint.

A discussion of the analysis of polarisation data is left for Sect. 9.2, but some aspects of how the data are used are relevant for consideration within the data production pipeline. Most polarimetric analysis focuses on the spectral behaviour of the Stokes parameters. Despite this, most polarimetry data products are stored with frequency as the slowest-varying axis. The two most common ways in which full Stokes image data are stored is to either separate the Stokes parameters into separate FITS files, each of which is a position-position-frequency cube, or to separate the channels as individual FITS files, each of which is a position-position-Stokes cube. Both formats make it difficult (and inefficient, in terms of disk I/O) to isolate a single position and extract data along the frequency and Stokes axes; this can make data access particularly slow on large datasets. Solutions to this have followed two broad types: to re-order the data before analysis (e.g., Sridhar et al. 2018) so that frequency is the fastest-varying axis, or to divide the data by sky area into smaller files that can be manipulated more easily.

References

E.A.K. Adams, B. Adebahr, W.J.G. de Blok, H. Dénes, K.M. Hess, J.M. van der Hulst, A. Kutkin, D.M. Lucero, R. Morganti, V.A. Moss, T.A. Oosterloo, E. Orrú, R. Schulz, A.S. van Amesfoort, A. Berger, O.M. Boersma, M. Bouwhuis, R. van den Brink, W.A. van Cappellen, L. Connor, A.H.W.M. Coolen, S. Damstra, G.N.J. van Diepen, T.J. Dijkema, N. Ebbendorf, Y.G. Grange, R. de Goei, A.W. Gunst, H.A. Holties, B. Hut, M.V. Ivashina, G.I.G. Józsa, J. van Leeuwen, G.M. Loose, Y. Maan, M. Mancini, Á. Mika, H. Mulder, M.J. Norden, A.R. Offringa, L.C. Oostrum, I. Pastor-Marazuela, D.J. Pisano, A.A. Ponomareva, J.W. Romein, M. Ruiter, A.P. Schoenmakers, D. van der Schuur, J.J. Sluman, R. Smits, K.J.C. Stuurwold, J. Verstappen, N.P.E. Vilchez, D. Vohl, K.J. Wierenga, S.J. Wijnholds, E.E.M. Woestenburg, A.W. Zanting, J. Ziemke, First release of Apertif imaging survey data. Astron. Astrophys. **667**, A38 (2022). https://doi.org/10.1051/0004-6361/202244007

I. Altintas, D. Crawl, C.J. Crosby, P. Cornillon, Scientific workflows for the geosciences: an emerging approach to building integrated data analysis systems, in *Geoinformatics: Cyber-infrastructure for the Solid Earth Sciences* (Cambridge University Press, Cambridge, 2011), pp. 237–250. https://doi.org/10.1017/CBO9780511976308.016

K.M.B. Asad, L.V.E. Koopmans, V. Jelić, V.N. Pandey, A. Ghosh, F.B. Abdalla, G. Bernardi, M.A. Brentjens, A.G. de Bruyn, S. Bus, B. Ciardi, E. Chapman, S. Daiboo, E.R. Fernandez, G. Harker, I.T. Iliev, H. Jensen, O. Martinez-Rubi, G. Mellema, M. Mevius, A.R. Offringa, A.H. Patil, J. Schaye, R.M. Thomas, S. van der Tol, H.K. Vedantham, S. Yatawatta, S. Zaroubi, Polarization leakage in epoch of reionization windows - I. Low Frequency Array observations of the 3C196 field. Mon. Not. R. Astron. Soc. **451**(4), 3709–3727 (2015). https://doi.org/10.1093/mnras/stv1107

Astropy Collaboration, A.M. Price-Whelan, P.L. Lim, N. Earl, N. Starkman, L. Bradley, D.L. Shupe, A.A. Patil, L. Corrales, C.E. Brasseur, M. Nöthe, A. Donath, E. Tollerud, B.M. Morris, A. Ginsburg, E. Vaher, B.A. Weaver, J. Tocknell, W. Jamieson, M.H. van Kerkwijk, T.P. Robitaille, B. Merry, M. Bachetti, H. Moritz Günther, T.L. Aldcroft, J.A. Alvarado-Montes, A.M. Archibald, A. Bódi, S. Bapat, G. Barentsen, J. Bazán, M. Biswas, M. Boquien, D.J. Burke, D. Cara, M. Cara, K.E. Conroy, S. Conseil, M.W. Craig, R.M. Cross, K.L. Cruz, F. D'Eugenio, N. Dencheva, H.A.R. Devillepoix, J.P. Dietrich, A.D. Eigenbrot, T. Erben, L. Ferreira, D. Foreman-Mackey, R. Fox, N. Freij, S. Garg, R. Geda, L. Glattly, Y. Gondhalekar, K.D. Gordon, D. Grant, P. Greenfield, A.M. Groener, S. Guest, S. Gurovich, R. Handberg, A. Hart, Z. Hatfield-Dodds, D. Homeier, G. Hosseinzadeh, T. Jenness, C.K. Jones, P. Joseph, J. Bryce Kalmbach, E. Karamehmetoglu, M. Kałuszyński, M.S.P. Kelley, N. Kern, W.E. Kerzendorf, E.W. Koch, S. Kulumani, A. Lee, C. Ly, Z. Ma, C. MacBride, J.M. Maljaars, D. Muna, N.A. Murphy, H. Norman, R. O'Steen, K.A. Oman, C. Pacifici, S. Pascual, J. Pascual-Granado, R.R. Patil, G.I. Perren, T.E. Pickering, T. Rastogi, B.R. Roulston, D.F. Ryan, E.S. Rykoff, J. Sabater, P. Sakurikar, J. Salgado, A. Sanghi, N. Saunders, V. Savchenko, L. Schwardt, M. Seifert-Eckert, A.Y. Shih, A.S. Jain, G. Shukla, J. Sick, C. Simpson, S. Singanamalla, L.P. Singer, J. Singhal, M. Sinha, B.M. Sipőcz, L.R. Spitler, D. Stansby, O. Streicher, J. Šumak, J.D. Swinbank, D.S. Taranu, N. Tewary, G.R. Tremblay, M. de Val-Borro, S.J. Van Kooten, Z. Vasović, S. Verma, J. Vinícius de Miranda Cardoso, P.K.G. Williams, T.J. Wilson, B. Winkel, W.M. Wood-Vasey, R. Xue, P. Yoachim, C. Zhang, A. Zonca, Astropy Project Contributors. The Astropy Project: sustaining and growing a community-oriented open-source project and the latest major release (v5.0) of the core package. Astrophys. J. **935**(2), 167 (2022). https://doi.org/10.3847/1538-4357/ac7c74

B.M. Gaensler, T.L. Landecker, A.R. Taylor, POSSUM Collaboration, Survey science with ASKAP: Polarization Sky Survey of the Universe's Magnetism (POSSUM), in *American Astronomical Society Meeting Abstracts #215*, vol. 215. American Astronomical Society Meeting Abstracts, p. 470.13 (January 2010)

M. Janssen, J.F. Radcliffe, J. Wagner, Software and techniques for VLBI data processing and analysis. Universe **8**(10), 527 (2022). https://doi.org/10.3390/universe8100527

M. Lacy, S.A. Baum, C.J. Chandler, S. Chatterjee, T.E. Clarke, S. Deustua, J. English, J. Farnes, B.M. Gaensler, N. Gugliucci, G. Hallinan, B.R. Kent, A. Kimball, C.J. Law, T.J.W. Lazio, J. Marvil, S.A. Mao, D. Medlin, K. Mooley, E.J. Murphy, S. Myers, R. Osten, G.T. Richards, E. Rosolowsky, L. Rudnick, F. Schinzel, G.R. Sivakoff, L.O. Sjouwerman, R. Taylor, R.L. White, J. Wrobel, H. Andernach, A.J. Beasley, E. Berger, S. Bhatnager, M. Birkinshaw, G.C. Bower, W.N. Brandt, S. Brown, S. Burke-Spolaor, B.J. Butler, J. Comerford, P.B. Demorest, H. Fu, S. Giacintucci, K. Golap, T. Güth, C.A. Hales, R. Hiriart, J. Hodge, A. Horesh, Ž. Ivezić, M.J. Jarvis, A. Kamble, N. Kassim, X. Liu, L. Loinard, D.K. Lyons, J. Masters, M. Mezcua, G.A. Moellenbrock, T. Mroczkowski, K. Nyland, C.P. O'Dea, S.P. O'Sullivan, W.M. Peters, K. Radford, U. Rao, J. Robnett, J. Salcido, Y. Shen, A. Sobotka, S. Witz, M. Vaccari, R.J. van Weeren, A. Vargas, P.K.G. Williams, I. Yoon, The Karl G. Jansky Very Large Array Sky Survey (VLASS). Science case and survey design. Publ. Astron. Soc. Pac. **132**(1009), 035001 (2020). https://doi.org/10.1088/1538-3873/ab63eb

I. Raicu, *Many-Task Computing: Bridging the Gap Between High Throughput Computing and High Performance Computing* (VDM Verlag, Saarbrücken, DEU, 2009). ISBN 3639156145

A.E.E. Rogers, S.S. Doeleman, J.M. Moran, Fringe detection methods for very long baseline arrays. Astron. J. **109**, 1391 (1995). https://doi.org/10.1086/117371

T.W. Shimwell, H.J.A. Röttgering, P.N. Best, W.L. Williams, T.J. Dijkema, F. de Gasperin, M.J. Hardcastle, G.H. Heald, D.N. Hoang, A. Horneffer, H. Intema, E.K. Mahony, S. Mandal, A.P. Mechev, L. Morabito, J.B.R. Oonk, D. Rafferty, E. Retana-Montenegro, J. Sabater, C. Tasse, R.J. van Weeren, M. Brüggen, G. Brunetti, K.T. Chyży, J.E. Conway, M. Haverkorn, N. Jackson, M.J. Jarvis, J.P. McKean, G.K. Miley, R. Morganti, G.J. White, M.W. Wise, I.M. van Bemmel, R. Beck, M. Brienza, A. Bonafede, G. Calistro Rivera, R. Cassano, A.O. Clarke,

D. Cseh, A. Deller, A. Drabent, W. van Driel, D. Engels, H. Falcke, C. Ferrari, S. Fröhlich, M.A. Garrett, J.J. Harwood, V. Heesen, M. Hoeft, C. Horellou, F.P. Israel, A.D. Kapińska, M. Kunert-Bajraszewska, D.J. McKay, N.R. Mohan, E. Orrú, R.F. Pizzo, I. Prandoni, D.J. Schwarz, A. Shulevski, M. Sipior, D.J.B. Smith, S.S. Sridhar, M. Steinmetz, A. Stroe, E. Varenius, P.P. van der Werf, J.A. Zensus, J.T.L. Zwart, The LOFAR two-metre sky survey. I. Survey description and preliminary data release. Astron. Astrophys. **598**, A104 (2017). https://doi.org/10.1051/0004-6361/201629313

S.S. Sridhar, G. Heald, J.M. van der Hulst, cuFFS: a GPU-accelerated code for Fast Faraday rotation measure Synthesis. Astron. Comput. **25**, 205–212 (2018). https://doi.org/10.1016/j.ascom.2018.10.002

A.R. Taylor, C.J. Salter, GALFACTS: the G-ALFA continuum transit survey, in *The Dynamic Interstellar Medium: A Celebration of the Canadian Galactic Plane Survey*, ed. by R. Kothes, T.L. Landecker, A.G. Willis, vol. 438. Astronomical Society of the Pacific Conference Series (December 2010), p. 402

The Galaxy Community, The Galaxy platform for accessible, reproducible and collaborative biomedical analyses: 2022 update. Nucleic Acids Res. **50**(W1), W345–W351 (2022). ISSN 0305-1048. https://doi.org/10.1093/nar/gkac247

A.R. Thompson, J.M. Moran, G.W. Swenson Jr., *Interferometry and Synthesis in Radio Astronomy*, 3rd edn. (2017). https://doi.org/10.1007/978-3-319-44431-4

R. Wang, R. Tobar, M. Dolensky, T. An, A. Wicenec, C. Wu, F. Dulwich, N. Podhorszki, V. Anantharaj, E. Suchyta, B. Lao, S. Klasky, Processing full-scale square kilometre array data on the summit supercomputer, in *SC20: International Conference for High Performance Computing, Networking, Storage and Analysis* (2020), pp. 1–12. https://doi.org/10.1109/SC41405.2020.00006

L. Wratten, A. Wilm, J. Göke, Reproducible, scalable, and shareable analysis pipelines with bioinformatics workflow managers. Nat. Methods **18**(10), 1161–1168 (2021). ISSN 1548-7091. https://doi.org/10.1038/s41592-021-01254-9

Chapter 6
Implementations of Data Processing

Danielle Fenech, Jordan D. Collier, Matthew Whiting, Yan G. Grange, Aidan Hotan, Marco Iacobelli, Vanessa A. Moss, Daizhong Liu, Michael Janssen, Hongming Tang, and Alexander Drabent

Existing as well as planned instruments are a powerful tool in understanding the challenges and potential solutions for dealing with large data volumes. In particular, with the advent of digital signal processing, broad-band receivers operating with a high frequency resolution, and the increasing amount of antennas or stations (and thus baselines) at modern interferometric radio facilities, these challenges came

D. Fenech (✉)
SKA Observatory, Jodrell Bank, Lower Withington, Macclesfield, Cheshire, UK

Astrophysics Group, Cavendish Laboratory, Cambridge, UK
e-mail: danielle.fenech@skao.int

J. D. Collier
The Inter-University Institute for Data Intensive Astronomy (IDIA), Department of Astronomy, University of Cape Town, Rondebosch, Cape Town, South Africa

School of Science, Western Sydney University, Penrith, NSW, Australia

CSIRO, Space and Astronomy, Australia Telescope National Facility, Bentley, WA, Australia
e-mail: jordan@idia.ac.za

M. Whiting · V. A. Moss
CSIRO Space and Astronomy, Epping, NSW, Australia
e-mail: Matthew.Whiting@csiro.au; Vanessa.Moss@csiro.au

Y. G. Grange · M. Iacobelli
ASTRON, Dwingeloo, The Netherlands
e-mail: grange@astron.nl; iacobelli@astron.nl

A. Hotan
CSIRO Space and Astronomy, Bentley, WA, Australia
e-mail: Aidan.Hotan@csiro.au

D. Liu
Max-Planck-Institut für extraterrestrische Physik, Garching, Germany

E. Vardoulaki et al. (eds.), *Data-Intensive Radio Astronomy*, Astrophysics and Space Science Library 472, https://doi.org/10.1007/978-3-031-58468-8_6

to light in the recent years. In many cases this is part of their own continued improvements to carry-out operations as dedicated instruments providing unique and important science. Many of the challenges they face are the same as for the SKA and they can provide invaluable input to the development process for this and future instruments. Thus, it is becoming increasingly urgent to find sustainable and forward-looking solutions.

Given the scale, building towards something like the SKA is essential—a number of pathfinder instruments like LOFAR (Sect. 6.7) and precursor instruments like MeerKAT (Sect. 6.8) or ASKAP (Sect. 6.9) fulfil this purpose. These help us identify our progress, as well as our limitations, and aid in providing solutions to the big data challenges we are facing. This chapter outlines the specific instrumental properties, their processing requirements, and solutions of a number of established research facilities and telescopes including both precursor and pathfinder instruments depending on how the instrument is designed and operated. As it will be highlighted in these upcoming sections, virtually all of the existing newly upgraded and new facilities are facing the same challenges in data processing. The paradigm has in many ways already shifted and high-performance high data-volume processing is becoming the norm.

6.1 VLA

The VLA (see Sect. 1.3) or JVLA (Karl G. Jansky VLA) accommodates broad-band receivers. Depending on the sampler, the receiver bandwidth can be 2 or 8 GHz per polarisation. The current Wideband Interferometric Digital ARchitecture (WIDAR) correlator can process 64 independent subbands (i.e., spectral windows) at 100 ms sampling rate, with 256/128/64 channels for each subband with single/dual/full polarisations. The special pulsar mode can even have a sampling rate of 15 μs with a decreased number of channels. The usual data rate is below 60 MB/s, above which the observing program would be identified as a Resident Shared Risk Observing (RSRO).

All current JVLA data reduction use the CASA software. Before that the Astronomical Image Processing System (AIPS) was used, which was developed by

M. Janssen
Department of Astrophysics, Institute for Mathematics, Astrophysics and Particle Physics (IMAPP), Radboud University, Nijmegen, The Netherlands
e-mail: m.janssen@astro.ru.nl

H. Tang
Department of Astronomy, Tsinghua University, Beijing, China
e-mail: hongmingt@mail.tsinghua.edu.cn

A. Drabent
Thüringer Landessternwarte, Tautenburg, Germany
e-mail: alex@tls-tautenburg.de

NRAO and migrated into CASA. The data reduction generally follows the standard way of using flux/bandpass/phase/pointing calibrators to calibrate the science source data. For high-frequency K–Q band observations, it is required that the interval time between phase calibrator and science target scans to be much shorter, e.g. slightly less than 10 min, so as to better calibrate the atmospheric phase change. RFI usually affect lower frequency observations, which sometimes cannot be well autoflagged but need manual flagging. Some relatively aggressive flagging technique based on sigma-clipping are also under development.

After applying calibration by the users, the MeasurementSet data can be obtained. Such data nowadays can be as large as a few hundred GB, depending on e.g., tens of hours of observing. Imaging can be challenging, especially given the large primary beam (field-of-view) of tens of arcmin scale and synthesised beam of sub-arcsec scale at these GHz frequencies. Mosaicking of multiple pointings makes a larger field-of-view, but is more computationally expensive to image as a whole. Among VLA scientific programs, the VLA-COSMOS 3 GHz survey ((Smolčić et al. 2017)) is one of the largest mosaics. The main $\sim$2 square degree area of the COSMOS deep field ((Scoville et al. 2007)) was imaged at $0.35''$ resolution (with a pixel size usually $\lesssim 1/5$ of the beam). Such large mosaic, high-resolution surveys have to deal with computationally intensive imaging with the CASA tclean task, i.e., taking days of computation time with hundreds of CPU cores.

The next-generation VLA (ngVLA) will be the next major upgrade of VLA, promisingly recommended by the US Astro 2020 Decadal Survey. A large number of new antennas (244 at 18 m size and 19 at 6 m size[1]) will be built by about 2030, extending the maximum baseline to 1000 km. It will also extend the frequency coverage up to 116 GHz. All these will make ngVLA a greatly needed, unique facility for millimetre/radio observations in the northern hemisphere, but also facing the challenge of dealing with at least $\sim$3000$\times$ bigger data, e.g., hundreds of TB per day of observing and for calibration and imaging ($\sim$100$\times$ in baseline number and $\sim$30$\times$ in resolution, not even accounting receiver upgrades).

6.2 ALMA

ALMA (see Sect. 1.3) operates a heterodyne receiver that is installed at each antenna for each observing band. Two new bands, Band 1 and Band 2, are under development to cover lower observing frequencies down to 35 GHz. Each heterodyne receiver has two sidebands, lower sideband (LSB) and upper sideband (USB), symmetrically distributed around the Local Oscillator (LO). ALMA Band 3 to 8 receivers are 2 SB type, i.e., both LSB and USB will be output to correlator separately. Band 9 and 10 receivers are DSB type, i.e., the sum of LSB and USB will be output, but one sideband can be blocked by LO offsetting so that only the other sideband is output.

[1] https://ngvla.nrao.edu/.

Then, four correlator basebands are defined within the bandwidths of LSB and USB, each having a bandwidth of 2 GHz per polarisation and 2 polarisations, totalling up to 8 GHz per polarisation or 16 GHz for dual/full polarisations. These basebands can have flexible, non-overlapping frequency ranges within the LSB and USB coverage. A flexible number of spectral windows can then be defined for high spectral resolution observations, with a channel width ranging from 7.6 kHz (15 kHz) to 7.8 MHz (15 MHz) and number of channels up to 8192 (4096) if in single (dual) polarisation mode. If in full polarisation mode, i.e., all four Stokes parameters are output, these numbers double/half correspondingly. In addition, for actual observations, the edge of the spectral windows are discarded, making the actual channel number and bandwidth 15/16 of the aforementioned nominal numbers (e.g., 7680 instead of 8192 channels).

The ALMA 12-m and 7-m arrays operate separately. The full 12-m array produces $54 \times 53 = 2862$ baselines. Taking a 2SB receiver's dual polarisation, 4 spectral window, lowest spectral resolution mode as an example, i.e., 128 channels (15.6 MHz channel width) $\times$ 2 polarisations, and a sampling rate of 2 s,[2] the data to be taken is 42 GB/hr. Using the highest resolution mode for all spectral windows will have a data rate of 1.27 TB hr^{-1}, with the corresponding 6 s sampling rate.

However, the latter is not usually reached because scientific program observers need to justify their data rate in their proposal, and scientifically the relatively small ALMA bandwidth usually only have one strong line that need very high spectral resolution. The ALMA Design Reference Science Plan (DRSP) has estimated a mean data rate of a few tens MB/s, and its operation plan is based on a data rate of 200 TB yr^{-1}.

The ALMA raw data model uses the Archival Science Data Model (ASDM; Sect. 3.5), which is commonly developed by ALMA and VLA for efficient data recording and transporting. This format is not easily read for calibration or further scientific analysis. The CASA MS file format (see Sect. 2.2.2) is the data format aiming for the latter purpose. The calibration of raw ALMA data in ASDM uses the CASA software, and the first step is importing the data into MeasurementSet format. Then the relatively standard calibration steps as described in Sect. 5.2 are applied.

ALMA dynamically schedules and carries out observations with the basic unit named Member ObsUnitSet (OUS). A member OUS then has one or more Scheduling Blocks (SBs), and each SB generates a Execution Block when it is executed. ALMA analysts will inspect each observation Member OUS data, make preliminary, scientific-alike analysis and produce the so-called QA2 products, which are usually images and/or data cubes for preview, and also ALMA Data Reduction scripts for the observers/users to reapply the calibrations if not generated by the pipeline. It is usually recommended for the users to download the raw data and run the ALMA Data Reduction scripts to (re)apply calibrations and obtain the calibrated MS data, then perform customizable imaging and subsequent scientific analysis.

---

[2] https://science.nrao.edu/facilities/alma/naasc-memo-series/naasc-memos/110.naasc-data-rates.

6.3 NOEMA

NOEMA (see Sect. 1.3) uses the GILDAS[3] software for the raw data calibration and uv-data processing and imaging. GILDAS is an ensemble of several data reduction and observation planning software packages for IRAM telescopes, i.e., NOEMA and IRAM 30 m single dish telescope. Its ASTRO package is for the observation planning, CLIC package is for the NOEMA raw data reduction, MAPPING is for interferometric uv data processing, and CLASS is for the IRAM 30 m or other similar single-dish data reduction.

6.4 FAST

In general, FAST (see Sect. 1.3) has adopted two novel approaches (Nan et al. 2011):

- An active refractive surface to correct spherical aberration and ensure parabolic shape of the dish in real time, which enables FAST to observe in much larger sky coverage.
- A lightweight driven antenna feed cabin driven by cables and servomechanism, allowing the feed to move and receive signals from different directions, and resulting in significant improvement of the telescope pointing accuracy.

Though FAST allows a comparatively wide observation frequency range, the so-called 19-beam receiver by far is the most frequently used one, covering a frequency range of 1.05–1.45 GHz. In accordance to the test results during the 'Shared-Risk' testing observation period, FAST pointing accuracy has standard deviation of $7.9''$. When using the 19-beam receiver, at around 1.4 GHz, the telescope has its aperture efficiency and central beam system temperature to be around 0.63 and 24 K if the zenith angle is smaller than $26^\circ.4$, respectively (Jiang et al. 2020). New receivers are under development or being tested, which shall come into practice in the future.

FAST supports up to 18 observation modes under 9 categories. Users could choose any of them when observing with the 19-beam receiver[4] (Jiang et al. 2020). Among all these modes, for instance, the *OnTheFlyMapping* mode has been used for the FAST Ultra-Deep Survey (FUDS; Xi et al. 2022); the *DriftScan* mode had supported the discovery of repeating FRBs (i.e., Niu et al. 2022) as part of the Commensal Radio Astronomy FAST Survey (CRAFTS; Li et al. 2018); observers could discover new pulsars through the *Tracking* mode (i.e., Wang et al. 2020); or mapping HI structure with the *On-Off* mode (Xu et al. 2022). If observers find the

[3] https://www.iram.fr/IRAMFR/GILDAS/.

[4] https://fast.bao.ac.cn/cms/article/24/.

available modes do not match their requirements, they can also use the *User-defined scan* mode to customize their observation strategies.[5]

After observers have their observation proposals approved and finish their observations, the resulting data would be primarily stored at the FAST Data Center.[6] The FAST Data Centre is located in Guizhou, China. The storage space of the FAST Data Center by now is 58 PB and may be improved when future funding is available. It provides a free data storage service and would periodically release lists of projects which recently went through their 12-month data protection periods. Other FAST users then could submit their data transmission requests on their official website and gain access to these data.[7]

Though FAST has made effort on providing data open access services, stable online data transmission still remains challenging. Depending on the selection of observation settings and observation modes, FAST low-level data product package (data went through ADC sampling processes and saved in FITS format) usually requires storage space from tens of GB to tens of TB. For instance, if a typical pulsar searching oriented observation using only a single horn (the central beam of the 19-beam receiver), 4096 frequency channels, 49.125 μs sampling time and request full polarisation data, then the observer shall receive data in size of around 1.2 TB per observation hour (22.8 TB per hour if all 19 beams are used). FUDS as another example, however, requiring its HI spectrum data sampling time at the second level and 65,536 frequency channels. A 1.2 h observation data file recorded single polarisation per beam observed then required only 2 GB storage space (Xi et al. 2022), and finishing FUDS would require data space in the scale of 100 TB.

Given the limited data transmission ability and relatively manageable individual data package volume, the data center for now encourages users to either access their data on FAST local clusters, or using the FAST data external transmission service to physically send them disks with the requested data. For those users who work on the FAST clusters, they can reduce and analyse their data via computation nodes of these clusters. Both CPU based computation nodes (26 nodes) and GPU based nodes (10 nodes) are available for FAST cluster users, along with some conventional and popular data analysis softwares (i.e., PRESTO for pulsar researcher[8] or Astropy for HI researcher and other astronomers) pre-installed and ready to use.

In terms of data reduction, multiple customized RFI correction algorithms (i.e., for HI observations, Wang et al. 2022; Zhang et al. 2022; Liu et al. 2022) and data reduction algorithms (i.e., HiFAST[9]) have been developed by previous FAST users to improve observation sensitivity and increase the technical accessibility of using FAST data for research. The FAST data center, on the other hand, supports these

[5] Archival FAST observation logs with technical details can be accessed through https://fast.bao.ac.cn/observation_log/search once people sign in to their FAST accounts (open for registration).

[6] https://fast.bao.ac.cn/cms/article/11/.

[7] https://fast.bao.ac.cn.

[8] https://github.com/scottransom/presto.

[9] https://github.com/ICRAR/FAST-HI.

developments by providing regular RFI environment monitoring reports and noise diode calibration reports. FAST users then could develop, modify or improve their RFI correction algorithms or data reduction pipelines accordingly.

6.5 EHT

The EHT (see Sect. 1.3) currently observes for a few days or nights in the first quarter of each year, collecting a few PB of raw baseband data. Based on the inaugural science observations in 2017, the first images of black holes—of Sgr A* and M87* (Event Horizon Telescope Collaboration et al. 2019b, 2022)—and the highest resolution images of the jet launching regions of 3C 279 (Kim et al. 2020), Centaurus A (Janssen et al. 2021), J1924-2914 (Issaoun et al. 2022), and NRAO530 have so far been made.

There are stringent requirements to make sufficiently smooth telescope surfaces that are reflective in the (sub)-millimetre range. Many EHT telescopes are only 10 m-class dishes. At the same time, the AGN synchrotron emission is weak at the high radio observing frequencies and water vapour in the turbulent troposphere often results in short (few seconds) coherence times combined with notable signal attenuation and sky noise (Thompson et al. 2017; Roelofs et al. 2020). Sufficient sensitivity to detect fringes and correct the atmospheric phase turbulence on short enough timescales is accumulated through the high 8 GHz aggregate recording bandwidth at 2-bit sampling and sensitive central stations, such as phased ALMA and the 50 m LMT dish.

Next to the phased sum data, the phased arrays in the EHT also record the connected-element interferometric data simultaneously. These measurements provide information about the large-scale structure of the observed sources and the corresponding total-flux light curves within the interferometer fields-of-view. Furthermore, the connected-element data is needed for the calibration of the phasing systems and an improved amplitude gain calibration of the co-located EHT stations through the usage of the very short baselines in the array (Event Horizon Telescope Collaboration et al. 2019a).

The 64 Gbps recording rates make the transfer of the data via internet to the two central correlation facilities at the Max Planck Institute for Radio Astronomy in Bonn, Germany and the MIT Haystack Observatory in Westford, Massachusetts, infeasible, particularly for remote sites like the South Pole Telescope (SPT). Instead, the data is stored locally on Mark 6 recording units (Whitney et al. 2013), which write to 128 hard drives in parallel. The complete set of full recording modules after an observing session are then physically shipped to the correlators. There, the data is played back from the media and correlated with DiFX (Deller et al. 2007, 2011) using the CALC correlator model. With PolConvert (Martí-Vidal et al. 2016), the linear-feed data from ALMA is converted to a circular polarisation basis, common with the other EHT stations. The correlator output is converted to the HOPS/MkIV and FITS-IDI formats. The corresponding station metadata needed for the a-priori flux density calibration is gathered in ANTAB table files (Janssen et al. 2019a).

The EHT data is processed in a two-step approach. "Level one" (L1) asserts the validity and completeness of the data for scientific analysis. Through the L1 process, all issues that would require a re-correlation are identified and fixed, so that the recording media can be recycled for future observing sessions. Next to the verified correlator output, the L1 package consists of a complete and validated set of a-priori calibration metadata for all stations.

The "Level two" (L2) process takes the L1 output as input and produces science-ready calibrated data for individual observed sources or projects that consists of multiple sources. The most important L2 steps are the removal of bad data (e.g., telescope was not on source and data drop-outs due to recorder issues), application of the a-priori metadata, fringe-fitting to remove post-correlation residual delays and rates, as well as the stabilisation of tropospheric phase turbulence. A careful processing of the data allows for substantial reductions of data volumes, from petabytes of the raw voltages, to terabytes of the L1 correlated data, to megabytes of the $\sim$10 s time- and fully frequency-averaged L2 visibilities, to a few kilobytes of a source image reconstruction within the small EHT field-of-view.

Multiple calibration pipelines are used in parallel by the EHT for cross-comparisons and an additional validation of the final scientific results in L2 (Event Horizon Telescope Collaboration et al. 2019a). For L1, the data is also checked with the different EHT calibration pipelines, utilising each's own internal diagnostic tools. Currently the EHT makes use of the HOPS-based (Whitney et al. 2004) EHT-HOPS (Blackburn et al. 2019) and CASA-based (The CASA Team et al. 2022; van Bemmel et al. 2022) rPICARD (Janssen et al. 2018, 2019b) pipelines. EHT-HOPS uses the MkIV data and is designed to handle mm-VLBI data. Due to its geodetic origin, HOPS is able to accurately measure delay and delay-rate errors in the data. Furthermore, the HOPS routines have been exercised in many early 1 mm-VLBI measurements and are thus well tested. For the EHT, this pipeline is run on the Google Cloud Platform. The on-demand availability of computing power is well suited for the current low-duty-cycle operational mode of the EHT. Through the *GNU parallel* tool, the visibility data is efficiently processed on a per-scan basis in multi-CPU virtual machines on the cloud. Moreover, the cloud was found to be a convenient data staging and sharing area. rPICARD uses the FITS-IDI data, internally converted to a MeasurementSet, and is a generic VLBI data reduction pipeline. For the EHT, rPICARD is run on ALMA Regional Centre (ARC) computing platforms. Here, we make use of the CASA *MPI* implementation, working with Multi-MS datasets split across VLBI scans, to distribute the computing across many CPU cores. Global fringe-fitting to solve for gross station-based delays and delay-rates is the most resource-intensive calibration step. It becomes memory-limited with >10 GB of RAM needed per task rather than CPU-limited on the ARC machines. The rPICARD development is based on a full-fledged VLBI upgrade of the CASA software, initiated by the ERC-funded BlackHoleCam project (Goddi et al. 2017) and maintained by JIVE and NRAO (van Bemmel et al. 2022). The widely used CASA software was chosen for its secure long-term user support and continuous upgrades to scale with the needs of future observatories as well as to lower to barrier of entry for VLBI data processing. For both the CASA and HOPS pathways, containerised software versions are used for scientific reproducibility and the common standard UVFITS file format (see Sect. 2.2.1) is adopted as final calibration output for downstream analysis.

6.6 e-MERLIN

One of the unique challenges for e-MERLIN (see Sect. 1.3) is the heterogeneity of the array with five different structures and sizes. The Lovell telescope is a 76 m steerable dish, the MK2 dish is a 25 m elliptical structure, there is a 32 m dish situated at Lords Bridge site in Cambridge and three 25 m "E-systems" telescopes and a former RADAR dish with a wire-mesh construction on the dish surface. The telescopes are fibre-linked and data is recorded and correlated at the observatory on a dedicated hardware correlator.

The range presents some challenges in imaging particularly when attempting to produce wide fields-of-view or mosaicked images to ensure the array primary beam is correctly considered.

Data is processed as MeasurementSet through a CASA-based pipeline which includes quality assessment. This is presented to the user as a web-based report that provides information about the observations and pipeline processing. The e-MERLIN pipeline[10] utilises CASA tasks (The CASA Team et al. 2022) and AOFlagger (Offringa et al. 2012) to perform RFI removal, instrumental calibration, plotting and imaging. The RFI removal includes stages to remove the data from the Lovell-MK2 baseline (the correlated data from the Lovell and MK2 telescopes), when present, as well as applying masks in frequency and time of known RFI. The calibration stages include flux density scaling, frequency-dependent bandpass and complex gain correction as well as statistical weighting to account for the significant range in dish sensitivities. Given the relatively small number of baselines, where appropriate, data is conserved as much as possible and the pipeline runs a number of the calibration tasks iteratively to incrementally improve the calibration and maximise sensitivity. Field and zoom images of the calibration and target sources are created using Tclean as part of the pipeline and are included in the report alongside plots of the data and calibration solutions derived as part of the processing.

Unlike a number of the telescopes discussed here, the e-MERLIN dishes are situated in relatively close proximity to towns, cities and airports with observations suffering from significant RFI, particularly at L-band where a large portion of telecommunications operate.

[10] https://github.com/e-merlin/eMERLIN_CASA_pipeline.

Given the challenges, the pipeline results are not always suitable for immediate scientific publication but used as input for further improvements to the data reduction by the user, particularly for further imaging and self-calibration.

Observations with e-MERLIN require relatively high time and frequency sampling given the high resolution as well as typically several hours of on-source observing time (typically 12 h) to ensure sufficient uv-coverage. Alongside the broad frequency coverage this results in datasets of 10–100 s GB for a standard observation. Whilst some averaging may be possible it is highly source dependent and is often minimally applied in order to preserve the unaberrated field-of-view. This can become increasingly challenging when imaging large fields-of-view which requires direction-dependent corrections including accounting for the primary beam response. This is complicated by the variation in size of the e-MERLIN dishes which also varies as a function of frequency.

The full e-MERLIN array is also now available to be used concurrently as part of EVN (European VLBI Network[11]) observations which is a VLBI facility with telescopes situated across Europe and Asia. Typical EVN observations require a larger number of telescopes (of order 10–20) to ensure sufficient uv-coverage, with potentially higher time and frequency resolution than for e-MERLIN alone, increasing the dataset sizes further.

6.7 LOFAR

LOFAR (see Sect. 1.3) offers the astronomical community a set of three main observing modes (beam-formed (BF), interferometric (IF), Transient Buffer Boards (TBB)), as well as any combination of these (e.g. simultaneous BF and IF). Such flexibility of the telescope configuration and its observing modes enables a large spectrum of science use cases (from cosmology to solar-physics) and is mirrored in both the large (~50 TB/day) data rate and the composition of the LOFAR archived data collection. The LOFAR Long Term Archive (LTA) consists of a central catalogue with related metadata services and a distributed hardware infrastructure and data services at three data centres (Surf SARA, FZJ and Poznan) spread through Europe. At the time of writing, the LOFAR catalogue amounts to about 57 PB of data, the largest astronomical data collection to date. However the vast majority of these data are in the form of low-level data products, which is sub-optimal for enabling an efficient discovery process although low-level data products allow for multi-purpose scientific exploitation (via different workflows).

Main causes for this status are the complexity of the processing strategies, the computing intensive nature of the processing and the large volumes of the archived datasets at metrewave regime, which call for large scale (and distributed) processing.

[11] https://www.evlbi.org/.

Indeed several big data problems posed by the explosion of data in radio astronomy are relevant in the context of LOFAR IF (visibility) data:

- How to calibrate and process these datasets?
- How to optimise the algorithms and the processing workflows?
- How to extract astronomical info from the final data products?

Not surprisingly, a large data rate observatory requires a high-throughput processing infrastructure. In the following of this section, to clarify these aspects we will focus on the LOFAR IF mode data flow.

To serve multiple science use cases and to enable calibration, LOFAR IF measurements are typically stored at high time and spectral resolutions, thus leading to sizes of tens of TBs. Such sizes are hardly manageable, thus a first (innovative) step in the data flow consists in a pre-processing (e.g. RFI excision, bright off-axis removal and averaging) step which includes the compression of the visibility data (see Offringa 2016). As a result, a multi-terabyte (<10 TB) dataset can be archived and delivered to the astronomer for further processing. Note that the compression step not only mitigate storage constraints but also speed up computations for its processing.

An accurate correction of both instrumental and ionospheric errors over a wide field-of-view is needed to achieve the desired scientific outcome. Since the beginning of this century it was realised that standard processing techniques where not adequate to achieve high dynamic range and resolving power at low frequency, therefore efforts started to develop more complex calibration strategies (see, e.g., Cotton et al. 2004). For LOFAR it was also clear the need of both dedicated software tools (and algorithms) as well as dedicated calibration and imaging strategies. In particular to manage the signal corruption imposed by the ionosphere on the low frequency radio waves, a set of innovative direction-dependent calibration (and imaging) strategies (see, e.g., Williams et al. 2016; van Weeren et al. 2016; de Gasperin et al. 2020; Tasse et al. 2021) and related algorithms (see, e.g., van Diepen et al. 2018; Offringa et al. 2014; Offringa and Smirnov 2017; van der Tol et al. 2018) have been developed since the start of the LOFAR operations in 2013. An ongoing joint effort of the LOFAR community (mostly on workflow side) and ASTRON (mainly on the software components and algorithm side) proved to be fruitful and has resulted in standard processing recipes, currently available for LBA (de Gasperin et al. 2019, 2020) and HBA (de Gasperin et al. 2019; van Weeren et al. 2016; Tasse et al. 2018, 2021) visibility data, while for the exploitation of VLBI capabilities the maturity of techniques has been achieved for HBA only (see, e.g., Morabito et al. 2022). Complex schemes aimed to track and correct a number of systematic effects as a function of (the pointing) direction, frequency, time are used in LOFAR data reduction workflows.

After pre-processing, a set of calibration workflows, like LINC,[12] or prefactor[13] is run on flux density calibrator(s) and target datasets to correct for Direction Independent Errors (DIE). During this calibration step, averaging can be done thus decreasing the dataset size down to a fraction ($\sim$0.4) of TB. For a typical (HBA) IF datasets about 2000 cpu hours are needed to complete this step.

The next step consists of correcting for Direction Dependent Errors (DDE) via either the faceting or the screen fitting approaches. With the first option the field-of-view is divided in sky regions around a set of suited calibrator sources (i.e. directions) and solutions are found and applied iteratively on each source (e.g. the Factor workflow) or at once (e.g. the SAGEcal[14] and DDF-pipeline[15] for HBA data and the LiLF[16] pipeline for LBA data). Note that in order to perform sub-arcsecond imaging, an in-field delay calibration step is performed in a dedicated VLBI-workflow[17] (see Morabito et al. 2022). With the second approach a set of spatially smooth solutions is found from the selected set of directions and applied during the full field-of-view imaging (e.g. the Rapthor workflow). At this data processing stage no further shrink of dataset size can be done and for a typical HBA IF dataset about 7000 CPU hours are needed. The exploitation of VLBI capabilities is the most challenging computing process because decorrelation effects strongly limits the possibility to average data (both in time and frequency) to decrease the volume of visibilities to be calibrated and imaged. Additionally, because of the high pixel resolution and the wide field-of-view the imaging process becomes prohibitively computationally expensive. As a result, a further increase of processing resources of a factor of a few is required.

Although the impressive progresses made with customising and developing software and workflows for LOFAR data processing, a few limitations should be mentioned:

- **I/O bottle-necks.** Not all software is equally well matured to handle I/O in the most optimized way. Unfortunately, a single solution to tackle I/O bottle-necks can not be achieved, because solutions much depend on the type of data editing (i.e. operations) that need to be performed. With ever-growing datasets, getting the data to the CPU/GPU becomes more and more challenging.
- **Exploitation of parallelism.** There's still room for improvement for exploiting opportunities to do parallel processing. To this aim different approaches can be considered; in some cases further decomposing the processing algorithm to allow massively parallel processing, e.g. on GPUs could be appropriate. But parallelism can also be achieved by farming out computing jobs to many compute nodes of e.g. a computing cluster. In this context old-school methods like job submission and MPI may get obsoleted by more modern approaches like e.g. Python Dask.

[12] https://git.astron.nl/RD/LINC.

[13] https://github.com/lofar-astron/prefactor.

[14] https://github.com/nlesc-dirac/sagecal.

[15] https://github.com/mhardcastle/ddf-pipeline.

[16] https://github.com/revoltek/LiLF.

[17] https://github.com/lmorabit/lofar-vlbi.

- **Workflow management.** Processing data to yield scientifically useful data requires complex workflows. In order to fully exploit parallelism, it is unavoidable to use dedicated workflow management systems. The problem is that there exist many hundreds of different workflow management systems. Most of these systems have a very small user community and often work only for a very limited use case. Here, standardisation efforts, like with the Common Workflow Language,[18] may provide a solution. However, approaching the complex processing requirements in terms of workflows, using existing workflow management systems is still rather new in radio astronomy.

To conclude, it is useful to briefly overview a few aspects regarding large scale processing of large datasets, which are important for next generation radio observatories in primis the SKA. Given the high rate of throughput and the complex structure of IF radio data, processing of LOFAR data is an I/O intensive process and it can not be performed on either single powerful machine or small computing clusters. Instead, high-throughput distributed computing systems (e.g. grid like) close to the storage sites are needed. Moreover a high level of parallelisation is possible for processing IF radio data and calibration workflows should try to maximise the degree of parallel computing. As a consequence, efficient processing platforms are required to handle bulk processing (see Mechev 2019). To enable processing on distributed computing systems, efforts to enforce and improve the portability of both software components and the full, complex processing software environment are fundamental. As the LOFAR use case demonstrates, powerful correlators and computing clusters are necessary in the framework of data-intensive radio astronomy. This poses sustainability challenges in terms of operational costs and environmental impact. Therefore developing and adopting more efficient algorithms and hardware will be a crucial goal for the bulk processing long term preservation of large astronomical data collections. Finally, a note should be made about the complexity of the many tools used to (post-)process radio astronomical data. The use of these tools require advanced knowledge of experienced radio astronomer but in order to engage a larger community of astronomers into using the available (LOFAR) data, processing needs to be streamlined. One way to achieve this is to provide default and robust processing workflows that can generate high quality outputs in most of the processing use cases.

6.8 MeerKAT

As outlined in Sect. 1.3, the MeerKAT radio telescope consists of 64 dishes with a 13.5 m diameter. A default observation will include all four correlations (XX,YY,XY,YX) and 8 s integration time, observed in 4k mode for continuum and

---

[18] https://www.commonwl.org/.

polarisation projects, 32k mode for spectral line projects, and in some cases, both narrowband and wideband coarse modes. A typical full synthesis track will span 8 h of observing; however, to address scheduling constraints across the diverse set of supported projects, observers are often requested to schedule a rising and setting track, typically spanning 4 h each.

Therefore, with up to 2016 baselines, MeerKAT's data rates are considerable, reaching 1.1 TB per hour during 32k mode, and 0.14 TB per hour during 4k mode. These data, stored in the native MVF format, can be converted into MeasurementSet format, following a KATDAL procedure in which the data are converted on the fly during a data transfer (see Sect. 2.3.3). A typical 8 h 32k MS is ∼12 TB in size, and therefore, processing these data requires considerable computing power, and ideally, software efficiently utilising this infrastructure.

Due to the diverse range of science projects supported by MeerKAT, including LSPs and OT proposals (see Sect. 2.3.3), a broad community of MeerKAT users make use of a number of computing facilities and software packages.

IDIA and ilifu

The primary third party facility, used as the processing facility for the majority of the MeerKAT imaging LSPs, is ilifu.[19] This facility is operated by the Inter-University Institute for Data-Intensive Astronomy (IDIA), and is a tier 2 joint cloud platform for Astronomy and Bioinformatics, emulating an SKA Regional Centre (SRC). The ilifu facility consists of a number of services, including a computing cluster, a JupyterHub interface, GPUs, development nodes and transfer nodes.

The primary resource of ilifu is its cluster, comprising >80 nodes,[20] consisting of 32 cores each. The cluster contains ∼80 nodes in its Main partition, with 232 GB of usable RAM, two nodes in its HighMem partition, with 480 GB of usable RAM, and five GPU nodes, consisting of Tesla P100 16 GB GPUs and one Tesla V100 32 GB GPU. Jupyter sessions are spawned within a dedicated SLURM partition that manages their resources, ensuring users have sole access to those resources.

To support MeerKAT's significant data volumes, ilifu contains ∼7 PB of usable storage, across a number of shared CephFS mounts, and consists of a 100 GB/s network to the South African National Research Network (SANReN).

IDIA provides systems and astronomy support across various aspects of scientific research, including purpose-built software containers, and services for data transport, access, processing, analysis and storage. LDAP is utilised to configure project and accounting groups for access to proprietary data. A typical software stack for MeerKAT processing on ilifu will include Linux VMs deployed by OpenStack, Singularity containers including CASA and other software dependencies, with jobs scheduled and resources managed by SLURM, and a wrapper software written and managed by those within the user community, often making use of OpenMPI.

[19] Xhosa word for 'cloud', https://docs.ilifu.ac.za.

[20] Within its cloud platform, ilifu nodes are regularly repurposed for custom services.

Science Data Processor
All observations, apart from extremely short snapshot observations, will be run through the SDP pipelines in real time, with flags, calibration solutions and images made available to the end users via the SARAO archive. Although only intended for quality assurance purposes, the SDP pipelines have become fully fledged pipelines in their own right.

The first part of the SDP pipeline that is run during a compatible MeerKAT observation is the SDP calibration (SDPcal) pipeline. This pipeline is run in real time on data streamed from telescope (see Sect. 2.3.3), and produces calibration solutions and RFI flags. When observing in 32k mode, the data are split into four frequency bands and processed in parallel on separate compute servers. Solutions are derived according to the tags given to each field within the observing schedule, and include delay, bandpass, gain and cross-hand delay calibration. The pipeline makes a smart selection of a clean section of the band unaffected by RFI in order to produce these solutions, other than for bandpass calibration, which solves across the entire band. Similarly, a smart selection of reference antenna is made during the start of an observation and carried through the observation and real-time processing. The visibilities to which SDPcal solutions are applied are referred to as level 1 data.

The SDPcal pipeline produces a large set of RFI flags according to separate categories, produced via advanced algorithms such as a version of AOFlagger rewritten via DASK to operate on MVF data. These flag categories are stored within individual bits in the flag byte, which allows end-users to choose which category of flag to apply to their data. The SDP calibration is presented within an HTML calibration report as a means of quality assessment. This report presents plots and basic metrics of the calibration solutions and flags, including antenna S/N, waterfall plots of the percentage flagged, and various additional plots, tables and metrics presenting the calibration solutions, according to field and calibration tag.

The next step of the SDP processing is the continuum pipeline, which is run for compatible observations in which a target is observed for over 15 min. This pipeline is run offline after observing and real-time calibration has taken place, and takes $\sim$10 h for a typical 8 h observation. The continuum pipeline builds on top of the calibration pipeline by using the level 1 data and producing level 2 data, in the form of self-calibration solutions, a continuum image and model for continuum-subtraction. The continuum pipeline is based on the Obit package (Cotton 2008), and images each target field within two rounds of phase-only self-calibration, with an initial high threshold of 1 mJy, and a final threshold around 0.1 mJy. Baseline-dependent averaging is used to reduce the data volumes, and widefield effects are solved by imaging in several sub-bands and many facets, including around bright sources outside the primary beam, as identified from SUMSS. A final image is produced by combining the facets from each sub-band, and applying primary beam correction. This image is then passed into a validation tool to produce a continuum image quality report, made available to the user alongside their data via the SARAO archive. This tool runs the Aegean source finder to produce a MeerKAT catalogue, and compares to an optional set of reference catalogues (e.g. ASKAP's RACS-low near-all-sky catalogue) for comparison of positions and fluxes, and derivation of spectral indices. These are presented within fully interactive plots, and the data quality is summarised within several key metrics, assessing the MeerKAT data quality internally, with metrics for its RMS, source counts, and integrated to peak flux densities, and externally, with metrics for its positional offset and flux density ratios.

The final step of the SDP processing is the spectral pipeline, which is run for compatible observations in which a target is observed for over 45 min. This pipeline is also run offline after the continuum pipeline has run, and takes ~5 h for a typical 8 h observation. The spectral pipeline builds on top of both the calibration and continuum pipelines by using the level 1 and level 2 data, including the continuum sky model, and producing individual channels of continuum-subtracted data. The spectral pipeline includes GPU support, and used a hybrid gridding algorithm combining both w-stacking and w-projection. The self-calibration solutions from the continuum pipeline are applied, continuum-subtraction is performed within the visibility domain using the continuum sky model, and the data are imaged, typically converging within three major cycles. Finally, primary beam correction is applied and the individual channel images are placed on the SARAO archive for downloading. A spectral imager report is produced and made available alongside the data via the SARAO archive. This report summarises the individual channels imaged, including which passed, failed, had no/flagged data, or were masked out, and includes several key plots of the individual channel images, including the flux density and S/N measured.

Making use of data processed via the SDP pipelines can be done in a number of ways. Users may simply wish to access the final images, which can be downloaded as FITS images directly from the SARAO archive. This includes continuum images before and after primary beam correction, containing individual sub-band images and a spectral index map, and individual spectral-line channel images. Users who wish to access the calibrated visibilities can interact with the data via KATDAL, either on-the-fly (i.e. directly access the data hosted on the archive) or using a downloaded copy of the MVF data, or can convert and transfer the data in MeasurementSet format. Solutions and flags may be applied to MS data, which are separately stored under different categories of flags and solutions.

Users who wish do their own processing can download raw MS data with no solutions applied, and a minimal set of flags applied. Typically the 'data_lost' and 'cam' (control and monitoring) flags will be applied to all MS data, as these flags are identified in real time by the telescope software during an observation, and cannot be identified during processing. Such users will process the data manually/interactively, via a custom set of scripts, of a pipeline. Although many scripts and software pipelines exist, the majority of approaches follow a similar two-round calibration approach, making use of CASA for at least the initial first generation calibration (1GC) steps. Users wishing to make use of a processing pipeline will generally make use of the IDIA pipeline, CARACal, or Oxkat, which we now describe in more detail.

IDIA Pipeline

The IDIA PROCESSMEERKAT pipeline is a fully automated end-to-end parallel processing pipeline for calibration and imaging of radio data, with a unique focus on performance, usability, stability, and polarisation calibration. The pipeline performs full-polarisation cross-calibration, self-calibration and science imaging, including automated flagging and automated masked imaging. The pipeline has a user-friendly interface that manages SLURM job submission, resources and dependencies, and offers convenience scripts for submitting the pipeline, cancelling it, seeing its status, and more. It makes use of off-the-shelf software, in particular, CASA 6, and other Python-based packages such as PYBDSF and ASTROPY. The pipeline avoids making use of other more experimental software and large parameter sets, but uses sensible defaults that result in good image quality the majority of the time, while offering the ability to edit pipeline scripts or insert one's own scripts, containers or MPI wrappers into the automated workflow.

The pipeline makes use of parallel processing via two parallelism schemes, achieved by partitioning the data and processing each partition in parallel, in the following way:

1. Partition in frequency via user-specified spectral windows (SPWs)
2. Partition in time (scan) via a CASA multi-MS (MMS)indexdata, format, MeasurementSet, MS

The result of this design is an ability to process the data within a short period of time, necessary for keeping up with MeerKAT's enormous data rates. Figure 6.1 shows the cumulative run-time for typical 32k data ($\sim$12 TB in volume) throughout the initial cross-calibration steps, from partitioning the data, through two rounds of flagging and calibration, and splitting out the data ready for imaging. A comparison is shown between a case where neither parallelism scheme is used, where each of them are used in turn, and where full parallelism is used. Using the IDIA pipeline, 32k data can be cross-calibrated in a few hours per spectral window, typically completed in several hours, less than the total observing time. Similar parallelism schemes are used during self-calibration and imaging, as well as additional algorithmic approaches to improve both performance and image fidelity, although benchmarking of these steps remains ongoing.

The IDIA pipeline is primarily designed to process MeerKAT data on the ilifu facility, although has been demonstrated to perform well for for other telescopes and on other HPC facilities, including uGMRT and VLA, using other SLURM facilities, such as the Pawsey Supercomputing Centre. A future release is planned in which other platforms will be supported, including PBS/Torque for the CHPC, and AW-projection will be supported via CASA's TCLEAN task, using a new telescope-agnostic algorithm based only upon input holography, implemented by the NRAO Algorithm Research and Development Group (ARDG).

CARACal

The Containerized Automated Radio Astronomy Calibration (CARACal) software is a pipeline for processing any radio interferometer data that is in MeasurementSet

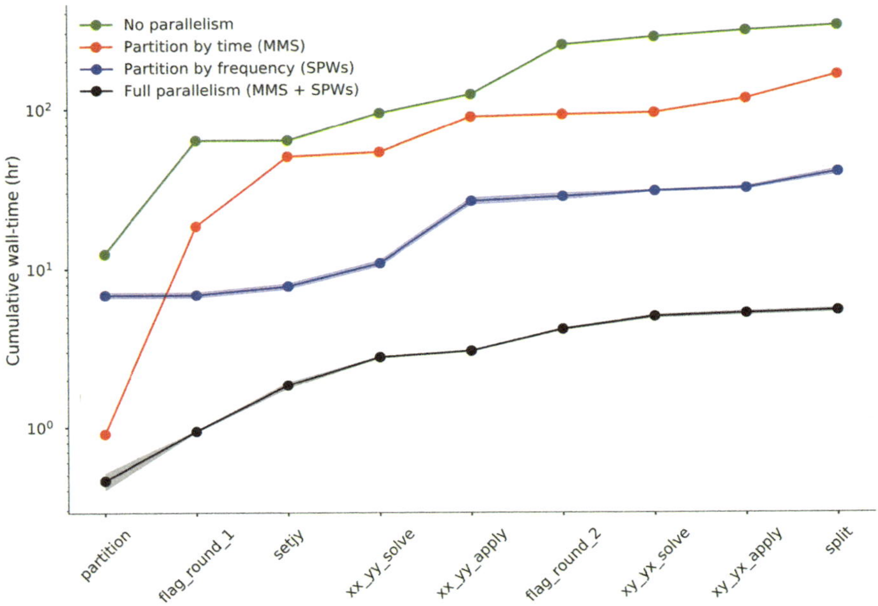

Fig. 6.1 The cumulative run-time of a typical IDIA pipeline cross-calibration run for typical 32k MeerKAT data (~12 TB), according to four parallelism schemes: without partitioning the data (i.e. no parallelism—green), and when partitioning the data in time (red), in frequency (blue), and in both time and frequency (black). The solid line and shaded areas show the median run-time and robust standard deviation over all spectral windows that were concurrently processed

format. CARACal makes use of Stimela, a Python scripting framework utilising container software such as Docker and Singularity, enabling the user to execute tasks from different software packages sequentially, without having to install those packages. CARACal is particularly focused on offering users the ability to process data with any desired software package, and unifying those steps into a single pipeline, including the basic flagging and cross-calibration steps, all the way through to direction-dependent calibration, configured via a single YAML file.

Oxkat

Oxkat (Heywood 2020) is a highly automated Python-based pipeline designed for MeerKAT processing, written for and commonly used on the ilifu facility via SLURM, and the CHPC facility via PBS/Torque. The pipeline processes data all the way through to 3GC (third-generation, direction-dependent calibration), making use of tools such as CubiCal, DDFacet, killMS and WSClean. Oxkat is particularly focused on high-fidelity imaging using a suite of available tools, achieved using tools designed for direction-dependent solving and imaging, including complex solvers and facet-based imagers.

6.9 ASKAP

Knowing that ASKAP (see Sect. 1.3) would push the limits of computational power, the telescope was designed to incorporate custom signal processing and supercomputing systems as a core part of its operational workflow. Instead of storing raw data for subsequent analysis, ASKAP data must be imaged and archived in less time than it took to observe, to avoid overflowing the limited disk space available. Similarly, the complexity of the overall system combined with high data rates, the need to survey the sky efficiently and the extremely remote site meant that science operations needed to adapt to best suit this context, shifting away from a reliance on human monitoring and control towards more automation and autonomy with human experts reacting where necessary.

Scheduling of the telescope is handled by SAURON (Scheduling Autonomously Under Reactive Observational Needs). This scheduler has visibility of the current observational pool, associated survey constraints, the environmental conditions of the site, the current status of the telescope system (including hardware availability or issues) and the status of Pawsey disks for ingest, storage and processing. This provides SAURON with the information required to dynamically choose an appropriate field to observe at any given moment without human input or supervision, also ensuring appropriate calibration is carried out. Enabling SAURON to work effectively required collaborative development work on the telescope system to automate processes that were originally expected to be carried out manually (e.g. telescope configuration changes, delay calibration, beamweight updates, etc.). Following a consolidation period of development in 2020 which involved upgrading automation and stability on the telescope, the first dedicated period of SAURON observing resulted in a jump from 50% to 80% observing efficiency and 70% to 95% successful observing rate. In preparation for full surveys, work is currently underway to extend existing automation to include an autonomous pipeline processing workflow and better link SAURON decisions to the status of processing, such that the full ASKAP operations workflow is as automated as possible for maximum efficiency. Alongside this, research work via the Collaborative Intelligence Future Science Platform of CSIRO will explore how machine learning and artificial intelligence can be used to augment existing workflows, improve data quality and better leverage human expertise in a collaborative way.

The antennas produce over 100 Tb s^{-1} of raw voltage data. This is reduced to about 20 Gb s^{-1} by signal processing hardware at the observatory site, which forms a set of beams and then correlates beams pointing in the same direction across all antennas, averaging the result over 10 s intervals. These visibilities are sent over a long-haul optical fibre network to the Pawsey supercomputing centre in the nearest capital city, Perth.

Flagging, calibration, imaging, mosaicking, source finding and archiving are all done within the Pawsey data ecosystem, using custom-made software designed specifically for parallel processing environments. Many different kinds of data products can be produced for each observation, from continuum images in all polarisation products, to spectral cubes and source catalogues. This ability to support multiple science goals from a single observation increases the scientific output of the telescope and makes the most efficient use of observing time.

When running at full spectral resolution, we permanently archive up to 50 TB each day. For the first 5 years of operation, around 90% of ASKAP's observing time will be dedicated to large-scale survey projects in conjunction with international survey science teams.

ASKAP Ingest

The ASKAP data ingest pipeline formats 20 Gb s^{-1} of raw visibility data (tagged with hardware-specific metadata and packetised according to the needs of the correlator at the observatory) into a file structure compatible with downstream imaging software (see also Sect. 2.3.4). Initially, it was hoped that we could keep all the data in memory without having to write anything to disk until the final archive-ready outputs, but this proved to be impractical. During commissioning and verification, it is essential to retain raw data for some time, so it can be investigated and checked for quality. We therefore developed software to write several CASA MeasurementSets in parallel, dividing up the data stream by beam and frequency on a Lustre filesystem.

Due to the performance criticality of this stage in ASKAP's workflow, a dedicated ingest cluster was built. The ingest cluster was isolated from other users at the supercomputing centre and fitted with an independent high-speed storage array and dedicated point-to-point links to the observatory. Initial trials with common storage proved unreliable, as unpredictable load spikes due to other user activity could cause the ingest buffers to overflow, resulting in loss of data. The need for an independent storage buffer necessitates an additional copy before processing can begin, but this overhead was the only way to ensure data integrity.

Data logistics and the processing workflow are managed by a dedicated event-based software layer that also tracks performance and processing status for each observation.

ASKAPsoft

The processing of ASKAP data is managed through a pipeline workflow running on a supercomputer at the Pawsey Centre. For the early phases of ASKAP through to the completion of the pilot surveys, this took place on the *Galaxy* Cray XC-30, although processing will move to the new *Setonix* Cray EX in 2023. The pipeline workflow takes the raw data written by the ingest pipeline and performs the necessary calibration, flagging, imaging, continuum-subtraction and source-finding to create a set of science-ready data products that are made available to end-users through the CSIRO ASKAP Science Data Archive (CASDA), see Sect. 3.3.

The workflow itself is based on custom-written shell (`bash`) and python scripts, that make use of the SLURM workload manager to run a large sequence of jobs with inter-dependencies. Each of these jobs makes use of the ASKAPsoft suite of tools—a package developed at CSIRO specifically to meet the challenges of ASKAP processing, with distributed processing (MPI and OpenMP) built in.

The multi-beam nature of ASKAP data, where different beams are essentially independent observations, means each beam can be processed in parallel (up to the point where they are mosaicked together). We are able to choose the number of compute nodes used by many of the jobs, such that we optimally use the available resources. This allows a single processing run for a dataset to fill the queue on the *Galaxy* machine.

The outline of the workflow is as follows:

1. Data preparation—a copy is made from the raw data in the local processing area to provide an optimised dataset for processing, and to prevent the raw data from being altered by the processing.
2. Bandpass calibration—frequency-dependent complex gain calibration is applied, having been determined for each beam from a separate observation of the primary southern flux density calibrator PKS B1934-638. This sets the overall flux density scale.
3. Flagging and averaging—RFI is excised from the visibility data, and full-spectral-resolution data is averaged to form 1 MHz-resolution "continuum" data.
4. Continuum imaging and self-calibration—the continuum dataset is imaged and used to further calibrate the data through an iterative self-calibration process. This step provides a fully-calibrated continuum image, as well as a time-dependent but frequency-independent complex gain calibration table. This is then applied to the full-resolution dataset.
5. Continuum cube imaging—the continuum dataset may be imaged to form coarse-resolution cubes, preserving the 1 MHz frequency channels. This is done in all four Stokes parameters to provide full polarisation information.
6. Continuum subtraction—if spectral processing of the full-resolution dataset is required, the continuum emission is subtracted from the visibilities by use of a model derived either from the continuum imaging or from fitting to the final image.
7. Spectral imaging—the spectral data is then imaged to form a large spectral cube. Residual continuum emission (that may not have been removed in the previous step due to an incomplete model) is removed by fitting each spectrum in the image domain.
8. Mosaicking—once all beams are complete, they are combined in the image plane through linear mosaicking. At this point we correct for the primary-beam attenuation by applying models of each beam determined from holography observations.
9. Cataloguing—source-finding is run on the continuum mosaic using `selavy`, creating catalogues of islands and the fitted components in each island.
10. Archiving—the images, calibration tables and catalogues are then uploaded to CASDA for access by astronomers.

6.10 SKA

As outlined in Sect. 1.3, the SKA Observatory is a radio facility comprising of two telescopes situated in South Africa and Western Australia, with a headquarters at Jodrell Bank Observatory in the UK,[21] due to become fully operational in 2028. It has been designed to address a broad spectrum of science challenges from the origins of life, planet and star formation, the evolution of galaxies to the Epoch of Reionisation.[22]

SKA-Low is sighted in the Murchison Radio astronomy Observatory with a 1 km-wide core of stations and a maximum baseline of 70 km with a total of 512 stations. Each station will comprise of 256 log-periodic antennas designed to enable a high sensitivity response over the broad operating frequency range of 50–350 MHz. For standard observations SKA-Low will have a 0.9 s integration time and channel width of 5.4 kHz.

SKA-Mid will consist of the existing 64 MeerKAT dishes (13.5-m diameter) with the addition of 133 SKA antennas (15-m diameter). These will be situated in a randomly distributed 2 km core in the Karoo desert and along three spiral arms out to a maximum baseline of 150 km. SKA-Mid will operate at 350 MHz to 15 GHz separated over five bands with a standard channel width of 13.75 kHz and integration time of 0.14 s.

When complete, the SKA will be the largest radio facility in the world providing unprecedented sensitivity and image fidelity at radio frequencies. This will be achieved through the large number of antennas/dishes and the broad frequency spectrum of the receivers, but also through the data processing facility and techniques. The SKA will see a number of roll-out stages during construction (known as Array Assemblies) but at full capability will produce 20 Tb/s from the Mid and 2 Pb s^{-1} and Low telescopes, with ingest for data processing 8 Tb s^{-1} data from each telescope, necessitating high-performance computing facilities and automated processing.

The digitised data from the array (in the form of a number of station beams) will be sent to the Central Processing Facility (CPF) for correlation and formation of multiple tied-array beams. High time and frequency resolution pulsar search and pulsar timing processing is also performed.

All further data processing for the SKA will be (automated and as a result of the sheer volume of data produced will be streamed into) performed by the Science Data Processor (SDP) subsystem in the Science Processing Centre (SPC), which receives the data post-correlation. This will include visibility data as well as the outputs from the pulsar timing and search (candidates and associated voltage data).

---

[21] https://www.skao.int/.

[22] https://www.skao.int/en/resources/402/key-documents.

The telescope will operate in a number of observing modes including imaging (continuum and spectral-line) and non-imaging (pulsar search and timing, VLBI and transient searching).

SDP will be responsible for the data capture, calibration, imaging and non-imaging processing and delivery of the 'science-ready' data products to the Science Regional Centres which will include image cubes, calibrated visibilities, source catalogues, pulsar candidates and timing solutions, and transient data.[23]

There are a number of real-time processing stages performed by SDP including providing gain solutions and pointing offsets to the rest of the system to enable pointing corrections and beam-forming.

Fast-imaging and transient pipelines will be used to provide quick processing to search for and provide automated transient notifications.

The majority of calibration and imaging stages to produce the expected data products will be performed as part of the batch processing. This will include (but is not limited to) pre-processing (RFI removal, source extraction and averaging as required), application of instrumental corrections, iterative self-calibration (to correct for atmospheric effects), polarisation and imaging as well as spectral-line processing.

Given the sheer volume of data produced, the majority of observing projects on the SKA will not have access to raw visibility data, requiring high-performance pipelines in order to provide data products including image cubes and source catalogues suitable for scientific analysis. This will necessitate appropriate algorithms to provide e.g. direction-dependent corrections and wide-field imaging with full polarisation calibration to minimise errors and maximise sensitivity. These will also require highly-parallelised implementations to maximise efficiency and ensure high throughput of data.

In order to prepare for this challenge, the SDP consortium formed an architectural design for the necessary platform and software.[24] The development of the processing software is now underway working under a Scaled Agile Framework (SAFe ®) with open-source access.[25] As has been discussed in previous sections, there are a number of existing instruments implementing high-performance pipelines on super-computing scales, driven by the same challenges the SKA will face, providing an ideal opportunity to learn and utilise the existing software, adapting as needed to meet the needs of the SKA.

Lessons Learned from Pathfinders and Precursors

The SKA will, in principle, meet many aspects of the challenges faced in the pathfinder and precursor instruments. Fundamentally, the data processing required is the same in terms of what corrections are necessary to the data to provide the output imaging and non-imaging data products needed. This means that existing

---

[23] https://www.skao.int/sites/default/files/documents/d7-SKA-TEL-SDP-0000027_01C_DRE_SDPPipelinesDesign-signed.pdf.

[24] https://developer.skao.int/projects/ska-sdp-integration/en/latest/.

[25] https://gitlab.com/ska-telescope/sdp.

approaches can be used and adapted from those currently in-working systems. For example, the imaging and deconvolution algorithms already employed in LOFAR data processing (as discussed in Sect. 6.7) provide fundamental wide-field capability and can be used to understand where the challenges are likely to arise in providing the equivalent scientific results for SKA-Low. Likewise, there are significant similarities with how the Science Data Processor will operate for SKA-Mid and that of MeerKAT (see Sect. 6.8).

Particularly for the precursor instruments a significant amount of the control systems needed to operate the telescope during observations and to provide the necessary planning, monitoring and feedback have also been implemented in facilities that will look very similar to the SKA. These instruments already share common infrastructure with the SKA. They will both deliver science on the way and help scale out to what's needed for the SKA telescopes, i.e., much has been learned about the required planning, construction and infrastructure needed to support these working instruments on the SKA site. Existing instruments also play a key role in testing the software being developed for the SKA. They can provide invaluable feedback on deploying and using the software components as they are developed. Especially while in the early stages of construction, the SKA will have comparatively little infrastructure (hardware and software) available for testing, meaning the precursor instruments offer the experience of on-sky testing in a working instrument, something that's crucial to ensure that the software is performing as it needs to.

The SKA has additional challenges in operating two independent telescopes and the need to perform all of the processing as part of the scheduled time and resource allocation on the instrument, but will seek to overcome these challenges with updated implementations and architecture that are based on existing practices of these precursor and pathfinder systems.

References

L. Blackburn, C.-k. Chan, G.B. Crew, V.L. Fish, S. Issaoun, M.D. Johnson, M. Wielgus, K. Akiyama, J. Barrett, K.L. Bouman, R. Cappallo, A.A. Chael, M. Janssen, C.J. Lonsdale, S.S. Doeleman, EHT-HOPS pipeline for millimeter VLBI data reduction. Astrophys. J. **882**(1), 23 (2019). https://doi.org/10.3847/1538-4357/ab328d

W.D. Cotton, Obit: a development environment for astronomical algorithms. Publ. Astron. Soc. Pac. **120**(866), 439 (2008). https://doi.org/10.1086/586754

W.D. Cotton, J.J. Condon, R.A. Perley, N. Kassim, J. Lazio, A. Cohen, W. Lane, W.C. Erickson, Beyond the isoplanatic patch in the VLA Low-frequency Sky Survey, in *Ground-Based Telescopes*, ed. by J.M. Oschmann Jr., Society of Photo-Optical Instrumentation Engineers (SPIE) Conference Series, vol. 5489 (2004), pp. 180–189. https://doi.org/10.1117/12.551298

F. de Gasperin, T.J. Dijkema, A. Drabent, M. Mevius, D. Rafferty, R. van Weeren, M. Brüggen, J.R. Callingham, K.L. Emig, G. Heald, H.T. Intema, L.K. Morabito, A.R. Offringa, R. Oonk, E. Orrù, H. Röttgering, J. Sabater, T. Shimwell, A. Shulevski, W. Williams, Systematic effects in LOFAR data: a unified calibration strategy. Astron. Astrophys. **622**, A5 (2019). https://doi.org/10.1051/0004-6361/201833867

F. de Gasperin, G. Brunetti, M. Brüggen, R. van Weeren, W.L. Williams, A. Botteon, V. Cuciti, T.J. Dijkema, H. Edler, M. Iacobelli, H. Kang, A. Offringa, E. Orrú, R. Pizzo, D. Rafferty, H. Röttgering, T. Shimwell, Reaching thermal noise at ultra-low radio frequencies. Toothbrush radio relic downstream of the shock front. Astron. Astrophys. **642**, A85 (2020). https://doi.org/10.1051/0004-6361/202038663

A.T. Deller, S.J. Tingay, M. Bailes, C. West, DiFX: a software correlator for very long baseline interferometry using multiprocessor computing environments. Publ. Astron. Soc. Pac. **119**, 318–336 (2007). https://doi.org/10.1086/513572

A.T. Deller, W.F. Brisken, C.J. Phillips, J. Morgan, W. Alef, R. Cappallo, E. Middelberg, J. Romney, H. Rottmann, S.J. Tingay, R. Wayth, DiFX-2: a more flexible, efficient, robust, and powerful software correlator. Publ. Astron. Soc. Pac. **123**, 275 (2011). https://doi.org/10.1086/658907

Event Horizon Telescope Collaboration, K. Akiyama, A. Alberdi, W. Alef, K. Asada, R. Azulay, A.-K. Baczko, D. Ball, M. Baloković, J. Barrett et al., First M87 event horizon telescope results. III. Data processing and calibration. Astrophys. J. Lett. **875**, L3 (2019a). https://doi.org/10.3847/2041-8213/ab0c57

Event Horizon Telescope Collaboration, K. Akiyama, A. Alberdi, W. Alef, K. Asada, R. Azulay, A.-K. Baczko, D. Ball, M. Baloković, J. Barrett, D. Bintley, L. Blackburn, W. Boland, K.L. Bouman, G.C. Bower, M. Bremer, C.D. Brinkerink, R. Brissenden, S. Britzen, A.E. Broderick, D. Broguiere, T. Bronzwaer, D.-Y. Byun, J.E. Carlstrom, A. Chael, C.-k. Chan, S. Chatterjee, K. Chatterjee, M.-T. Chen, Y. Chen, I. Cho, P. Christian, J.E. Conway, J.M. Cordes, G.B. Crew, Y. Cui, J. Davelaar, M. De Laurentis, R. Deane, J. Dempsey, G. Desvignes, J. Dexter, S.S. Doeleman, R.P. Eatough, H. Falcke, V.L. Fish, E. Fomalont, R. Fraga-Encinas, W.T. Freeman, P. Friberg, C.M. Fromm, J.L. Gómez, P. Galison, C.F. Gammie, R. García, O. Gentaz, B. Georgiev, C. Goddi, R. Gold, M. Gu, M. Gurwell, K. Hada, M.H. Hecht, R. Hesper, L.C. Ho, P. Ho, M. Honma, C.-W. L. Huang, L. Huang, D.H. Hughes, S. Ikeda, M. Inoue, S. Issaoun, D.J. James, B.T. Jannuzi, M. Janssen, B. Jeter, W. Jiang, M.D. Johnson, S. Jorstad, T. Jung, M. Karami, R. Karuppusamy, T. Kawashima, G.K. Keating, M. Kettenis, J.-Y. Kim, J. Kim, J. Kim, M. Kino, J.Y. Koay, P.M. Koch, S. Koyama, M. Kramer, C. Kramer, T.P. Krichbaum, C.-Y. Kuo, T.R. Lauer, S.-S. Lee, Y.-R. Li, Z. Li, M. Lindqvist, K. Liu, E. Liuzzo, W.-P. Lo, A.P. Lobanov, L. Loinard, C. Lonsdale, R.-S. Lu, N.R. MacDonald, J. Mao, S. Markoff, D.P. Marrone, A.P. Marscher, I. Martí-Vidal, S. Matsushita, L.D. Matthews, L. Medeiros, K.M. Menten, Y. Mizuno, I. Mizuno, J.M. Moran, K. Moriyama, M. Moscibrodzka, C. Müller, H. Nagai, N.M. Nagar, M. Nakamura, R. Narayan, G. Narayanan, I. Natarajan, R. Neri, C. Ni, A. Noutsos, H. Okino, H. Olivares, G.N. Ortiz-León, T. Oyama, F. Özel, D.C.M. Palumbo, N. Patel, U.-L. Pen, D.W. Pesce, V. Piétu, R. Plambeck, A. PopStefanija, O. Porth, B. Prather, J.A. Preciado-López, D. Psaltis, H.-Y. Pu, V. Ramakrishnan, R. Rao, M.G. Rawlings, A.W. Raymond, L. Rezzolla, B. Ripperda, F. Roelofs, A. Rogers, E. Ros, M. Rose, A. Roshanineshat, H. Rottmann, A.L. Roy, C. Ruszczyk, B.R. Ryan, K.L.J. Rygl, S. Sánchez, D. Sánchez-Arguelles, M. Sasada, T. Savolainen, F. Peter Schloerb, K.-F. Schuster, L. Shao, Z. Shen, D. Small, B.W. Sohn, J. SooHoo, F. Tazaki, P. Tiede, R.P.J. Tilanus, M. Titus, K. Toma, P. Torne, T. Trent, S. Trippe, S. Tsuda, I. van Bemmel, H.J. van Langevelde, D.R. van Rossum, J. Wagner, J. Wardle, J. Weintroub, N. Wex, R. Wharton, M. Wielgus, G.N. Wong, Q. Wu, K. Young, A. Young, Z. Younsi, F. Yuan, Y.-F. Yuan, J.A. Zensus, G. Zhao, S.-S. Zhao, Z. Zhu, J.-C. Algaba, A. Allardi, R. Amestica, J. Anczarski, U. Bach, F.K. Baganoff, C. Beaudoin, B.A. Benson, R. Berthold, J.M. Blanchard, R. Blundell, S. Bustamente, R. Cappallo, E. Castillo-Domínguez, C.-C. Chang, S.-H. Chang, S.-C. Chang, C.-C. Chen, R. Chilson, T.C. Chuter, R. Córdova Rosado, I.M. Coulson, T.M. Crawford, J. Crowley, J. David, M. Derome, M. Dexter, S. Dornbusch, K.A. Dudevoir, S.A. Dzib, A. Eckart, C. Eckert, N.R. Erickson, W.B. Everett, A. Faber, J.R. Farah, V. Fath, T.W. Folkers, D.C. Forbes, R. Freund, A.I. Gómez-Ruiz, D.M. Gale, F. Gao, G. Geertsema, D.A. Graham, C.H. Greer, R. Grosslein, F. Gueth, D. Haggard, N.W. Halverson, C.-C. Han, K.-C. Han, J. Hao, Y. Hasegawa, J.W. Henning, A. Hernández-Gómez, R. Herrero-Illana, S. Heyminck, A. Hirota, J. Hoge, Y.-D. Huang, C.M. Violette Impellizzeri, H. Jiang, A. Kamble, R. Keisler, K. Kimura, Y. Kono, D. Kubo, J. Kuroda, R. Lacasse, R.A.

Laing, E.M. Leitch, C.-T. Li, L.C.C. Lin, C.-T. Liu, K.-Y. Liu, L.-M. Lu, R.G. Marson, P.L. Martin-Cocher, K.D. Massingill, C. Matulonis, M.P. McColl, S.R. McWhirter, H. Messias, Z. Meyer-Zhao, D. Michalik, A. Montaña, W. Montgomerie, M. Mora-Klein, D. Muders, A. Nadolski, S. Navarro, J. Neilsen, C.H. Nguyen, H. Nishioka, T. Norton, M.A. Nowak, G. Nystrom, H. Ogawa, P. Oshiro, T. Oyama, H. Parsons, S.N. Paine, J. Peñalver, N.M. Phillips, M. Poirier, N. Pradel, R.A. Primiani, P.A. Raffin, A.S. Rahlin, G. Reiland, C. Risacher, I. Ruiz, A.F. Sáez-Madaín, R. Sassella, P. Schellart, P. Shaw, K.M. Silva, H. Shiokawa, D.R. Smith, W. Snow, K. Souccar, D. Sousa, T.K. Sridharan, R. Srinivasan, W. Stahm, A.A. Stark, K. Story, S.T. Timmer, L. Vertatschitsch, C. Walther, T.-S. Wei, N. Whitehorn, A.R. Whitney, D.P. Woody, J.G.A. Wouterloot, M. Wright, P. Yamaguchi, C.-Y. Yu, M. Zeballos, S. Zhang, L. Ziurys, First M87 event horizon telescope results. I. The shadow of the supermassive Black Hole. Astrophys. J. Lett. **875**(1), L1 (2019b). https://doi.org/10.3847/2041-8213/ab0ec7

Event Horizon Telescope Collaboration, K. Akiyama, A. Alberdi, W. Alef, J.C. Algaba, R. Anantua, K. Asada, R. Azulay, U. Bach, A.-K. Baczko, D. Ball, M. Baloković, J. Barrett, M. Bauböck, B.A. Benson, D. Bintley, L. Blackburn, R. Blundell, K.L. Bouman, G.C. Bower, H. Boyce, M. Bremer, C.D. Brinkerink, R. Brissenden, S. Britzen, A.E. Broderick, D. Broguiere, T. Bronzwaer, S. Bustamante, D.-Y. Byun, J.E. Carlstrom, C. Ceccobello, A. Chael, C.-k. Chan, K. Chatterjee, S. Chatterjee, M.-T. Chen, Y. Chen, X. Cheng, I. Cho, P. Christian, N.S. Conroy, J.E. Conway, J.M. Cordes, T.M. Crawford, G.B. Crew, A. Cruz-Osorio, Y. Cui, J. Davelaar, M. De Laurentis, R. Deane, J. Dempsey, G. Desvignes, J. Dexter, V. Dhruv, S.S. Doeleman, S. Dougal, S.A. Dzib, R.P. Eatough, R. Emami, H. Falcke, J. Farah, V.L. Fish, E. Fomalont, H. Alyson Ford, R. Fraga-Encinas, W.T. Freeman, P. Friberg, C.M. Fromm, A. Fuentes, P. Galison, C.F. Gammie, R. García, O. Gentaz, B. Georgiev, C. Goddi, R. Gold, A.I. Gómez-Ruiz, J.L. Gómez, M. Gu, M. Gurwell, K. Hada, D. Haggard, K. Haworth, M.H. Hecht, R. Hesper, D. Heumann, L.C. Ho, P. Ho, M. Honma, C.-W.L. Huang, L. Huang, D.H. Hughes, S. Ikeda, C.M. Violette Impellizzeri, M. Inoue, S. Issaoun, D.J. James, B.T. Jannuzi, M. Janssen, B. Jeter, W. Jiang, A. Jiménez-Rosales, M.D. Johnson, S. Jorstad, A.V. Joshi, T. Jung, M. Karami, R. Karuppusamy, T. Kawashima, G.K. Keating, M. Kettenis, D.-J. Kim, J.-Y. Kim, J. Kim, J. Kim, M. Kino, J. Y. Koay, P. Kocherlakota, Y. Kofuji, P.M. Koch, S. Koyama, C. Kramer, M. Kramer, T.P. Krichbaum, C.-Y. Kuo, N. La Bella, T.R. Lauer, D. Lee, S.-S. Lee, P.K. Leung, A. Levis, Z. Li, R. Lico, G. Lindahl, M. Lindqvist, M. Lisakov, J. Liu, K. Liu, E. Liuzzo, W.-P. Lo, A.P. Lobanov, L. Loinard, C.J. Lonsdale, R.-S. Lu, J. Mao, N. Marchili, S. Markoff, D.P. Marrone, A.P. Marscher, I. Martí-Vidal, S. Matsushita, L.D. Matthews, L. Medeiros, K.M. Menten, D. Michalik, I. Mizuno, Y. Mizuno, J.M. Moran, K. Moriyama, M. Moscibrodzka, C. Müller, A. Mus, G. Musoke, I. Myserlis, A. Nadolski, H. Nagai, N.M. Nagar, M. Nakamura, R. Narayan, G. Narayanan, I. Natarajan, A. Nathanail, S.N. Fuentes, J. Neilsen, R. Neri, C. Ni, A. Noutsos, M.A. Nowak, J. Oh, H. Okino, H. Olivares, G.N. Ortiz-León, T. Oyama, F. Özel, D.C.M. Palumbo, G.F. Paraschos, J. Park, H. Parsons, N. Patel, U.-L. Pen, D.W. Pesce, V. Piétu, R. Plambeck, A. PopStefanija, O. Porth, F.M. Pötzl, B. Prather, J.A. Preciado-López, D. Psaltis, H.-Y. Pu, V. Ramakrishnan, R. Rao, M.G. Rawlings, A.W. Raymond, L. Rezzolla, A. Ricarte, B. Ripperda, F. Roelofs, A. Rogers, E. Ros, C. Romero-Cañizales, A. Roshanineshat, H. Rottmann, A.L. Roy, I. Ruiz, C. Ruszczyk, K.L.J. Rygl, S. Sánchez, D. Sánchez-Argüelles, M. Sánchez-Portal, M. Sasada, K. Satapathy, T. Savolainen, F. Peter Schloerb, J. Schonfeld, K.-F. Schuster, L. Shao, Z. Shen, D. Small, B.W. Sohn, J. SooHoo, K. Souccar, H. Sun, F. Tazaki, A.J. Tetarenko, P. Tiede, R.P.J. Tilanus, M. Titus, P. Torne, E. Traianou, T. Trent, S. Trippe, M. Turk, I. van Bemmel, H.J. van Langevelde, D.R. van Rossum, J. Vos, J. Wagner, D. Ward-Thompson, J. Wardle, J. Weintroub, N. Wex, R. Wharton, M. Wielgus, K. Wiik, G. Witzel, M.F. Wondrak, G.N. Wong, Q. Wu, P. Yamaguchi, D. Yoon, A. Young, K. Young, Z. Younsi, F. Yuan, Y.-F. Yuan, J. Anton Zensus, S. Zhang, G.-Y. Zhao, S.-S. Zhao, C. Agurto, A. Allardi, R. Amestica, J.P. Araneda, O. Arriagada, J.L. Berghuis, A. Bertarini, R. Berthold, J. Blanchard, K. Brown, M. Cárdenas, M. Cantzler, P. Caro, E. Castillo-Domínguez, T.L. Chan, C.-C. Chang, D.O. Chang, S.-H. Chang, S.-C. Chang, C.-C. Chen, R. Chilson, T.C. Chuter, M. Ciechanowicz, E. Colin-Beltran, I.M. Coulson, J. Crowley, N. Degenaar, S. Dornbusch, C.A. Durán, W.B. Everett, A. Faber, K. Forster, M.M. Fuchs, D.M.

Gale, G. Geertsema, E. González, D. Graham, F. Gueth, N.W. Halverson, C.-C. Han, K.-C. Han, Y. Hasegawa, J. Luis Hernández-Rebollar, C. Herrera, R. Herrero-Illana, S. Heyminck, A. Hirota, J. Hoge, S.R. Hostler Schimpf, R.E. Howie, Y.-D. Huang, H. Jiang, H. Jinchi, D. John, K. Kimura, T. Klein, D. Kubo, J. Kuroda, C. Kwon, R. Lacasse, R. Laing, E.M. Leitch, C.-T. Li, C.-T. Liu, K.-Y. Liu, L.C.C. Lin, L.-M. Lu, F. Mac-Auliffe, P. Martin-Cocher, C. Matulonis, J.K. Maute, H. Messias, Z. Meyer-Zhao, A. Montaña, F. Montenegro-Montes, W. Montgomerie, M.E. Moreno Nolasco, D. Muders, H. Nishioka, T.J. Norton, G. Nystrom, H. Ogawa, R. Olivares, P. Oshiro, J.P. Pérez-Beaupuits, R. Parra, N.M. Phillips, M. Poirier, N. Pradel, R. Qiu, P.A. Raffin, A.S. Rahlin, J. Ramírez, S. Ressler, M. Reynolds, I. Rodríguez-Montoya, A.F. Saez-Madain, J. Santana, P. Shaw, L.E. Shirkey, K.M. Silva, W. Snow, D. Sousa, T.K. Sridharan, W. Stahm, A.A. Stark, J. Test, K. Torstensson, P. Venegas, C. Walther, T.-S. Wei, C. White, G. Wieching, R. Wijnands, J.G.A. Wouterloot, C.-Y. Yu, W. Yu, M. Zeballos, First Sagittarius A* event horizon telescope results. I. The shadow of the supermassive Black Hole in the center of the Milky Way. Astrophys. J. Lett. **930**(2), L12 (2022). https://doi.org/10.3847/2041-8213/ac6674

C. Goddi, H. Falcke, M. Kramer, L. Rezzolla, C. Brinkerink, T. Bronzwaer, J.R.J. Davelaar, R. Deane, M. de Laurentis, G. Desvignes, R.P. Eatough, F. Eisenhauer, R. Fraga-Encinas, C.M. Fromm, S. Gillessen, A. Grenzebach, S. Issaoun, M. Janßen, R. Konoplya, T.P. Krichbaum, R. Laing, K. Liu, R.S. Lu, Y. Mizuno, M. Moscibrodzka, C. Müller, H. Olivares, O. Pfuhl, O. Porth, F. Roelofs, E. Ros, K. Schuster, R. Tilanus, P. Torne, I. van Bemmel, H. J. van Langevelde, N. Wex, Z. Younsi, A. Zhidenko, BlackHoleCam: fundamental physics of the galactic center. Int. J. Mod. Phys. D **26**(2), 1730001-239 (2017). https://doi.org/10.1142/S0218271817300014

I. Heywood, oxkat: semi-automated imaging of MeerKAT observations. Astrophysics Source Code Library, record ascl:2009.003 (September 2020)

S. Issaoun, M. Wielgus, S. Jorstad, T.P. Krichbaum, L. Blackburn, M. Janssen, C.-k. Chan, D.W. Pesce, J.L. Gómez, K. Akiyama, M. Mościbrodzka, I. Martí-Vidal, A. Chael, R. Lico, J. Liu, V. Ramakrishnan, M. Lisakov, A. Fuentes, G.-Y. Zhao, K. Moriyama, A.E. Broderick, P. Tiede, N.R. MacDonald, Y. Mizuno, E. Traianou, L. Loinard, J. Davelaar, M. Gurwell, R.-S. Lu, A. Alberdi, W. Alef, J.C. Algaba, R. Anantua, K. Asada, R. Azulay, U. Bach, A.-K. Baczko, D. Ball, M. Baloković, J. Barrett, M. Bauböck, B.A. Benson, D. Bintley, R. Blundell, W. Boland, K.L. Bouman, G.C. Bower, H. Boyce, M. Bremer, C.D. Brinkerink, R. Brissenden, S. Britzen, D. Broguiere, T. Bronzwaer, S. Bustamante, D.-Y. Byun, J.E. Carlstrom, C. Ceccobello, K. Chatterjee, S. Chatterjee, M.-T. Chen, Y. Chen, I. Cho, P. Christian, N.S. Conroy, J.E. Conway, J.M. Cordes, T.M. Crawford, G.B. Crew, A. Cruz-Osorio, Y. Cui, M.D. Laurentis, R. Deane, J. Dempsey, G. Desvignes, J. Dexter, S.S. Doeleman, V. Dhruv, S.A. Dzib Quijano, R.P. Eatough, R. Emami, H. Falcke, J. Farah, V.L. Fish, E. Fomalont, H. Alyson Ford, R. Fraga-Encinas, W.T. Freeman, P. Friberg, C.M. Fromm, P. Galison, C.F. Gammie, R. García, O. Gentaz, B. Georgiev, C. Goddi, R. Gold, A.I. Gómez-Ruiz, M. Gu, K. Hada, D. Haggard, M.H. Hecht, R. Hesper, L.C. Ho, P. Ho, M. Honma, C.-W.L. Huang, L. Huang, D.H. Hughes, S. Ikeda, C.M. Violette Impellizzeri, M. Inoue, D.J. James, B.T. Jannuzi, B. Jeter, W. Jiang, A. Jimenez-Rosales, M.D. Johnson, A.V. Joshi, T. Jung, M. Karami, R. Karuppusamy, T. Kawashima, G.K. Keating, M. Kettenis, D.-J. Kim, J.-Y. Kim, J. Kim, J. Kim, M. Kino, J.Y. Koay, P. Kocherlakota, Y. Kofuji, P.M. Koch, S. Koyama, C. Kramer, M. Kramer, C.-Y. Kuo, N.L. Bella, T.R. Lauer, D. Lee, S.-S. Lee, P.K. Leung, A. Levis, Z. Li, R. Lico, G. Lindahl, M. Lindqvist, K. Liu, E. Liuzzo, W.-P. Lo, A.P. Lobanov, C. Lonsdale, J. Mao, N. Marchili, S. Markoff, D.P. Marrone, A.P. Marscher, S. Matsushita, L.D. Matthews, L. Medeiros, K.M. Menten, D. Michalik, I. Mizuno, Y. Mizuno, J.M. Moran, C. Müller, A. Mus, G. Musoke, I. Myserlis, A. Nadolski, H. Nagai, N.M. Nagar, M. Nakamura, R. Narayan, G. Narayanan, I. Natarajan, A. Nathanail, J. Neilsen, R. Neri, C. Ni, A. Noutsos, M.A. Nowak, J. Oh, H. Okino, H. Olivares, G.N. Ortiz-León, T. Oyama, F. Özel, D.C.M. Palumbo, G.F. Paraschos, J. Park, H. Parsons, N. Patel, U.-L. Pen, V. Piétu, R. Plambeck, A. PopStefanija, O. Porth, F.M. Pötzl, B. Prather, J.A. Preciado-López, D. Psaltis, H.-Y. Pu, R. Rao, M.G. Rawlings, A.W. Raymond, L. Rezzolla, A. Ricarte, B. Ripperda, F. Roelofs, A. Rogers, E. Ros, C. Romero-Canizales, A. Roshanineshat, H. Rottmann, A.L. Roy,

I. Ruiz, C. Ruszczyk, K.L.J. Rygl, S. Sánchez, D. Sánchez-Arguelles, M. Sanchez-Portal, M. Sasada, K. Satapathy, T. Savolainen, F. Peter Schloerb, K.-F. Schuster, L. Shao, Z. Shen, D. Small, B.W. Sohn, J. SooHoo, K. Souccar, H. Sun, F. Tazaki, A.J. Tetarenko, P. Tiede, R.P.J. Tilanus, M. Titus, P. Torne, T. Trent, S. Trippe, I. van Bemmel, H.J. van Langevelde, D.R. van Rossum, J. Vos, J. Wagner, D. Ward-Thompson, J. Wardle, J. Weintroub, N. Wex, R. Wharton, K. Wiik, G. Witzel, M. Wondrak, G.N. Wong, Q. Wu, P. Yamaguchi, D. Yoon, A. Young, K. Young, Z. Younsi, F. Yuan, Y.-F. Yuan, J. Anton Zensus, S. Zhang, S.-S. Zhao, Resolving the inner Parsec of the Blazar J1924-2914 with the event horizon telescope. Astrophys. J. **934**(2), 145 (2022). https://doi.org/10.3847/1538-4357/ac7a40

M. Janssen, C. Goddi, H. Falcke, D. van Rossum, I. van Bemmel, M. Kettenis, D. Small, I. Marti-Vidal, RPICARD: a CASA-based calibration pipeline for VLBI data, in *14th European VLBI Network Symposium & Users Meeting (EVN 2018)* (November 2018), p. 80

M. Janssen, L. Blackburn, S. Issaoun, M. Wielgus, T.P. Krichbaum, H. Falcke, A priori calibration of eht stations. EHT Memo Series, 2019-CE-01 (https://eventhorizontelescope.org/for-astronomers/memos) (March 2019a)

M. Janssen, C. Goddi, I.M. van Bemmel, M. Kettenis, D. Small, E. Liuzzo, K. Rygl, I. Martí-Vidal, L. Blackburn, M. Wielgus, H. Falcke, rPICARD: a CASA-based calibration pipeline for VLBI data. Calibration and imaging of 7 mm VLBA observations of the AGN jet in M 87. Astron. Astrophys. **626**, A75 (2019b). https://doi.org/10.1051/0004-6361/201935181

M. Janssen, H. Falcke, M. Kadler, E. Ros, M. Wielgus, K. Akiyama, M. Baloković, L. Blackburn, K.L. Bouman, A. Chael, C.-k. Chan, K. Chatterjee, J. Davelaar, P.G. Edwards, C.M. Fromm, J.L. Gómez, C. Goddi, S. Issaoun, M.D. Johnson, J. Kim, J.Y. Koay, T.P. Krichbaum, J. Liu, E. Liuzzo, S. Markoff, A. Markowitz, D.P. Marrone, Y. Mizuno, C. Müller, C. Ni, D.W. Pesce, V. Ramakrishnan, F. Roelofs, K.L.J. Rygl, I. van Bemmel, Event Horizon Telescope Collaboration, A. Alberdi, W. Alef, J.C. Algaba, R. Anantua, K. Asada, R. Azulay, A.-K. Baczko, D. Ball, D. Ball, J. Barrett, B.A. Benson, D. Bintley, D. Bintley, R. Blundell, W. Boland, W. Boland, G.C. Bower, H. Boyce, M. Bremer, C.D. Brinkerink, R. Brissenden, S. Britzen, A.E. Broderick, D. Broguiere, T. Bronzwaer, D.-Y. Byun, J.E. Carlstrom, S. Chatterjee, M.-T. Chen, Y. Chen, P.M. Chesler, I. Cho, P. Christian, J.E. Conway, J.M. Cordes, T.M. Crawford, G.B. Crew, A. Cruz-Osorio, Y. Cui, Y. Cui, M. De Laurentis, R. Deane, J. Dempsey, G. Desvignes, J. Dexter, S.S. Doeleman, R.P. Eatough, J. Farah, J. Farah, V.L. Fish, E. Fomalont, H. Alyson Ford, R. Fraga-Encinas, P. Friberg, P. Friberg, A. Fuentes, P. Galison, C.F. Gammie, R. García, Z. Gelles, O. Gentaz, B. Georgiev, B. Georgiev, R. Gold, R. Gold, A.I. Gómez-Ruiz, M. Gu, M. Gurwell, K. Hada, D. Haggard, M.H. Hecht, R. Hesper, E. Himwich, L.C. Ho, P. Ho, M. Honma, C.-W.L. Huang, L. Huang, D.H. Hughes, S. Ikeda, M. Inoue, M. Inoue, D.J. James, B.T. Jannuzi, B.T. Jannuzi, B. Jeter, W. Jiang, A. Jimenez-Rosales, A. Jimenez-Rosales, S. Jorstad, T. Jung, M. Karami, R. Karuppusamy, T. Kawashima, G.K. Keating, M. Kettenis, D.-J. Kim, J.-Y. Kim, J.-Y. Kim, J. Kim, M. Kino, M. Kino, Y. Kofuji, S. Koyama, M. Kramer, C. Kramer, C. Kramer, C.-Y. Kuo, T.R. Lauer, S.-S. Lee, A. Levis, Y.-R. Li, Z. Li, M. Lindqvist, R. Lico, G. Lindahl, G. Lindahl, K. Liu, K. Liu, W.-P. Lo, A.P. Lobanov, L. Loinard, C. Lonsdale, R.-S. Lu, N.R. MacDonald, J. Mao, N. Marchili, N. Marchili, N. Marchili, A.P. Marscher, I. Martí-Vidal, S. Matsushita, L.D. Matthews, L. Medeiros, K.M. Menten, I. Mizuno, I. Mizuno, J.M. Moran, K. Moriyama, M. Moscibrodzka, M. Moscibrodzka, G. Musoke, A.M. Mejías, H. Nagai, N.M. Nagar, M. Nakamura, R. Narayan, G. Narayanan, I. Natarajan, A. Nathanail, J. Neilsen, R. Neri, R. Neri, A. Noutsos, M.A. Nowak, H. Okino, H. Olivares, G.N. Ortiz-León, T. Oyama, F. Özel, D.C.M. Palumbo, J. Park, N. Patel, U.-L. Pen, U.-L. Pen, V. Piétu, R. Plambeck, A. PopStefanija, O. Porth, F.M. Pötzl, B. Prather, J.A. Preciado-López, D. Psaltis, H.-Y. Pu, H.-Y. Pu, R. Rao, M.G. Rawlings, A.W. Raymond, L. Rezzolla, A. Ricarte, B. Ripperda, B. Ripperda, A. Rogers, A. Rogers, M. Rose, A. Roshanineshat, H. Rottmann, A.L. Roy, C. Ruszczyk, C. Ruszczyk, S. Sánchez, D. Sánchez-Arguelles, M. Sasada, T. Savolainen, F.P. Schloerb, K.-F. Schuster, L. Shao, Z. Shen, D. Small, B.W. Sohn, J. SooHoo, H. Sun, F. Tazaki, A.J. Tetarenko, P. Tiede, R.P.J. Tilanus, M. Titus, P. Torne, T. Trent, E. Traianou, S. Trippe, I. van Bemmel, H.J. van Langevelde, D.R. van Rossum, J. Wagner, D. Ward-Thompson, J. Wardle, J. Weintroub, N. Wex, R. Wharton, R. Wharton,

G.N. Wong, Q. Wu, D. Yoon, A. Young, K. Young, Z. Younsi, F. Yuan, Y.-F. Yuan, J. Anton Zensus, G.-Y. Zhao, S.-S. Zhao. Event horizon telescope observations of the jet launching and collimation in Centaurus A. Nat. Astron. **5**, 1017–1028 (2021). https://doi.org/10.1038/s41550-021-01417-w

P. Jiang, N.-Y. Tang, L.-G. Hou, M.-T. Liu, M. Krčo, L. Qian, J.-H. Sun, T.-C. Ching, B. Liu, Y. Duan, Y.-L. Yue, H.-Q. Gan, R. Yao, H. Li, G.-F. Pan, D.-J. Yu, H.-F. Liu, D. Li, B. Peng, J. Yan, FAST Collaboration, The fundamental performance of FAST with 19-beam receiver at L band. Res. Astron. Astrophys. **20**(5), 064 (2020). https://doi.org/10.1088/1674-4527/20/5/64

J.-Y. Kim, T.P. Krichbaum, A.E. Broderick, M. Wielgus, L. Blackburn, J.L. Gómez, M.D. Johnson, K.L. Bouman, A. Chael, K. Akiyama, S. Jorstad, A.P. Marscher, S. Issaoun, M. Janssen, C.-k. Chan, T. Savolainen, D.W. Pesce, F. Özel, A. Alberdi, W. Alef, K. Asada, R. Azulay, A.-K. Baczko, D. Ball, M. Baloković, J. Barrett, D. Bintley, W. Boland, G.C. Bower, M. Bremer, C.D. Brinkerink, R. Brissenden, S. Britzen, D. Broguiere, T. Bronzwaer, D.-Y. Byun, J.E. Carlstrom, S. Chatterjee, K. Chatterjee, M.-T. Chen, Y. Chen, I. Cho, P. Christian, J.E. Conway, J.M. Cordes, G.B. Crew, Y. Cui, J. Davelaar, M. De Laurentis, R. Deane, J. Dempsey, G. Desvignes, J. Dexter, S.S. Doeleman, R.P. Eatough, H. Falcke, V.L. Fish, E. Fomalont, R. Fraga-Encinas, P. Friberg, C.M. Fromm, P. Galison, C.F. Gammie, R. García, O. Gentaz, B. Georgiev, C. Goddi, R. Gold, A.I. Gómez-Ruiz, M. Gu, M. Gurwell, K. Hada, M.H. Hecht, R. Hesper, L.C. Ho, P. Ho, M. Honma, C.-W.L. Huang, L. Huang, D.H. Hughes, S. Ikeda, M. Inoue, D.J. James, B.T. Jannuzi, B. Jeter, W. Jiang, A. Jimenez-Rosales, T. Jung, M. Karami, R. Karuppusamy, T. Kawashima, G.K. Keating, M. Kettenis, J. Kim, J. Kim, M. Kino, J.Y. Koay, P.M. Koch, S. Koyama, M. Kramer, C. Kramer, C.-Y. Kuo, T.R. Lauer, S.-S. Lee, Y.-R. Li, Z. Li, M. Lindqvist, R. Lico, K. Liu, E. Liuzzo, W.-P. Lo, A.P. Lobanov, L. Loinard, C. Lonsdale, R.-S. Lu, N.R. MacDonald, J. Mao, S. Markoff, D.P. Marrone, I. Martí-Vidal, S. Matsushita, L.D. Matthews, L. Medeiros, K.M. Menten, Y. Mizuno, I. Mizuno, J.M. Moran, K. Moriyama, M. Moscibrodzka, G. Musoke, C. Müller, H. Nagai, N.M. Nagar, M. Nakamura, R. Narayan, G. Narayanan, I. Natarajan, R. Neri, C. Ni, A. Noutsos, H. Okino, H. Olivares, G.N. Ortiz-León, T. Oyama, D.C.M. Palumbo, J. Park, N. Patel, U.-L. Pen, V. Piétu, R. Plambeck, A. PopStefanija, O. Porth, B. Prather, J.A. Preciado-López, D. Psaltis, H.-Y. Pu, V. Ramakrishnan, R. Rao, M.G. Rawlings, A.W. Raymond, L. Rezzolla, B. Ripperda, F. Roelofs, A. Rogers, E. Ros, M. Rose, A. Roshanineshat, H. Rottmann, A.L. Roy, C. Ruszczyk, B.R. Ryan, K.L.J. Rygl, S. Sánchez, D. Sánchez-Arguelles, M. Sasada, F.P. Schloerb, K.-F. Schuster, L. Shao, Z. Shen, D. Small, B.W. Sohn, J. SooHoo, F. Tazaki, P. Tiede, R.P.J. Tilanus, M. Titus, K. Toma, P. Torne, T. Trent, E. Traianou, S. Trippe, S. Tsuda, I. van Bemmel, H.J. van Langevelde, D.R. van Rossum, J. Wagner, J. Wardle, D. Ward-Thompson, J. Weintroub, N. Wex, R. Wharton, G.N. Wong, Q. Wu, D. Yoon, A. Young, K. Young, Z. Younsi, F. Yuan, Y.-F. Yuan, J.A. Zensus, G. Zhao, S.-S. Zhao, Z. Zhu, J.-C. Algaba, A. Allardi, R. Amestica, J. Anczarski, U. Bach, F.K. Baganoff, C. Beaudoin, B.A. Benson, R. Berthold, J.M. Blanchard, R. Blundell, S. Bustamente, R. Cappallo, E. Castillo-Domínguez, C.-C. Chang, S.-H. Chang, S.-C. Chang, C.-C. Chen, R. Chilson, T.C. Chuter, R.C. Rosado, I.M. Coulson, J. Crowley, M. Derome, M. Dexter, S. Dornbusch, K.A. Dudevoir, S.A. Dzib, A. Eckart, C. Eckert, N.R. Erickson, W.B. Everett, A. Faber, J.R. Farah, V. Fath, T.W. Folkers, D.C. Forbes, R. Freund, D.M. Gale, F. Gao, G. Geertsema, D.A. Graham, C.H. Greer, R. Grosslein, F. Gueth, D. Haggard, N.W. Halverson, C.-C. Han, K.-C. Han, J. Hao, Y. Hasegawa, J.W. Henning, A. Hernández-Gómez, R. Herrero-Illana, S. Heyminck, A. Hirota, J. Hoge, Y.-D. Huang, C.M. Violette Impellizzeri, H. Jiang, D. John, A. Kamble, R. Keisler, K. Kimura, Y. Kono, D. Kubo, J. Kuroda, R. Lacasse, R.A. Laing, E.M. Leitch, C.-T. Li, L.C.C. Lin, C.-T. Liu, K.-Y. Liu, L.-M. Lu, R.G. Marson, P.L. Martin-Cocher, K.D. Massingill, C. Matulonis, M.P. McColl, S.R. McWhirter, H. Messias, Z. Meyer-Zhao, D. Michalik, A. Montaña, W. Montgomerie, M. Mora-Klein, D. Muders, A. Nadolski, S. Navarro, J. Neilsen, C.H. Nguyen, H. Nishioka, T. Norton, M.A. Nowak, G. Nystrom, H. Ogawa, P. Oshiro, T. Oyama, H. Parsons, J. Peñalver, N.M. Phillips, M. Poirier, N. Pradel, R.A. Primiani, P.A. Raffin, A.S. Rahlin, G. Reiland, C. Risacher, I. Ruiz, A.F. Sáez-Madaín, R. Sassella, P. Schellart, P. Shaw, K.M. Silva, H. Shiokawa, D.R. Smith, W. Snow, K. Souccar, D. Sousa, T.K. Sridharan, R. Srinivasan, W. Stahm, A.A. Stark, K. Story, S.T.

Timmer, L. Vertatschitsch, C. Walther, T.-S. Wei, N. Whitehorn, A.R. Whitney, D.P. Woody, J.G.A. Wouterloot, M. Wright, P. Yamaguchi, C.-Y. Yu, M. Zeballos, S. Zhang, L. Ziurys, Event Horizon Telescope Collaboration, Event Horizon Telescope imaging of the archetypal Blazar 3C 279 at an extreme 20 microarcsecond resolution. Astron. Astrophys. **640**, A69 (2020). https://doi.org/10.1051/0004-6361/202037493

D. Li, P. Wang, L. Qian, M. Krco, P. Jiang, Y. Yue, C. Jin, Y. Zhu, Z. Pan, R. Nan, A. Dunning, FAST in Space: considerations for a multibeam, multipurpose survey using China's 500-m aperture spherical radio telescope (FAST). IEEE Microwave Mag. **19**(3), 112–119 (2018). https://doi.org/10.1109/MMM.2018.2802178

D.-J. Liu, Y. Xu, Y.-J. Li, Z.-H. Lin, S.-B. Bian, C.-J. Hao, A baseline correction algorithm for FAST. Res. Astron. Astrophys. **22**(8), 081001 (2022). https://doi.org/10.1088/1674-4527/ac7d43

I. Martí-Vidal, A. Roy, J. Conway, A.J. Zensus, Calibration of mixed-polarization interferometric observations. Tools for the reduction of interferometric data from elements with linear and circular polarization receivers. Astron. Astrophys. **587**, A143 (2016). https://doi.org/10.1051/0004-6361/201526063

A.P. Mechev, Orchestration of Distributed LOFAR Workflows. PhD thesis, University of Leiden, Leiden (December 2019)

L.K. Morabito, N.J. Jackson, S. Mooney, F. Sweijen, S. Badole, P. Kukreti, D. Venkattu, C. Groeneveld, A. Kappes, E. Bonnassieux, A. Drabent, M. Iacobelli, J.H. Croston, P.N. Best, M. Bondi, J.R. Callingham, J.E. Conway, A.T. Deller, M.J. Hardcastle, J.P. McKean, G.K. Miley, J. Moldon, H.J.A. Röttgering, C. Tasse, T.W. Shimwell, R.J. van Weeren, J.M. Anderson, A. Asgekar, I.M. Avruch, I.M. van Bemmel, M.J. Bentum, A. Bonafede, W.N. Brouw, H.R. Butcher, B. Ciardi, A. Corstanje, A. Coolen, S. Damstra, F. de Gasperin, S. Duscha, J. Eislöffel, D. Engels, H. Falcke, M.A. Garrett, J. Griessmeier, A.W. Gunst, M.P. van Haarlem, M. Hoeft, A.J. van der Horst, E. Jütte, M. Kadler, L.V.E. Koopmans, A. Krankowski, G. Mann, A. Nelles, J.B.R. Oonk, E. Orru, H. Paas, V.N. Pandey, R.F. Pizzo, M. Pandey-Pommier, W. Reich, H. Rothkaehl, M. Ruiter, D.J. Schwarz, A. Shulevski, M. Soida, M. Tagger, C. Vocks, R.A.M.J. Wijers, S.J. Wijnholds, O. Wucknitz, P. Zarka, P. Zucca, Sub-arcsecond imaging with the International LOFAR Telescope. I. Foundational calibration strategy and pipeline. Astron. Astrophys. **658**, A1 (2022). https://doi.org/10.1051/0004-6361/202140649

R. Nan, D. Li, C. Jin, Q. Wang, L. Zhu, W. Zhu, H. Zhang, Y. Yue, L. Qian, The five-hundred aperture spherical radio telescope (fast) project. Int. J. Mod. Phys. D **20**(6), 989–1024 (2011). https://doi.org/10.1142/S0218271811019335

C.H. Niu, K. Aggarwal, D. Li, X. Zhang, S. Chatterjee, C.W. Tsai, W. Yu, C.J. Law, S. Burke-Spolaor, J.M. Cordes, Y.K. Zhang, S.K. Ocker, J.M. Yao, P. Wan, Y. Feng, Y. Niino, C. Bochenek, M. Cruces, L. Connor, J.A. Jiang, S. Dai, R. Luo, G.D. Li, C.C. Miao, J.R. Niu, R. Anna-Thomas, J. Sydnor, D. Stern, W.Y. Wang, M. Yuan, Y.L. Yue, D.J. Zhou, Z. Yan, W.W. Zhu, B. Zhang, A repeating fast radio burst associated with a persistent radio source. Nature **606**(7916), 873–877 (2022). https://doi.org/10.1038/s41586-022-04755-5

A.R. Offringa, Compression of interferometric radio-astronomical data. Astron. Astrophys. **595**, A99 (2016). https://doi.org/10.1051/0004-6361/201629565

A.R. Offringa, O. Smirnov, An optimized algorithm for multiscale wideband deconvolution of radio astronomical images. Mon. Not. R. Astron. Soc. **471**(1), 301–316 (2017). https://doi.org/10.1093/mnras/stx1547

A.R. Offringa, J.J. van de Gronde, J.B.T.M. Roerdink, A morphological algorithm for improved radio-frequency interference detection. Astron. Astrophys. **539**, 10 pp. (2012)

A.R. Offringa, B. McKinley, Hurley-Walker et al., WSClean: an implementation of a fast, generic wide-field imager for radio astronomy. Mon. Not. R. Astron. Soc. **444**(1), 606–619 (2014). https://doi.org/10.1093/mnras/stu1368

F. Roelofs, M. Janssen, I. Natarajan, R. Deane, J. Davelaar, H. Olivares, O. Porth, S.N. Paine, K.L. Bouman, R.P.J. Tilanus, I.M. van Bemmel, H. Falcke, K. Akiyama, A. Alberdi, W. Alef, K. Asada, R. Azulay, A. Baczko, D. Ball, M. Baloković, J. Barrett, D. Bintley, L. Blackburn,

W. Boland, G.C. Bower, M. Bremer, C.D. Brinkerink, R. Brissenden, S. Britzen, A.E. Broderick, D. Broguiere, T. Bronzwaer, D. Byun, J.E. Carlstrom, A. Chael, C. Chan, S. Chatterjee, K. Chatterjee, M. Chen, Y. Chen, I. Cho, P. Christian, J.E. Conway, J.M. Cordes, G.B. Crew, Y. Cui, M. De Laurentis, J. Dempsey, G. Desvignes, J. Dexter, S.S. Doeleman, R.P. Eatough, V.L. Fish, E. Fomalont, R. Fraga-Encinas, P. Friberg, C.M. Fromm, J.L. Gómez, P. Galison, C.F. Gammie, R. García, O. Gentaz, B. Georgiev, C. Goddi, R. Gold, M. Gu, M. Gurwell, K. Hada, M.H. Hecht, R. Hesper, L.C. Ho, P. Ho, M. Honma, C.L. Huang, L. Huang, D.H. Hughes, S. Ikeda, M. Inoue, S. Issaoun, D.J. James, B.T. Jannuzi, B. Jeter, W. Jiang, M.D. Johnson, S. Jorstad, T. Jung, M. Karami, R. Karuppusamy, T. Kawashima, G.K. Keating, M. Kettenis, J. Kim, J. Kim, J. Kim, M. Kino, J. Yi Koay, P.M. Koch, S. Koyama, M. Kramer, C. Kramer, T.P. Krichbaum, C. Kuo, T.R. Lauer, S. Lee, Y. Li, Z. Li, M. Lindqvist, R. Lico, K. Liu, E. Liuzzo, W. Lo, A.P. Lobanov, L. Loinard, C. Lonsdale, R. Lu, N.R. MacDonald, J. Mao, S. Markoff, D.P. Marrone, A.P. Marscher, I. Martí-Vidal, S. Matsushita, L.D. Matthews, L. Medeiros, K.M. Menten, Y. Mizuno, I. Mizuno, J.M. Moran, K. Moriyama, M. Moscibrodzka, C. Müller, H. Nagai, N.M. Nagar, M. Nakamura, R. Narayan, G. Narayanan, R. Neri, C. Ni, A. Noutsos, H. Okino, G.N. Ortiz-León, T. Oyama, F. Özel, D.C.M. Palumbo, N. Patel, U. Pen, D.W. Pesce, V. Piétu, R. Plambeck, A. PopStefanija, B. Prather, J.A. Preciado-López, D. Psaltis, H. Pu, V. Ramakrishnan, R. Rao, M.G. Rawlings, A.W. Raymond, L. Rezzolla, B. Ripperda, A. Rogers, E. Ros, M. Rose, A. Roshanineshat, H. Rottmann, A.L. Roy, C. Ruszczyk, B.R. Ryan, K.L.J. Rygl, S. Sánchez, D. Sánchez-Arguelles, M. Sasada, T. Savolainen, F. Peter Schloerb, K. Schuster, L. Shao, Z. Shen, D. Small, B. Won Sohn, J. SooHoo, F. Tazaki, P. Tiede, M. Titus, K. Toma, P. Torne, T. Trent, S. Trippe, S. Tsuda, H.J. van Langevelde, D.R. van Rossum, J. Wagner, J. Wardle, D. Ward-Thompson, J. Weintroub, N. Wex, R. Wharton, M. Wielgus, G.N. Wong, Q. Wu, A. Young, K. Young, Z. Younsi, F. Yuan, Y. Yuan, J.A. Zensus, G. Zhao, S. Zhao, Z. Zhu, SYMBA: an end-to-end VLBI synthetic data generation pipeline. Astron. Astrophys. **636**, A5 (2020)

N. Scoville, R. G. Abraham, H. Aussel, J. E. Barnes, A. Benson, A. W. Blain, D. Calzetti, A. Comastri, P. Capak, C. Carilli, J. E. Carlstrom, C. M. Carollo, J. Colbert, E. Daddi, R. S. Ellis, M. Elvis, S. P. Ewald, M. Fall, A. Franceschini, M. Giavalisco, W. Green, R. E. Griffiths, L. Guzzo, G. Hasinger, C. Impey, J.-P. Kneib, J. Koda, A. Koekemoer, O. Lefevre, S. Lilly, C. T. Liu, H. J. McCracken, R. Massey, Y. Mellier, S. Miyazaki, B. Mobasher, J. Mould, C. Norman, A. Refregier, A. Renzini, J. Rhodes, M. Rich, D. B. Sanders, D. Schiminovich, E. Schinnerer, M. Scodeggio, K. Sheth, P. L. Shopbell, Y. Taniguchi, N. D. Tyson, C. M. Urry, L. Van Waerbeke, P. Vettolani, S. D. M. White, L. Yan, COSMOS: Hubble Space Telescope Observations. Astrophys. J. Suppl. Ser. **172**(1), 38–45 (2007). https://doi.org/10.1086/516580

V. Smolčić, M. Novak, M. Bondi, P. Ciliegi, K. P. Mooley, E. Schinnerer, G. Zamorani, F. Navarrete, S. Bourke, A. Karim, E. Vardoulaki, S. Leslie, J. Delhaize, C. L. Carilli, S. T. Myers, N. Baran, I. Delvecchio, O. Miettinen, J. Banfield, M. Baloković, F. Bertoldi, P. Capak, D. A. Frail, G. Hallinan, H. Hao, N. Herrera Ruiz, A. Horesh, O. Ilbert, H. Intema, V. Jelić, H.-R. Klöckner, J. Krpan, S. R. Kulkarni, H. McCracken, C. Laigle, E. Middleberg, E. J. Murphy, M. Sargent, N. Z. Scoville, K. Sheth, The VLA-COSMOS 3 GHz Large Project: Continuum data and source catalog release. American Acad. Pediatr. **602**, A1 (2017). https://doi.org/10.1051/0004-6361/201628704

C. Tasse, B. Hugo, M. Mirmont, O. Smirnov, M. Atemkeng, L. Bester, M.J. Hardcastle, R. Lakhoo, S. Perkins, T. Shimwell, Faceting for direction-dependent spectral deconvolution. Astron. Astrophys. **611**, A87 (2018). https://doi.org/10.1051/0004-6361/201731474

C. Tasse, T. Shimwell, M.J. Hardcastle, S.P. O'Sullivan, R. van Weeren, P.N. Best, L. Bester, B. Hugo, O. Smirnov, J. Sabater, G. Calistro-Rivera, F. de Gasperin, L.K. Morabito, H. Röttgering, W.L. Williams, M. Bonato, M. Bondi, A. Botteon, M. Brüggen, G. Brunetti, K.T. Chyży, M.A. Garrett, G. Gürkan, M.J. Jarvis, R. Kondapally, S. Mandal, I. Prandoni, A. Repetti, E. Retana-Montenegro, D.J. Schwarz, A. Shulevski, Y. Wiaux, The LOFAR Two-meter Sky Survey: deep fields data release 1. I. Direction-dependent calibration and imaging. Astron. Astrophys. **648**, A1 (2021). https://doi.org/10.1051/0004-6361/202038804

The CASA Team, B. Bean, S. Bhatnagar, S. Castro, J. Donovan Meyer, B. Emonts, E. Garcia, R. Garwood, K. Golap, J. Gonzalez Villalba, P. Harris, Y. Hayashi, J. Hoskins, M. Hsieh, P. Jagannathan, W. Kawasaki, A. Keimpema, M. Kettenis, J. Lopez, J. Marvil, J. Masters, A. McNichols, D. Mehringer, R. Miel, G. Moellenbrock, F. Montesino, T. Nakazato, J. Ott, D. Petry, M. Pokorny, R. Raba, U. Rau, D. Schiebel, N. Schweighart, S. Sekhar, K. Shimada, D. Small, J.-W. Steeb, K. Sugimoto, V. Suoranta, T. Tsutsumi, I.M. van Bemmel, M. Verkouter, A. Wells, W. Xiong, A. Szomoru, M. Griffith, B. Glendenning, J. Kern, CASA, the Common Astronomy Software Applications for Radio Astronomy. arXiv e-prints, art. arXiv:2210.02276 (October 2022)

A.R. Thompson, J.M. Moran, G.W. Swenson Jr., *Interferometry and Synthesis in Radio Astronomy*, 3rd edn. (2017). https://doi.org/10.1007/978-3-319-44431-4

I.M. van Bemmel, M. Kettenis, D. Small, M. Janssen, G.A. Moellenbrock, D. Petry, C. Goddi, J.D. Linford, K.L.J. Rygl, E. Liuzzo, B. Marcote, O.S. Bayandina, N. Schweigart, M. Verkouter, A. Keimpema, A. Szomoru, H.J. van Langevelde, CASA on the fringe – development of VLBI processing capabilities for CASA. arXiv e-prints, art. arXiv:2210.02275 (October 2022)

S. van der Tol, B. Veenboer, A.R. Offringa, Image domain gridding: a fast method for convolutional resampling of visibilities. Astron. Astrophys. **616**, A27 (2018). https://doi.org/10.1051/0004-6361/201832858

G. van Diepen, T.J. Dijkema, A. Offringa, DPPP: Default Pre-Processing Pipeline. Astrophysics Source Code Library, record ascl:1804.003 (April 2018)

R.J. van Weeren, W.L. Williams, M.J. Hardcastle, T.W. Shimwell, D.A. Rafferty, J. Sabater, G. Heald, S.S. Sridhar, T.J. Dijkema, G. Brunetti, M. Brüggen, F. Andrade-Santos, G.A. Ogrean, H.J.A. Röttgering, W.A. Dawson, W.R. Forman, F. de Gasperin, C. Jones, G.K. Miley, L. Rudnick, C.L. Sarazin, A. Bonafede, P.N. Best, L. Bîrzan, R. Cassano, K.T. Chyży, J.H. Croston, T. Ensslin, C. Ferrari, M. Hoeft, C. Horellou, M.J. Jarvis, R.P. Kraft, M. Mevius, H.T. Intema, S.S. Murray, E. Orrú, R. Pizzo, A. Simionescu, A. Stroe, S. van der Tol, G.J. White, LOFAR facet calibration. Astrophys. J. Suppl. **223**(1), 2 (2016). https://doi.org/10.3847/0067-0049/223/1/2

L. Wang, B. Peng, B.W. Stappers, K. Liu, M.J. Keith, A.G. Lyne, J. Lu, Y.-Z. Yu, F. Kou, J. Yan, P. Jiang, C. Jin, D. Li, Q. Li, L. Qian, Q. Wang, Y. Yue, H. Zhang, S. Zhang, Y. Zhu, FAST Collaboration, Discovery and timing of pulsars in the globular cluster M13 with FAST. Astrophys. J. **892**(1), 43 (2020). https://doi.org/10.3847/1538-4357/ab76cc

Y. Wang, Z. Zhang, H. Zhang, W. Zhu, D. Li, P. Wang, Description of a RFI mitigation pipeline for the FAST 19-beam receiver. Astron. Comput. **39**, 100568 (2022). https://doi.org/10.1016/j.ascom.2022.100568

A.R. Whitney, R. Cappallo, W. Aldrich, B. Anderson, A. Bos, J. Casse, J. Goodman, S. Parsley, S. Pogrebenko, R. Schilizzi, D. Smythe, Mark 4 vlbi correlator: architecture and algorithms. Radio Sci. **39**(1), RS1007 (2004). https://doi.org/10.1029/2002RS002820

A.R. Whitney, C.J. Beaudoin, R.J. Cappallo, B.E. Corey, G.B. Crew, S.S. Doeleman, D.E. Lapsley, A.A. Hinton, S.R. McWhirter, A.E. Niell, A.E.E. Rogers, C.A. Ruszczyk, D.L. Smythe, J. SooHoo, M.A. Titus, Demonstration of a 16 Gbps station^{-1} broadband-RF VLBI system. Publ. Astron. Soc. Pac. **125**, 196 (2013). https://doi.org/10.1086/669718

W.L. Williams, R.J. van Weeren, H.J.A. Röttgering, P. Best, T.J. Dijkema, F. de Gasperin, M.J. Hardcastle, G. Heald, I. Prandoni, J. Sabater, T.W. Shimwell, C. Tasse, I.M. van Bemmel, M. Brüggen, G. Brunetti, J.E. Conway, T. Enßlin, D. Engels, H. Falcke, C. Ferrari, M. Haverkorn, N. Jackson, M.J. Jarvis, A.D. Kapińska, E.K. Mahony, G.K. Miley, L.K. Morabito, R. Morganti, E. Orrú, E. Retana-Montenegro, S.S. Sridhar, M.C. Toribio, G.J. White, M.W. Wise, J.T.L. Zwart, LOFAR 150-MHz observations of the Boötes field: catalogue and source counts. Mon. Not. R. Astron. Soc. **460**(3), 2385–2412 (2016). https://doi.org/10.1093/mnras/stw1056

H. Xi, B. Peng, L. Staveley-Smith, B.-Q. For, B. Liu, The FAST Ultra-Deep Survey (FUDS): observational strategy, calibration and data reduction. Publ. Astron. Soc. Aust. **39**, e019 (2022). https://doi.org/10.1017/pasa.2022.16

C.K. Xu, C. Cheng, P.N. Appleton, P.A. Duc, Y. Gao, N.Y. Tang, M. Yun, Y.S. Dai, J.S. Huang, U. Lisenfeld, F. Renaud, A 0.6 Mpc HI structure associated with Stephan's Quintet. arXiv e-prints, art. arXiv:2208.04870 (August 2022)

C.-P. Zhang, J.-L. Xu, J. Wang, Y. Jing, Z. Liu, M. Zhu, P. Jiang, Radio frequency interference mitigation and statistics in the spectral observations of FAST. Res. Astron. Astrophys. **22**(2), 025015 (2022). https://doi.org/10.1088/1674-4527/ac3f2d

Chapter 7
Co-design and Software Architecture

Claudio Gheller, Manolis Marazakis, Estela Suarez, and Giuliano Taffoni

7.1 Co-design Motivations and Challenges

During the next decade we expect novel radio astronomy instruments to come online and start producing datasets of high volume and complexity at a rate that will be orders of magnitude above any existing radio telescope. This development necessitates the adoption of new data processing approaches that take advantage of latest-generation computing facilities and analysis algorithms. In this chapter, we consider associated challenges from the viewpoint of both software architecture and scientific data processing workflow. We advocate taking a co-design approach, whereby hardware and software are developed and validated in close coordination, to ensure that HPC-based workflows meet performance, cost, and reliability goals as necessitated by the overwhelming data flow expected by different observational experiments. We consider software stack development following on the principle of separation of concerns, and with reproducibility and performance portability as

C. Gheller (✉)
Institute of Radioastronomy, INAF, Bologna, Italy
e-mail: claudio.gheller@inaf.it

M. Marazakis
Institute of Computer Science, Foundation for Research and Technology - Hellas (FORTH), Heraklion, Crete, Greece
e-mail: maraz@ics.forth.gr

E. Suarez
Jülich Supercomputing Centre (JSC), Forschungszentrum Jülich, Jülich, Germany

Computer Science Department, University of Bonn, Bonn, Germany
e-mail: e.suarez@fz-juelich.de

G. Taffoni
INAF - Astronomical Observatory of Trieste, Trieste, Italy
e-mail: giuliano.taffoni@inaf.it

E. Vardoulaki et al. (eds.), *Data-Intensive Radio Astronomy*, Astrophysics and Space Science Library 472, https://doi.org/10.1007/978-3-031-58468-8_7

primary design goals, for systems built on top of heterogeneous hardware platforms. Finally, motivated by the particular case of the LOFAR radio interferometer, we present a study on how to exploit large-scale parallel architectures for efficiently solving the gridding processing step.

The last decade has seen the advent of a multitude of radio astronomy instruments, in particular the SKA precursors and pathfinders, e.g., LOFAR (Sect. 6.7) or MeerKAT (Sect. 6.8), which produce terabytes of data, often with tens of measured parameters associated to each observed object. This is due to the combined effects of extreme high sensitivity and wider fields that led radio astronomy to the cusp of a big-data revolution.

Moreover, during the next decade, new highly complex and massively large data sets are expected by novel and more complex scientific instruments whose rate of data production will be orders of magnitude higher than any existing radio telescope. Handling and exploring these new data volumes, and actually making real scientific discoveries, poses a considerable technical challenge. This requires the adoption of new approaches that overcome the traditional research methods and requires the access to a new generation of computing facilities and computational High-Performance Computing algorithms.

The efficient and effective exploitation of large computing capabilities is fundamental in order to obtain new scientific results, combined with the use of high-performance data processing and visualisation to reduce, analyse and explore large volumes of data. Data processing codes must be able to efficiently use heterogeneous systems that combine general-purpose processors with accelerators, high-speed networks, and composite storage devices that represent the state of the art and future trends in supercomputing resources.

The opportunity of using cutting edge HPC systems is not anymore only a way to reduce the time-to-solution, but the sole approach to process datasets of the expected size and complexity. This has a huge impact both on the software architecture, leading to the re-design and refactoring of large portions of data processing/analysis codes, and on the sophisticated workflows, which combine different pieces of software to perform the various steps of the radio-data processing/analysis pipeline.

The related challenges can be summarised as follows:

- the memory challenge. Huge datasets cannot be loaded in the memory of a single CPU and cannot be handled by a single processor but by distributed memory systems. Distributed computing, based on the adoption of the MPI standard, represents a feasible and effective solution. Data are split among multiple memories and the computing effort is distributed among multiple processors, speeding up the computation and maintaining the time to solution reasonably low. Furthermore parallel I/O solutions (based, e.g. on the MPI I/O library or higher level HDF5 or ADIOS solutions) will be tested, together with the usage of high-throughput, low-latency devices like specialised SSD solutions.
- the acceleration challenge. In order to improve the codes' performance, multi (multi-core CPUs) and many cores (GPUs) architectures have to be exploited, This requires a reimplementation of large fractions of the codes adopting suitable

programming models like OpenMP, P-threads or C++11 for multithreading on the GPU and CUDA, ROCm, OpenCL, OpenACC, OpenMP for the GPU.
- the data challenge. This addresses the management, archiving and access of the raw data, the science data products, and the final outcomes of data processing and analysis. It requires capable and reliable long term storage facilities and fast and efficient on-line storage integrated into an HPC architecture capable of providing the required computing power "in–situ", where the data is stored, avoiding any bottlenecks related to the movement of large data volumes. A science analysis platform has to be deployed to users in order to provide an easy and effective access to the data and the computing services.

To face and overcome those challenges, it is crucial to implement a co-design approach. The hardware and software can be designed together to ensure that the implementation not only functions properly but also meets performance, cost, and reliability goals as necessitated by the overwhelming data flow expected by different experiments.

7.2 Software Trends in the Path to Exascale

As stated in Sect. 4.1, the increasing complexity and heterogeneity of modern HPC systems pose enormous challenges to system software and application developers, especially in terms of efficiency, performance, portability, and long-term code maintainability. A trade-off between portability and performance is unavoidable: achieving the maximum performance on a given system requires hardware-specific optimisations, which are not directly transferable onto a different platform.

The software stack must support an increasing diversity of hardware devices and employ all their newest features, in order to get as close as possible to the theoretical peak performance. At the same time, to safeguard portability one strives to hide the hardware complexity from the users as much as possible, so that the application codes do not need to be fully rewritten to run on any other system. It is preferred to let the middleware deal with the hardware complexity, while keeping the application programming interfaces (API) as unmodified as possible.

Within the software stack an approach based on the separation of concerns is taken, to facilitate manageability and long-term maintenance. Each of the main functionalities is implemented by a specific software package, which interact with the rest via (ideally) standardised interfaces. For example, installation packages make the necessary compilers and libraries available in the system, jobs are submitted via the scheduler, the distribution of the binaries to the different compute nodes is done by the resource manager, and the distribution of data to processes is done message-passing libraries. The combination of all these elements and many more is a multi-layer, modularised software stack, in which its lower layers implement specific hardware adaptations, while the higher layers facing the user are hardware-agnostic and therefore portable between platforms. However, the resulting application performance is less predictable when using such deep software stacks, as it becomes much more difficult to understand where within the numerous layers the performance gains/losses are coming from. A strategy to at least achieve reproducible results across platforms is virtualisation, i.e., packing all necessary software modules and dependencies of a given code into a container, which runs directly on top of the operating system and therefore has minimal external dependencies (Sect. 4.3).

One first step towards improving portability between different processing units (CPUs, GPUs, other accelerators) is ensuring that applications can run on them using the same programming interface. For the last decades, the de-facto standard in HPC has been the combination of OpenMP[1] (for shared memory) and MPI[2] (for distributed memory) programming. Today, they co-exist with vendor specific interfaces such as CUDA[3] or HIP, which achieve maximum performance on the GPUs of the respective vendors. Alternative, platform-agnostic programming models for accelerators exist (e.g. OpenCL,[4] OpenACC[5]) but, as usual, improved portability comes at the cost of some performance loss.

The holy grail of *performance portability* is sought through high-level programming models that abstract the user even further, so that he/she does not need to know what kind of hardware is underneath, or even does not need to explicitly write parallel code. Examples of portability libraries are Kokkos,[6] Raja,[7] and DaCE (Ziogas et al. 2021), which are built on top of standard programming languages (C++ or Python). It is worth noting that users increasingly use machine and deep learning as part of their applications, so that frameworks such as TensorFlow (Abadi et al. 2015) and PyTorch[8] are very widely employed on HPC systems. These are ported to all major compute devices, which guarantees the portability of the codes using them.

However, even if the syntax of programming languages and parallelisation libraries are portable across platforms, still some changes might be needed to run on different kinds of hardware, if one really needs to achieve close to peak performance. For instance, changing process pinning or data layouts might be beneficial when the processor microarchitecture uses different number of cores and/or memory channels. In order to take away this effort from the user, much more *wisdom* needs to be given to the runtime systems. Hope is put on the use of artificial intelligence

---

[1] https://www.openmp.org.

[2] https://www.mpi-forum.org/docs/mpi-4.0/mpi40-report.pdf.

[3] https://developer.nvidia.com/cuda-toolkit.

[4] https://www.khronos.org/opencl.

[5] https://www.openacc.org.

[6] https://kokkos.org/about.

[7] https://github.com/LLNL/RAJA.

[8] https://pytorch.org.

to help the runtime systems in taking the right decisions for each application on any given hardware configuration and specific situation.

7.3 Toward the HPC Enabling of LOFAR Data Processing Pipelines

In the coming years, current and upcoming radio-interferometers, like LOFAR (Cornwell et al. 2008), will produce huge volumes of data, related, in particular, to high resolution low-frequency radio surveys. The LOFAR international baselines could enable mapping the entire northern low-frequency sky at sub-arcsecond resolution.

A number of technical difficulties make this kind of observation an outstanding challenge, such as the limited number of available calibrators and the long observing times required. A further limiting factor is the amount of computational resources, in terms of both storage and computing, necessary to produce such images.

The expected data volume, of the order of tens of Petabytes per year, in fact, is hard to manage with traditional approaches. In addition, data size and complexity requires the adoption of highly computational demanding algorithms for the processing and the analysis. Data has to be stored in dedicated facilities, providing the necessary capacity at the highest performance. Data processing has to be performed local to the data, exploiting available HPC resources.

In modern HPC systems, performance is being achieved through many-core and accelerated computing (based, for instance, on GPUs), and current trends suggest that some form of heterogeneous computing will continue to be prevalent in emerging architectures. Therefore, the ability to fully exploit new heterogeneous and many-core solutions is of paramount importance towards achieving optimal performance. On the other hand, with the increasing size and complexity of observational data, it is of primary importance for scientists to be able to exploit all available hardware in emerging HPC environments. Exploiting these novel hybrid architectures is non-trivial however, due to the challenges presented by diverse hardware computing technologies and the increasing levels of architectural parallelism. New algorithms, together with numerical and computational solutions, are required.

In radio interferometry, data processing is generally performed in various steps (often repeated iteratively). Imaging represents one of the most computational demanding, both in terms of memory request and in terms of computing time, associated to operations like gridding, i.e. the convolutional resampling of the observed data on a computational mesh, or FFT transforming, to move from Fourier to real space and vice-versa. The computational requirements increase further when observations with large fields of view (as typically happens with current radio interferometers and at the lowest frequencies) are considered, since curvature effects cannot be neglected and the problem becomes fully three dimensional, with the introduction of the *w-term* correction (Cornwell et al. 2008; Offringa et al. 2014).

Here, we discuss several solutions toward the exploitation of large parallel architectures to the gridding operation on a Cartesian 3D computational mesh, that can be effectively distributed and efficiently managed across different processing units, potentially leading to a good scalability on large HPC systems.

7.3.1 The Parallel Gridding Algorithm

The gridding algorithm consists in the convolution of the point like observed visibilities on a regular Cartesian mesh:

$$\tilde{V}(u_i, v_j, w_k) = \sum_{m \in \text{measures}} V_m G((u_m, v_m, w_m), (u_i, v_j, w_k)), \tag{7.1}$$

where m is the m-th measurement, (u_m, v_m, w_m) are its coordinates, (u_i, v_j, w_k) is a computational grid point, V_m is the measured visibility and $\tilde{V}$ is the visibility convolved on the mesh. The G kernel is the convolution function. In this work, for simplicity we have adopted a Gaussian function.

7.3.1.1 Data Distribution

Two distinct data structures are involved in the gridding operation. The first is an unstructured dataset storing the (u, v, w) coordinates of the array baselines at each measurement time. Each baseline has a number of associated visibilities, which is determined by the frequency bandwidth, the frequency resolution and the number of polarisations. A further quantity, the weight, is also assigned to each measurement and polarisation. The second, is a Cartesian computational mesh where the convolved visibilities and their FFT transformed counterpart are calculated. The two data structures determine the memory request of the algorithm.

To achieve an effective memory utilisation, the two datasets must be distributed among different processors. The concurrent presence of unstructured point-like data (the observed visibilities), together with regular mesh-based data (the convolved visibilities) requires some care in order to obtain an efficient data distribution.

Point-like visibilities have a spatial number density that varies dramatically depending on the arrangement of the antennas of the interferometer and on the observation time. For instance, areas around the mesh centre are way more dense than peripheral areas. Also radial number density is not uniform, changing with the orientation. On the other hand, in memory there is no correlation between the memory location of a measurement and its physical spatial position, i.e. measurements that are close in memory can be spatially distant (and vice-versa). As a result, their distribution can be considered as almost-random.

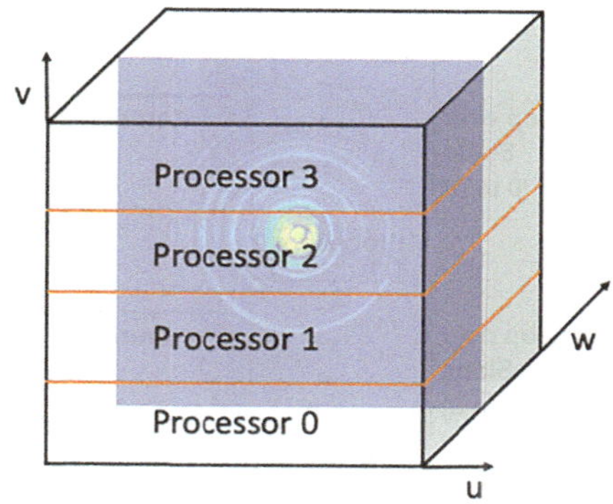

Fig. 7.1 Computational mesh data distribution. Each computing unit owns a rectangular *resident slab*. In this example data are split among four computing units

In this situation, data can be distributed among disjoint memories assigning to each computing unit a consecutive fraction of the measurements. This has several advantages. First and foremost, data distribution is balanced. Each computing unit owns the same number of visibilities. Second, data can be distributed already when they are read from the disk, without communication. This, of course, requires a parallel file system (which we assume to be available in an HPC facility) to be efficient. Third, data in the same area of the computational mesh are distributed among different computing units, reducing the computational pressure and potential race conditions on that specific area. Main drawbacks are represented by the non-locality of the data, that are spread all over the spatial domain, hence the computational mesh, and possible imbalances for short observations.

The data defined on the computational grid have been partitioned among the computing elements adopting a rectangular slab-like decomposition (Fig. 7.1), assigning to each task a rectangular region of $N_u \times N_w \times (N_v/N_s)$ cells, where N_u and N_w are the mesh size in the u and in the w directions respectively. N_v is the mesh size in the v direction, which is split in N_s parts, equal to the number of computing tasks.

Such domain decomposition allows storing on each memory only a fraction of the whole mesh, avoiding replicas. Hence, each different computing element is in charge of storing and managing a specific part of the mesh (hereafter the *resident slab*). The main drawbacks of this approach are that (1) the maximum number of computing units that can be used is equal to the 1D v-size of the mesh; (2) a suitable communication scheme has to be implemented in order to compose the images. Since we deal with rather large $N_v > 1000$ meshes, the former is not expected to be a substantial limiting factor.

7.3.1.2 Parallel Processing

Data distribution implies additional management, of the work performed by multiple computing units and some communication to exchange data needed by remote processors. Furthermore, synchronisation has to be properly enforced in order

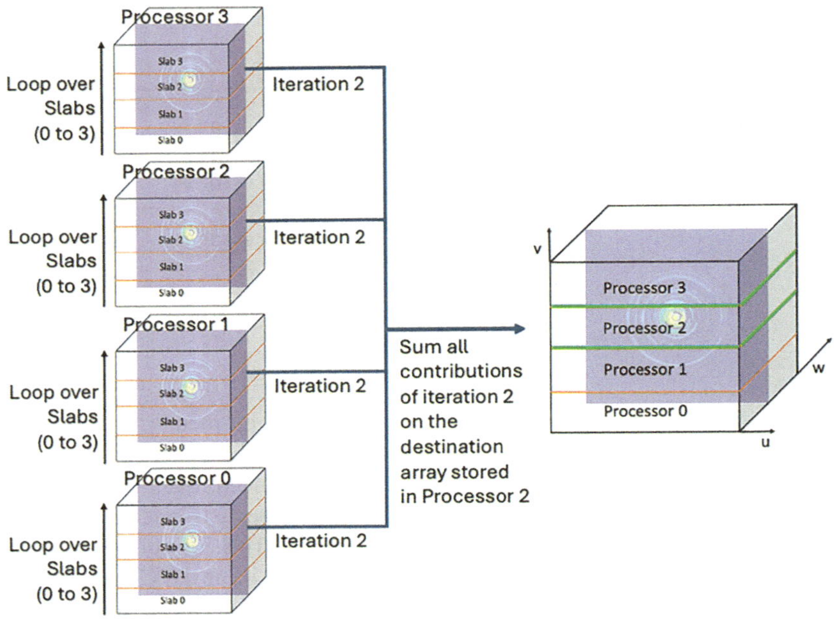

Fig. 7.2 The different computing units (four units in this example) perform the gridding operation on its fraction of visibilities slab by slab. The contribution of each computing unit to a slab is finally *accumulated* on the resident slab

to ensure that communication and work are properly completed at specific steps of code execution, when this is necessary (e.g. before writing results on disk). Communication and synchronisation limit the effectiveness of the parallel execution of an algorithm, i.e. its scalability, which measures the ability to handle more work as the size of the computer or of the data grows. Therefore, they both have to be minimised in order to exploit larger and larger computing resources, supporting bigger data and reducing the time to solution.

The parallel implementation of the gridding algorithm is accomplished by exploiting the Message Passing Interface (MPI, Gropp et al. 1994), the de-facto standard for parallel computing.

As described above, each CPU holds the same fraction of visibility data, each fraction covering the whole computational mesh. Since the mesh cannot be replicated, the following procedure, sketched in Fig. 7.2, is implemented:

- N_s linked lists are created, each collecting the measurements lying on a different slab in which the mesh is divided. Boundary conditions are managed including in the list of a given slab also measurements lying in neighbouring slabs but still contributing to the processed one.
- An auxiliary slab per computing unit of size equal to the resident one is created.

- A N_s iterations loop is run by each computing unit, processing, one after the other, the various slabs. At each iteration, exploiting the corresponding linked list, the measurements lying on the specific slab are processed to calculate the contribution of that computing unit to that slab according to Eq. 7.1.
- At the end of each single iteration the processed slab is sent to the processor that holds the corresponding resident slab and summed to it. This has been implemented adopting two different strategies: (1) using a blocking *MPI_Reduce* function from all the MPI tasks to that owning the resident slab, that implies that all communication has to be completed before the next iteration starts; (2) using a non-blocking one-side *MPI_Accumulate* operation, designed to perform concurrent reduce operations on the memory of a target processor. Synchronisation is imposed only at the end of the loop over slabs, calling a collective *MPI_Win_fence* function.

At the end of the loop each CPU has its own slab of the image up to date, ready for the next step. Notice that only the last step requires MPI communication, all the rest being completely local.

7.3.1.3 Multi/Many-Threads parallelism

For each MPI task, the time to solution of the gridding algorithm can be considerably reduced by accelerating the code exploiting either multi-core CPUs (multithreads parallelism) or by *accelerators* like Graphic Processing Units (GPU, many threads parallelism).

Multithreading for multi-core CPUs has been implemented by using the OpenMP[9] API. For the gridding algorithm, OpenMP is used to parallelize the gridding loop over the visibilities. Loop iterations are independent from one another and they can be trivially distributed among OpenMP threads. When it comes to accumulation of the contributions from different threads, one has to take care about race conditions (i.e. several threads updating the same memory location simultaneously). We then use the OpenMP built-in *atomic* clause.

The approach adopted for the GPU is similar to that described for multithreading. The code implementation adopts two different approaches, the first based on CUDA,[10] the second on the offload constructs of OpenMP.

[9] https://www.openmp.org.

[10] https://developer.nvidia.com/cuda-zone.

Table 7.1 Main characteristics of the LOFAR HBA observation adopted for the tests. Coordinates of the target are given in the first two rows

LOFAR HBA inner station	
Right ascension	15:58:18.96
Declination	+27:29:19.2
Observation time	8 h
Integration time	4 s
N. antennas	62
Bandwidth	120–170 MHz
No. of subbands	25
No. of channels per sub-band	20
Subband bandwidth	2 MHz
MaxUVdist (kλ)	58
MaxUVdist (km)	120
MinUVdist (wavelength)	13.60
MinUVdist (m)	27.82
Field of view	$\sim$12 deg^2
Angular resolution	$\sim$6 arcsec
N. visibilities (approx)	0.54×10^9
Input data size (GB)	4.4
Mesh size	$4096 \times 4096 \times 16$
Mesh size (GB)	4.0

7.3.2 Performance Analysis

In order to analyse the performance and the scalability of the parallel implementation of our gridding algorithm, we have performed a number of tests and benchmarks exploiting the different state-of-the-art computation resources offered by the Marconi 100 (hereafter M100) supercomputer, available at CINECA HPC centre.[11] The systems deploy an IBM Spectrum Scale parallel filesystem.

The code, all written adopting the C standard, has been compiled using the GCC compiler and the available MPI library (Spectrum MPI 10.4.0).

The input dataset used for the tests and the benchmarks comes from a LOFAR HBA 8 h observation. Its main characteristics are presented in Table 7.1. The dataset is a single sub-band at 146 MHz with a bandwidth of 2 MHz (such narrow bandwidth is peculiar of the specific observation, typical HBA observations having a 48 MHz bandwidth) split into 20 channels. It has been used for benchmarking the single node/GPU implementation of the code and for strong scalability tests.

We have first benchmarked our algorithm at the node level, measuring the performance achievable by a code capable of exploiting all the computational resources of the available CPU. The results are presented in Table 7.2.

[11] https://www.hpc.cineca.it/hardware/marconi100.

Table 7.2 Timings for processing the test dataset generating a 4096 × 4096 × 16 pixels image for the different code kernels on different devices (core, GPU, node) of the Marconi 100 system. Errors are calculated as the standard deviation over ten measurements

	Gridding (s)
CORE	81.71 ± 0.03
GPU CUDA	3.25 ± 0.01
GPU OMP	2.64 ± 0.02
NODE (cores)	2.88 ± 0.01
NODE (GPUs CUDA)	1.79 ± 0.06
NODE (GPUs OMP)	0.98 ± 0.02

The time needed to complete the calculation on a single core is between 25 and 80 times slower than that obtained exploiting the full CPU in different configurations. It is therefore critical for the algorithm to exploit some levels of parallelism to exploit at its full capabilities the computing devices (CPU, GPU or both). Using a single GPU, the time to solution is dramatically reduced. The same holds when all the 32 cores of the CPU are used, adopting the OpenMP implementation. The best results, however are obtained when the computing node is exploited at its full power, with all four available GPUs contributing to the calculation. In that case the time to solution is approximately three times faster than using the CPU's cores.

When multiple CPUs are used, strong scalability estimates the performance of the code keeping the data and the mesh size constant and progressively increasing the adopted computational resources. Strong scalability of an algorithm is measured by the *Speed-up* parameter, calculated as:

$$S = \frac{T_1}{T_n} \tag{7.2}$$

where T_n is the wall clock time measured N MPI tasks (one core or one GPU) and T_1 is the corresponding time measured on the same system using a baseline configuration (e.g. a single MPI task). Given the limited size of the input datasets, tests have been performed up to 4 nodes.

Figure 7.3 presents the speed-up of the gridding algorithm using different parallel solutions. In the left panel, the speed-up of the pure MPI implementation is shown, from 1 to 128 MPI tasks (corresponding to the same number of cores). Almost ideal linear scalability is achieved up to the maximum number of tasks, showing that communication overheads are negligible compared to computing. When 128 cores are used, the code runs 128 times faster than on one core. Similar conclusions can be drawn analysing the performance of the OpenMP CPU implementation, presented in the right panel. In this case however, timings can be collected only up to the 16 cores lying on a single CPU socket, the performance degrading substantially when two sockets are used. Multi-node tests cannot be run with the pure OpenMP implementation. The central panel of Fig. 7.3 shows the speed-up of the GPU implementations (CUDA and OpenMP), from 1 to 16 GPUs (4 GPUs per node are

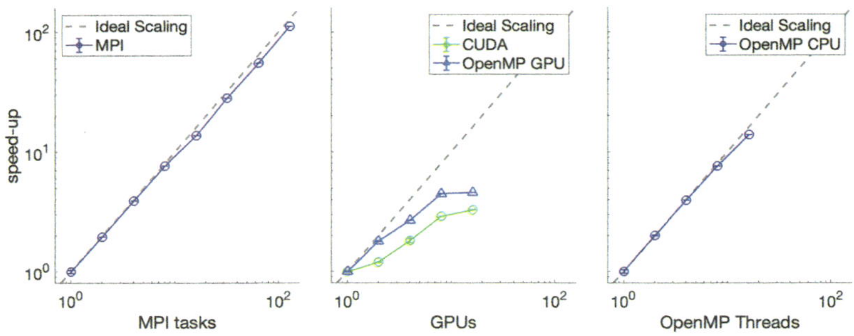

Fig. 7.3 Strong scalability of the gridding algorithm. Pure MPI run, up to 128 tasks, is shown in the left panel, GPU runs, up to 16 GPUs, are shown in the centre panel (with CUDA and OpenMP the corresponding GPU implementations are indicated) and the pure CPU OpenMP run (limited to a single socket, 16 cores) is shown on the right panel. Strong scaling is measured in terms of speed-up (see Eq. 7.2). Ideal linear scaling is shown as a grey, dashed line

available). The GPU version of our algorithm shows worse scalability than the CPU one. This is due to the much faster gridding kernel execution of the GPU, that leads to increase the relative weight of communication, that, although negligible on the CPU, clearly affects the scaling properties of the GPU algorithm. Therefore, using 16 GPUs, the time to solution is only 3–5 times shorter than with a single GPU.

7.3.2.1 Conclusions from the Performance Analysis

The huge data volumes that will be generated in the coming decade by LOFAR and other astronomical observatories requires the development of effective solutions for their efficient processing. The exploitation of state-of-the-art, hybrid HPC architectures is key to this endeavour. We have shown how the combination of parallel distributed computing, multithreading and accelerated computing can be effectively used for gridding radio-interferometry data, leading not only to a substantial reduction of the time to solution, but also to support increasingly larger datasets and images, overcoming the limits imposed by centralised architectures.

References

M. Abadi, A. Agarwal, P. Barham, E. Brevdo, Z. Chen, C. Citro, G.S. Corrado, A. Davis, J. Dean, M. Devin, S. Ghemawat, I. Goodfellow, A. Harp, G. Irving, M. Isard, Y. Jia, R. Jozefowicz, L. Kaiser, M. Kudlur, J. Levenberg, D. Mané, R. Monga, S. Moore, D. Murray, C. Olah, M. Schuster, J. Shlens, B. Steiner, I. Sutskever, K. Talwar, P. Tucker, V. Vanhoucke, V. Vasudevan, F. Viégas, O. Vinyals, P. Warden, M. Wattenberg, M. Wicke, Y. Yu, X. Zheng, TensorFlow: Large-scale machine learning on heterogeneous systems (2015). https://www.tensorflow.org/. Software available from tensorflow.org

T.J. Cornwell, K. Golap, S. Bhatnagar, The noncoplanar baselines effect in radio interferometry: the W-projection algorithm. IEEE J. Sel. Top. Signal Process. **2**(5), 647–657 (2008). https://doi.org/10.1109/JSTSP.2008.2005290

W. Gropp, E. Lusk, A. Skjellum, *Using MPI: Portable Parallel Programming with the Message-Passing Interface* (MIT Press, Cambridge, MA, 1994). ISBN 0-262-57104-8

A.R. Offringa, B. McKinley, N. Hurley-Walker, et al. WSClean: an implementation of a fast, generic wide-field imager for radio astronomy. Mon. Not. R. Astron. Soc. **444**(1), 606–619 (2014). https://doi.org/10.1093/mnras/stu1368

A.N. Ziogas, T. Schneider, T. Ben-Nun, A. Calotoiu, T. De Matteis, J. de Fine Licht, L. Lavarini, T. Hoefler, Productivity, portability, performance: Data-centric python. CoRR, abs/2107.00555 (2021). https://arxiv.org/abs/2107.00555

Part III
Post-processing and Data Analysis

Chapter 8
Continuum Source Identification and Measurement

Andrew M. Hopkins, Yjan Gordon, Martin J. Hardcastle, Daizhong Liu, David A. Rafferty, Michelle M. Boyce, Hongming Tang, Joshua Marvil, Wendy Williams, Biny Sebastian, Christopher O'Dea, Jack Radcliffe, Lawrence Rudnick, and Mattia Vaccari

A. M. Hopkins (✉)
School of Mathematical and Physical Sciences, Macquarie University, Sydney, NSW, Australia
e-mail: andrew.hopkins@mq.edu.au

Y. Gordon
Department of Physics, University of Wisconsin-Madison, Madison, WI, USA
e-mail: yjan.gordon@wisc.edu

M. J. Hardcastle
Department of Physics, Astronomy and Mathematics, University of Hertfordshire, Hatfield, UK
e-mail: m.j.hardcastle@herts.ac.uk

D. Liu
Max-Planck-Institut für extraterrestrische Physik, Garching, Germany

D. A. Rafferty
Hamburger Sternwarte, Universität Hamburg, Hamburg, Germany
e-mail: drafferty@hs.uni-hamburg.de

M. M. Boyce
Department of Physics and Astronomy, University of Manitoba, Winnipeg, MB, Canada
e-mail: michelle.boyce2@umanitoba.ca

H. Tang
Department of Astronomy, Tsinghua University, Beijing, China
e-mail: hongmingt@mail.tsinghua.edu.cn

J. Marvil
National Radio Astronomy Observatory, Socorro, NM, USA
e-mail: jmarvil@nrao.edu

W. Williams
SKA Observatory, Jodrell Bank, Macclesfield, UK
e-mail: wendy.williams@skao.int

B. Sebastian · C. O'Dea
Department of Physics and Astronomy, University of Manitoba, Winnipeg, MB, Canada
e-mail: Biny.Sebastian@umanitoba.ca; odeac@umanitoba.ca

J. Radcliffe
Department of Physics, University of Pretoria, Hatfield, Pretoria, South Africa

E. Vardoulaki et al. (eds.), *Data-Intensive Radio Astronomy*, Astrophysics and Space Science Library 472, https://doi.org/10.1007/978-3-031-58468-8_8

In this chapter we illustrate the scientific information that is commonly extracted from processed radio data, and some of the commonly used tools. As described in Part II, the radio data is first calibrated and then imaged. This results in a 4D image dataset, with the first two dimensions being the spatial axes, the third dimension the frequency axis, and the fourth dimension the polarisation (Stokes; see Sect. 1.2) axis. This 4D data is commonly averaged along the stokes and frequency axes to make a 2D continuum image, if the polarisation or spectral properties of the target astronomical objects are not the primary goal. The natural next step is to identify continuum sources, measure their coordinates and fluxes, and then, where possible, identify their physical nature.

The source detection and measurement stage, even for 2D images, is becoming a progressively data intensive challenge as image sizes continue to grow with the latest generation of radio telescope facilities. Such images may individually be many gigabytes in size, and contain tens of thousands of sources, or more. In large existing, ongoing, or planned surveys with telescopes including the MWA, VLA, LOFAR, MeerKAT, and ASKAP, many hundreds or thousands of such images require analysis, and there is a strong push toward real-time or near-real-time processing to manage the data volume and flow. This will be even more significant in the SKA era.

Astronomical objects in radio data can be quite diverse depending on the scientific goals of each observation, e.g., from pulsars to supernova remnants, planets or planetary systems, through Galactic emission such as molecular clouds, to external galaxies including AGN/QSOs/quasars to radio galaxies and star-forming galaxies (SFGs), and more. Depending on the extent of the object and the angular resolution of the radio data, i.e., the (synthesised) beam or point spread function (PSF), an object can either be resolved so that its spatial extent can be measured in the data, or unresolved so that it appears in the image at the size and shape of the beam (PSF).

Origin of Radio Emission

As introduced in Chap. 1, the continuum emission in the radio observations covering centimetre to submillimetre wavelengths mainly originates from the synchrotron emission from relativistic electrons, and from dust thermal emission, the thermal electron bremsstrahlung emission which is also known as free-free emission (Condon 1992). Radio continuum observations in galaxies can be used to constrain these properties, and indirectly the properties of the ISM and star formation, or the supermassive black holes hosted in galaxy centres. Spatially resolved observations

L. Rudnick
Minnesota Institute for Astrophysics, University of Minnesota, Minneapolis, MN, USA
e-mail: odeac@umanitoba.ca

M. Vaccari
Department of Astronomy, University of Cape Town, Cape Town, South Africa

Institute of Radioastronomy, INAF, Bologna, Italy
e-mail: mattia.vaccari@uct.ac.za

can further reveal the distribution of the ISM and star formation in galaxies or star-forming regions.

Finding Astronomical Objects in 2D Images
The primary task in analysing the continuum data is finding astronomical objects in the 2D image, which are referred to as "sources" This process resembles (although significantly predates) the object detection tasks in modern computer vision (CV), where AI progressively plays a more significant role, as mentioned in other chapters. There are many efforts to apply such approaches to radio astronomy data (Lao et al. 2021), but none are yet mature. This chapter consequently focuses on traditional and current methods of source identification and measurement. These all essentially follow a similar process: identifying emission peaks above some intensity threshold (equivalently, these could also be absorption below some threshold) followed by a measurement or fitting step to obtain the source position, flux and morphological properties. In practice this is complicated by the fact that sources are not all simple or compact, many instead showing extended structure, or being composed of adjacent but separate components (such as the lobes and possibly core of a radio galaxy). Such sources may also overlap on the sky, due to projection effects or to real physical proximity, further confusing the link between emission peaks and true astrophysical sources. The term "island" is often used to refer to a distribution of flux lying above a threshold, and which may be composed of one or more components belonging to one or more intrinsic astrophysical source. In general, the term "component" is used to describe (elliptical) Gaussian fits to flux peaks, including those (possibly multiple peaks) within an island, that may either be independent sources, or true components of a complex source.

Ultimately, this process aims to provide a quantitative description of each identified object, and with a performance that can be robustly characterised in terms of the completeness, C, and reliability, R, (or its complement, the false detection rate, FDR; Hopkins et al. 2002), and the measurement uncertainties. Here completeness (often called "recall" in the machine learning literature) is the fraction of real sources as a function of signal-to-noise, or flux density, that are successfully recovered. The reliability (referred to as "precision" in the machine learning literature) is the fraction of measured sources that are real, as a function of signal-to-noise, or flux density. Its complement, the False detection rate FDR, ($\mathrm{FDR} = 1 - R$) is the probability that any measured source is not real, rather being produced by the noise characteristics or imaging artefacts in the data (Hopkins et al. 2002).

In the following sections we divide the continuum source extraction topic into blind extraction methods using source finders (Sect. 8.1); source fitting methods using known source prior information (Sect. 8.2); and measuring sources with complex morphology (Sect. 8.3). We then discuss various ways to identify the nature of radio sources, such as cross-matching blindly-extracted sources to known source catalogues (Sect. 8.4), and classifying and cataloguing radio sources in a Big Data fashion (Sect. 8.5). Finally, some telescope specific approaches and "lessons learned" perspectives are detailed (Sect. 8.7).

8.1 Source Finders

There is now a large suite of source finders available for analysing radio continuum images, having been developed over the past two decades and more. Many of these have been developed specifically to analyse radio data, but almost as many originate from the similar needs for source detection in optical or infrared images, and even other domains beyond astronomy. Many of these have featured in data challenges to help characterise their performance on radio data as the community works toward developing optimised source finding tools for the SKA (Hopkins et al. 2015; Bonaldi et al. 2021).

The early generation of source finders focused on detection and measurement of compact (point or point-like) sources, using a thresholding and Gaussian-fitting approach, and included SAD (Condon et al. 1998) and its variant HAPPY (White et al. 1997), and SFIND (Hopkins et al. 2002). Natural evolution of these approaches led to tools like Aegean (Hancock et al. 2012), aiming to better characterise overlapping emission peaks within islands, BLOBCAT (Hales et al. 2012), incorporating measurement of polarised sources as well as continuum, SOURCE_FIND (AMI Consortium et al. 2011) initially for high-frequency radio images, PyBDSF (Mohan and Rafferty 2015), initially for low-frequency radio images, and Selavy (Whiting and Humphreys 2012), using a 3D thresholding source finder (Duchamp; Whiting 2012) at its core. More recent developments use a staged approach for characterising complex source structures, such as Caesar (Riggi et al. 2016, 2019). Aegean, Caesar, and Selavy are also multiprocessor based, i.e. NxGen source finders, with high data throughput in mind. Other approaches include the use of curvature thresholds and adaptive or matched filters (e.g., CuTEx; Molinari et al. 2011, IFCA BAF; López-Caniego et al. 2006, IFCA MF; López-Caniego and Vielva 2012), and those developed for optical images (e.g., PySE; Spreeuw 2010; Swinbank et al. 2015, an evolution of SExtractor; Bertin and Arnouts 1996, and ProFound; Robotham et al. 2018; Hale et al. 2019).

There are, of course, tools developed as well for identifying sources in 3D images, for HI measurements for example, such as Duchamp (Whiting 2012) and SoFiA (Serra et al. 2015). These are discussed in more detail in the next chapter.

An important aspect of source finders is background and noise estimation, a required step before island detection and component determination. Radio source finders, such as, Aegean, PyBDSF, and Selavy, use a sliding box method to estimate noise, wherein the box size is crucial in terms of being not too small around bright sources and not too large around faint sources (Huynh et al. 2012), with a step size chosen so as to provide a statistically uniform image map. Box-wise, background noise maps are commonly estimated through the mean (μ), median (m), or mode of the image pixel intensities, with appropriate RMS (σ), MADFM, or IRQ noise map statistics, linked from box-to-box with bilinear/bicubic interpolation.

Median/MADFM statistics, when implemented in a computationally efficient fashion, are often referred to as robust statistics, as they tend to filter out source contributions (i.e., outliers), but at a time-complexity (Knuth 2000) cost of $O(n \log n)$

as compared to $O(n)$ for μ/σ statistics. Optical source finders, such as ProFound, PySE, and SExtractor, use σ-clipping (Akhlaghi and Ichikawa 2015) to trim source contributions beforehand, which is of $O(n)$.

Islands are typically determined by setting RMS threshold limits with respect to the background noise, estimated from the images themselves as described above. Some source finders, such as ProFound, Selavy, and SExtractor, use a watershed approach, scanning downwards from image peaks to some threshold, forming islands along with their component positions (i.e. local island peaks) for deblending. ProFound takes this one step further, by dilating the islands according to a Kron (Kron 1980)/Petrosian (Petrosian 1976) like profile, down to a lower threshold limit. Source finders such as Aegean, BLOBCAT, Caesar, PySE, and PyBDSF, use a flood-fill approach, using a set of rules to grow islands outwards from an initial (detection) to a lower (measurement) threshold.

Component extraction, or "deblending," is perhaps the most varied step in terms of radio source characterisation. In more traditional source finders, flux density peaks within islands are characterised by elliptical Gaussians. The number and location of such peaks are determined through various methods, optimised through some metric, such as χ^2. Selavy determines the position and number of peaks detected, to within some tolerance, during a raster-scan, for constraining the fit. Aegean uses curvature (i.e. local maxima) to define the component number and location for the fits. PyBDSF and PySE use less constrained global techniques. ProFound, by contrast, uses a watershed deblending (Beucher and Lantuejoul 1979) technique, wherein islands are separated into segments along "valley" boundaries; such segments being akin to components, in that they can be characterised by flux-weighted Gaussians. This process has the advantage of measuring and characterising extended sources with diffuse emission. Caesar uses watershed deblending, treating segments as sub-islands for local Gaussian fits, also making it more sensitive to complex structures with diffuse emission. A difference between ProFound and those tools that use local Gaussian fits to characterise sources is in the approach to recovering flux below the threshold or noise level. Gaussian fitting implicitly assumes that the a source has a Gaussian extent and estimates the total flux accordingly. With ProFound's "aperture" approach, it requires a correction when applied on radio images to account for this low level flux, referred to as a "beam correction" (Hale et al. 2019). Table 8.1 lists some general design characteristics of various source finders.

Source finders have their strengths and weaknesses, depending on the task at hand, and various statistics are used in order to quantify their performance. These typically include obvious metrics such as position accuracy, flux recovery, completeness, and reliability (Hopkins et al. 2015; Riggi et al. 2019; Hale et al. 2019; Boyce et al. 2023b). Often such metrics are calculated through running a finder on a simulated image whose source properties are known. Most source finders perform well in determining position and fluxes for point sources as the deblending is trivial, although finders developed for optical images, such as ProFound, may require corrections for accurate radio flux estimation (Hale et al. 2019). In terms of completeness and reliability, carefully tuned source finders perform close to

Table 8.1 Design characteristics of various 2D source finders, used in Radio astronomy

Source finder	Band	Targeted source morphology			NxGen
		Point	Extended	Diffuse	
Aegean[a,b,c,d]	Radio	✓			✓
APEX[b]	Radio	✓			
BLOBCAT[b]	Radio	✓	✓		
Caesar[c,e]	Radio	✓	✓	✓	✓
CuTEx[b]	Optical	✓			
HAPPY	Radio	✓			
IFCA BAF[b]	Radio		✓		
IFCA MF[b]	Radio		✓		
IMSAD[d]	Radio	✓			
Selavy[b,c,d]	Radio	✓			✓
ProFound[a,c]	Optical	✓	✓	✓	
PyBDSF[a,b,c]	Radio	✓	✓		
PySE[b]	Radio	✓			
SAD[b]	Radio	✓			
SExtractor[b,d]	Optical		✓		
SFIND[d]	Radio	✓			
SOURCE_FIND[b]	Radio		✓		

Performance Analysis Papers
[a] Radio source extraction with ProFound (Hale et al. 2019)
[b] The ASKAP/EMU Source Finding Data Challenge (Hopkins et al. 2015)
[c] Hydra II: Characterisation of Aegean, Caesar, ProFound, PyBDSF and Selavy source finders (Boyce et al. 2023b)
[d] Compact continuum source finding for next generation radio surveys (Hancock et al. 2012)
[e] Caesar source finder: Recent developments and testing (Riggi et al. 2019)

100% above $\approx 10\sigma$, dropping to 10–50% by $\approx 5\sigma$, with significant variation between finders (Hopkins et al. 2015; Boyce et al. 2023a). There is a trade-off in terms of a low detection threshold leading to high completeness at the expense of poor reliability (i.e. high FDR), and vice versa. Optimisation between these statistics depends on the science goals of particular investigations, and also requires expert knowledge for a given source finder, as they often have many parameters that can be tuned.

Source finders developed to primarily measure compact sources, unsurprisingly, perform less well with extended sources, both in terms of detection and characterisation. This can be reflected in poorer completeness and reliability statistics (Hopkins et al. 2015). Some finders, such as PyBDSF and more recently Caesar aim at explicitly handling extended emission more effectively. Some optical source finders (such as ProFound) also fare better with extended radio sources, as they are often designed to handle extended galaxy images in optical data.

In terms of characterising sources with diffuse emission, Caesar and ProFound perform well (Riggi et al. 2019; Boyce et al. 2023b), whereas point-source based finders, such as, Aegean, PyBDSF, and Selavy, can be biased away from accurate source positions, as diffuse emission may affect the estimate of source locations. It can also limit accurate flux recovery, as the diffuse emission blends into the noise with hard to model spatial structure (Boyce et al. 2023b). In general, there has been very little in terms of statistical assessment of source finder performance using simulated images of complex sources with diffusion emission. Caesar is one of the few that has been characterised in this fashion (Riggi et al. 2019).

Recently, an extensible source finder comparison tool, Hydra, has been developed that can measure comparison statistics based on simulated and real images (Boyce et al. 2023a,b). In the former case, detections are made in shallow images generated from real, or deep, images by adding noise. The point here is to enable detections in the original (deep) images to be used as a reference (assuming them to be accurate for at least the high S/N sources). The performance statistics can then be measured using shallow-deep comparisons without the need for simulated images or artificial (injected) sources. Hydra currently supports Aegean, Caesar, ProFound, PyBDSF, and Selavy. These were chosen to provide a good cross-section of the current range of source finders available.

8.2 Source Extraction with Prior Information

In many cases the astronomical objects to extract from the radio data are already known at other wavelengths, such as optical or infrared. An illustration of such a case is shown in Fig. 8.1, where the optical image of the Hubble Ultra Deep Field (HUDF) observed by the Hubble Space Telescope (HST) is compared with the sub-millimetre (sub-mm) image observed by ALMA in the same field. In these cases, photometry in the radio (here the sub-mm) image can be directly measured at known spatial locations. This will typically enable the detection of fainter objects than blind source finding, such as that discussed above, since the faintest such detections are dominated by spurious objects (noise spikes) in the blind case, while a known a priori object ensures that the measurement is more likely to be real. Also, even if a known source is not measurable (below the noise level) in the radio data, an upper limit can still be obtained, which is useful in other ways, e.g., constraining galaxy spectral energy distributions (SEDs).

There are two broad approaches that can be taken here. A simple and direct approach is referred to as "forced photometry", in which the flux is measured directly in the radio image at the location of a known prior source. This is often done assuming the radio source is point-like, fitting a Gaussian of the same size as the beam or PSF. A more sophisticated approach is a forward-modelling approach, where each prior source is assigned some flux, which is then integrated and compared against the radio image, until a suitable match is achieved (measured, for example, through a chi-squared minimisation). This approach is sometimes referred to as "prior fitting".

Prior fitting usually requires a prior source position catalogue (or for simplicity, just "prior catalogue") in addition to the science and RMS images commonly used

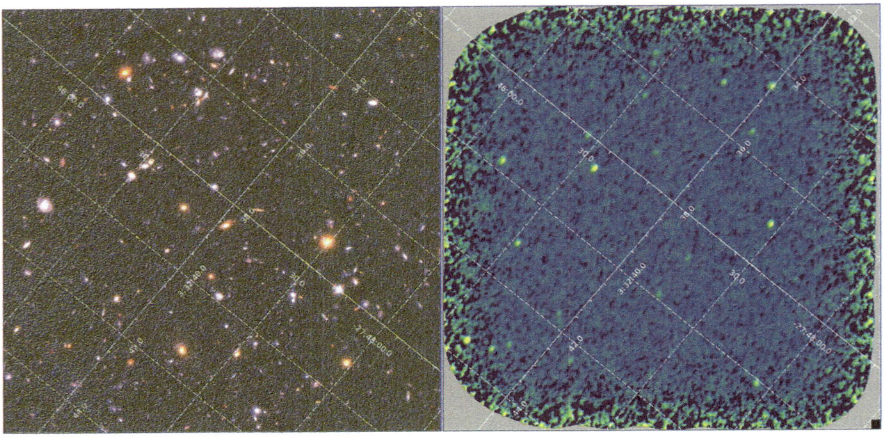

Fig. 8.1 *Hubble* Space Telescope optical to near-infrared RGB image (*left*) and ALMA 1mm continuum image (*right*) of the Hubble Ultra Deep Field, based on the public science data of HUDF (Beckwith et al. 2006), CANDELS (Grogin et al. 2011; Koekemoer et al. 2011) and 3D-HST (Brammer et al. 2012) (https://archive.stsci.edu/prepds/3d-hst/) surveys and the ALMA ASPECS large program (Walter et al. 2016) (https://almascience.eso.org/alma-data/lp/ASPECS). Images are aligned at the same World Coordinate System (WCS) with North at 50° counterclockwise from up and East to the lower left. WCS axes are shown as white lines. The noisy boundaries in the right panel indicates the edge of primary beams of the ALMA mosaic where primary beam attenuation becomes large. The ALMA image has a synthesised beam FWHM of $1.5''$. The field of view is about $2.2' \times 2.2'$

by source finders. A prior catalogue then should contain the spatial coordinates of all prior sources. These coordinates are usually taken from optical or near-infrared catalogues, with any astrometric offsets between those and the radio images corrected. There are two potential issues with a prior catalogue, namely the prior completeness and prior density. We return to this point later.

With a prior catalogue, what prior photometry does is simply modelling source shapes, convolving with the PSF, then comparing with the data to find the best source shape models that minimise the chi-square or equivalently maximise the likelihood. Below we first discuss a few widely used prior photometry software packages, then note some challenges in the prior photometry method itself.

A simple, widely-used source fitting tool is GALFIT (Peng et al. 2002). It was originally designed for use with optical images, but has also been used in sub-mm/radio studies. It has several advantages: it is simple and fast on small images, can fit multiple sources simultaneously, can fit various source models including point source, Sérsic, Gaussian and exponential 2D profiles, etc. It also has several disadvantages, making it less directly usable for the big data future: It is much slower (or has a memory issue) when dealing with large images (e.g. 10k × 10k pixels), has no parallelisation support, its chi-square minimisation algorithm sometimes does not converge, requiring careful manual oversight, and it cannot probe the posterior distributions well enough for a large number of priors (which is a known issue for the chi-square minimisation, but which can be dealt with by MCMC algorithms).

Similar simple source fitting tools or scripts exist in the literature and can be quickly written (e.g. in Python with ASTROPY.MODELING package), which is adequate when only small numbers of sources need handling. However, there are no mature prior fitting or forced photometry tools that have been developed to support the scale of the next generation of surveys. The key points to consider in developing such a prior fitting pipeline are that it should be able to:

- be parallel (or parallelisable), enabling it to process any size of images by internally cutting an image into tiles and performing source fitting inside each tile,
- process any number of input images given a prior catalogue even if some prior sources fall outside of the image,
- automatically select point and extended source models and handle non-convergence if using the chi-square minimisation algorithm,
- optimise the prior density by selective fitting, and improve the prior completeness by adding unknown sources in the residual image,
- routinely run for new incoming data to update the final prior photometry catalogue.

The A3COSMOS survey (Liu et al. 2019) has implemented such a prior photometry pipeline for routinely extracting (sub-)mm photometry from the ever-increasing public science archive of ALMA for high-redshift galaxies in the 2 sq. deg. COSMOS deep field (Scoville et al. 2007). Individual ALMA observations usually only cover a small field with a limited number of galaxies and a specific band. However, it is typical to detect galaxies other than the primary target even in these small fields. Information about these serendipitous galaxies are critical to constrain the sub-mm galaxy number count, luminosity functions, cosmic star-formation rate (SFR) density and dust/gas-SFR scaling relations. Therefore, systematically and routinely exploiting these pieces of information becomes an important task.

The A3COSMOS prior photometry pipeline uses galfit (Peng et al. 2002) as its internal fitting tool, and uses shell scripts to provide wrappers and achieve the goals of recursive, smart, selective source fitting. It uses an optical/near-IR source catalogue actively maintained and updated by the optical/near-IR community as the input prior catalogue, and fits thousands of ALMA images. Clever choice of priors has been implemented to avoid needlessly fitting to an arbitrarily large set of known objects, keeping the prior density to a manageable level. Similarly, automatically adding unknown sources from the fitting residual is not implemented given that the prior catalogue for the COSMOS field is already very dense (more than two million galaxies over 2 sq. deg., or about 300 per sq. arcmin). If there is any sub-mm/radio-only source that is not in the prior catalogue, the sub-mm/radio emission will be wrongly assigned to some priors, which can be identified subsequently using a galaxy SED fitting diagnostic (see also Sect. 8.4).

Next, we discuss why we care about the prior density, prior completeness, photometry completeness, and photometry FDR. The prior density affects the uncertainty of the photometry. If priors are too crowded given the resolution of the sub-mm or radio image, the internal source fitting will simply end up with a larger uncertainty because of the blending. Reducing the prior density is typically more important for low-frequency (GHz) radio surveys which often have larger beam sizes (e.g., $8''$–$15''$) than for the much higher frequency ALMA images. In the former case, in addition to a pre-step fixed-position fitting as implemented in A3COSMOS, a pre-selection based on the redshift or mass or some SED predicted flux for the prior source could be very useful.

The prior completeness describes how many sources in the sub-mm/radio data are actually in the prior catalogue. If the optical surveys used for prior catalogue construction are not deep enough, for example, dusty high-redshift galaxies can be missed, leading to sources of radio emission in the image lacking a prior counterpart. Future large radio surveys may require post-processing to implement a systematic source finding step to extract additional priors in the residual image after a prior source fitting step is necessary. Some analyses with Herschel photometry have already adopted such an algorithm, where sources in the residual are identified with the SExtractor software, or manually added to the prior catalogue for refitting (Liu et al. 2018; Jin et al. 2018).

The photometry completeness and FDR, as discussed in the previous section, also applies here. The photometry completeness describes, in a Monte Carlo simulation, how many input sources can be recovered from the photometry process, and the FDR describes how many recovered sources are actually not input sources, i.e., falsely detected. As for source finders, the photometry completeness and FDR can also be estimated by Monte Carlo simulations for prior photometry. The A3COSMOS survey did such calculations and compared the completeness and FDRs of blind photometry from PyBDSF compared with A3COSMOS prior photometry, finding that the latter achieves a 40% higher completeness and $3\times$ lower FDR at a given signal-to-noise ratio (S/N) (peak/rms $\sim$ 5). These quantities also depend on the source size relative to the beam size (Liu et al. 2019; Jiménez-Andrade et al. 2019).

In summary, prior photometry is a technique used to extract the deepest photometry from sub-mm/radio data, taking advantage of sources known to exist at other wavelengths. There are numerous source fitting tools, and their principal ideas are similar. However, there are few pipeline-level tools that can systematically process large datasets for on-going and future sub-mm/radio surveys. Comprehensive, modern pipelines and actively maintained prior catalogues are still rather preliminary for the future systematic exploitation of large radio survey and archival datasets.

8.3 Multi-Component Sources

While the source finders described in Sect. 8.1 can identify compact or slightly resolved sources with high completeness and reliability, the accurate identification and characterisation of extended sources can be problematic. This is true both

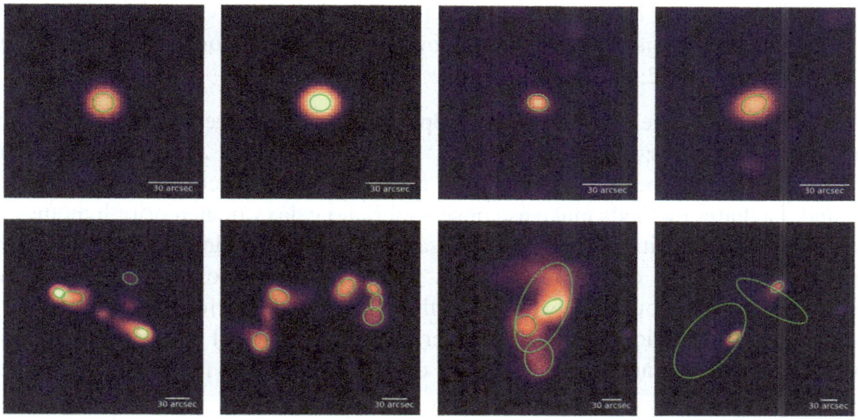

Fig. 8.2 Top row: simple (single-component) morphology radio sources from the EMU Pilot Survey (Norris et al. 2021b). Bottom row: complex (multi-component) radio sourced from the EMU Pilot Survey. The green ellipses show the Gaussian components fit by the Selavy source finder

for sources with complex morphologies as well as those whose radio emission is split into multiple components (such as the lobes and core of AGN (see Fig. 8.2 for examples). Identifying which independently measured component in an image might be associated with others and need to be grouped together as part of a single astrophysical source is not a trivial problem. Identifying radio loud AGN and well-resolved SFGs typically requires their radio emission to be cross-matched with optical/infrared images in order to identify their host galaxies. This links the radio component grouping problem together with both the cross-matching and source classification challenges (Sects. 8.4 and 8.5).

When a survey has its expected number of multi-component sources on the scale of hundreds of thousands (see Fig. 8.2), such as MIGHTEE (Sect. 8.7.2), LoTSS DR1 (Sect. 8.7.3), EMU, or VLA-COSMOS (Smolčić et al. 2017; Vardoulaki et al. 2019), the traditional approach of visual inspection to group and cross-match radio components associated with the same host galaxy becomes challenging to infeasible. Even existing large-scale surveys such as EMU, LoTSS and VLASS, are now measuring millions of sources, of which 5–10% are typically multi-component objects, which either require crowd-sourced visual inspection (Banfield et al. 2015; Alegre et al. 2022) or more advanced source finder algorithms. Automated or semi-automated approaches are clearly necessary. Several technical routes have been proposed:

1. Source finder algorithms developed to handle extended objects with diffuse emission (i.e., Fan et al. 2015; Robotham et al. 2018; Hale et al. 2019; Fan et al. 2020).
2. Public human-powered visual inspection projects, or citizen science, to perform manual multi-component source finding (i.e., Banfield et al. 2015).

3. Machine learning based algorithms to lower the burden of visual inspection (i.e., Proctor 2016; Segal et al. 2023) or provide end-to-end automated solutions (i.e., Alegre et al. 2022; Gupta et al. 2022).

Representative methods of the first approach include image segmentation methods (i.e., ProFound; Robotham et al. 2018; Hale et al. 2019) and Bayesian hypothesis testing methods (i.e., Fan et al. 2015, 2020). In comparison with ProFound, the latter methods make use of several radio galaxy related prior hypotheses, such as possible positions of the hosting galaxy relative to radio components, source bending angles and core-lobe distances to cross identify components belonging to the same radio source. It was found that this method performed similarly to a manually identification approach; further improvements will rely on the increase of data sample size and the introduction of other parameters for describing source properties (Fan et al. 2020).

Citizen science can both supplement grouping of components of complex or well-extended sources when the reliability of source finder algorithms like PyBDSF or Selavy drop (see Sect. 8.7.3) and pave ways for developing machine-learning-based source finders that usually require well-labelled source data entries (see Sect. 10.2). In the field of multi-component source finding, representative projects include Radio Galaxy Zoo (RGZ; Banfield et al. 2015), Radio Galaxy Zoo (LOFAR; Hardcastle et al. 2023) and Radio Galaxy Zoo (EMU). RGZ, for instance, is an EMU initiated citizen science project asking involved citizen scientists to (a) group or assemble islands that belong to the same parent source and (b) cross-match the source with its infrared counterparts. The project was seen as the pathfinder of EMU data mining and catalogue generation pipelines. Since launch in 2013, the project attracted over 12,000 volunteers and helped with over 100,000 source identifications, and further supported development of ClaRAN—an end-to-end machine learning based source finder algorithm (Wu et al. 2019). Radio Galaxy Zoo LOFAR, on the other hand, incorporated PyBDSF and machine learning into project sample image generation and subject data selection that go to users (Alegre et al. 2022, see Sect. 8.7.3 for further details).

While machine learning approaches will be discussed in Sects. 10.2 and 10.4 with more detail, there are alternative methods developed to lower the workload of extracting complex (not limited to multi-component sources) radio sources by applying a decision tree selection pipeline (Williams et al. 2019) or computing typical parameters (Segal et al. 2019, 2023). One of these applications, for instance, identifies complex radio sources by measuring the coarse-grained complexity (see Sect. 10.6.2) of their image cutouts (Segal et al. 2019, 2023). Ideally, an image cutout would have high complexity if the image contains a source with complex morphology, while the value would be low if there is either no emission or completely random emission. The combination of complexity measure and experienced expert validation then should help identify sources with complex morphology. In fact, a recent study has shown that by applying a complexity-based approach on source image cutouts, experts were able to find unusual objects such as Odd Radio

Circles (ORCs; Norris et al. 2021a) in the EMU Pilot Survey data (Segal et al. 2023).

8.4 Source Cross-Matching and Multi-Wavelength Catalogues

At least conceptually, the process of radio catalogue generation is carried out independently of the multi-wavelength data available in the field, although in practice the multi-wavelength data can be used to assist the decomposition of the radio sky into sources (see Sects. 8.2 and 8.3). The majority of large-scale radio catalogue work, however, will initially generate a catalogue with no optical identifications. Early radio catalogues were published in this state and, due to their low resolution compared to optical data-sets, typically required significant investment in high-resolution radio followup in order to generate optical IDs at all. The NVSS and TGSS-ADR catalogues (Condon et al. 1998; Intema et al. 2017) are examples of data products from large surveys with low resolution which, while valuable for the community, cannot be matched to optical data without additional radio information.

Current-generation radio surveys typically have resolution that in principle at least does allow the identification of radio sources, and often explicitly aim to produce a multi-wavelength catalogue in which the properties of the optical counterparts of the radio sources, where available, are compiled. For a high-resolution survey, the vast majority of the discrete sources identified in the radio will be galaxies, generating radio emission either from star formation or as a result of AGN activity, and in this chapter we focus on matching with galaxies, while noting that other objects (individual stars, supernova remnants, clusters of galaxies) should in principle also appear in these multi-wavelength catalogues.

The mainstay of large-volume cross-matching is the likelihood ratio method (Richter 1975; Sutherland and Saunders 1992) which allows the selection of the best counterpart, if any, from optical/IR catalogues to a given compact radio source. This technique is well suited to the regime in which most radio surveys operate: the radio resolution is usually low compared to that of the optical/IR catalogues and at present the density of radio sources on the sky is always much smaller than the number of possible galaxy counterparts, so that we expect typically multiple possible optical counterparts to any given radio source. This technique tends to break down for resolved or multi-component sources, where the number of possible optical counterparts becomes too great and where it is necessary to use morphological information from the radio to select the correct optical identification. The problem is more acute for surveys at low frequency (where flat-spectrum radio cores for AGN are often not visible) and surveys with high sensitivity to extended structure (where many faint diffuse sources are seen). The largest radio sources in angular size terms can extend to degree scales, and for these there is currently no substitute for human visual inspection to find the optical counterpart, which, however, is time-consuming and potentially error-prone. It is also necessary in some cases to re-analyse the radio catalogue in the light of potential optical identifications, for example in deblending multi-component radio sources where multiple components have an optical ID, or associating distinct sources from the radio catalogue. The approaches taken by different current-generation surveys are discussed in detail in Sect. 8.7.

Once cross-matching has been carried out, the goal of surveys is often to generate a 'value-added' catalogue which links the original (or modified) radio catalogue to the properties of the optical and multi-wavelength counterparts. Typically these might include, at a minimum, the position, catalogue identifier and optical/IR magnitude of the matched source; for galaxies, it is important to have the best possible spectroscopic or photometric redshift; and for objects with redshifts the radio sources can then be linked to, for example, galaxy stellar masses or star-formation rates if the multi-wavelength data is good enough to allow these to be calculated reliably. Links between the optical hosts and other wavelengths (e.g. X-rays) can be provided in the catalogue or through Virtual Observatory tools.

8.5 Source Classification

Exploiting the catalogues derived from a radio survey will almost always involve some element of *classification* of the radio sources, i.e. dividing them using their radio and multi-wavelength properties into groups that are intended to have some physical significance. Even without optical identifications, morphological classification of radio sources is possible: some classifications can be extracted directly from the catalogue (e.g. compact versus extended) while others depend on further image analysis—for example, the Fanaroff-Riley class of an object (Fanaroff and Riley 1974) can be estimated by eye (e.g., Vardoulaki et al. 2021) or in an automated way (e.g., Mingo et al. 2019; Scaife and Porter 2021) and similar methods applied to resolved sources can be used to detect bent sources in the narrow-angle tail or wide-angle tail classes, or other rare AGN types such as restarting or X-shaped sources. Additional physical information can be derived from broad-band spectral indices, if available, which can be used to select candidate remnant sources where particle acceleration may have ceased (Jurlin et al. 2020) or from spectral index or polarisation mapping. In the context of morphological classification, it is important to note that traditional terminology is both subjective and non-exclusive, and so best practice is to make use of multiple non-exclusive tags (Rudnick 2021) that can be used to select samples of interest, rather than to rigidly force morphological classes into the parent catalogue.

Optical identification, when possible, often provides a redshift which allows the physical properties of the radio source to be estimated—so, for example, highly luminous sources or candidate giant radio galaxies (GRGs) above some physical projected size threshold can be selected directly from the data. For galaxies associated with unresolved radio emission, a key question is whether the radio is predominantly associated with star formation or AGN activity. If the galaxy's star-formation rate is known, radio AGN can be selected on the basis of an excess of luminosity over the expected luminosity from star formation (Hardcastle et al. 2016; Smolčić et al. 2017; Delvecchio et al. 2017) but the SFR may not be reliably derived from the available ancillary data, in which case less direct methods may be used (Sabater et al. 2019). All of these methods rely on pre-existing source classifications, as well as spectral index information to measure the luminosity for high-z sources, and the only absolutely model-independent method for assessing the contribution from SF or AGN activity is to resolve the source, e.g. to show that the radio emission comes from a galactic disk rather than the jets and lobes of a radio AGN (Muxlow et al. 2020). If VLBI observations are available, AGN can be selected by looking for brightness temperatures in excess of 10^5 K (Radcliffe et al. 2021), but not all radio AGN have compact features exceeding this brightness temperature.

Once a host galaxy has been found, the physical classification of that host galaxy can then be applied to the radio source: in this way hosts of radio AGN selected from surveys can be classified as quasars, blazars, Seyfert galaxies and so on if the appropriate optical and/or emission-line data are available. Mid-IR colour-colour plots and spectral line diagnostic diagrams, if available, offer options for selecting different types of AGN, though it is important to note that many radio AGN hosts appear as quiescent galaxies using these selection methods.

8.6 Radio Source Disentangling via Wide-Field VLBI

One method that can routinely distinguish between AGN and SFG is the use of Very Long Baseline Interferometry (VLBI). Due to the long baselines and low sensitivity to diffuse emission, VLBI will typically only detect compact, high brightness temperature objects (often $>10^5$ K) such as supernovae, pulsars, radios stars and AGN. In the distant Universe ($z > 0.1$), the only objects typically energetic and compact enough to be detected by VLBI are AGN (e.g., Kewley et al. 2000; Middelberg et al. 2013). This makes VLBI a useful tool to investigate the nature and number of AGN across cosmic time without the use of multi-wavelength data. In particular, it has been shown that a VLBI-selected AGN sample cannot be classified as AGN using a single multi-wavelength identifier (Radcliffe et al. 2021).

However, typical VLBI observations have historically been restricted to a small field-of-view meaning that obtaining a significant sized sample required for such studies is difficult and expensive. This was because the averaging required to the smearing of the images would create large terabyte data-rates that were unfeasible until recently, and nonsensical if only concerned with the source at the pointing centre. In recent years, the renaissance in computational capabilities have made wide-field VLBI across the whole primary beam possible. Initially, only a few sources were targeted, either by multiple passes in the correlator (e.g., Garrett et al. 2005), or through a fine frequency and time averaging (e.g., Chi et al. 2013). These methods were unwieldy, either producing a monolithic large single data-set and/or impractical correlation times.

In recent years, through the development of new powerful and expandable software correlators, a new correlation mode to image the entire primary beam was developed. Called 'multiple phase centre observing' (Deller et al. 2011), this method takes advantage of the modern parallel computing architecture to allow hundreds of sources to be targeted in a single observation. This method is more practical, producing multiple small data-sets for which calibration was identical for each. This embarrassingly parallel process means that subsequent reduction and analysis is more straightforward for the user. When this new correlation method is coupled with new calibration algorithms (such as multi-source self-calibration, Radcliffe et al. 2016 and heterogeneous primary beam corrections, Middelberg et al. 2013; Radcliffe et al. 2018), imaging degrees of the sky at VLBI resolutions is now feasible.

This has been applied to many different science cases including probing the interstellar medium (ISM) of nearby galaxies (Morgan et al. 2013), investigating young stellar objects (Forbrich et al. 2021; Dzib et al. 2021), searching for gravitational lens systems (Spingola et al. 2019), supernovae (Radcliffe et al. 2019), and binary supermassive black-holes, and of course investigating the incidence of AGN and their milliarcsecond scale structure (Deller and Middelberg 2014; Middelberg et al. 2013; Herrera Ruiz et al. 2017; Radcliffe et al. 2018).

Returning to AGN, now that wide-field VLBI is a distinct possibility, we can obtain statistically complete samples of thousands AGN across cosmic time, providing a dust-free window on the milliarcsecond non-thermal emission from these objects. For example, the mJIVE-20 survey (Deller and Middelberg 2014) targeted over 20,000 sources from the FIRST survey with the VLBA, detecting 4336 sources and revealing that the radio emission becomes more compact towards lower flux densities. At fainter flux densities, there have been surveys of multiple deep-fields, such as the CDF-S, Lockman Hole (Middelberg et al. 2011), XMM-LSS (Middelberg et al. 2013), GOODS-N (Chi et al. 2013; Radcliffe et al. 2018) and COSMOS (Herrera Ruiz et al. 2017, 2018) fields. The COSMOS survey with the VLBA was the largest of these, targeting 2865 sources across a 2×2 deg area, resulting in 468 detections (Herrera Ruiz et al. 2017).

Recently, there has been the prospect of commensal surveys which have begun on the EVN and VLBA telescopes. These experiments piggyback on standard VLBI observing to search for compact emission in radio sources across the primary beam. It is prohibitively expensive to image the whole primary beam therefore existing surveys are used to pre-select sources for correlation.

At lower frequencies, VLBI is now possible with the International LOFAR telescope (ILT). Here the combination of a wider field-of-view and lower resolution required for a high brightness temperature AGN, makes this an attractive option in the hunt for AGN. The proof-of-concept came from the ILT survey of the 2 degree Lockman Hole field (Sweijen et al. 2022), where 940 AGN were identified out of 2316 detections (Morabito et al. 2022b). These were validated with multi-wavelength identifiers which classified 83% of the sources meaning that 160 were new AGN detections. To summarise, wide-field VLBI at cm and metre wavelengths now provides an attractive and dust-independent classification option to identify AGN.

In the future, the capabilities of wide-field VLBI will be vastly improved through the planned widening bandwidths and data-rates of receivers and the inclusion of new sensitive elements (including phased-up SKA and the 500 m FAST telescope). This will provide its own challenge in terms of data-sizes, as the required data-rates for wide-field VLBI are much larger. For example, the current largest wide-field VLBI experiment, the 72 h observation of COSMOS (PI Radcliffe) which consists of ~20 antennae, 1024 Mbps data-rate and targets 700 sources, has a total raw data-size of 34.3 TB. If this were to be observed at 32 Mbps, the total data-size would be 1.1 PB. While computational and storage capabilities will improve and become cheaper in the future, the majority of VLBI data processing effort is still borne by the user. This means that many institutions may be disadvantaged by their inability to process such volumes of data. This necessitates investment into cloud/server/compute availability to VLBI users which could follow a model such as that offered to MeerKAT users with the ilifu computing cluster (see Sect. 6.8).

8.6.1 VLBI with the SKA

The SKA has been designed to be able to operate in VLBI experiments. For the initial phase of the SKA, the observatory will provide a number of tied-array 'beams' that are achieved by phasing up various sub-arrays of SKA antennae. In the case of SKA-Low, these sub-arrays will be formed from a sample of stations, and, for SKA-Mid, they will comprise samples of antennae. These sub-arrays can be customised to vary the sensitivity, and effective field-of-view of the phased beam. Depending on the observing frequency, this VLBI mode will be able to operate commensally with other observing modes such as continuum imaging and pulsar timing.

Once the tied-array beams are formed, each one effectively acts as a normal 'VLBI' station where the visibilities are exported to external correlation facilities outside of the SKA. While the increase of a few telescopes will not drastically increase the data-rates, the incorporation of such a sensitive element along with the restrictive field-of-view imposed by phased array beams provides extra challenges to the various science cases and technical challenges to data-processing pipelines. It is for this reason that significant effort is underway to simulate SKA-VLBI observations in order to gauge the impact of these restrictions.

Going forward, there is a significant push to increase the data-rates and the number of telescopes in the next decade. For example, the EVN 2020–2030 vision (Venturi et al. 2020) states the desire for 32 Gbps capabilities and the development of broad-band receivers to match the SKA receivers. With a host of new possible stations, such as the African VLBI Network and SKA, there will be a step-change in data-size. When combined with specialised modes, such as wide-field, spectral line, or many SKA phased beams, a vast improvement in sensitivity is possible but also the data storage, and processing requirements that will need to be conducted by the user will increase. Furthermore, for a wide-field observing mode, there will be the extra complication of imaging a large field-of-view at such high resolutions.

Should the final SKA vision be achieved, the future SKA-Mid instrument will eventually span large swathes of sub-Saharan Africa and it will truly become a VLBI instrument in its own right. This will bring extra complications both pre- (e.g., clock synchronisation) and post-correlation (e.g., extra calibration steps, larger pixel images), especially if the data is to be calibrated and correlated in real time. However, it is hoped that the computational capacities at this time will be capable of overcoming these obstacles.

8.7 Perspective from Different Surveys

So far in this chapter we have discussed various approaches to identifying and measuring the properties of sources in radio images. The latest generation of wide-field radio continuum sky surveys represent the largest current datasets to which these methods are applied, with millions of sources being detected. These surveys are often carried out with different telescopes and with various scientific goals, resulting in distinct characteristics for each survey in terms of, e.g., angular resolution and sensitivity (see Table 8.2). Subsequently, any given survey will have source-finding, measurement, and cross-identification challenges that are unique to that specific survey. In the final part of this chapter, we discuss the approaches to tackle such challenges that have been adopted by four different large survey projects being undertaken by ASKAP, MeerKAT, LOFAR and the enhanced VLA.

Table 8.2 Comparison of the basic survey parameters for MIGHTEE, EMU, LoTSS and VLASS

	MIGHTEE	EMU	LoTSS	VLASS
Frequency (MHz)	1300	943	144	3000
Beam size (arcsec)	5	15	6	2.5
RMS (μJy beam^{-1})	2	20	70	70[d]
Footprint (deg^2)	20	20,000	20,000	34,000
Expected source detections (10^6)	0.12[a]	30[b]	16[c]	5[d]
Multi-component sources[e] (10^3)	0.2	1000	25	10

[a] Extrapolation from early science data release to full survey (Heywood et al. 2022)
[b] Prediction based on the Tiered Radio Extragalactic Continuum Simulation (Bonaldi et al. 2019)
[c] Extrapolating from LoTSS DR2 to the entire survey footprint (Shimwell et al. 2022)
[d] Predicted values for the cumulative VLASS imaging after all three epochs have been observed (Lacy et al. 2020)
[e] Predictions extrapolated from previous surveys accounting for resolution and sensitivity (Vardoulaki et al. 2019; Gordon et al. 2021)

8.7.1 EMUCAT

The Evolutionary Map of the Universe survey (EMU; Norris et al. 2011, 2021b) is an ongoing (2022–2027) all-southern-sky continuum survey with the Australian Square Kilometre Array Pathfinder (ASKAP; Hotan et al. 2021). With a targeted rms noise level of $\sim$20 μJy/beam and typical beam size of $\sim 15''$, EMU will be the deepest $\nu \sim 1$ GHz continuum survey to date and is expected to detect $\sim$1400 sources per square degree. This depth, combined with a survey footprint of 2π Sr, is expected to produce nearly 3×10^7 individual source detections, making it the largest SKA pathfinder survey in terms of sheer number counts. As such, EMU is at the forefront of data-intensive radio astronomy, and will be pioneering new approaches to producing both basic and enhanced catalogues for the astronomical community.

The EMU 'value-added' catalogue (EMUCAT) is a sophisticated resource being developed by EMU team members in coordination with the Australian SKA Regional Center (AusSRC). The primary aims of EMUCAT are to combine all individual ASKAP EMU observations into a single cohesive all-sky data product and then add value through the inclusion of other multi-wavelength survey data. EMUCAT will substantially improve the scientific value of the ASKAP data by mosaicking adjacent observations to optimise the overall sensitivity and uniformity. Source finding is then performed on these new images using the source-finder Selavy (Whiting and Humphreys 2012) to produce an accurate, de-duplicated radio catalogue. EMUCAT will also greatly enhance the scientific utility of the ASKAP data through tasks including cross-matching, classification, redshift identification and estimation, in order to provide a fully science-ready resource.

EMUCAT pipelines are automatically triggered as new EMU data appears in the CSIRO ASKAP Science Data Archive (CASDA) through the VOEvent service. These pipelines perform a series of post-processing and cross-identification tasks and compute various derived astrophysical properties, updating the EMUCAT database at each stage. The value-added data produced includes identification of simple (single-component) and complex (multi-component) radio sources (see Fig. 8.2) through cross-matching to infrared source catalogues from the Wide-field Infrared Survey Explorer telescope (WISE Wright et al. 2010). Following this initial multi-wavelength cross identification, additional value is added in the form of:

- multi-band photometry and known redshifts obtained via catalogue matching,
- derivation of redshifts where otherwise unavailable through, e.g., photo-z and machine learning techniques,
- providing derived properties such as colours, sizes, morphologies, AGN or star-forming classification, etc.

The EMUCAT data products are served through a VO-compliant Table Access Protocol (TAP) server hosted at the AusSRC. These are accessible through TAP-aware tools such as TOPCAT (Taylor 2005), through python notebooks running at the AusSRC co-located with the data, and via a user portal that will facilitate basic database queries. Over the course of the EMU survey, the team will produce sequential EMUCAT data releases which will be exported to CASDA.

8.7.2 MeerKAT and MIGHTEE

The MeerKAT telescope (Jonas and MeerKAT Team 2016) is carrying out a number of surveys, with MIGHTEE (Jarvis et al. 2016) being the principal extragalactic continuum survey. MIGHTEE aims to carry out deep surveys covering $\sim$20 deg^2 in L-band over four extragalactic fields (COSMOS, E-CDFS, ELAIS-S1, XMM-LSS) with a central frequency of 1284 MHz, a maximum resolution around 5 arcsec, and a target confusion-limited noise level of 2 μJy beam^{-1} at 8-arcsec resolution, comparable to the depth expected from SKA all-sky surveys. An S-band survey of $\sim$4 deg^2 over COSMOS and E-CDFS will also be carried out. Polarisation as well as HI data products will also be produced, and a complementary uGMRT survey of the same sky area, known as superMIGHTEE, is in progress. At the time of writing, early science data in the COSMOS and XMM-LSS data have been made public (Heywood et al. 2022) and the survey team is preparing for a first data release that will include wider areas and increased depths in COSMOS and XMM-LSS and the first data in E-CDFS. Radio catalogues for the released fields are produced using PyBDSF (Sect. 8.1).

MIGHTEE benefits from extremely good multi-wavelength data over its well-studied target fields (Hudelot et al. 2012; McCracken et al. 2012; Jarvis et al. 2013; Vaccari 2015; Vaccari et al. 2016; Aihara et al. 2018), including far-IR data from Spitzer and Herschel collated in a systematic way by the HELP project (Shirley et al. 2019, 2021) and the aim is to release science-ready catalogues that will incorporate these data; high-quality homogenised photometry, photometric and spectroscopic redshifts as well as source classifications are already available for some of the fields (Adams et al. 2021; Hatfield et al. 2022; Vaccari 2022a,b; Weaver et al. 2022; Zou et al. 2022). As with LOFAR (Sect. 8.7.3), the identification of optical/near-infrared counterpart of MIGHTEE radio sources is carried out via a combination of likelihood-ratio cross-matching for compact sources and visual inspection for association and identification of extended sources. Given the limited sky area, the fraction of extended sources will be comparatively low and the effort needed for visual inspection small. Source classification can then be carried out with a substantial degree of accuracy using the extensive multi-wavelength data available across the majority of the MIGHTEE footprint, and the fraction of different radio source populations can thus be studied in detail (Hale et al. 2022; Whittam et al. 2022).

8.7.3 LOFAR

To date, LOFAR cataloguing efforts have been primarily based on source catalogues produced by PyBDSF (Sect. 8.1). PyBDSF is generally well suited to this purpose, as it was specifically designed to deal with the most important source-finding issues encountered with LOFAR images. Broadly, these issues are the result of LOFAR's low observing frequency and high resolution, which result in images that can suffer from position-dependent artefacts (and point spread function) yet also have a significant fraction of sources that are resolved. LOFAR cataloguing efforts must take into account these issues to produce reliable source catalogues.

Basic LOFAR source catalogues made with PyBDSF are derived by fitting Gaussians to sources found via blind searches that identify emission through thresholding, with a typical peak flux density threshold of five times the standard deviation of the local background noise. The background noise is estimated in a position-dependent manner by taking the median of pixel values inside a rectangular box that is moved across the image. This approach is necessitated by the position-dependent nature of artefacts in LOFAR images (due to ionospheric and beam effects that cannot be entirely corrected for during calibration): the background noise maps must account for this position-dependent source of noise to avoid false detections near bright sources. In the case of LOFAR data, it is often beneficial to use the option to derive the background noise maps using two scales for the median box, with a smaller scale used near the brightest sources (which tend to be compact) and a larger one used elsewhere, where a small box might result in noise that is biased high due to extended emission with a scale similar to the size of the box or larger. Lastly, for LOFAR catalogues, the thresholding is often done on the so-called "flat-noise" image, the image that has not been corrected for the attenuation of the LOFAR primary beam. Doing so alleviates the need to account for the rising noise as a function of distance from the phase centre that occurs in primary-beam corrected images, resulting in somewhat more robust background noise maps.

Once sources have been identified through thresholding, a decomposition is done by fitting Gaussians to the emission. This approach is well suited to point sources, which are generally Gaussian-like due to the typical restoration process used in LOFAR imaging. However, extended, resolved sources are usually not well modelled by simple Gaussians. PyBDSF mitigates this issue to some extent by doing a wavelet decomposition of the image and fitting scale-dependent Gaussians to each wavelet image. The resulting Gaussians are then grouped together into sources by a simple algorithm that estimates whether the dip in surface brightness between two Gaussians implies that they are likely members of a single source or independent sources. Nevertheless, in general accurate source association is a challenging task, and considerable further effort has often been necessary to reliably associate source components, as discussed below.

Catalogues produced by PyBDSF fall into two basic types: Gaussian or source catalogues. Gaussian catalogues include information on each Gaussian component (its position, peak flux density, size, source association, etc.). These catalogues preserve the most information about the source structure, and are therefore useful for use as basic calibration models (for example, as a starting model for self calibration). However, they are not generally useful for cross matching with catalogues at other wavelengths as often only a single Gaussian component of a source will be matched (rather that the entire source). For such purposes the source-type catalogue is more useful, where the component Gaussians are summed into a single entry for each source. For these multi-component sources, a moment analysis is done to produce estimates of the aggregate source position, size, and shape. The source-type catalogues include information on whether the source is composed of a single Gaussian component or multiple ones (which may be useful for further analysis), and of course the two catalogue types (Gaussian and source) can be used together to understand which specific Gaussians are included in a multi-component source. As mentioned above, these basic catalogues may need considerable further processing to allow the full scientific exploitation of the data.

PyBDSF can be, and is, used for individual LOFAR observations, but its most intensive use comes from wide-area sky surveys that need to generate large-volume catalogue. The LOFAR all-sky survey LoTSS (Shimwell et al. 2017) aims to survey the entire northern sky (above $\delta = 0°$) at a resolution of 6 arcsec with a typical *rms* noise level of 100 μJy beam^{-1} at 144 MHz. In addition, deep observations of selected famous fields are being carried out with the aim to reach *rms* noise levels of 10–15 μJy beam^{-1}. So far the all-sky survey has had a first (Shimwell et al. 2019) and second (Shimwell et al. 2022) data release: DR2 covers around 5700 degree of sky away from the Galactic plane and is the largest radio sky survey so far carried out by number of sources, with 4.4 million radio sources detected. In addition a first deep-field data release consisting of observations of the ELAIS-N1, Boötes and Lockman Hole fields is available (Sabater et al. 2021; Tasse et al. 2021). Although LoTSS is currently largely generating images using only the Dutch baselines of LOFAR, with a typical resolution of 6 arcsec, a stretch goal of the project is to exploit the much higher resolution provided by the full International LOFAR Telescope (ILT), which can be ~0.3 arcsec at 144 MHz (Morabito et al. 2022a), over large areas of the sky. A low-frequency counterpart of LoTSS, LoLSS, is being carried out at a central frequency of 57 MHz (de Gasperin et al. 2021).

The basic LoTSS data product from the wide-area surveys is a mosaiced Stokes I radio image from each pointing of LOFAR (841 pointings in DR2) and a source catalogue for that mosaic generated by PyBDSF. The source catalogues from the different mosaics are then de-duplicated to form a single catalogue for a given data release. A number of other per-pointing data products, including low-resolution Stokes I mosaics and Stokes QU cubes, Stokes V images, and sub-band imaging, are not discussed further here: see Shimwell et al. (2022) for more details.

LOFAR's combination of low frequency, high surface brightness sensitivity and short baselines (with sensitivity to structure on scales up to 1° in the default imaging pipeline for DR2) means that the LoTSS sky is unusually complex compared to that seen by VLA surveys where extended structure on scales ~10 times the beam size is filtered out by the lack of short baselines. Therefore the decomposition of the radio sky into 'sources' by PyBSDF is only ever approximate—for example, a classical double radio source can easily be seen by PyBDSF as two or more separate objects (which need to be 'associated' in post-processing so that their status as a single physical object can be recorded). More rarely, PyBDSF may erroneously combine emission from two physically distinct sources into one (the sources then need to be 'deblended').

A goal of the LoTSS project since its earliest days has been to provide not just the radio images but the best available identification with an optical source for each radio object. The resolution of the survey, at 6 arcsec, is good enough to get reliable identifications in most cases when combined with all-sky optical and infrared data. For DR2 the DESI Legacy Survey (Dey et al. 2019) and *WISE* (Wright et al. 2010) data are used for optical and infrared respectively: for the deep fields deeper dedicated optical and infrared datasets were used. The optical IDs will form the basis of spectroscopic followup with WEAVE-LOFAR (Smith et al. 2016).

For compact radio sources (which remain the vast majority of the objects in the survey) simple likelihood-ratio cross-matching can be used to select the most plausible optical counterpart. For more complex sources a more complex approach is needed, described for DR1 by Williams et al. (2019), and for the deep fields by Kondapally et al. (2021). Essentially the radio and optical properties of the source, including whether the PyBDSF source is made up of multiple Gaussians and whether those Gaussians themselves have plausible optical IDs, are used to decide how its optical ID should be allocated. For a small but significant number of sources, human visual inspection is required, which allows the team to combine the tasks of optical identification and source association making use of all the available data. This was done using the Zooniverse platform—for the wide-area and deep fields DR1 LoTSS used a private Zooniverse instance while DR2 made use of a public citizen science project, Radio Galaxy Zoo (LOFAR) in which nearly 177,000 sources were inspected by around 11,000 citizen science volunteers. In all cases expert classifiers inspected objects where processing or the input from citizen scientists indicated some problem with the data, and manually corrected blended sources (Fig. 8.3). For DR2, a version of the ridge line algorithm of Barkus et al. (2022) was run at the end of the process to find identifications for large objects which had not been successfully identified by the volunteers. DR2 also benefited from the use of machine-learning-based selection of some of the targets sent to visual inspections (Alegre et al. 2022). These automated approaches, together with machine learning association of complex sources (Mostert et al. 2022), are likely to be increasingly important in future larger-scale surveys.

A full optical ID catalogue, together with spectroscopic and photometric redshifts (Duncan 2022) enables a large amount of other science, including, for example, systematic morphological analysis and more detailed measurements of size and radio flux (Mingo et al. 2019) or AGN selection (Hardcastle et al. 2019), the results of which can be provided as supplementary catalogues. For DR2 over 80% of sources have candidate optical IDs from either *WISE* or the Legacy survey or both, and around 50% have spectroscopic or photometric redshift estimates. For sources with redshift estimates, value-added physical quantities such as luminosity and host galaxy mass were provided for DR1 and the deep fields.

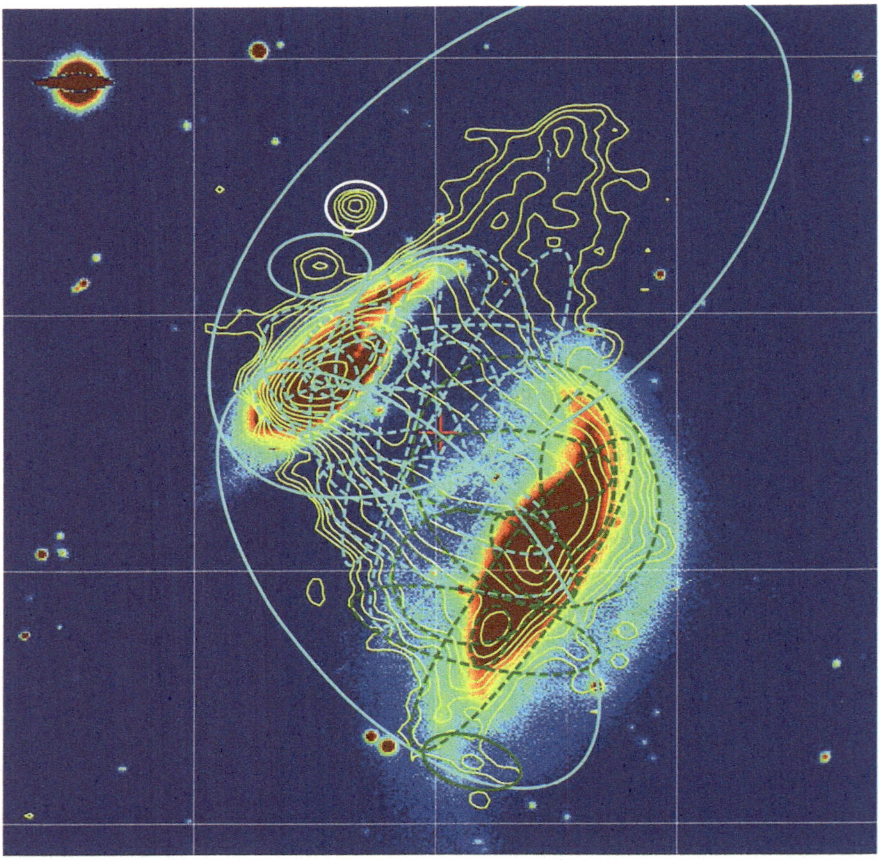

Fig. 8.3 Manual deblending in action. A complex radio source from LoTSS DR2 with contributions from two physical objects, in this case interacting galaxies, has been broken down into the component Gaussians from PyBDSF which are manually being assigned to two different catalogued sources. Radio contours, PyBDSF sources (solid lines) and Gaussians (dashed lines) are shown over an optical image, in this case from the Legacy survey. Grid lines are spaced at 1 arcmin

The planned processing of long-baseline data (LOFAR VLBI) will add interesting complexity to the production of catalogues, since the existing wide-field imaging, (e.g. Sweijen et al. 2022) show that the long-baseline images can give a very different view of the sky to the 6-arcsec ones made with the same data. Extended structures seen at 6 arcsec may well not be detectable at 0.3 arcsec because of the very different surface brightness sensitivity of the images. Generating a full sky catalogue will likely require the combination of different *radio* catalogues before multi-wavelength cross-matching begins.

8.7.4 VLASS

The Karl G. Jansky Very Large Array (VLA) Sky Survey (VLASS; Lacy et al. 2020) is an S-band ($2 < \nu < 4$ GHz) continuum survey of the entire sky north of $\delta = -40°$. VLASS is undertaken using the B- and BnA-configurations of the VLA, resulting in a typical synthesised beam size of $\sim 2''.5$, making it the highest angular resolution near-all-sky radio continuum survey yet undertaken. One of the primary survey goals of VLASS is to explore the population of transient objects that exhibit radio variability on timescales of a few years. To this end, VLASS is observing its entire footprint on three separate occasions (epochs) to an *rms* noise level of ~130μJy/beam, with each observing epoch separated by ~2 years. When all three of these epochs are combined, VLASS can reach an rms noise level of ~70 μJy/beam. A comparison of VLASS to three of the other big SKA-precursor surveys is shown in Table 8.2.

The basic science products produced by NRAO include images and cubes for each epoch. In order to enable immediate follow-up of potential newly identified radio transient sources, rapidly produced *Quick Look* images that have only been cleaned to first order are made available shortly after observation. These *Quick Look* images are released months, or sometimes years, ahead of what is possible for the fully cleaned *Single Epoch* images. Therefore, understanding and dealing with data quality is a key aspect of detecting components/sources in VLASS.

On-the-Fly Mapping

The VLASS on-sky footprint of ~34,000 $\deg^2$ represents more than 80% of the sky. In order to facilitate observing such a wide area in the lowest possible time, the VLA makes use of a new on-the-fly mapping (OTFM) strategy. In short, OTFM allows the VLA to continuously scan large patches of the sky at a constant slew rate (3.31 arcsec/min in the case of VLASS (Lacy et al. 2020)). Here, integration and mosaicking is performed while scanning rather than requiring a number of individual pointed observations to map the same region that are mosaicked together post-hoc.

For a sufficiently small time period, Δt, the phase centre of the scan is effectively constant while the pointing centre varies. Smearing across even a small time interval introduces amplitude errors that must be compensated for with an *effective* primary beam, $\boldsymbol{B}_{\nu,\mathbf{eff}}$. This time-dependent effective primary beam is related to the primary beam of a static pointing, $\boldsymbol{B}_\nu$, by:

$$\boldsymbol{B}_{\nu,\mathbf{eff}}(\boldsymbol{s} - \hat{\boldsymbol{s}}_p; \Delta t) = \frac{1}{\Delta t} \int_{-\Delta t/2}^{+\Delta t/2} \boldsymbol{B}_\nu[\boldsymbol{s} - \boldsymbol{s}_p(t)]dt. \tag{8.1}$$

Here, $\boldsymbol{s}$ is the sky coordinate, $\boldsymbol{s}_p$ is the pointing center, and $\hat{\boldsymbol{s}}_p$ is the average pointing center over Δt (Mooley et al. 2019).

The advantage of using OTFM is that observing overheads, particularly the fixed 'settling time' that is required with each individual VLA pointing, are

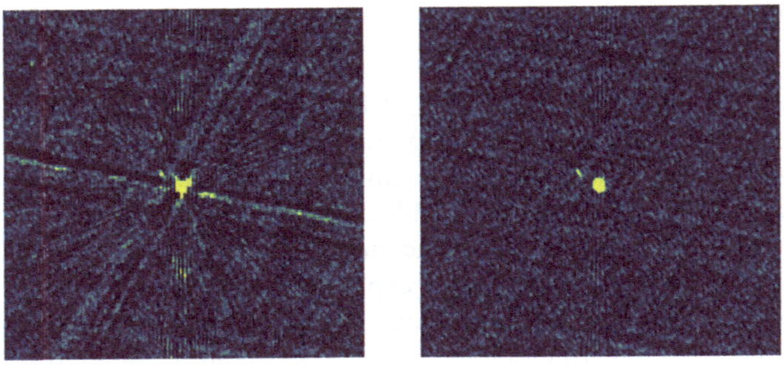

Fig. 8.4 Comparison of the *Quick Look* (left) and *Single Epoch* (right) images of a ~10 mJy source observed by VLASS. The more rapid processing needed to produce *Quick Look* in a timely manner leaves residual image artefacts around the source (especially visible to the east, west and south of the source)

substantially reduced. As such overheads can account for a significant fraction (a few tens of percent) of observing time with relatively shallow observations, OTFM results in a $>10\%$ increase in survey efficiency where the rms noise level is $\gtrsim$ a few 10s μJy/beam (Mooley et al. 2018), as is the case for VLASS. However, a consequence of OTFM, is that imaging artefacts may be more frequent and require additional processing to remove, many of which are still present in the VLASS *Quick Look* images (see Fig. 8.4).

Source Finding Challenges in VLASS Quick Look Images

Although the VLASS *Quick Look* images are not designed to be the final, science-ready, VLASS images, their rapid availability is nonetheless a scientifically useful resource to radio astronomers. That is to say, *limited usefulness* is not the same as *useless*. Figure 8.4 shows a comparison between a *Quick Look* image and a finalised *Single Epoch* image of the same source. The impact of not performing self-calibration and only carrying out limited cleaning in the *Quick Look* image result in image artefacts that are absent in the *Single Epoch* image. These strong image artefacts have the potential to be detected by source-finding algorithms, and thus contaminate resultant source catalogues with spurious detections.

In order to produce a scientifically useful catalogue, false-positive detections need to be identified. In practice this can be achieved by any algorithm that segregates the parameter space of sources into regions occupied by real and spurious detections. However, it is likely that whatever algorithm is used to identify contaminant sources in a catalogue will be imperfect, and this should be considered when deciding how to address spurious detections in a catalogue. Once identified, contaminant sources may be dealt with by adopting one of two philosophies. A *search and destroy* approach can be used to remove detections in a catalogue that have been identified as spurious. An alternative method is to *flag and retain*. Here, instead of removing the identified contaminants, they are left in the catalogue, but with the addition of one or more flags that distinguishes them from detections that are believed to be real.

For VLASS *Quick Look*, PyBDSF was used to produce a catalogue of source detections. The next step undertaken was to *flag and retain* those detections that were likely to be spurious (Gordon et al. 2021). The advantages of this approach are (a) that the robust sample produced by the *search and destroy* approach is an easily recoverable subset of the catalogue, and (b) it provides a more complete sample for people to use if they disagree that a given source is spurious. The main disadvantage of a *flag and retain* philosophy is that it passes an additional burden to the end user, i.e. they must make a decision on which data they wish to use. It is therefore of paramount importance to clearly define the flag(s) used and communicate that using these flags is left to the end user when adopting the *flag and retain* philosophy.

Spectral Index Measurements

The broad bandwidth of VLASS (2 GHz) is well suited to measuring the radio spectral indices of detected components. Spectral index values based on a single VLASS observation have the advantage of being simultaneous measurements, rather than being determined from multi-band observations taken at different times. Such simultaneous spectral index measurements present the true value of spectral index, unbiased by any variability of the source over time. Making use of the CASA algorithm MTMFS (Multi-Term Multi Frequency Synthesis; Rau and Cornwell 2011), the image at a given frequency, $\boldsymbol{I}_\nu$ is modelled by:

$$\boldsymbol{I}_\nu = \boldsymbol{I}_{\nu_0}\left(\frac{\nu}{\nu_0}\right)^\alpha, \tag{8.2}$$

relative to a reference frequency, ν_0, and assuming a power law with gradient α and no curvature (Sohn et al. 2003). VLASS *Single Epoch* imaging produces coefficient images for both the zeroth- ($\boldsymbol{TT}_0$) and first-order ($\boldsymbol{TT}_1$) terms of the Taylor expansion of Eq. 8.2, such that:

$$\boldsymbol{TT}_0 = \boldsymbol{I}_{\nu_0},\ \boldsymbol{TT}_1 = \alpha \boldsymbol{I}_{\nu_0}. \tag{8.3}$$

Here, I is the image intensity, and α is the spectral index across the bandwidth of the image. Hence, the spectral index of a source in VLASS can be determined from *Single Epoch* images by:

$$\alpha = \frac{\boldsymbol{TT}_1}{\boldsymbol{TT}_0}. \tag{8.4}$$

It is worth remembering that beam size is dependent on observing frequency, i.e., the angular resolution of VLASS at 2 GHz will be poorer than at 4 GHz. Furthermore, the signal-to-noise of a source will often vary with frequency, and reliable spectral index measurements may be unable to be obtained for faint sources. Spectral index measurements from Taylor term images are therefore best used

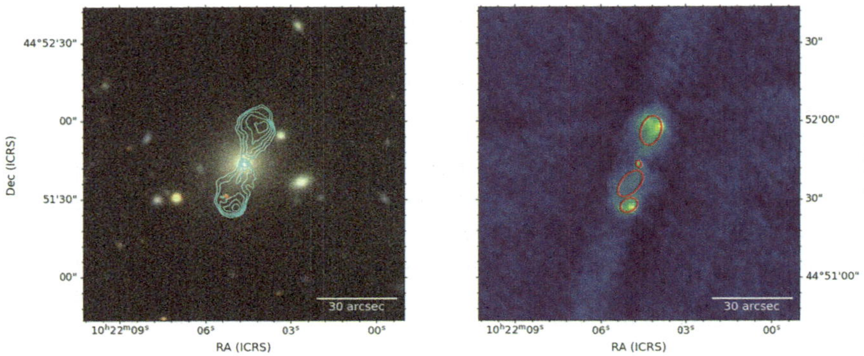

Fig. 8.5 The left-hand panel shows radio contours from VLASS (cyan) overlaid on an optical image from the Sloan Digital Sky Survey (York et al. 2000). These indicate the radio source is a single physical object associated with a typical early-type host galaxy. The right-hand panel shows that this single radio galaxy is detected as multiple 'components' by the source-finder PyBDSF

for bright, unresolved sources. For identified sources in VLASS, spectral index measurements will be obtained in the first instance by measuring $\boldsymbol{TT}_1/\boldsymbol{TT}_0$ at the peak brightness of the detection.

Multi-Component Sources in VLASS

The high angular resolution of VLASS relative to other wide-field continuum surveys has a number of effects with respect to source identification and measurements. For many extended radio AGN, high resolution observations will split a source into multiple components, each with its own entry in a catalogue (see Chap. 8.3). For example double radio sources that are unresolved in a poorer resolution survey may be split into two (or more) distinct components by VLASS (see Fig. 8.5). Indeed, comparisons with the Faint Images of the Radio Sky at Twenty cm survey (FIRST, $5.4''$ beam (Becker et al. 1995)) show that $\sim$20% of AGN seen as a single source in FIRST are split into multiple components by VLASS (Gordon et al. 2021). Furthermore, in VLASS, the lack of short baselines in the array configuration results in diffuse structures larger than $\sim 20''$–$30''$ being resolved out (Lacy et al. 2020). Thus, for VLASS in particular, large structures may be broken up into multiple components, necessitating the grouping of components in physical sources (see Sects. 8.3 and 10.2).

Host Identification

One key advantage of the high resolution of VLASS compared to other wide-field continuum surveys is the improved ability to correctly identify the host galaxy associated with a radio source in multi-wavelength data. Historically, even for unresolved sources, the relatively poor resolution of radio with respect to optical observations would often result in a number of candidate cross-identifications for a given radio source. In such cases, statistical methods can be used to identify the most likely real host from a number of candidates (McAlpine et al. 2012), although these

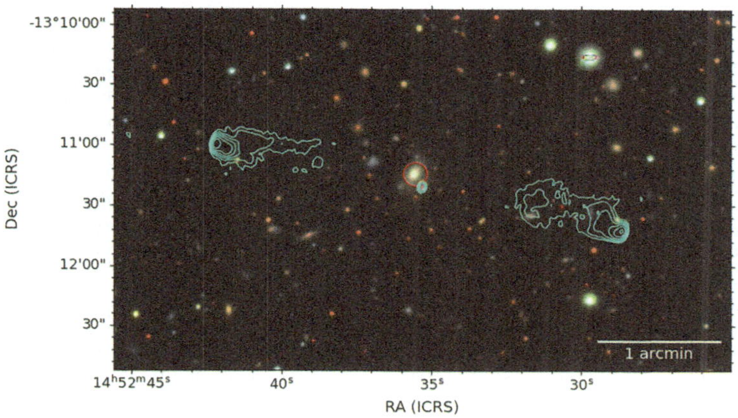

Fig. 8.6 VLASS epoch 1 contours (cyan) of the giant radio galaxy PKS1449-129 overlaid on an optical image from the Dark Energy Camera Legacy Survey (Dey et al. 2019). The high resolution observation of a radio core by VLASS allows the radio galaxy to be correctly cross identified with the quasar PSO J145235.301-131120.98 (coincident with the cyan contours at the centre of the image; Villarreal Hernández and Andernach 2018). Previously it had erroneously been matched with the bright galaxy $\sim 8''$ to the north-east of the quasar (shown by the red circle in the above image)

inevitably suffer some level of contamination from misidentifications. The point spread function of VLASS is only a factor of ~ 3 (linearly) larger than that of most wide field optical surveys, therefore reducing the likelihood of host identification errors for unresolved sources (see Fig. 8.6). For VLASS the Canadian Initiative for Radio Astronomy Data Analysis (CIRADA) are providing likelihood ratio derived infrared host identifications from WISE.

References

N.J. Adams, R.A.A. Bowler, M.J. Jarvis, B. Häußler, C.D.P. Lagos, Evolution of the galaxy stellar mass function: evidence for an increasing M* from z = 2 to the present day. Mon. Not. R. Astron. Soc. **506**(4), 4933–4951 (2021). https://doi.org/10.1093/mnras/stab1956

H. Aihara, N. Arimoto, R. Armstrong, S. Arnouts, N.A. Bahcall, S. Bickerton, J. Bosch, K. Bundy, P.L. Capak, J.H.H. Chan, M. Chiba, J. Coupon, E. Egami, M. Enoki, F. Finet, H. Fujimori, S. Fujimoto, H. Furusawa, J. Furusawa, T. Goto, A. Goulding, J.P. Greco, J.E. Greene, J.E. Gunn, T. Hamana, Y. Harikane, Y. Hashimoto, T. Hattori, M. Hayashi, Y. Hayashi, K.G. Hełminiak, R. Higuchi, C. Hikage, P.T.P. Ho, B.-C. Hsieh, K. Huang, S. Huang, H. Ikeda, M. Imanishi, A.K. Inoue, K. Iwasawa, I. Iwata, A.T. Jaelani, H.-Y. Jian, Y. Kamata, H. Karoji, N. Kashikawa, N. Katayama, S. Kawanomoto, I. Kayo, J. Koda, M. Koike, T. Kojima, Y. Komiyama, A. Konno, S. Koshida, Y. Koyama, H. Kusakabe, A. Leauthaud, C.-H. Lee, L. Lin, Y.-T. Lin, R.H. Lupton, R. Mandelbaum, Y. Matsuoka, E. Medezinski, S. Mineo, S. Miyama, H. Miyatake, S. Miyazaki, R. Momose, A. More, S. More, Y. Moritani, T.J. Moriya, T. Morokuma, S. Mukae, R. Murata, H. Murayama, T. Nagao, F. Nakata, M. Niida, H. Niikura, A.J. Nishizawa, Y. Obuchi, M. Oguri, Y.

Oishi, N. Okabe, S. Okamoto, Y. Okura, Y. Ono, M. Onodera, M. Onoue, K. Osato, M. Ouchi, P.A. Price, T.-S. Pyo, M. Sako, M. Sawicki, Takatoshi Shibuya, K. Shimasaku, A. Shimono, M. Shirasaki, J.D. Silverman, M. Simet, J. Speagle, D.N. Spergel, M.A. Strauss, Y. Sugahara, N. Sugiyama, Y. Suto, S.H. Suyu, N. Suzuki, P.J. Tait, M. Takada, T. Takata, N. Tamura, M.M. Tanaka, M. Tanaka, M. Tanaka, Y. Tanaka, T. Terai, Y. Terashima, Y. Toba, N. Tominaga, J. Toshikawa, E.L. Turner, T. Uchida, H. Uchiyama, K. Umetsu, F. Uraguchi, Y. Urata, T. Usuda, Y. Utsumi, S.-Y. Wang, W.-H. Wang, K.C. Wong, K. Yabe, Y. Yamada, H. Yamanoi, N. Yasuda, S. Yeh, A. Yonehara, S. Yuma, The hyper Suprime-Cam SSP survey: overview and survey design. Publ. Astron. Soc. Japan **70**, S4 (2018). https://doi.org/10.1093/pasj/psx066

M. Akhlaghi, T. Ichikawa, Noise-based detection and segmentation of nebulous objects. Astrophys. J. Suppl. **220**, 1 (2015). https://doi.org/10.1088/0067-0049/220/1/1

L. Alegre, J. Sabater, P. Best, R.I.J. Mostert, W.L. Williams, G. Gürkan, M.J. Hardcastle, R. Kondapally, T.W. Shimwell, D.J.B. Smith, A machine learning classifier for LOFAR radio galaxy cross-matching techniques. arXiv e-prints, art. arXiv:2207.01645 (2022)

AMI Consortium, T.M.O. Franzen, M.L. Davies, E.M. Waldram, K.J.B. Grainge, M.P. Hobson, N. Hurley-Walker, A. Lasenby, M. Olamaie, G.G. Pooley, C. Rodríguez-Gonzálvez, R.D.E. Saunders, A.M.M. Scaife, M.P. Schammel, P.F. Scott, T.W. Shimwell, D.J. Titterington, J.T.L. Zwart, 10C survey of radio sources at 15.7 GHz - I. Observing, mapping and source extraction. Mon. Not. R. Astron. Soc. **415**, 2699 (2011). https://doi.org/10.1111/j.1365-2966.2011.18887.x

J.K. Banfield, O.I. Wong, K.W. Willett, R.P. Norris, L. Rudnick, S.S. Shabala, B.D. Simmons, C. Snyder, A. Garon, N. Seymour, E. Middelberg, H. Andernach, C.J. Lintott, K. Jacob, A.D. Kapińska, M.Y. Mao, K.L. Masters, M.J. Jarvis, K. Schawinski, E. Paget, R. Simpson, H.R. Klöckner, S. Bamford, T. Burchell, K.E. Chow, G. Cotter, L. Fortson, I. Heywood, T.W. Jones, S. Kaviraj, Á.R. López-Sánchez, W.P. Maksym, K. Polsterer, K. Borden, R.P. Hollow, L. Whyte, Radio Galaxy Zoo: host galaxies and radio morphologies derived from visual inspection. Mon. Not. R. Astron. Soc. **453**(3), 2326–2340 (2015). https://doi.org/10.1093/mnras/stv1688

B. Barkus, J.H. Croston, J. Piotrowska, B. Mingo, P.N. Best, M.J. Hardcastle, R.I.J. Mostert, H.J.A. Röttgering, J. Sabater, B. Webster, W.L. Williams, The application of ridgelines in extended radio source cross-identification. Mon. Not. R. Astron. Soc. **509**(1), 1–15 (2022). https://doi.org/10.1093/mnras/stab2952

R.H. Becker, R.L. White, D.J. Helfand, The FIRST survey: faint images of the radio sky at twenty centimeters. Astrophys. J. **450**, 559 (1995). https://doi.org/10.1086/176166

S.V.W. Beckwith, M. Stiavelli, A.M. Koekemoer, J.A.R. Caldwell, H.C. Ferguson, R. Hook, R.A. Lucas, L.E. Bergeron, M. Corbin, S. Jogee, N. Panagia, M. Robberto, P. Royle, R.S. Somerville, M. Sosey, The hubble ultra deep field. Astron. J. **132**(5), 1729–1755 (2006). https://doi.org/10.1086/507302

E. Bertin, S. Arnouts, Sextractor: Software for source extraction. Astron. Astrophys. Suppl. Ser. **117**, 393 (1996) https://doi.org/10.1051/aas:1996164

S. Beucher, C. Lantuejoul, *International Workshop on Image Processing: Real-time Edge and Motion detection/estimation, Rennes, France*, Sept. (1979), p. 12

A. Bonaldi, M. Bonato, V. Galluzzi, I. Harrison, Marcella Massardi, S. Kay, G. De Zotti, and M.L. Brown. The tiered radio extragalactic continuum simulation (T-RECS). Mon. Not. R. Astron. Soc. **482**(1), 2–19 (2019). https://doi.org/10.1093/mnras/sty2603

A. Bonaldi, T. An, M. Brüggen, S. Burkutean, B. Coelho, H. Goodarzi, P. Hartley, P.K. Sandhu, C. Wu, L. Yu, M.H. Z.Haghighi, S. Antón, Z. Bagheri, D. Barbosa, J.P. Barraca, D. Bartashevich, M. Bergano, M. Bonato, J. Brand, A. de Gasperin, F. Giannetti, R. Dodson, P. Jain, S. Jaiswal, B. Lao, B. Liu, E. Liuzzo, Y. Lu, V. Lukic, D. Maia, N. Marchili, M. Massardi, P. Mohan, J.B. Morgado, M. Panwar, P. Prabhakar, V.A.R.M. Ribeiro, K.L.J. Rygl, V.S. Ali, E. Saremi, E. Schisano, S. Sheikhnezami, A.V. Sadr, A. Wong, O.I. Wong, Square kilometre array science data challenge 1: analysis and results. Mon. Not. R. Astron. Soc. **500**, 3821 (2021) https://doi.org/10.1093/mnras/staa3023

M.M. Boyce, A.M. Hopkins, S. Riggi, L. Rudnick, M. Ramsay, C.L. Hale, J. Marvil, M. Whiting, P. Venkataraman, C.P. O'Dea, S.A. Baum, Y.A. Gordon, A.N. Vantyghem, M. Dionyssiou, H. Andernach, J.D. Collier, J. English, B.S. Koribalski, D. Leahy, M.J. Michałowski, S. Safi-Harb, M. Vaccari, E. Alexander, M. Cowley, A.D. Kapinska, A.S.G. Robotham, H.. Tang, Hydra I: An extensible multi-source-finder comparison and cataloguing tool. Publ. Astron. Soc. Austral. **40**, e028 (2023a)

M.M. Boyce, A.M. Hopkins, S. Riggi, L. Rudnick, M. Ramsay, C.L. Hale, J. Marvil, M. Whiting, P. Venkataraman, C.P. O'Dea, S.A. Baum, Y.A. Gordon, A.N. Vantyghem, M. Dionyssiou, H. Andernach, J.D. Collier, J. English, B.S. Koribalski, D. Leahy, M.J. Michałowski, S. Safi-Harb, M. Vaccari, E. Alexander, M. Cowley, A.D. Kapinska, A.S.G. Robotham, H. Tang, Hydra II: Characterisation of Aegean, Caesar, ProFound, PyBDSF, and Selavy. Publ. Astron. Soc. Austral. **40**, e027 (2023b)

G.B. Brammer, R. Sánchez-Janssen, I. Labbé, E. da Cunha, D.K. Erb, M. Franx, M. Fumagalli, B. Lundgren, D. Marchesini, I. Momcheva, E. Nelson, S. Patel, R. Quadri, Hans-W. Rix, R.E. Skelton, K.B. Schmidt, A. van der Wel, P.G. van Dokkum, D.A. Wake, K.E. Whitaker, 3D-HST Grism spectroscopy of a gravitationally lensed, low-metallicity starburst galaxy at z = 1.847. Astrophys. J. Lett. **758**(1), L17 (2012). https://doi.org/10.1088/2041-8205/758/1/L17

S. Chi, P.D. Barthel, M.A. Garrett, Deep, wide-field, global VLBI observations of the Hubble deep field north (HDF-N) and flanking fields (HFF). Astron. Astrophys. **550**, A68 (2013). https://doi.org/10.1051/0004-6361/201220783

J.J. Condon, Radio emission from normal galaxies. Annu. Rev. Astron. Astrophys. **30**, 575–611 (1992). https://doi.org/10.1146/annurev.aa.30.090192.003043

J.J. Condon, W.D. Cotton, E.W. Greisen, Q.F. Yin, R.A. Perley, G.B. Taylor, J.J. Broderick, The NRAO VLA sky survey. Astron. J. **115**, 1693 (1998). https://doi.org/10.1086/300337

F. de Gasperin, W.L. Williams, P. Best, M. Brüggen, G. Brunetti, V. Cuciti, T.J. Dijkema, M.J. Hardcastle, M.J. Norden, A. Offringa, T. Shimwell, R. van Weeren, D. Bomans, A. Bonafede, A. Botteon, J.R. Callingham, R. Cassano, K.T. Chyży, K.L. Emig, H. Edler, M. Haverkorn, G. Heald, V. Heesen, M. Iacobelli, H.T. Intema, M. Kadler, K. Małek, M. Mevius, G. Miley, B. Mingo, L.K. Morabito, J. Sabater, R. Morganti, E. Orrú, R. Pizzo, I. Prandoni, A. Shulevski, C. Tasse, M. Vaccari, P. Zarka, H. Röttgering, The LOFAR LBA Sky Survey. I. Survey description and preliminary data release. Astron. Astrophys. **648**, A104 (2021). https://doi.org/10.1051/0004-6361/202140316

A.T. Deller, E. Middelberg, mJIVE-20: A survey for compact mJy radio objects with the very long baseline array. Astron. J. **147**(1), 14 (2014). https://doi.org/10.1088/0004-6256/147/1/14

A.T. Deller, W.F. Brisken, C.J. Phillips, J. Morgan, W. Alef, R. Cappallo, E. Middelberg, J. Romney, H. Rottmann, S.J. Tingay, R. Wayth, DiFX-2: A more flexible, efficient, robust, and powerful software correlator. Publ. Astron. Soc. Pac. **123**(901), 275 (2011). https://doi.org/10.1086/658907

I. Delvecchio, V. Smolčić, G. Zamorani, C.D.P. Lagos, S. Berta, J. Delhaize, N. Baran, D.M. Alexander, D.J. Rosario, V. Gonzalez-Perez, O. Ilbert, C.G. Lacey, O. Le Fèvre, O. Miettinen, M. Aravena, M. Bondi, C. Carilli, P. Ciliegi, K. Mooley, M. Novak, E. Schinnerer, P. Capak, F. Civano, N. Fanidakis, N. Herrera Ruiz, A. Karim, C. Laigle, S. Marchesi, H.J. McCracken, E. Middleberg, M. Salvato, L. Tasca, The VLA-COSMOS 3 GHz large project: AGN and host-galaxy properties out to $z \lesssim 6$. Astron. Astrophys. **602**, A3 (2017). https://doi.org/10.1051/0004-6361/201629367

A. Dey, D.J. Schlegel, D. Lang, R. Blum, K. Burleigh, X. Fan, J.R. Findlay, D. Finkbeiner, D. Herrera, Stéphanie Juneau, M. Landriau, M. Levi, I. McGreer, A. Meisner, A.D. Myers, J. Moustakas, P. Nugent, A. Patej, E.F. Schlafly, A.R. Walker, F. Valdes, B.A. Weaver, C. Yèche, H. Zou, X. Zhou, B. Abareshi, T.M.C. Abbott, B. Abolfathi, C. Aguilera, S. Alam, L. Allen, A. Alvarez, J. Annis, B. Ansarinejad, M. Aubert, J. Beechert, E.F. Bell, S.Y. BenZvi, F. Beutler, R.M. Bielby, A.S. Bolton, C. Briceño, E.J. Buckley-Geer, K. Butler, A. Calamida, R.G. Carlberg, P. Carter, R. Casas, F.J. Castander, Y. Choi, J. Comparat, E. Cukanovaite, T. Delubac, K. DeVries, S. Dey, G. Dhungana, M. Dickinson, Z. Ding, J.B. Donaldson, Y. Duan, C.J. Duckworth, S. Eftekharzadeh, D.J. Eisenstein, T. Etourneau, P.A. Fagrelius, J.

Farihi, M. Fitzpatrick, A. Font-Ribera, L. Fulmer, B.T. Gänsicke, E. Gaztanaga, K. George, D.W. Gerdes, S.G.A. Gontcho, C. Gorgoni, G. Green, J. Guy, D. Harmer, M. Hernandez, K. Honscheid, L.W. Huang, D.J. James, B.T. Jannuzi, L. Jiang, R. Joyce, A. Karcher, S. Karkar, R. Kehoe, J.-P. Kneib, A. Kueter-Young, T.-W. Lan, T.R. Lauer, L. Le Guillou, A. Le Van Suu, J.H. Lee, M. Lesser, L. Perreault Levasseur, T.S. Li, J.L. Mann, R. Marshall, C.E. Martínez-Vázquez, P. Martini, H. du Mas des Bourboux, S. McManus, T.G. Meier, B. Ménard, N. Metcalfe, A. Muñoz-Gutiérrez, J. Najita, K. Napier, G. Narayan, J.A. Newman, J. Nie, B. Nord, D.J. Norman, K.A.G. Olsen, A. Paat, N. Palanque-Delabrouille, X. Peng, C.L. Poppett, M.R. Poremba, A. Prakash, D. Rabinowitz, A. Raichoor, M. Rezaie, A.N. Robertson, N.A. Roe, A.J. Ross, N.P. Ross, G. Rudnick, S. Safonova, A. Saha, F.J. Sánchez, E. Savary, H. Schweiker, A. Scott, Hee-J. Seo, H. Shan, D.R. Silva, Z. Slepian, C. Soto, D. Sprayberry, R. Staten, C.M. Stillman, R.J. Stupak, D.L. Summers, S. Sien Tie, H. Tirado, M. Vargas-Magaña, A.K. Vivas, R.H. Wechsler, D. Williams, J. Yang, Q. Yang, T. Yapici, D. Zaritsky, A. Zenteno, K. Zhang, T. Zhang, R. Zhou, Z. Zhou, Overview of the DESI legacy imaging surveys. Astron. J. **157**(5), 168 (2019). https://doi.org/10.3847/1538-3881/ab089d

K.J. Duncan, All-purpose, all-sky photometric redshifts for the Legacy Imaging Surveys Data Release 8. Mon. Not. R. Astron. Soc. **512**(3), 3662–3683 (2022). https://doi.org/10.1093/mnras/stac608

S.A. Dzib, J. Forbrich, M.J. Reid, K.M. Menten, A VLBA survey of radio stars in the Orion Nebula Cluster. II. Astrometry. Astrophys. J. **906**(1), 24 (2021). https://doi.org/10.3847/1538-4357/abc68f

D. Fan, T. Budavári, R.P. Norris, A.M. Hopkins, Matching radio catalogues with realistic geometry: application to SWIRE and ATLAS. Mon. Not. R. Astron. Soc. **451**(2), 1299–1305 (2015). https://doi.org/10.1093/mnras/stv994

D. Fan, T. Budavári, R.P. Norris, A. Basu, Optimal probabilistic catalogue matching for radio sources. Mon. Not. R. Astron. Soc. **498**(1), 565–573 (2020). https://doi.org/10.1093/mnras/staa2447

B.L. Fanaroff, J.M. Riley, The morphology of extragalactic radio sources of high and low luminosity. Mon. Not. R. Astron. Soc. **167**, 31P–36P (1974). https://doi.org/10.1093/mnras/167.1.31P

J. Forbrich, S.A. Dzib, M.J. Reid, K.M. Menten, A VLBA survey of radio stars in the Orion Nebula Cluster. I. The nonthermal radio population. Astrophys. J. **906**(1), 23 (2021). https://doi.org/10.3847/1538-4357/abc68e

M.A. Garrett, J.M. Wrobel, R. Morganti, 21st Century VLBI: deep wide-field surveys, in *Future Directions in High Resolution Astronomy*, ed. by J. Romney, M. Reid, volume 340 of Astronomical Society of the Pacific Conference Series (2005), p. 600. https://doi.org/10.48550/arXiv.astro-ph/0403642

Y.A. Gordon, M.M. Boyce, C.P. O'Dea, Lawrence Rudnick, H. Andernach, A.N. Vantyghem, S.A. Baum, Jean-P. Bui, M. Dionyssiou, S. Safi-Harb, and I. Sander. A quick look at the 3 GHz radio sky. I. Source statistics from the very large array sky survey. Astrophys. J. Suppl. **255**(2), 30 (2021). https://doi.org/10.3847/1538-4365/ac05c0

N.A. Grogin, D.D. Kocevski, S.M. Faber, H.C. Ferguson, A.M. Koekemoer, A.G. Riess, V. Acquaviva, D.M. Alexander, O. Almaini, Matthew L.N. Ashby, M. Barden, E.F. Bell, Frédéric Bournaud, T.M. Brown, K.I. Caputi, S. Casertano, P. Cassata, M. Castellano, P. Challis, Ranga-R. Chary, E. Cheung, M. Cirasuolo, C.J. Conselice, A. Roshan Cooray, D.J. Croton, E. Daddi, T. Dahlen, R. Davé, D.F. de Mello, A. Dekel, M. Dickinson, T. Dolch, J.L. Donley, J.S. Dunlop, A.A. Dutton, D. Elbaz, G.G. Fazio, A.V. Filippenko, S.L. Finkelstein, A. Fontana, J.P. Gardner, P.M. Garnavich, E. Gawiser, M. Giavalisco, A. Grazian, Y. Guo, N.P. Hathi, B. Häussler, P.F. Hopkins, J.-S. Huang, K.-H. Huang, S.W. Jha, J.S. Kartaltepe, R.P. Kirshner, D.C. Koo, K. Lai, K.-S. Lee, W. Li, J.M. Lotz, R.A. Lucas, P. Madau, P.J. McCarthy, E.J. McGrath, D.H. McIntosh, R.J. McLure, B. Mobasher, L.A. Moustakas, M. Mozena, K. Nandra, J.A. Newman, S.-M. Niemi, K.G. Noeske, C.J. Papovich, L. Pentericci, A. Pope, J.R. Primack, A. Rajan, S. Ravindranath, N.A. Reddy, A. Renzini, H.-W. Rix, A.R. Robaina, S.A. Rodney, D.J. Rosario, P. Rosati, S. Salimbeni, C. Scarlata, B. Siana, L. Simard, J. Smidt, R.S. Somerville, H. Spinrad,

A.N. Straughn, L.-G. Strolger, O. Telford, H.I. Teplitz, J.R. Trump, A. van der Wel, C. Villforth, R.H. Wechsler, B.J. Weiner, T. Wiklind, V. Wild, G. Wilson, S. Wuyts, H.-J. Yan, M.S. Yun, CANDELS: The cosmic assembly near-infrared deep extragalactic legacy survey. Astrophys. J. Suppl. **197**(2), 35 (2011). https://doi.org/10.1088/0067-0049/197/2/35

N. Gupta, M. Huynh, R.P. Norris, R. Wang, A.M. Hopkins, H. Andernach, B.S. Koribalski, T.J. Galvin, Discovery of peculiar radio morphologies with ASKAP using unsupervised machine learning. arXiv e-prints, art. arXiv:2208.13997 (2022)

C.L. Hale, A.S.G. Robotham, L.J.M. Davies, M.J. Jarvis, S.P. Driver, I. Heywood, Radio source extraction with profound. Mon. Not. R. Astron. Soc. **487**, 1486 (2019). https://doi.org/10.1093/mnras/stw982

C.L. Hale, I.H. Whittam, M.J. Jarvis, P.N. Best, N.L. Thomas, I. Heywood, M. Prescott, N. Adams, J. Afonso, F. An, R.A.A. Bowler, J.D. Collier, R.H.W. Cook, R. Davé, B.S. Frank, M. Glowacki, P.W. Hatfield, S. Kolwa, C.C. Lovell, N. Maddox, L. Marchetti, L.K. Morabito, E. Murphy, I. Prandoni, Z. Randriamanakoto, A.R. Taylor, MIGHTEE: deep 1.4 GHz source counts and the sky temperature contribution of star forming galaxies and active galactic nuclei. Mon. Not. R. Astron. Soc. (2022). https://doi.org/10.1093/mnras/stac3320

C.A. Hales, T. Murphy, J.R. Curran, E. Middelberg, B.M. Gaensler, R.P. Norris, Blobcat: software to catalogue flood-filled blobs in radio images of total intensity and linear polarization. Mon. Not. R. Astron. Soc. **425**, 979 (2012). https://doi.org/10.1111/j.1365-2966.2012.21373.x

P.J. Hancock, T. Murphy, B.M. Gaensler, A. Hopkins, J.R. Curran, Compact continuum source finding for next generation radio surveys. Mon. Not. R. Astron. Soc. **422**, 1812 (2012). https://doi.org/10.1111/j.1365-2966.2012.20768.x

M.J. Hardcastle, G. Gürkan, R.J. van Weeren, W.L. Williams, P.N. Best, F. de Gasperin, D.A. Rafferty, S.C. Read, J. Sabater, T.W. Shimwell, D.J.B. Smith, C. Tasse, N. Bourne, M. Brienza, M. Brüggen, G. Brunetti, K.T. Chyży, J. Conway, L. Dunne, S.A. Eales, S.J. Maddox, M.J. Jarvis, E.K. Mahony, R. Morganti, I. Prandoni, H.J.A. Röttgering, E. Valiante, G.J. White, LOFAR/H-ATLAS: a deep low-frequency survey of the Herschel-ATLAS North Galactic Pole field. Mon. Not. R. Astron. Soc. **462**(2), 1910–1936 (2016). https://doi.org/10.1093/mnras/stw1763

M.J. Hardcastle, W.L. Williams, P.N. Best, J.H. Croston, K.J. Duncan, H.J.A. Röttgering, J. Sabater, T.W. Shimwell, C. Tasse, J.R. Callingham, R.K. Cochrane, F. de Gasperin, G. Gürkan, M.J. Jarvis, V. Mahatma, G.K. Miley, B. Mingo, S. Mooney, L.K. Morabito, S.P. O'Sullivan, I. Prandoni, A. Shulevski, D.J.B. Smith, Radio-loud AGN in the first LoTSS data release. The lifetimes and environmental impact of jet-driven sources. Astron. Astrophys. **622**, A12 (2019). https://doi.org/10.1051/0004-6361/201833893

M.J. Hardcastle, M.A. Horton, W.L. Williams, K.J. Duncan, L. Alegre, B. Barkus, J.H. Croston, H. Dickinson, E. Osinga, H.J.A. Röttgering, J. Sabater, T.W. Shimwell, D.J.B. Smith, P.N. Best, A. Botteon, M. Brüggen, A. Drabent, F. de Gasperin, G. Gürkan, M. Hajduk, C.L. Hale, M. Hoeft, M. Jamrozy, M. Kunert-Bajraszewska, R. Kondapally, M. Magliocchetti, V.H. Mahatma, R.I.J. Mostert, S.P. O'Sullivan, U. Pajdosz-Śmierciak, J. Petley, J.C.S. Pierce, I. Prandoni, D.J. Schwarz, A. Shulewski, T.M. Siewert, J.P. Stott, H. Tang, M. Vaccari, X. Zheng, T. Bailey, S. Desbled, A. Goyal, V. Gonano, M. Hanset, W. Kurtz, S.M. Lim, L. Mielle, C.S. Molloy, R. Roth, I.A. Terentev, M. Torres, The LOFAR two-metre sky survey. VI. Optical identifications for the second data release. Astron. Astrophys. **678**, A151 (2023). https://doi.org/10.1051/0004-6361/202347333

P.W. Hatfield, M.J. Jarvis, N. Adams, R.A.A. Bowler, B. Häußler, K.J. Duncan, Hybrid photometric redshifts for sources in the COSMOS and XMM-LSS fields. Mon. Not. R. Astron. Soc. **513**(3), 3719–3733 (2022). https://doi.org/10.1093/mnras/stac1042

N. Herrera Ruiz, E. Middelberg, A. Deller, R.P. Norris, P.N. Best, W. Brisken, E. Schinnerer, V. Smolčić, I. Delvecchio, E. Momjian, D. Bomans, N.Z. Scoville, C. Carilli, The faint radio sky: VLBA observations of the COSMOS field. Astron. Astrophys. **607**, A132 (2017). https://doi.org/10.1051/0004-6361/201731163

N. Herrera Ruiz, E. Middelberg, A. Deller, V. Smolčić, R.P. Norris, M. Novak, I. Delvecchio, P.N. Best, E. Schinnerer, E. Momjian, R.J. Dettmar, W. Brisken, A.M. Koekemoer, N.Z. Scoville,

VLBA+GBT observations of the COSMOS field and radio source counts at 1.4 GHz. Astron. Astrophys. **616**, A128 (2018). https://doi.org/10.1051/0004-6361/201832969

I. Heywood, M.J. Jarvis, C.L. Hale, I.H. Whittam, H.L. Bester, B. Hugo, J.S. Kenyon, M. Prescott, O.M. Smirnov, C. Tasse, J.M. Afonso, P.N. Best, J.D. Collier, R.P. Deane, B.S. Frank, M.J. Hardcastle, K. Knowles, N. Maddox, E.J. Murphy, I. Prandoni, S.M. Randriamampandry, M.G. Santos, S. Sekhar, F. Tabatabaei, A.R. Taylor, K. Thorat, MIGHTEE: total intensity radio continuum imaging and the COSMOS/XMM-LSS Early Science fields. Mon. Not. R. Astron. Soc. **509**(2), 2150–2168 (2022). https://doi.org/10.1093/mnras/stab3021

A.M. Hopkins, C.J. Miller, A.J. Connolly, C. Genovese, R.C. Nichol, L. Wasserman, A new source detection algorithm using the false-discovery rate. Astron. J. **123**(2), 1086–1094 (2002). https://doi.org/10.1086/338316

A.M. Hopkins, M.T. Whiting, N. Seymour, K.E. Chow, R.P. Norris, L. Bonavera, R. Breton, D. Carbone, C. Ferrari, T.M.O. Franzen, H. Garsden, J. González-Nuevo, C.A. Hales, P.J. Hancock, G. Heald, D. Herranz, M. Huynh, R.J. Jurek, M. López-Caniego, N. Massardi, M. Mohan, S. Molinari, E. Orrù, R. Paladino, M. Pestalozzi, R. Pizzo, D. Rafferty, H.J.A. Röttgering, L. Rudnick, E. Schisano, A. Shulevski, J. Swinbank, R. Taylor, A.J. van der Horst, The ASKAP/EMU source finding data challenge. Publ. Astron. Soc. Austral. **32**, E037 (2015). https://doi.org/10.1017/pasa.2015.37

A.W. Hotan, J.D. Bunton, A.P. Chippendale, M. Whiting, J. Tuthill, V.A. Moss, D. McConnell, S.W. Amy, M.T. Huynh, J.R. Allison, C.S. Anderson, K.W. Bannister, E. Bastholm, R. Beresford, D.C.J. Bock, R. Bolton, J.M. Chapman, K. Chow, J.D. Collier, F.R. Cooray, T.J. Cornwell, P.J. Diamond, P.G. Edwards, I.J. Feain, T.M.O. Franzen, D. George, N. Gupta, G.A. Hampson, L. Harvey-Smith, D.B. Hayman, I. Heywood, C. Jacka, C.A. Jackson, S. Jackson, K. Jeganathan, S. Johnston, M. Kesteven, D. Kleiner, B.S. Koribalski, K. Lee-Waddell, E. Lenc, E.S. Lensson, S. Mackay, E.K. Mahony, N.M. McClure-Griffiths, R. McConigley, P. Mirtschin, A.K. Ng, R.P. Norris, S.E. Pearce, C. Phillips, M.A. Pilawa, W. Raja, J.E. Reynolds, P. Roberts, D.N. Roxby, E.M. Sadler, M. Shields, A.E.T. Schinckel, P. Serra, R.D. Shaw, T. Sweetnam, E.R. Troup, A. Tzioumis, M.A. Voronkov, T. Westmeier, Australian square kilometre array pathfinder: I. System description. Publ. Astron. Soc. Austral. **38**, e009 (2021). https://doi.org/10.1017/pasa.2021.1

P. Hudelot, J.Ch. Cuillandre, K. Withington, Y. Goranova, H. McCracken, F. Magnard, Y. Mellier, N. Regnault, M. Betoule, H. Aussel, J.J. Kavelaars, P. Fernique, F. Bonnarel, F. Ochsenbein, O. Ilbert, VizieR online data catalog: The CFHTLS survey (T0007 release) (Hudelot+ 2012). *VizieR Online Data Catalog*, art. II/317 (2012)

M.T. Huynh, A. Hopkins, R. Norris, P. Hancock, T. Murphy, R. Jurek, M. Whitting, The completeness and reliability of threshold and false-discovery rate source extraction algorithms for compact continuum sources. Publ. Astron. Soc. Austral. **29**, 229 (2012). https://doi.org/10.1071/AS11026

H.T. Intema, P. Jagannathan, K.P. Mooley, D.A. Frail, The GMRT 150 MHz all-sky radio survey. First alternative data release TGSS ADR1. Astron. Astrophys. **598**, A78 (2017). https://doi.org/10.1051/0004-6361/201628536

M.J. Jarvis, D.G. Bonfield, V.A. Bruce, J.E. Geach, K. McAlpine, R.J. McLure, E. González-Solares, M. Irwin, J. Lewis, A.K. Yoldas, S. Andreon, N.J.G. Cross, J.P. Emerson, G. Dalton, J.S. Dunlop, S.T. Hodgkin, F.O. Le, M. Karouzos, K. Meisenheimer, S. Oliver, S. Rawlings, C. Simpson, I. Smail, D.J.B. Smith, M. Sullivan, W. Sutherland, S.V. White, J.T.L. Zwart, The VISTA deep extragalactic observations (VIDEO) survey. Mon. Not. R. Astron. Soc. **428**(2), 1281–1295 (2013). https://doi.org/10.1093/mnras/sts118

M. Jarvis, R. Taylor, I. Agudo, J.R. Allison, R.P. Deane, B. Frank, N. Gupta, I. Heywood, N. Maddox, K. McAlpine, M. Santos, A.M.M. Scaife, M. Vaccari, J.T.L. Zwart, E. Adams, D.J. Bacon, A.J. Baker, B.A. Bassett, P.N. Best, R. Beswick, S. Blyth, M.L. Brown, M. Bruggen, M. Cluver, S. Colafrancesco, G. Cotter, C. Cress, R. Davé, C. Ferrari, M.J. Hardcastle, C.L. Hale, I. Harrison, P.W. Hatfield, H.R. Klockner, S. Kolwa, E. Malefahlo, T. Marubini, T. Mauch, K. Moodley, R. Morganti, R.P. Norris, J.A. Peters, I. Prandoni, M. Prescott, S. Oliver, N. Oozeer, H.J.A. Rottgering, N. Seymour, C. Simpson, O. Smirnov, D.J.B. Smith, The

MeerKAT international GHz tiered extragalactic exploration (MIGHTEE) survey, in *MeerKAT Science: On the Pathway to the SKA* (2016, January), p. 6. https://doi.org/10.22323/1.277.0006

E.F. Jiménez-Andrade, B. Magnelli, A. Karim, G. Zamorani, M. Bondi, E. Schinnerer, M. Sargent, E. Romano-Díaz, M. Novak, P. Lang, F. Bertoldi, E. Vardoulaki, S. Toft, V. Smolčić, K. Harrington, S. Leslie, J. Delhaize, D. Liu, C. Karoumpis, J. Kartaltepe, A.M. Koekemoer, Radio continuum size evolution of star-forming galaxies over $0.35 < z < 2.25$. Astron. Astrophys. **625**, A114 (2019). https://doi.org/10.1051/0004-6361/201935178

S. Jin, E. Daddi, D. Liu, V. Smolčić, E. Schinnerer, A. Calabrò, Q. Gu, J. Delhaize, I. Delvecchio, Y. Gao, M. Salvato, A. Puglisi, M. Dickinson, F. Bertoldi, M. Sargent, M. Novak, G. Magdis, I. Aretxaga, G.W. Wilson, P. Capak, "Super-deblended" Dust emission in galaxies. II. Far-IR to (Sub)millimeter photometry and high-redshift galaxy candidates in the full COSMOS field. Astrophys. J. **864**(1), 56 (2018). https://doi.org/10.3847/1538-4357/aad4af

J. Jonas, MeerKAT Team, The MeerKAT radio telescope, in *MeerKAT Science: On the Pathway to the SKA* (2016), p. 1. https://doi.org/10.22323/1.277.0001

N. Jurlin, R. Morganti, M. Brienza, S. Mandal, N. Maddox, K.J. Duncan, S.S. Shabala, M.J. Hardcastle, I. Prandoni, H.J.A. Röttgering, V. Mahatma, P.N. Best, B. Mingo, J. Sabater, T.W. Shimwell, C. Tasse, The life cycle of radio galaxies in the LOFAR Lockman Hole field. Astron. Astrophys. **638**, A34 (2020). https://doi.org/10.1051/0004-6361/201936955

L.J. Kewley, C.A. Heisler, M.A. Dopita, R. Sutherland, R.P. Norris, J. Reynolds, S. Lumsden, Compact radio emission from warm infrared galaxies. Astrophys. J. **530**(2), 704–718 (2000). https://doi.org/10.1086/308397

D.E. Knuth, *The Art of Computer Programming* (Addison-Wesley, 2000). ISBN 0-201-89683-4

A.M. Koekemoer, S.M. Faber, H.C. Ferguson, N.A. Grogin, D.D. Kocevski, D.C. Koo, K. Lai, J.M. Lotz, R.A. Lucas, E.J. McGrath, S. Ogaz, A. Rajan, A.G. Riess, S.A. Rodney, L. Strolger, S. Casertano, M. Castellano, T. Dahlen, M. Dickinson, T. Dolch, A. Fontana, M. Giavalisco, A. Grazian, Y. Guo, N.P. Hathi, K.-H. Huang, A. van der Wel, H.-J. Yan, V. Acquaviva, D.M. Alexander, O. Almaini, M.L.N. Ashby, M. Barden, E.F. Bell, F. Bournaud, T.M. Brown, K.I. Caputi, P. Cassata, P.J. Challis, R.-R. Chary, E. Cheung, M. Cirasuolo, C.J. Conselice, A. Roshan Cooray, D.J. Croton, E. Daddi, R. Davé, D.F. de Mello, L. de Ravel, A. Dekel, J.L. Donley, J.S. Dunlop, A.A. Dutton, D. Elbaz, G.G. Fazio, A.V. Filippenko, S.L. Finkelstein, C. Frazer, J.P. Gardner, P.M. Garnavich, E. Gawiser, R. Gruetzbauch, W.G. Hartley, B. Häussler, J. Herrington, P.F. Hopkins, J.-S. Huang, S.W. Jha, A. Johnson, J.S. Kartaltepe, A.A. Khostovan, R.P. Kirshner, C. Lani, K.-S. Lee, W. Li, P. Madau, P.J. McCarthy, D.H. McIntosh, R.J. McLure, C. McPartland, B. Mobasher, H. Moreira, A. Mortlock, L.A. Moustakas, M. Mozena, K. Nandra, J.A. Newman, J.L. Nielsen, S. Niemi, K.G. Noeske, C.J. Papovich, L. Pentericci, A. Pope, J.R. Primack, S. Ravindranath, N.A. Reddy, A. Renzini, H.-W. Rix, A.R. Robaina, D.J. Rosario, P. Rosati, S. Salimbeni, C. Scarlata, B. Siana, L. Simard, J. Smidt, D. Snyder, R.S. Somerville, H. Spinrad, A.N. Straughn, O. Telford, H.I. Teplitz, J.R. Trump, C. Vargas, C. Villforth, C.R. Wagner, P. Wandro, R.H. Wechsler, B.J. Weiner, T. Wiklind, V. Wild, G. Wilson, S. Wuyts, M.S. Yun, CANDELS: The cosmic assembly near-infrared deep extragalactic legacy survey—The Hubble space telescope observations, imaging data products, and mosaics. Astrophys. J. Suppl. **197**(2), 36 (2011). https://doi.org/10.1088/0067-0049/197/2/36

R. Kondapally, P.N. Best, M.J. Hardcastle, D. Nisbet, M. Bonato, J. Sabater, K.J. Duncan, I. McCheyne, R.K. Cochrane, R.A.A. Bowler, W.L. Williams, T.W. Shimwell, C. Tasse, J.H. Croston, A. Goyal, M. Jamrozy, M.J. Jarvis, V.H. Mahatma, H.J.A. Röttgering, D.J.B. Smith, A. Wołowska, M. Bondi, M. Brienza, M.J.I. Brown, M. Brüggen, K. Chambers, M.A. Garrett, G. Gürkan, M. Huber, M. Kunert-Bajraszewska, E. Magnier, B. Mingo, R. Mostert, B. Nikiel-Wroczyński, S.P. O'Sullivan, R. Paladino, T. Ploeckinger, I. Prandoni, M.J. Rosenthal, D.J. Schwarz, A. Shulevski, J.D. Wagenveld, L. Wang, The LOFAR two-meter sky survey: deep fields data release 1. III. Host-galaxy identifications and value added catalogues. Astron. Astrophys. **648**, A3 (2021). https://doi.org/10.1051/0004-6361/202038813

R.G. Kron, Photometry of a complete sample of faint galaxies. Astrophys. J. Suppl. **43**, 305 (1980). https://doi.org/10.1086/190669

M. Lacy, S.A. Baum, C.J. Chandler, S. Chatterjee, T.E. Clarke, S. Deustua, J. English, J. Farnes, B.M. Gaensler, N. Gugliucci, G. Hallinan, B.R. Kent, A. Kimball, C.J. Law, T.J.W. Lazio, J. Marvil, S.A. Mao, D. Medlin, K. Mooley, E.J. Murphy, S. Myers, R. Osten, G.T. Richards, E. Rosolowsky, L. Rudnick, F. Schinzel, G.R. Sivakoff, L.O. Sjouwerman, R. Taylor, R.L. White, J. Wrobel, H. Andernach, A.J. Beasley, E. Berger, S. Bhatnager, M. Birkinshaw, G.C. Bower, W.N. Brandt, S. Brown, S. Burke-Spolaor, B.J. Butler, J. Comerford, P.B. Demorest, H. Fu, S. Giacintucci, K. Golap, T. Güth, C.A. Hales, R. Hiriart, J. Hodge, A. Horesh, Ž. Ivezić, M.J. Jarvis, A. Kamble, N. Kassim, X. Liu, L. Loinard, D.K. Lyons, J. Masters, M. Mezcua, G.A. Moellenbrock, T. Mroczkowski, K. Nyland, C.P. O'Dea, S.P. O'Sullivan, W.M. Peters, K. Radford, U. Rao, J. Robnett, J. Salcido, Y. Shen, A. Sobotka, S. Witz, M. Vaccari, R.J. van Weeren, A. Vargas, P.K.G. Williams, I. Yoon, The Karl G. Jansky Very Large Array Sky Survey (VLASS). Science case and survey design. Publ. Astron. Soc. Pac. **132**(1009), 035001 (2020). https://doi.org/10.1088/1538-3873/ab63eb

B. Lao, T. An, A. Wang, Z. Xu, S. Guo, W. Lv, X. Wu, Y. Zhang, Artificial intelligence for celestial object census: the latest technology meets the oldest science. Sci. Bull. **66**(21), 2145–2147 (2021). https://doi.org/10.1016/j.scib.2021.07.015

D. Liu, E. Daddi, M. Dickinson, F. Owen, M. Pannella, M. Sargent, M. Béthermin, G. Magdis, Y. Gao, X. Shu, T. Wang, S. Jin, H. Inami, "Super-deblended" Dust emission in galaxies. I. The GOODS-North catalog and the cosmic star formation rate density out to redshift 6. Astrophys. J. **853**(2), 172 (2018). https://doi.org/10.3847/1538-4357/aaa600

D. Liu, P. Lang, B. Magnelli, E. Schinnerer, S. Leslie, Y. Fudamoto, M. Bondi, B. Groves, E. Jiménez-Andrade, K. Harrington, A. Karim, P.A. Oesch, M. Sargent, E. Vardoulaki, T. Bădescu, L. Moser, F. Bertoldi, A. Battisti, E. da Cunha, J. Zavala, M. Vaccari, I. Davidzon, D. Riechers, M. Aravena, Automated mining of the ALMA archive in the COSMOS field (A^3COSMOS). I. Robust ALMA continuum photometry catalogs and stellar mass and star formation properties for $\sim$700 galaxies at z = 0.5-6. Astrophys. J. Suppl. **244**(2), 40 (2019). https://doi.org/10.3847/1538-4365/ab42da

M. López-Caniego, P. Vielva, Biparametric adaptive filter: detection of compact sources in complex microwave backgrounds. Mon. Not. R. Astron. Soc. **421**, 2139 (2012). https://doi.org/10.1111/j.1365-2966.2012.20444.x

M. López-Caniego, D. Herranz, J. González-Nuevo, J.L. Sanz, R.B. Barreiro, P. Vielva, F. Argüeso, L. Toffolatti, Comparison of filters for the detection of point sources in planck simulations. Mon. Not. R. Astron. Soc. **370**, 2047 (2006). https://doi.org/10.1111/j.1365-2966.2006.10639.x

K. McAlpine, D.J.B. Smith, M.J. Jarvis, D.G. Bonfield, S. Fleuren, The likelihood ratio as a tool for radio continuum surveys with Square Kilometre Array precursor telescopes. Mon. Not. R. Astron. Soc. **423**(1), 132–140 (2012). https://doi.org/10.1111/j.1365-2966.2012.20715.x

H.J. McCracken, B. Milvang-Jensen, J. Dunlop, M. Franx, J.P.U. Fynbo, O. Le Fèvre, J. Holt, K.I. Caputi, Y. Goranova, F. Buitrago, J.P. Emerson, W. Freudling, P. Hudelot, C. López-Sanjuan, F. Magnard, Y. Mellier, P. Møller, K.K. Nilsson, W. Sutherland, L. Tasca, J. Zabl, UltraVISTA: a new ultra-deep near-infrared survey in COSMOS. Astron. Astrophys. **544**, A156 (2012). https://doi.org/10.1051/0004-6361/201219507

E. Middelberg, A. Deller, J. Morgan, H. Rottmann, W. Alef, S. Tingay, R. Norris, U. Bach, W. Brisken, E. Lenc, Wide-field VLBA observations of the Chandra deep field South. Astron. Astrophys. **526**, A74 (2011). https://doi.org/10.1051/0004-6361/201015406

E. Middelberg, A.T. Deller, R.P. Norris, S. Fotopoulou, M. Salvato, J.S. Morgan, W. Brisken, D. Lutz, E. Rovilos, Mosaiced wide-field VLBI observations of the Lockman Hole/XMM. Astron. Astrophys. **551**, A97 (2013). https://doi.org/10.1051/0004-6361/201220374

B. Mingo, J.H. Croston, M.J. Hardcastle, P.N. Best, K.J. Duncan, R. Morganti, H.J.A. Rottgering, J. Sabater, T.W. Shimwell, W.L. Williams, M. Brienza, G. Gurkan, V.H. Mahatma, L.K. Morabito, I. Prandoni, M. Bondi, J. Ineson, S. Mooney, Revisiting the Fanaroff-Riley dichotomy and radio-galaxy morphology with the LOFAR Two-Metre Sky Survey (LoTSS). Mon. Not. R. Astron. Soc. **488**(2), 2701–2721 (2019). https://doi.org/10.1093/mnras/stz1901

N. Mohan, D. Rafferty, Pybdsf: Python blob detection and source finder. *Astrophysics Source Code Library*, ascl:1107.013 (2015). https://ui.adsabs.harvard.edu/abs/2015ascl.soft02007M/abstract

S. Molinari, E. Schisano, F. Faustini, M. Pestalozzi, A.M. di Giorgio, S. Liu, Source extraction and photometry for the far-infrared and sub-millimeter continuum in the presence of complex backgrounds. Astron. Astrophys. **530**, A133 (2011). https://doi.org/10.1051/0004-6361/201014752

K.P. Mooley, D.A. Frail, S.T. Myers, S.R. Kulkarni, K. Hotokezaka, L.P. Singer, A. Horesh, M.M. Kasliwal, S.B. Cenko, G. Hallinan, a case study of on-the-fly wide-field radio imaging applied to the gravitational wave event GW151226. Astrophys. J. **857**(2), 143 (2018). https://doi.org/10.3847/1538-4357/aab7f3

K.P. Mooley, S.T. Myers, D.A. Frail, G. Hallinan, B. Butler, A. Kimball, K. Golap, The Caltech-NRAO Stripe 82 Survey (CNSS). II. On-the-fly mosaicking methodology. Astrophys. J. **870**(1), 25 (2019). https://doi.org/10.3847/1538-4357/aaef7c

L.K. Morabito, N.J. Jackson, S. Mooney, F. Sweijen, S. Badole, P. Kukreti, D. Venkattu, C. Groeneveld, A. Kappes, E. Bonnassieux, A. Drabent, M. Iacobelli, J.H. Croston, P.N. Best, M. Bondi, J.R. Callingham, J.E. Conway, A.T. Deller, M.J. Hardcastle, J.P. McKean, G.K. Miley, J. Moldon, H.J.A. Röttgering, C. Tasse, T.W. Shimwell, R.J. van Weeren, J.M. Anderson, A. Asgekar, I.M. Avruch, I.M. van Bemmel, M.J. Bentum, A. Bonafede, W.N. Brouw, H.R. Butcher, B. Ciardi, A. Corstanje, A. Coolen, S. Damstra, F. de Gasperin, S. Duscha, J. Eislöffel, D. Engels, H. Falcke, M.A. Garrett, J. Griessmeier, A.W. Gunst, M.P. van Haarlem, M. Hoeft, A.J. van der Horst, E. Jütte, M. Kadler, L.V.E. Koopmans, A. Krankowski, G. Mann, A. Nelles, J.B.R. Oonk, E. Orru, H. Paas, V.N. Pandey, R.F. Pizzo, M. Pandey-Pommier, W. Reich, H. Rothkaehl, M. Ruiter, D.J. Schwarz, A. Shulevski, M. Soida, M. Tagger, C. Vocks, R.A.M.J. Wijers, S.J. Wijnholds, O. Wucknitz, P. Zarka, P. Zucca, Sub-arcsecond imaging with the International LOFAR Telescope. I. Foundational calibration strategy and pipeline. Astron. Astrophys. **658**, A1 (2022a). https://doi.org/10.1051/0004-6361/202140649

L.K. Morabito, F. Sweijen, J.F. Radcliffe, P.N. Best, R. Kondapally, M. Bondi, M. Bonato, K.J. Duncan, I. Prandoni, T.W. Shimwell, W.L. Williams, R.J. van Weeren, J.E. Conway, G. Calistro Rivera, Identifying active galactic nuclei via brightness temperature with sub-arcsecond international LOFAR telescope observations. Mon. Not. R. Astron. Soc. **515**(4), 5758–5774 (2022b). https://doi.org/10.1093/mnras/stac2129

J.S. Morgan, M.K. Argo, C.M. Trott, J.-P. Macquart, A. Deller, E. Middelberg, J. Miller-Jones, S.J. Tingay, Wide-field VLBI Observations of M31: A unique probe of the ionized interstellar medium of a nearby galaxy. Astrophys. J. **768**(1), 12 (2013). https://doi.org/10.1088/0004-637X/768/1/12

R.I.J. Mostert, K.J. Duncan, L. Alegre, H.J.A. Röttgering, W.L. Williams, P.N. Best, M.J. Hardcastle, R. Morganti, Radio source-component association for the LOFAR Two-metre Sky Survey with region-based convolutional neural networks. arXiv e-prints, art. arXiv:2209.14226 (2022)

T.W.B. Muxlow, A.P. Thomson, J.F. Radcliffe, N.H. Wrigley, R.J. Beswick, I. Smail, I.M. McHardy, S.T. Garrington, R.J. Ivison, M.J. Jarvis, I. Prandoni, M. Bondi, D. Guidetti, M.K. Argo, D. Bacon, P.N. Best, A.D. Biggs, S.C. Chapman, K. Coppin, H. Chen, T.K. Garratt, M.A. Garrett, E. Ibar, J.-P. Kneib, K.K. Knudsen, L.V.E. Koopmans, L.K. Morabito, E.J. Murphy, A. Njeri, C. Pearson, M.A. Pérez-Torres, A.M.S. Richards, H.J.A. Röttgering, M.T. Sargent, S. Serjeant, C. Simpson, J.M. Simpson, A.M. Swinbank, E. Varenius, T. Venturi, The e-MERGE Survey (e-MERLIN Galaxy Evolution Survey): overview and survey description. Mon. Not. R. Astron. Soc. **495**(1), 1188–1208 (2020). https://doi.org/10.1093/mnras/staa1279

R.P. Norris, A.M. Hopkins, J. Afonso, S. Brown, J.J. Condon, L. Dunne, I. Feain, R. Hollow, M. Jarvis, M. Johnston-Hollitt, E. Lenc, E. Middelberg, P. Padovani, I. Prandoni, L. Rudnick, N. Seymour, G. Umana, H. Andernach, D.M. Alexander, P.N. Appleton, D. Bacon, J. Banfield, W. Becker, M.J.I. Brown, P. Ciliegi, C. Jackson, S. Eales, A.C. Edge, B.M. Gaensler, G. Giovannini, C.A. Hales, P. Hancock, M.T. Huynh, E. Ibar, R.J. Ivison, R. Kennicutt, A.E. Kimball, A.M. Koekemoer, B.S. Koribalski, Á. R. López-Sánchez, M.Y. Mao, T. Murphy,

H. Messias, K.A. Pimbblet, A. Raccanelli, K.E. Randall, T.H. Reiprich, I.G. Roseboom, H. Röttgering, D.J. Saikia, R.G. Sharp, O.B. Slee, I. Smail, M.A. Thompson, J.S. Urquhart, J.V. Wall, G.B. Zhao, EMU: evolutionary map of the universe. Publ. Astron. Soc. Austral. **28**(3), 215–248 (2011). https://doi.org/10.1071/AS11021

R.P. Norris, H.T. Intema, A.D. Kapińska, B.S. Koribalski, E. Lenc, L. Rudnick, R.Z.E. Alsaberi, C. Anderson, G.E. Anderson, E. Crawford, R. Crocker, J. English, M.D. Filipović, T.J. Galvin, A.M. Hopkins, N. Hurley-Walker, S. Inoue, K. Luken, P.J. Macgregor, P. Manojlović, J. Marvil, A.N. O'Brien, L. Park, W. Raja, D. Shobhana, T. Venturi, J.D. Collier, C. Hale, A. Hotan, V. Moss, M. Whiting, Unexpected circular radio objects at high Galactic latitude. Publ. Astron. Soc. Austral. **38**, e003 (2021a). https://doi.org/10.1017/pasa.2020.52

R.P. Norris, J. Marvil, J.D. Collier, A.D. Kapińska, A.N. O'Brien, L. Rudnick, H. Andernach, J. Asorey, M.J.I. Brown, M. Brüggen, E. Crawford, J. English, S.F.u. Rahman, M.D. Filipović, Y. Gordon, G. Gürkan, C. Hale, A.M. Hopkins, M.T. Huynh, K. HyeongHan, M. James Jee, B.S. Koribalski, E. Lenc, K. Luken, D. Parkinson, I. Prandoni, W. Raja, T.H. Reiprich, C.J. Riseley, S.S. Shabala, J.R. Sheil, T. Vernstrom, M.T. Whiting, J.R. Allison, C.S. Anderson, L. Ball, M. Bell, J. Bunton, T.J. Galvin, N. Gupta, A. Hotan, C. Jacka, P.J. Macgregor, E.K. Mahony, U. Maio, V. Moss, M. Pandey-Pommier, M.A. Voronkov, The evolutionary map of the universe pilot survey. Publ. Astron. Soc. Austral. **38**, e046 (2021b). https://doi.org/10.1017/pasa.2021.42

C.Y. Peng, L.C. Ho, C.D. Impey, H.-W. Rix, Detailed structural decomposition of galaxy images. Astron. J. **124**(1), 266–293 (2002). https://doi.org/10.1086/340952

V. Petrosian, Surface brightness and evolution of galaxies. Astrophys. J. Lett. **210**, L53 (1976). https://doi.org/10.1086/182301

D.D. Proctor, A selection of giant radio sources from NVSS. Astrophys. J. Suppl. **224**(2), 18 (2016). https://doi.org/10.3847/0067-0049/224/2/18

J.F. Radcliffe, M.A. Garrett, R.J. Beswick, T.W.B. Muxlow, P.D. Barthel, A.T. Deller, E. Middelberg, Multi-source self-calibration: Unveiling the microJy population of compact radio sources. Astron. Astrophys. **587**, A85 (2016). https://doi.org/10.1051/0004-6361/201527980

J.F. Radcliffe, M.A. Garrett, T.W.B. Muxlow, R.J. Beswick, P.D. Barthel, A.T. Deller, A. Keimpema, R.M. Campbell, N. Wrigley, Nowhere to hide: Radio-faint AGN in GOODS-N field. I. Initial catalogue and radio properties. AAP **619**, A48 (2018). https://doi.org/10.1051/0004-6361/201833399

J.F. Radcliffe, R.J. Beswick, A.P. Thomson, M.A. Garrett, P.D. Barthel, T.W.B. Muxlow, An insight into the extragalactic transient and variable microJy radio sky across multiple decades. Mon. Not. R. Astron. Soc. **490**(3), 4024–4039 (2019). https://doi.org/10.1093/mnras/stz2748

J.F. Radcliffe, P.D. Barthel, A.P. Thomson, M.A. Garrett, R.J. Beswick, T.W.B. Muxlow, Nowhere to hide: Radio-faint AGN in the GOODS-N field. II. Multi-wavelength AGN selection techniques and host galaxy properties. Astron. Astrophys. **649**, A27 (2021). https://doi.org/10.1051/0004-6361/202038591

U. Rau, T.J. Cornwell, A multi-scale multi-frequency deconvolution algorithm for synthesis imaging in radio interferometry. Astron. Astrophys. **532**, A71 (2011). https://doi.org/10.1051/0004-6361/201117104

G.A. Richter. Search for optical identifications in the 5C3 radio survey. II. Statistical treatment and results. Astron. Nachr. **296**(2), 65 (1975). https://doi.org/10.1002/asna.19752960203

S. Riggi, A. Ingallinera, P. Leto, F. Cavallaro, F. Bufano, F. Schilliró, C. Trigilio, G. Umana, C.S. Buemi, R.P. Norris, Automated detection of extended sources in radio maps: progress from the scorpio survey. Mon. Not. R. Astron. Soc. **460**, 1486 (2016). https://doi.org/10.1093/mnras/stw982

S. Riggi, F. Vitello, U. Becciani, F. Buemi, A. Calanducci, F. Cavallaro, A. Costa, A. Ingallinera, P. Leto, R.P. Norris, F. Schillirò, E. Sciacca, C. Trigilio, G. Umana, Caesar source finder: Recent developments and testing. Publ. Astron. Soc. Austral. **36**, 29 (2019). https://doi.org/10.1017/pasa.2019.29

A.S.G. Robotham, L.J.M. Davies, S.P. Driver, S.P. Koushan, D.S. Taranu, S. Casura, J. Liske, Profound: Source extraction and application to modern survey data. Mon. Not. R. Astron. Soc. **476**, 3137 (2018). https://doi.org/10.1093/mnras/sty440

L. Rudnick, Radio galaxy classification: #Tags, not boxes. Galaxies **9**(4), 85 (2021). https://doi.org/10.3390/galaxies9040085

J. Sabater, P.N. Best, M.J. Hardcastle, T.W. Shimwell, C. Tasse, W.L. Williams, M. Brüggen, R.K. Cochrane, J.H. Croston, F. de Gasperin, K.J. Duncan, G. Gürkan, A.P. Mechev, L.K. Morabito, I. Prandoni, H.J.A. Röttgering, D.J.B. Smith, J.J. Harwood, B. Mingo, S. Mooney, A. Saxena, The LoTSS view of radio AGN in the local Universe. The most massive galaxies are always switched on. Astron. Astrophys. **622**, A17 (2019). https://doi.org/10.1051/0004-6361/201833883

J. Sabater, P.N. Best, C. Tasse, M.J. Hardcastle, T.W. Shimwell, D. Nisbet, V. Jelic, J.R. Callingham, H.J.A. Röttgering, M. Bonato, M. Bondi, B. Ciardi, R.K. Cochrane, M.J. Jarvis, R. Kondapally, L.V.E. Koopmans, S.P. O'Sullivan, I. Prandoni, D.J. Schwarz, D.J.B. Smith, L. Wang, W.L. Williams, S. Zaroubi, The LOFAR two-meter sky survey: deep fields data release 1. II. The ELAIS-N1 LOFAR deep field. Astron. Astrophys. **648**, A2 (2021). https://doi.org/10.1051/0004-6361/202038828

A.M.M. Scaife, F. Porter, Fanaroff-Riley classification of radio galaxies using group-equivariant convolutional neural networks. Mon. Not. R. Astron. Soc. **503**(2), 2369–2379 (2021). https://doi.org/10.1093/mnras/stab530

N. Scoville, H. Aussel, M. Brusa, P. Capak, C.M. Carollo, M. Elvis, M. Giavalisco, L. Guzzo, G. Hasinger, C. Impey, J.P. Kneib, O. LeFevre, S.J. Lilly, B. Mobasher, A. Renzini, R.M. Rich, D.B. Sanders, E. Schinnerer, D. Schminovich, P. Shopbell, Y. Taniguchi, N.D. Tyson, The Cosmic Evolution Survey (COSMOS): overview. Astrophys. J. Suppl. **172**(1), 1–8 (2007). https://doi.org/10.1086/516585

G. Segal, D. Parkinson, R.P. Norris, J. Swan, Identifying complex sources in large astronomical data using a coarse-grained complexity measure. Publ. Astron. Soc. Pac. **131**(1004), 108007 (2019). https://doi.org/10.1088/1538-3873/ab0068

G. Segal, D. Parkinson, R. Norris, A.M. Hopkins, H. Andernach, E.L. Alexander, E. Carretti, B.S. Koribalski, L.S. Legodi, S. Leslie, Y. Luo, J.C.S. Pierce, H. Tang, E. Vardoulaki, T. Vernstrom, Identifying anomalous radio sources in the Evolutionary Map of the Universe Pilot Survey using a complexity-based approach. Mon. Not. R. Astron. Soc. **521**(1), 1429–1447 (2023). https://doi.org/10.1093/mnras/stad537

P. Serra, T. Westmeier, N. Giese, R. Jurek, L. Flöer, A. Popping, B. Winkel, T. van der Hulst, M. Meyer, B.S. Koribalski, L. Staveley-Smith, H. Courtois, SoFiA: a flexible source finder for 3D spectral line data. Mon. Not. R. Astron. Soc. **448**(2), 1922 (2015). ISSN 0035-8711. https://doi.org/10.1093/mnras/stv079

T.W. Shimwell, H.J.A. Röttgering, P.N. Best, W.L. Williams, T.J. Dijkema, F. de Gasperin, M.J. Hardcastle, G.H. Heald, D.N. Hoang, A. Horneffer, H. Intema, E.K. Mahony, S. Mandal, A.P. Mechev, L. Morabito, J.B.R. Oonk, D. Rafferty, E. Retana-Montenegro, J. Sabater, C. Tasse, R.J. van Weeren, M. Brüggen, G. Brunetti, K.T. Chyży, J.E. Conway, M. Haverkorn, N. Jackson, M.J. Jarvis, J.P. McKean, G.K. Miley, R. Morganti, G.J. White, M.W. Wise, I.M. van Bemmel, R. Beck, M. Brienza, A. Bonafede, G. Calistro Rivera, R. Cassano, A.O. Clarke, D. Cseh, A. Deller, A. Drabent, W. van Driel, D. Engels, H. Falcke, C. Ferrari, S. Fröhlich, M.A. Garrett, J.J. Harwood, V. Heesen, M. Hoeft, C. Horellou, F.P. Israel, A.D. Kapińska, M. Kunert-Bajraszewska, D.J. McKay, N.R. Mohan, E. Orrú, R.F. Pizzo, I. Prandoni, D.J. Schwarz, A. Shulevski, M. Sipior, D.J.B. Smith, S.S. Sridhar, M. Steinmetz, A. Stroe, E. Varenius, P.P. van der Werf, J.A. Zensus, J.T.L. Zwart, The LOFAR Two-metre Sky Survey. I. Survey description and preliminary data release. Astron. Astrophys. **598**, A104 (2017). https://doi.org/10.1051/0004-6361/201629313

T.W. Shimwell, C. Tasse, M.J. Hardcastle, A.P. Mechev, W.L. Williams, P.N. Best, H.J.A. Röttgering, J.R. Callingham, T.J. Dijkema, F. de Gasperin, D.N. Hoang, B. Hugo, M. Mirmont, J.B.R. Oonk, I. Prandoni, D. Rafferty, J. Sabater, O. Smirnov, R.J. van Weeren, G.J. White, M. Atemkeng, L. Bester, E. Bonnassieux, M. Brüggen, G. Brunetti, K.T. Chyży, R. Cochrane,

J.E. Conway, J.H. Croston, A. Danezi, K. Duncan, M. Haverkorn, G.H. Heald, M. Iacobelli, H.T. Intema, N. Jackson, M. Jamrozy, M.J. Jarvis, R. Lakhoo, M. Mevius, G.K. Miley, L. Morabito, R. Morganti, D. Nisbet, E. Orrú, S. Perkins, R.F. Pizzo, C. Schrijvers, D.J.B. Smith, R. Vermeulen, M.W. Wise, L. Alegre, D.J. Bacon, I.M. van Bemmel, R.J. Beswick, A. Bonafede, A. Botteon, S. Bourke, M. Brienza, G. Calistro Rivera, R. Cassano, A.O. Clarke, C.J. Conselice, R.J. Dettmar, A. Drabent, C. Dumba, K.L. Emig, T.A. Enßlin, C. Ferrari, M.A. Garrett, R.T. Génova-Santos, A. Goyal, G. Gürkan, C. Hale, J.J. Harwood, V. Heesen, M. Hoeft, C. Horellou, C. Jackson, G. Kokotanekov, R. Kondapally, M. Kunert-Bajraszewska, V. Mahatma, E.K. Mahony, S. Mandal, J.P. McKean, A. Merloni, B. Mingo, A. Miskolczi, S. Mooney, B. Nikiel-Wroczyński, S.P. O'Sullivan, J. Quinn, W. Reich, C. Roskowiński, A. Rowlinson, F. Savini, A. Saxena, D.J. Schwarz, A. Shulevski, S.S. Sridhar, H.R. Stacey, S. Urquhart, M.H.D. van der Wiel, E. Varenius, B. Webster, A. Wilber, The LOFAR Two-metre Sky Survey. II. First data release. Astron. Astrophys. **622**, A1 (2019). https://doi.org/10.1051/0004-6361/201833559

T.W. Shimwell, M.J. Hardcastle, C. Tasse, P.N. Best, H.J.A. Röttgering, W.L. Williams, A. Botteon, A. Drabent, A. Mechev, A. Shulevski, R.J. van Weeren, L. Bester, M. Brüggen, G. Brunetti, J.R. Callingham, K.T. Chyży, J.E. Conway, T.J. Dijkema, K. Duncan, F. de Gasperin, C.L. Hale, M. Haverkorn, B. Hugo, N. Jackson, M. Mevius, G.K. Miley, L.K. Morabito, R. Morganti, A. Offringa, J.B.R. Oonk, D. Rafferty, J. Sabater, D.J.B. Smith, D.J. Schwarz, O. Smirnov, S.P. O'Sullivan, H. Vedantham, G.J. White, J.G. Albert, L. Alegre, B. Asabere, D.J. Bacon, A. Bonafede, E. Bonnassieux, M. Brienza, M. Bilicki, M. Bonato, G. Calistro Rivera, R. Cassano, R. Cochrane, J.H. Croston, V. Cuciti, D. Dallacasa, A. Danezi, R.J. Dettmar, G. Di Gennaro, H.W. Edler, T.A. Enßlin, K.L. Emig, T.M.O. Franzen, C. García-Vergara, Y.G. Grange, G. Gürkan, M. Hajduk, G. Heald, V. Heesen, D.N. Hoang, M. Hoeft, C. Horellou, M. Iacobelli, M. Jamrozy, V. Jelić, R. Kondapally, P. Kukreti, M. Kunert-Bajraszewska, M. Magliocchetti, V. Mahatma, K. Małek, S. Mandal, F. Massaro, Z. Meyer-Zhao, B. Mingo, R.I.J. Mostert, D.G. Nair, S.J. Nakoneczny, B. Nikiel-Wroczyński, E. Orrú, U. Pajdosz-Śmierciak, T. Pasini, I. Prandoni, H.E. van Piggelen, K. Rajpurohit, E. Retana-Montenegro, C.J. Riseley, A. Rowlinson, A. Saxena, C. Schrijvers, F. Sweijen, T.M. Siewert, R. Timmerman, M. Vaccari, J. Vink, J.L. West, A. Wołowska, X. Zhang, J. Zheng, The LOFAR Two-metre Sky Survey. V. Second data release. Astron. Astrophys. **659**, A1 (2022). https://doi.org/10.1051/0004-6361/202142484

R. Shirley, Y. Roehlly, P.D. Hurley, V. Buat, M.C. Campos Varillas, S. Duivenvoorden, K.J. Duncan, A. Efstathiou, D. Farrah, E. González Solares, K. Malek, L. Marchetti, I. McCheyne, A. Papadopoulos, E. Pons, R. Scipioni, M. Vaccari, S. Oliver, HELP: a catalogue of 170 million objects, selected at 0.36-4.5 μm, from 1270 deg^2 of prime extragalactic fields. Mon. Not. R. Astron. Soc. **490**(1), 634–656 (2019). https://doi.org/10.1093/mnras/stz2509

R. Shirley, K. Duncan, M.C. Campos Varillas, P.D. Hurley, K. Małek, Y. Roehlly, M.W.L. Smith, H. Aussel, T. Bakx, V. Buat, D. Burgarella, N. Christopher, S. Duivenvoorden, S. Eales, A. Efstathiou, E.A. González Solares, M. Griffin, M. Jarvis, B.L. Faro, L. Marchetti, I. McCheyne, A. Papadopoulos, K. Penner, E. Pons, M. Prescott, E. Rigby, H. Rottgering, A. Saxena, J. Scudder, M. Vaccari, L. Wang, S.J. Oliver, HELP: the Herschel extragalactic legacy project. Mon. Not. R. Astron. Soc. **507**(1), 129–155 (2021). https://doi.org/10.1093/mnras/stab1526

D.J.B. Smith, P.N. Best, K.J. Duncan, N.A. Hatch, M.J. Jarvis, H.J.A. Röttgering, C.J. Simpson, J.P. Stott, R.K. Cochrane, K.E. Coppin, H. Dannerbauer, T.A. Davis, J.E. Geach, C.L. Hale, M.J. Hardcastle, P.W. Hatfield, R.C.W. Houghton, N. Maddox, S.L. McGee, L. Morabito, D. Nisbet, M. Pandey-Pommier, I. Prandoni, A. Saxena, T.W. Shimwell, M. Tarr, I. van Bemmel, A. Verma, G.J. White, W.L. Williams, The WEAVE-LOFAR survey, in *SF2A-2016: Proceedings of the Annual meeting of the French Society of Astronomy and Astrophysics*, ed. by C. Reylé, J. Richard, L. Cambrésy, M. Deleuil, E. Pécontal, L. Tresse, I. Vauglin (2016, December), pp. 271–280

V. Smolčić, M. Novak, I. Delvecchio, L. Ceraj, M. Bondi, J. Delhaize, S. Marchesi, E. Murphy, E. Schinnerer, E. Vardoulaki, G. Zamorani, The VLA-COSMOS 3 GHz large project: cosmic evolution of radio AGN and implications for radio-mode feedback since z 5. Astron. Astrophys. **602**, A6 (2017). https://doi.org/10.1051/0004-6361/201730685

B.W. Sohn, U. Klein, K.H. Mack, The spectral-curvature parameter: An alternative tool for the analysis of synchrotron spectra. Astron. Astrophys. **404**, 133–144 (2003). https://doi.org/10.1051/0004-6361:20030435

C. Spingola, J.P. McKean, M. Lee, A. Deller, J. Moldon, A novel search for gravitationally lensed radio sources in wide-field VLBI imaging from the mJIVE-20 survey. Mon. Not. R. Astron. Soc. **483**(2), 2125–2153 (2019). https://doi.org/10.1093/mnras/sty3189

J.N. Spreeuw, Search and detection of low frequency radio transients. Dissertation, Astronomical Institute Anton Pannekoek, University of Amsterdam, ISBN 978-90-9024055-8, 2010

W. Sutherland, W. Saunders, On the likelihood ratio for source identification. Mon. Not. R. Astron. Soc. **259**, 413–420 (1992). https://doi.org/10.1093/mnras/259.3.413

F. Sweijen, R.J. van Weeren, H.J.A. Röttgering, L.K. Morabito, N. Jackson, A.R. Offringa, S. van der Tol, B. Veenboer, J.B.R. Oonk, P.N. Best, M. Bondi, T.W. Shimwell, C. Tasse, A.P. Thomson, Deep sub-arcsecond wide-field imaging of the Lockman Hole field at 144 MHz. Nat. Astron. **6**, 350–356 (2022). https://doi.org/10.1038/s41550-021-01573-z

J.D. Swinbank, T.D. Staley, G.J. Molenaar, E. Rol, A. Rowlinson, B. Scheers, H. Spreeuw, M.E. Bell, J.W. Broderick, D. Carbone, H. Garsden, A.J. van der Horst, C.J. Law, M. Wise, R.P. Breton, Y. Cendes, S. Corbel, J. Eislöffel, H. Falcke, J.-M. Fender, R. Grießmeier, J.W.T. Hessels, B.W. Stappers, A.J. Stewart, R.A.M.J. Wijers, R. Wijnands, P. Zarka, The lofar transients pipeline. Astron. Comput. **11**, 25 (2015). https://doi.org/10.1016/j.ascom.2015.03.002

C. Tasse, T. Shimwell, M.J. Hardcastle, S.P. O'Sullivan, R. van Weeren, P.N. Best, L. Bester, B. Hugo, O. Smirnov, J. Sabater, G. Calistro-Rivera, F. de Gasperin, L.K. Morabito, H. Röttgering, W.L. Williams, M. Bonato, M. Bondi, A. Botteon, M. Brüggen, G. Brunetti, K.T. Chyży, M.A. Garrett, G. Gürkan, M.J. Jarvis, R. Kondapally, S. Mandal, I. Prandoni, A. Repetti, E. Retana-Montenegro, D.J. Schwarz, A. Shulevski, Y. Wiaux, The LOFAR two-meter sky survey: deep fields data release 1. I. Direction-dependent calibration and imaging. Astron. Astrophys. **648**, A1 (2021). https://doi.org/10.1051/0004-6361/202038804

M.B. Taylor, TOPCAT & STIL: Starlink Table/VOTable processing software, in *Astronomical Data Analysis Software and Systems XIV*, ed. by P. Shopbell, M. Britton, R. Ebert, volume 347 of Astronomical Society of the Pacific Conference Series (2005), p. 29

M. Vaccari, The Spitzer data fusion: contents, construction and applications to galaxy evolution studies, in *The Many Facets of Extragalactic Radio Surveys: Towards New Scientific Challenges* (2015), p. 27

M. Vaccari, The Spitzer data fusion, in *Zenodo Dataset*, vol. 61 (2022a), p. 6120914. https://doi.org/10.5281/zenodo.6120914

M. Vaccari, The Spitzer spectroscopic data fusion - merged spectroscopic redshift catalogs in Spitzer fields, in *Zenodo Dataset*, vol. 63 (2022b), p. 6368348. https://doi.org/10.5281/zenodo.6368348

M. Vaccari, G. Covone, M. Radovich, A. Grado, L. Limatola, M.T. Botticella, E. Cappellaro, M. Paolillo, G. Pignata, D. De Cicco, S. Falocco, L. Marchetti, M. Brescia, S. Cavuoti, G. Longo, M. Capaccioli, N. Napolitano, P. Schipani, The VOICE Survey: VST optical imaging of the CDFS and ES1 fields, in *The 4th Annual Conference on High Energy Astrophysics in Southern Africa (HEASA 2016)*, January (2016), p. 26. https://doi.org/10.22323/1.275.0026

E. Vardoulaki, E.F. Jiménez Andrade, A. Karim, M. Novak, S.K. Leslie, K. Tisanić, V. Smolčić, E. Schinnerer, M.T. Sargent, M. Bondi, G. Zamorani, B. Magnelli, F. Bertoldi, N. Herrera Ruiz, K.P. Mooley, J. Delhaize, S.T. Myers, S. Marchesi, A.M. Koekemoer, G. Gozaliasl, A. Finoguenov, E. Middleberg, P. Ciliegi, A closer look at the deep radio sky: Multi-component radio sources at 3 GHz VLA-COSMOS. Astron. Astrophys. **627**, A142 (2019). https://doi.org/10.1051/0004-6361/201832982

E. Vardoulaki, E.F. Jiménez Andrade, I. Delvecchio, V. Smolčić, E. Schinnerer, M.T. Sargent, G. Gozaliasl, A. Finoguenov, M. Bondi, G. Zamorani, T. Badescu, S.K. Leslie, L. Ceraj, K. Tisanić, A. Karim, B. Magnelli, F. Bertoldi, E. Romano-Diaz, K. Harrington, FR-type radio sources at 3 GHz VLA-COSMOS: Relation to physical properties and large-scale environment. Astron. Astrophys. **648**, A102 (2021). https://doi.org/10.1051/0004-6361/202039488

T. Venturi, Z. Paragi, M. Lindqvist, A. Bartkiewicz, R. Beswick, T. Bogdanović, W. Brisken, P. Charlot, F. Colomer, J. Conway, S. Frey, J.C. Guirado, L. Gurvits, H. van Langevelde, A. Lobanov, J. McKean, R. Morganti, T. Muxlow, M. Pérez-Torres, K. Rygl, R. Schulz, A. Szomoru, P. de Vicente, T. An, G. Anglada, M. Argo, R. Azulay, I. van Bemmel, T. Bocanegra, B. Boccardi, P. Castangia, J. Chibueze, G. Cimò, J.-B. Climent, R. Deane, A. Deller, R. Dodson, D. Duev, S. Etoka, D. Fenech, K. Gabányi, D. Gabuzda, M. Garrett, M. Gawroński, G. Ghirlanda, M. Giroletti, C. Goddi, J.L. Gómez, M. Gray, J. Greaves, J. Hessels, A. van der Horst, T. Hunter, R. Laing, D. Vir Lal, S. Lambert, L. Loinard, B. Marcote, A. Merloni, J. Miller-Jones, G. Molera Calvés, L. Moscadelli, H. Olofsson, L. Petrov, R. Pizzo, A. Possenti, L.H. Quiroga-Nuñez, C. Reynolds, A. Richards, M. Rioja, A. Sanna, T. Savolainen, T. Sbarrato, C. Spingola, G. Surcis, C. Trigilio, E. Varenius, W. Vlemmings, S. van Velzen, J. van der Walt, VLBI20-30: a scientific roadmap for the next decade – The future of the European VLBI Network. arXiv e-prints, art. arXiv:2007.02347 (2020). https://doi.org/10.48550/arXiv.2007.02347

A.C. Villarreal Hernández, H. Andernach, A search for extended radio sources in 1.3 sr of the VLA sky survey (VLASS). arXiv e-prints, art. arXiv:1808.07178 (2018)

F. Walter, R. Decarli, M. Aravena, C. Carilli, R. Bouwens, E. da Cunha, E. Daddi, R.J. Ivison, D. Riechers, I. Smail, M. Swinbank, A. Weiss, T. Anguita, R. Assef, R. Bacon, F. Bauer, E.F. Bell, F. Bertoldi, S. Chapman, L. Colina, P.C. Cortes, P. Cox, M. Dickinson, D. Elbaz, J. Gónzalez-López, E. Ibar, H. Inami, L. Infante, J. Hodge, A. Karim, O. Le Fevre, B. Magnelli, R. Neri, P. Oesch, K. Ota, G. Popping, H.-W. Rix, M. Sargent, K. Sheth, A. van der Wel, P. van der Werf, J. Wagg, ALMA spectroscopic survey in the hubble ultra deep field: survey description. Astrophys. J. **833**(1), 67 (2016). https://doi.org/10.3847/1538-4357/833/1/67

J.R. Weaver, O.B. Kauffmann, O. Ilbert, H.J. McCracken, A. Moneti, S. Toft, G. Brammer, M. Shuntov, I. Davidzon, B.C. Hsieh, C. Laigle, A. Anastasiou, C.K. Jespersen, J. Vinther, P. Capak, C.M. Casey, C.J.R. McPartland, B. Milvang-Jensen, B. Mobasher, D.B. Sanders, L. Zalesky, S. Arnouts, H. Aussel, J.S. Dunlop, A. Faisst, M. Franx, L.J. Furtak, J.P.U. Fynbo, K.M.L. Gould, T.R. Greve, S. Gwyn, J.S. Kartaltepe, D. Kashino, A.M. Koekemoer, V. Kokorev, O. Le Fèvre, S. Lilly, D. Masters, G. Magdis, V. Mehta, Y. Peng, D.A. Riechers, M. Salvato, M. Sawicki, C. Scarlata, N. Scoville, R. Shirley, J.D. Silverman, A. Sneppen, V. Smolćić, C. Steinhardt, D. Stern, M. Tanaka, Y. Taniguchi, H.I. Teplitz, M. Vaccari, W.H. Wang, G. Zamorani, COSMOS2020: A panchromatic view of the universe to z 10 from two complementary catalogs. Astrophys. J. Suppl. **258**(1), 11 (2022). https://doi.org/10.3847/1538-4365/ac3078

R.L. White, R.H. Becker, D.J. Helfand, M.D. Gregg, A catalog of 1.4 GHz radio sources from the FIRST survey. Astrophys. J. **475**(2), 479 (1997). https://doi.org/10.1086/303564

M. Whiting, Duchamp: a 3d source finder for spectral-line data. Mon. Not. R. Astron. Soc. **421**, 3242 (2012). https://doi.org/10.1111/j.1365-2966.2012.20548.x

M. Whiting, B. Humphreys, Source-finding for the Australian square kilometre array pathfinder. Publ. Astron. Soc. Austral. **29**, 371 (2012). https://doi.org/10.1071/AS12028

I.H. Whittam, M.J. Jarvis, C.L. Hale, M. Prescott, L.K. Morabito, I. Heywood, N.J. Adams, J. Afonso, F. An, Y. Ao, R.A.A. Bowler, J.D. Collier, R.P. Deane, J. Delhaize, B. Frank, M. Glowacki, P.W. Hatfield, N. Maddox, L. Marchetti, A.M. Matthews, I. Prandoni, S. Randriamampandry, Z. Randriamanakoto, D.J.B. Smith, A.R. Taylor, N.L. Thomas, M. Vaccari, MIGHTEE: the nature of the radio-loud AGN population. Mon. Not. R. Astron. Soc. **516**(1), 245–263 (2022). https://doi.org/10.1093/mnras/stac2140

W.L. Williams, M.J. Hardcastle, P.N. Best, J. Sabater, J.H. Croston, K.J. Duncan, T.W. Shimwell, H.J.A. Röttgering, D. Nisbet, G. Gürkan, L. Alegre, R.K. Cochrane, A. Goyal, C.L. Hale, N. Jackson, M. Jamrozy, R. Kondapally, M. Kunert-Bajraszewska, V.H. Mahatma, B. Mingo, L.K. Morabito, I. Prandoni, C. Roskowinski, A. Shulevski, D.J.B. Smith, C. Tasse, S. Urquhart, B. Webster, G.J. White, R.J. Beswick, J.R. Callingham, K.T. Chyży, F. de Gasperin, J.J. Harwood, M. Hoeft, M. Iacobelli, J.P. McKean, A.P. Mechev, G.K. Miley, D.J. Schwarz,

R.J. van Weeren, The LOFAR Two-metre Sky Survey. III. First data release: Optical/infrared identifications and value-added catalogue. Astron. Astrophys. **622**, A2 (2019). https://doi.org/10.1051/0004-6361/201833564

E.L. Wright, P.R.M. Eisenhardt, A.K. Mainzer, M.E. Ressler, R.M. Cutri, T. Jarrett, J.D. Kirkpatrick, D. Padgett, R.S. McMillan, M. Skrutskie, S.A. Stanford, M. Cohen, R.G. Walker, J.C. Mather, D. Leisawitz, T.N. Gautier, III, I. McLean, D. Benford, C.J. Lonsdale, A. Blain, B. Mendez, W.R. Irace, V. Duval, F. Liu, D. Royer, I. Heinrichsen, J. Howard, M. Shannon, M. Kendall, A.L. Walsh, M. Larsen, J.G. Cardon, S. Schick, M. Schwalm, M. Abid, B. Fabinsky, L. Naes, C.-W. Tsai, The Wide-field Infrared Survey Explorer (WISE): mission description and initial on-orbit performance. Astron. J. **140**, 1868–1881 (2010). https://doi.org/10.1088/0004-6256/140/6/1868

C. Wu, O.I. Wong, L. Rudnick, S.S. Shabala, M.J. Alger, J.K. Banfield, C.S. Ong, S.V. White, A.F. Garon, R.P. Norris, H. Andernach, J. Tate, V. Lukic, H. Tang, K. Schawinski, F.I. Diakogiannis, Radio Galaxy Zoo: CLARAN - a deep learning classifier for radio morphologies. Mon. Not. R. Astron. Soc. **482**(1), 1211–1230 (2019). https://doi.org/10.1093/mnras/sty2646

D.G. York, J. Adelman, J.E. Anderson Jr., S.F. Anderson, J. Annis, N.A. Bahcall, J.A. Bakken, R. Barkhouser, S. Bastian, E. Berman, W.N. Boroski, S. Bracker, C. Briegel, J.W. Briggs, J. Brinkmann, R. Brunner, S. Burles, L. Carey, M.A. Carr, F.J. Castander, B. Chen, P.L. Colestock, A.J. Connolly, J.H. Crocker, I. Csabai, P.C. Czarapata, J.E. Davis, M. Doi, T. Dombeck, D. Eisenstein, N. Ellman, B.R. Elms, M.L. Evans, X. Fan, G.R. Federwitz, L. Fiscelli, S. Friedman, J.A. Frieman, M. Fukugita, B. Gillespie, J.E. Gunn, V.K. Gurbani, E. de Haas, M. Haldeman, F.H. Harris, J. Hayes, T.M. Heckman, G.S. Hennessy, R.B. Hindsley, S. Holm, D.J. Holmgren, C.-H. Huang, C. Hull, D. Husby, S.-I. Ichikawa, T. Ichikawa, Ž. Ivezić, S. Kent, R.S.J. Kim, E. Kinney, M. Klaene, A.N. Kleinman, S. Kleinman, G.R. Knapp, J. Korienek, R.G. Kron, P.Z. Kunszt, D.Q. Lamb, B. Lee, R.F. Leger, S. Limmongkol, C. Lindenmeyer, D.C. Long, C. Loomis, J. Loveday, R. Lucinio, R.H. Lupton, B. MacKinnon, E.J. Mannery, P.M. Mantsch, B. Margon, P. McGehee, T.A. McKay, A. Meiksin, A. Merelli, D.G. Monet, J.A. Munn, V.K. Narayanan, T. Nash, E. Neilsen, R. Neswold, H.J. Newberg, R.C. Nichol, T. Nicinski, M. Nonino, N. Okada, S. Okamura, J.P. Ostriker, R. Owen, A.G. Pauls, J. Peoples, R.L. Peterson, D. Petravick, J.R. Pier, A. Pope, R. Pordes, A. Prosapio, R. Rechenmacher, T.R. Quinn, G.T. Richards, M.W. Richmond, C.H. Rivetta, C.M. Rockosi, K. Ruthmansdorfer, D. Sandford, D.J. Schlegel, D.P. Schneider, M. Sekiguchi, G. Sergey, K. Shimasaku, W.A. Siegmund, S. Smee, J.A. Smith, S. Snedden, R. Stone, C. Stoughton, M.A. Strauss, C. Stubbs, M. SubbaRao, A.S. Szalay, I. Szapudi, G.P. Szokoly, A.R. Thakar, C. Tremonti, D.L. Tucker, A. Uomoto, D. Vanden Berk, M.S. Vogeley, P. Waddell, S.-I. Wang, M. Watanabe, D.H. Weinberg, B. Yanny, N. Yasuda, SDSS Collaboration, The Sloan digital sky survey: technical summary. Astron. J. **120**(3), 1579–1587 (2000). https://doi.org/10.1086/301513

F. Zou, W.N. Brandt, C.-T. Chen, J. Leja, Q. Ni, W. Yan, G. Yang, S. Zhu, B. Luo, K. Nyland, F. Vito, Y. Xue, Spectral energy distributions in three deep-drilling fields of the Vera C. Rubin Observatory legacy survey of space and time: source classification and galaxy properties. Astrophys. J. Suppl. **262**(1), 15 (2022). https://doi.org/10.3847/1538-4365/ac7bdf

Chapter 9
Other Types of Source Extraction and Identification

Daizhong Liu, Prajwal Voraganti Padmanabh, Cameron L. Van Eck, Robert Main, Tim Sprenger, Michael Kramer, and Vishnu Balakrishnan

In Chap. 8, we mentioned that the radio data after imaging usually has 4D (5D if accounting for time domain), where only 2D information is relevant in the continuum source extraction. Here we discuss the commonly used extraction methods of other information like spectral lines (only if within the bandwidth), polarisation (if emission is polarised), and time domain properties (if the source has time variation, e.g. pulsar, fast radio burst).

9.1 Spectral Line

Spectral lines of astronomical objects in the submm/radio regime mostly originate from the absorption or emission of a species, either atoms (e.g. hydrogen) or molecules (e.g. carbon monoxide, CO), at a specific quantum state. Such species' particles will absorb/emit photons at a specific line centre frequency and change

D. Liu (✉)
Max-Planck-Institut für extraterrestrische Physik, Garching, Germany

P. V. Padmanabh
Max Planck Institute for Gravitational Physics (Albert Einstein Institute), Hannover, Germany
e-mail: prajwal.voraganti.padmanabh@aei.mpg.de

C. L. Van Eck
Research School of Astronomy & Astrophysics, The Australian National University, Canberra, ACT, Australia
e-mail: cameron.vaneck@anu.edu.au

R. Main · T. Sprenger · M. Kramer · V. Balakrishnan
Max-Planck-Institut für Radioastronomie, Bonn, Germany
e-mail: ramain@mpifr-bonn.mpg.de; tsprenger@mpifr-bonn.mpg.de; mkramer@mpifr-bonn.mpg.de; vishnu@mpifr-bonn.mpg.de

E. Vardoulaki et al. (eds.), *Data-Intensive Radio Astronomy*, Astrophysics and Space Science Library 472, https://doi.org/10.1007/978-3-031-58468-8_9

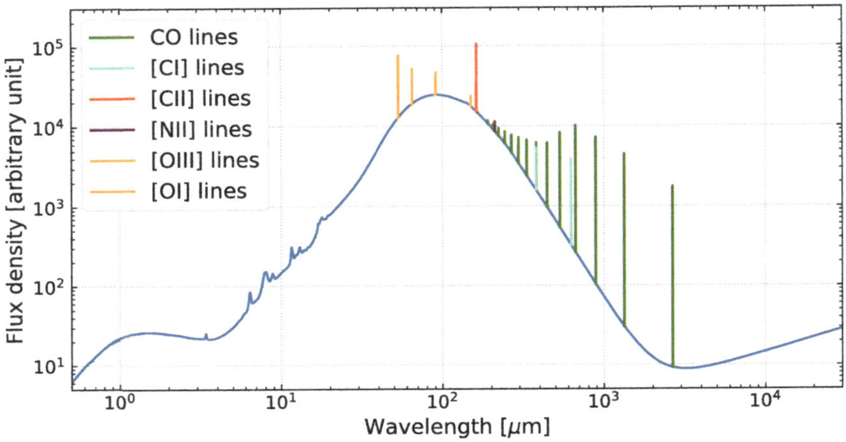

Fig. 9.1 Example galaxy optical-to-radio spectral energy distribution (SED) with major (sub-)mm/radio ISM cooling lines overlaid (Liu et al., 2021a). The galaxy SED is taken from the Arp 193 galaxy's model SED (Liu et al., 2021a), and the line strengths are simply calculated based on line luminosity versus infrared (IR) luminosity correlations (Liu et al., 2015; De Looze et al., 2011; Sargent et al., 2014; Zhao et al., 2016). An $L_{\mathrm{IR}} = 10^{12}$ $\mathrm{L}_{\odot}$ is used (which is 2× larger than Arp 193's true IR luminosity for this illustration purpose)

their quantum states following certain quantum rules. This transition creates a spectral line at the line centre frequency, with certain line width due to broadening, and peak height relating to the amount of the species and other properties like optical depth.

Spectral line and continuum usually coexist in the data. For example, Fig. 9.1 shows a sketch plot of a luminous infrared galaxy's spectral energy distribution (SED) and major (sub-)mm/radio ISM cooling lines on top of it (Liu et al., 2021a). The [CII] fine structure line at 158 μm is typically the strongest cooling line in molecular gas—star formation complexes (Madden et al., 2020). The [OIII] fine structure lines at 51 μm and 88 μm and [OI] fine structure lines at 63 μm and 145 μm are next strong cooling lines. At longer wavelengths, [NII], [CI] and CO lines are the mostly studied cooling lines. The CO lines are especially the most essential tool to probe the cold molecular gas in galaxies where molecular hydrogen (H_2) dominates but do not have detectable emission due to the lack of permanent dipole moment (Bolatto et al., 2013). These emission lines are inevitable ways to the studies of ISM, star formation, galaxy evolution and galaxy formation.

Spectral lines in the (sub-)mm/radio data can either be in emission or absorption, depending on the line-of-sight physical condition and strength of the continuum. An absorption line is seen only with a strong continuum, whereas an emission line can usually be seen with or without a detectable continuum. A continuum subtraction step is needed during the data processing if a noticeable continuum exists.

Before extracting the centre, width, peak height and integrated flux of a spectral line, we need to measure a spectrum from the 3D data cube, where the third axis

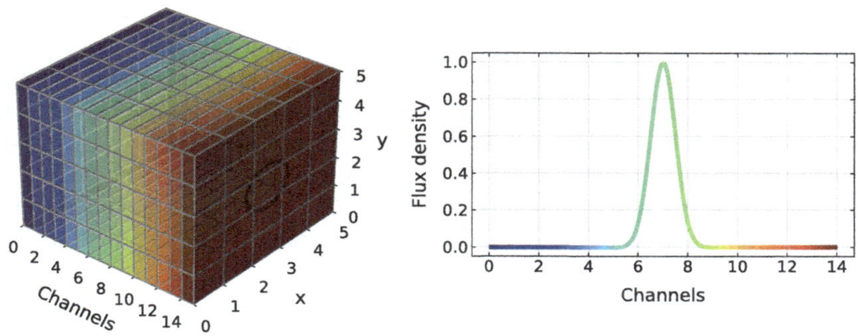

Fig. 9.2 Example 3D cube. x–y axes are spatial axes corresponding to the Right Ascendant and Declination on the sky, and channel axis is the spectral axis, which can be velocity, frequency or wavelength. The circle on the x–y plane represents an aperture. An 1D spectrum is usually extracted within an aperture (or a single spatial pixel; spaxel), e.g. as illustrated in the right panel

is frequency channels, or more commonly converted to radio velocity with respect to the line centre frequency ($v_{RAD} = (\nu_0 - \nu)/\nu_0 \times c$, where ν_0 is the line centre frequency, ν is the frequency axis, and c is the speed of light). Figure 9.2 shows such an example 3D cube. There are several ways to extract the 1D spectrum: at a fixed pixel (or also called spaxel because it's spatial coordinate is fixed) if the source is unresolved, or in an aperture, or performing source fitting with convolution of the beam. These extractions are done in the image cubes, which we refer to as image-plane extraction. There are also methods that directly extract spectrum from calibrated but not imaged uv-data, so that the measurements do not depend on the quality of CLEANing, which we refer to as uv-plane extraction hereafter.

The above extraction methods all need a known source position à priori. Since most targeted spectral line observations are aimed for detecting lines in known astronomical objects, the source spatial coordinate and redshift should all be known. However, there are also studies that aim at blindly finding line emitters in the radio data. We refer to this topic as blind line emitter detection. It can be done either in 2D by averaging chunks of channels in the spectral axis, or in 3D imaged cubes with certain image processing technique.

Then, in the later part of this section, we will illustrate a few high-level science examples, and list a few tools used in the community of astrophysics. They are likely incomplete but could be useful resources for the readers, and hopefully be inspiring for systematic, large-data processing in the future SKA era. For instance, SKA-Low (low-frequency) will be able to observe the redshifted HI 21 cm hyperfine structure line—the famous spectral line widely used to trace galaxy atomic gas content, its kinematics and even dark matter—beyond $z \sim 3$ way into the early Universe; SKA-Mid (mid-frequency) will be able to observe not only HI 21 cm but also OH 18 cm, H_2O maser and many other fainter lines in lower-redshift galaxies at an unprecedented angular resolution matching that of large submillimeter to optical telescopes; further with SKA-Upgrade (or SKA2) extending to frequencies

>15 GHz, the well-known molecular gas tracer CO(1-0) and dense gas tracer HCN(1-0) will be observable in distant galaxies—making systematic surveying of these gas contents finally possible way into the first generation of galaxies and black holes. The world-largest submillimeter array ALMA will continue being the major working force for observing more line traces, e.g., high-J CO, [CI] and [CII], highly complementary to SKA line tracer studies. How to best study the physics of the tracer lines is clearly based on how to best exploit the observational data, thus here we go into technical, broad views of exploiting spectral line data in different cases.

9.1.1 Detecting/Imaging a Spectral Line in a Galaxy with a Known Redshift

In the high-redshift community, it is very common to target a galaxy or a sample of galaxies with a known spectroscopic redshift (spec-z) for mm/radio spectral-line follow up. A spec-z can usually be obtained from optical/near-infrared spectroscopic surveys or from mm/radio spectroscopic scans for extremely sub-mm bright galaxies, i.e., submillimetre galaxies (SMGs). The former is much more efficient than the latter in identifying galaxy redshifts. There are numerous large surveys, e.g., the Sloan Digital Sky Survey[1] (SDSS; York et al., 2000; Abazajian et al., 2009), VIMOS VLT Deep Survey (VVDS; Le Fèvre et al., 2013; Garilli et al., 2008), zCOSMOS survey (Lilly et al., 2007, 2009), VIMOS Public Extragalactic Redshift Survey (VIPERS; Guzzo et al., 2014; Garilli et al., 2014; Scodeggio et al., 2018), VIMOS Ultra-Deep Survey (VUDS; Le Fèvre et al., 2015), and so on, have obtained optical spectra for millions of galaxies over the sky. Only a tiny fraction of them have (sub-)mm/radio spectroscopic follow up for the cold ISM in these galaxies. There are way fewer galaxies having spectroscopic redshifts from (sub-)mm/radio surveys. Only perhaps tens to hundreds of SMGs prominently detected in (sub-)mm bolometer array imaging surveys (Hughes et al., 1998; Borys et al., 2003; Perera et al., 2008; Penner et al., 2011) or from large-area (hundreds of $\deg^2$) *Herschel* Space Observatory imaging surveys (Oliver et al., 2010; Elbaz et al., 2011; Valiante et al., 2016; Smith et al., 2017; Liu et al., 2018; Jin et al., 2018) have subsequent (sub-)mm spectroscopic follow up to search for their redshifts. The Redshift Search Receiver (RSR) on the Large Millimeter Telescope (LMT) and the Z-Spec on the Atacama Pathfinder Experiment telescope (APEX) have been the most powerful instruments identifying redshifts for SMGs, confirming SMGs out to redshift of $z = 6.34$ (Riechers et al., 2013). ALMA has also been widely used for such purpose especially for scanning the brightest cooling line, [C II] 158 μm in the Epoch of Reionisation ($z > 6$; Bouwens et al., 2022). We will cover this topic in the next section.

[1] https://www.sdss.org/.

Once the redshift is known, it is fairly straightforward to detect the spectral line at the redshifted frequency ($\nu = \nu_0/(1+z)$, where ν_0 is the rest-frame frequency of the line). Single dish (sub-)mm telescopes usually do not have the capability to spatially resolve distant galaxies. Therefore, most science programs aim at a single line detection, that is, measuring the line profile integrated flux. For example, the Extended CO Legacy Database for GASS (xCOLD GASS) survey (Saintonge et al., 2011, 2017) is an IRAM 30 m large program making a large census of 366 galaxies' CO(1–0) or CO(2–1) emission. Molecular gas masses can be inferred from the CO line fluxes (Bolatto et al., 2013). JINGLE (Saintonge et al., 2018) is a 780-hour James Clerk Maxwell Telescope (JCMT) large program targeting the CO(2–1) emission in 193 far-infrared selected galaxies. These single dish surveys still only have the sensitivity to probe relatively low-z star-forming galaxies or only a handful of the brightest SMGs. Interferometers like ALMA in the southern hemisphere and NOEMA and VLA in the northern hemisphere have been the major forces observing higher-redshift normal SFGs, which are the bulk of galaxy population forming most of the stars in the Universe (Förster Schreiber and Wuyts, 2020). Due to the large cosmological distance and intrinsic faintness, detecting mm/radio lines in a large sample (e.g. hundreds) of SFGs has been extremely challenging mostly because the large time investment is usually not approved as Principle Investigator (PI) programs by the observatories.

In addition to large-sample galaxy evolution studies, high angular resolution imaging of (sub-)mm/radio spectral line has also been very valuable but also challenging. The challenge includes uv-coverage, surface brightness, and calibration of long baseline visibilities. Imaging itself is not an issue, given that these single target observations do not have a large mosaic. But whether one can recover the faint surface brightness at the outskirt of high-redshift SFGs is an issue. For example, when the synthesised beam FWHM becomes N times smaller, the beam area is N^2 smaller. The required sensitivity and on-source integration time to detect certain surface brightness threshold are thus N^2 lower and N^4 longer, respectively. This produces a large number of visibilities even for one galaxy, e.g., making the MeasurementSet data over a hundred GB. Currently, only very few high-redshift galaxies have such high-resolution, long-integration ALMA observations, e.g., a $z \sim 7$ QSO J234833.34-305410.0 imaged at ~35 mas (Walter et al., 2022), a $z \sim 4.7$ massive rotating SMG ALESS73.1 imaged at ~90 mas in dust (Hodge et al., 2019) to $\sim 0.1''$ in [C II] 158 μm line (Lelli et al., 2021), and another SMG AzTEC-1 at $z \sim 4.3$ imaged at ~ 70 mas (Tadaki et al., 2018).

In practice, to image some ALMA MeasurementSet (after calibration), one need to first identify the frequency channels that contain the expected lines, i.e., line channels, then run CASA `uvcontsub` task with those line channels excluded for the continuum subtraction of each visibility. The next imaging is fairly straightforward with the CASA `tclean` task. One needs to set a proper pixel size which is usually 1/5 of the synthesised beam size, an image size that matches the field of view, and a CLEANing threshold usually down to a few times the RMS noise of the data. The 'restoringbeam' parameter can be set to 'common' if there is no need to do per-channel beam. However, in the case when multiple MeasurementSets are jointly

CLEANed and they do not have exactly the same spectral window setups, then it is important to set 'restoringbeam' to empty to let 'tclean' use an independent beam per channel. In this case, the output image file (when exported as FITS-format files with the CASA 'exportfits' task) will contain a MultipleBeam extension in addition to the image data cube.

There are several tools or packages to analyze 3D image data cubes. For example, software with a Graphic User Interface (GUI) include: CARTA[2] (see also Sect. 11.2.1), DS9[3], QFitsView,[4] etc. The Python package `spectral_cube`[5] has also been more and more popular and powerful in dealing with image data cubes, computing moment maps, etc.

9.1.2 Spectral Line Scan for Redshift Determination

Galaxy redshifts are a particularly important property. Redshift affects the estimates of galaxy mass, physical size, and any absolute luminosity quantities. It is also a requirement for any spectral line imaging observation given that the current (sub-)mm/radio facilities (especially interferometers) usually do not have a large bandwidth to cover the whole probability distribution range of photometric redshifts (photo-z).

Optical spectroscopic surveys have contributed the vast majority of distant galaxies' redshift identification. However, at higher redshifts when galaxies' rest-frame optical emission red-shifts to the infrared part of the electromagnetic spectrum, it becomes progressively harder to identify their redshifts. Nevertheless, sub-mm strong cooling lines, as shown in Fig. 9.1, red-shift to much more observable windows, thus becoming potentially powerful tools to identify high-redshift ($z > 4$) galaxies.

The [C II] 158 μm line has been the most promising tool to identify galaxies as early as in the Epoch of Reionisation. The ALMA ALPINE large program (Le Fèvre et al., 2020), and the ALMA REBELS large program (Bouwens et al., 2022), have used this tool to survey in total over a hundred SFGs at redshifts $4 < z < 8$ with probably 200 hours of observing time.

Other lines like mid-to-high-J CO, [C I], H_2O, [N I] lines, etc., are also red-shifted to favourable observing windows at $2 \lesssim z \lesssim 5$, nearly 1/4 of the Universe's age ago. They are key tracers of the ISM properties (Cañameras et al., 2018; Yang et al., 2019; Harrington et al., 2021; Liu et al., 2021a; Dunne et al., 2022; Riechers et al., 2022). However, these lines are much fainter than the [C II] 158 μm line and other [O I] and [O III] lines at shorter rest-frame wavelengths. They have been used

[2] https://cartavis.github.io/.

[3] https://ds9.si.edu/.

[4] http://www.mpe.mpg.de/~ott/QFitsView/.

[5] https://github.com/radio-astro-tools/spectral-cube.

for redshift determination mostly only in a small number of strongly lensed and/or dusty galaxies (Brisbin et al., 2017; Jin et al., 2019, 2022; Reuter et al., 2020; Neri et al., 2020) (see also earlier studies in the review Carilli and Walter (2013)). Both the ALMA in the southern hemisphere and NOEMA in the northern hemisphere have been actively contributing to these efforts.

Technically, spectral line scan simply means using multiple, contiguous spectral window setups to cover a wide frequency range that cannot be fully covered by the bandwidth of the instrument. Since recent years, the ALMA Observing Tool (OT) has added a spectral line scan mode that allows the proposers to easily setup such a redshift scan observing program. At most five spectral setups can be set in one scheduling block to cover about 30–40 GHz bandwidth (given the ALMA single bandwidth of 8 GHz). Also a few years ago, NOEMA interferometer's new correlator PolyFiX has been commissioned and has twice the bandwidth of ALMA. Two NOEMA spectral setups can cover a contiguous 32 GHz range. Usually it is needed to have a small fraction of overlap between these spectral setups so as to avoid spectral window edge effects.

Spectral line scan programs are usually observed, calibrated and imaged as normal ones. Multiple MeasurementSets or UVTables can be obtained and CLEANed, leading to several image cubes covering different frequency ranges. Then line extraction can be performed for each cube to obtain the 1D spectrum. Concatenating the spectra will give us the full spectrum. Both the ALMA official software, CASA, and NOEMA official software, GILDAS can be used to make image cubes. Then users will need their own analysis on the spectra.

The spectral line scan is not always successful if the covered bandwidth is not wide enough, or the line is fainter than the proposed sensitivity. The former case can happen if one targets a high-redshift candidate which is however at a lower redshift, say $z \sim 2$, then two neighbouring CO lines will be more separated, e.g., CO(4–3) and CO(5–4) will be separated by $\sim$40 GHz at 2 mm, beyond the scanning bandwidth. The later case will happen easily if the target is simply too faint, for example, not having enough star formation rate or gas mass as expected, thus leading to no detectable line. Another practical issue is that there are a large number of noise peaks in the data that can be mistaken as science signal. This is because the large scanning bandwidth is much wider than the typical line width, e.g., 40 GHz bandwidth for a 400 km/s line at 2 mm means 200 independent searching chunks. Each chunk will also have noise peaks in the map that can reach 3–4-σ. It is particularly problematic when the source spatial coordinate is uncertain, e.g., based on single-dish positions. A detailed analysis of the noise properties in spectral line scans has been studied by Jin et al. (2019), finding that the false detection probability is more than one order of magnitude compared to pure Gaussian distribution statistics. Therefore, spectral line scan observation program should caution about not very high S/N detections.

9.1.3 Blind Line Emitter Detection

Blindly detecting line emitters in image data cubes is also an important topic in the context of data mining. How to systematically find and catalogue line emitters in 3D data is still less widely studied. It, however, will likely attract more attentions in the future large field of view (sub-)mm/radio surveys.

Several scientific works have addressed the blind line emitter detection techniques. For example, the ALMA large program ASPECS[6] (Walter et al., 2016; Decarli et al., 2016, 2019) and its pilot program with IRAM PdBI (Decarli et al., 2014; Walter et al., 2014), are the first blind deep field line survey, aiming at obtaining a cosmological volume-limited survey of high-redshift mm line emitter galaxies. Similar programs have also been conducted using the VLA, i.e., the VLA CO Luminosity Density at High Redshift (COLDz) survey (Pavesi et al., 2018; Riechers et al., 2019), and the VLA-ALMA Spectroscopic Survey in the Hubble Ultra Deep Field (VLASPECS) (Riechers et al., 2020). These surveys targeted the HUDF field (Fig. 8.1) and cover a contiguous area of about 50 arcmin^2 (e.g., 7 ALMA Band 3 pointings) and a wide frequency range, i.e., the full ALMA Band 3 (84–115 GHz) and Band 6 (212–272 GHz) coverages. In one of their work aiming at finding blind $z > 5.5$ [C II] line emitters (Aravena et al., 2016), the identifying of line emitters is made with the Astronomical Image Processing System (AIPS)'s SERCH task (Aravena et al., 2016), which convolves the data cube along the frequency axis with a Gaussian kernel with the expected line width as the FWHM then reports all channels and pixels above centre S/N threshold (e.g., 3.5). Then the IDL routine `CLUMPFIND` (Williams et al., 2011) is used to catalogue the line peaks. This, however, gives a large number of potential sources (e.g., thousands) where most of them should be boosted by noise. Aravena et al. (2016) analysed the statistics of negative peaks in their data cubes, and derived a fidelity, which they define as $P = 1 - N_{\text{negative}}/N_{\text{positive}}$ and approximates the probability of a positive detection being unlikely boosted by noise (and seen in intensity-reversed data cube). They found a 70% fidelity level at $\text{S/N} = 5.3$ for a blind line emitter without optical counterparts, and a 40% fidelity at $\text{S/N} = 4.5$ for the line candidates associated with optical dropouts at $z > 5.5$. In this way, a few more than ten $z > 5.5$ [C II] line emitters have been considered reliably detected in their survey. Their other papers also analysed CO and [C I] line emitters with similar methods (González-López et al., 2019).

The Australia Telescope Compact Array (ATCA) large program, CO ATCA Legacy Archive of Star-forming galaxies (COALAS[7]) (Jin et al., 2021), is an 820-hour program mapping the CO(1–0) line emission in several high-redshift (proto-)clusters. About half of the time was spent on the Spiderweb protocluster at $z = 2.16$, covering a contiguous 25.8 arcmin^2 or ~60 cMpc^2 area. With a bandwidth

[6] https://aspecs.info/; https://almascience.eso.org/alma-data/lp/ASPECS.

[7] http://research.iac.es/proyecto/COALAS/pages/en/home.php.

of 2 GHz at the observing frequency of 36.5 GHz, the observations cover a redshift range of $2.12 < z < 2.21$, or equivalently ±7000 km/s for the red-shifted CO(1–0) line. Jin et al. (2021) developed the MaxFinder code to automate the source finding and multi-wavelength counterpart cross-matching. The code iterates every 2D pixel in the map, takes the pixel's spectrum and calculates a weighted mean over channels between each pair of two channels n_i–n_j (with fixed channel width of 40 km/s). The channel pair that maximises the weighted mean is then taken as the signal, and an error is calculated over the channel range accordingly. In this way, a 2D line S/N map can be created. In their study, all lines with $S/N > 5$, and those with $S/N > 4$ and meanwhile having an optical counterpart are selected as line emitters.

Jin et al. (2021) also compared the Source Finding Application[8] (SoFiA; Serra et al., 2015b), which is widely used (Serra et al., 2015a; Kamphuis et al., 2015; Koribalski et al., 2020; Klitsch et al., 2022; Zheng et al., 2022). SoFiA is designed to search for line emitters in 3D across different spatial scales, and has a GUI. It takes the error and weights into account, and calculates a reliability for individual detection, $R = T/(T + F)$, where T is the number of positive peaks and F the negative ones. As pointed out by Jin et al. (2021), this is identical to the "fidelity" in Aravena et al. (2016) and González-López et al. (2019), and "purity" in Béthermin et al. (2020). Jin et al. (2021) find that in general SoFiA and MaxFinder agree well on the majority of sources, however, there are a small number of sources close to negative noise peaks or blended missed by SoFiA, and some sources with low $S/N \sim 3.5$ are in SoFiA but not in MaxFinder catalogues. There are even cases where a few known sources selected from optical emission line surveys that have relatively low-to-moderate $S/N \sim 3.8$–4.4 but missed by both SoFiA and MaxFinder. Therefore, it seems having multiple approaches and both blind and prior based line searches are indeed necessary to cover all cases.

The above cases mostly have not involved in a pre-processing of the data cube before searching. However, a matched-filter technique is sometimes also adopted for 3D line search. The matched filter method is basically computing the cross-correlation of the data and an expected model, to find the similarity of the data and model at each data position. That is, after matched filtering, the portion of the data more similar to the model will have a stronger signal, thus more easily to be picked out. Pavesi et al. (2018) applied the 3D matched-filter technique to blindly detect line emitters in the VLA COLDz survey data. Their code named MF3D is available in github.[9]

The Line Source Detection and Cataloguing (LSDCat; Herenz and Wisotzki, 2017) is also a software package that adopts 3D matched-filtering technique to detect line emitters in 3D. It is coded in Python and is developed for efficiently processing large 3D data cubes. LSDCat has been maintained and used by the MUSE consortium to search for high-z faint emission line galaxies (Herenz et al., 2017; Caruana et al., 2018; Urrutia et al., 2019).

[8] https://www.atnf.csiro.au/people/Tobias.Westmeier/tools_software_sofia.php.

[9] https://github.com/pavesiriccardo/MF3D.

The above contents are very likely incomplete and miss many significant codes or works, but hopefully can demonstrate the basics and varieties of blind line emitter detection. This will also be a rapidly evolving field with future technique like AI/ML and much larger data cubes obtained with larger telescopes/instruments.

9.1.4 High-Resolution Mapping of Spectral Lines in Large Nearby Galaxies

High-resolution mapping of (sub-)mm/mm spectral lines in large nearby galaxies, e.g., CO lines at sub-arcsec scales, have only been made within the last decade with ALMA and NOEMA. In the coming future, SKA and its precursor arrays, and ngVLA, will also be able to push H I 21 cm line mappings down to sub-arcsec scales, thus matching the angular resolutions of *HST*, *JWST* and other optical/near-infrared facilities.

The PHANGS-ALMA survey is an ALMA large program mapping a large number of 70+ nearby galaxies at $\sim 1''$ resolution in CO(2–1) (Leroy et al., 2021b). These mappings cover the full molecular gas disks of these galaxies with up to $\sim$23 arcmin^2 mosaic observations. To fully solve the missing flux issue, PHANGS-ALMA survey used both the ALMA 12 m main array and the Atacama Compact Array (ACA) with 7 m antennae covering shorter uv distances and 12 m total power observations filling in the zero uv space. A publicly available, comprehensive, flexible imaging and post-processing pipeline has been developed[10] (Leroy et al., 2021a) to best CLEAN the complicated nearby galaxy line emission and to combine interferometric and total-power data. In slightly more detail, the pipeline has the following stages: (1) preparing configuration keys; (2) uv data staging, (3) imaging, (4) post-processing and short-spacing correction, and (5) line map product creation. Stages (2) to (4) are to be run inside CASA, and stage (5) is to be executed outside CASA with some additional Python packages, e.g., the `spectral_cube` package.

The stage (3) imaging includes the key improvements of this pipeline, most suitable for the imaging of large-mosaic interferometric datasets. The pipeline iteratively calls the CASA `tclean` task each time with a gradually refined CLEAN mask, i.e., a data cube that defines which spaxels to CLEAN, and allows `tclean` to run for only a few major cycles so that the pipeline can frequently refine the mask during the whole process. The pipeline also conducts two types of CLEANing, firstly a 'multiscale' CLEANing, which allows the tclean to assume extended CLEAN models instead of the traditional point models, then a 'singlescale' traditional CLEANing. Each type of CLEANing is run for sufficient number of major cycles to reach a convergence on the CLEAN residual and model fluxes. In this way, the very complicated morphology of large nearby galaxies can be well modelled by the CLEAN algorithm, and the CLEAN residual will show no artefacts like negative

[10] https://github.com/akleroy/phangs_imaging_scripts.

bowling features or strong positive remaining signals in every channel map. This stage of pipeline is computationally expensive for moderately large data cubes, e.g., could be several days or even weeks for $3000\times3000\times300$ or larger data cubes, when without parallelisation. A future update with CASA parallelisation[11] implemented would be very promising.

The stage (4) post-processes the imaged data cubes and corrects the short spacing by running the CASA `feather` task to combing the interferometric imaged data cube and a single-dish cube. The latter can be obtained from single dish data reduction then imaged with the CASA `sdimaging` task. The PHANGS-ALMA pipeline will take care of image cube alignment before running the `feather` task. The `feather` task then performs Fourier transforms to both interferometric and single-dish image cubes, then combines the data in the Fourier space with some proper weighting (e.g., a low-pass filter and a high-pass filter depending on the single-dish primary beam), and finally does an inverse Fourier transform to get the feathered image cube. This is done channel by channel in the data cube. More detailed description of the task, the pipeline and the feathered versus non-feathered line maps are given in the relevant papers (Leroy et al., 2021a,b).

There are also other methods combining an interferometric dataset and a single-dish dataset thus correcting for the short spacing. For example, in the Plateau de Bure Arcsecond Whirlpool Survey (PAWS) (Schinnerer et al., 2013; Pety et al., 2013), a Hybrid Synthesis method has been developed. It firstly converts the single-dish image cube into pseudo uv visibilities by deconvolving with the single-dish primary beam and convolving with the interferometric synthesised beam in the Fourier space, then jointly CLEANing all visibilities with a proper weighting (Pety et al., 2013).

In another series of works using the Combined Array for Research in Millimeter Astronomy (CARMA) and the Nobeyama 45 m telescope (NRO 45m), and ALMA data, Koda et al. (2011, 2019) developed a Total Power Map to Visibilities ('tp2vis') code[12] to combine the interferometric and single dish data also in the pseudo visibility approach. A comprehensive comparison of short spacing corrections is given in Koda et al. (2019), where they classify the algorithms as Combination After Deconvolution (CASA feather), Combination During Deconvolution (Dirienzo et al., 2015), and Combination Before Deconvolution (the PAWS method and tp2vis).

To shortly summarise this part, high-resolution interferometric plus total-power mapping in nearby galaxies, and maybe also large-area Galactic regions, have gained major progresses and gathered a wealth of tools and experience. There are still challenging facts in efficiently performing the imaging with HPC and combining short-spacing datasets.

---

[11] https://casadocs.readthedocs.io/en/stable/notebooks/parallel-processing.html.

[12] https://github.com/tp2vis/distribute.

9.2 Polarisation

9.2.1 Methods

Polarimetry covers all analysis that involves the polarisation state of the signal, beyond the total intensity (Stokes I). This is typical divided into analysis of the linear polarisation (Stokes Q and U) and the circular polarisation (Stokes V). Polarisation is generally associated with magnetic fields, which provide a preferred axis for symmetry breaking in the physics of radio emission and radiative transfer.

Linear polarisation, in radio astronomy, is most often associated with the physical processes of synchrotron radiation and Faraday rotation. Synchrotron radiation comes from cosmic rays being accelerated by cosmic magnetic fields; it is linearly polarised perpendicular to the magnetic field orientation in the emitting volume. The synchrotron intensity and spectrum contains information on the magnetic field strength and cosmic ray properties, while the linear polarisation contains information on the orientation and degree of order of the magnetic field.

Faraday rotation is a radiative transfer process where linearly polarised emission passing through magnetised plasma experiences birefringence, causing the plane of polarisation to change. The strength of the Faraday rotation, which is typically called the rotation measure (RM) or Faraday depth, contains information on the magnetic field component parallel to the line of sight and the electron density, integrated over the radiation path. Synchrotron radiation and Faraday rotation are very often observed together: the synchrotron emission provides a source of linearly polarised radiation, which then experiences Faraday rotation as it propagates to the observer.

The change in the polarisation angle (Ψ) by Faraday rotation is proportional to the square of the wavelength (equivalently, the inverse square of frequency),

$$\Delta\Psi = \lambda^2 RM. \tag{9.1}$$

For polarised emission originating from a distance d, the RM is defined as

$$\mathrm{RM} = 0.812\,\mathrm{rad\,m^{-2}} \int_0^d \frac{n_e}{\mathrm{cm^{-3}}} \frac{B_{||}}{\mu\mathrm{G}} \frac{ds}{\mathrm{pc}} \tag{9.2}$$

where n_e is the free electron density, $B_{||}$ is the parallel component of the magnetic field (with positive being magnetic fields directed towards the observer), and ds is an element of the radiation path towards the observer (Ferrière et al., 2021).

The frequency dependence of linearly polarised signals after Faraday rotation dominates the analysis of polarisation data. Since the signal of interest is both frequency-dependent and broadband, it is not possible to average over frequency, as is typically done in total intensity analysis, nor to isolate small spectral regions of interest and discard unneeded channels, as is typically done in spectral line analysis.

While there are some analyses (mostly older, from a time where telescopes had comparatively little bandwidth and limited computing for processing multiple channels) that focus on spatial variation in the polarisation signal observed at a single frequency, the vast majority of polarimetric analysis focuses on the frequency variation of the polarised signal. The simplest, and earliest, form of spectropolarimetric analysis is to fit a linear relationship to the polarisation angle as a function of λ^2. The slope of this relationship gives the RM, while the intercept is the polarisation angle prior to Faraday rotation (i.e., the polarisation angle intrinsic to the synchrotron emission).

The linear fitting method is only effective for lines of sight with a single source of polarised emission with no internal Faraday rotation (Faraday rotation in the same volume as the polarised emission originates from); it also struggles to when the signal-to-noise ratio in polarisation is low in the individual channels. More sophisticated analysis techniques have been developed to overcome these limitations, the two most commonly used of which are RM synthesis and QU-fitting. Both methods can be used to analyse more complex lines of sight where there are either multiple emitting regions (with different degrees of Faraday rotation) or emission with internal Faraday rotation; such lines of sight are called Faraday-complex.

Rotation measure synthesis (Brentjens and de Bruyn, 2005) builds a Fourier formalism around the polarisation behaviour as a function of λ^2. Since polarisation angle is periodic (with a domain of 180°), a Fourier transform relationship can be defined between the polarisation as a function of λ^2 and the polarisation as a function of Faraday depth (often called the Faraday depth function (FDF), the RM[13] spectrum, or the Faraday spectrum); this is shown schematically in Fig. 9.3. In this way, all the Fourier properties we are comfortable with in other circumstances, such as time-frequency transforms or radio aperture synthesis, apply to RM synthesis. The polarisation as a function of frequency (or λ^2) acts analogously to the visibilities of aperture synthesis (including not being measured/gridded uniformly); after a Fourier transform the resulting polarisation as a function of Faraday depth is uniformly sampled, has a point spread function dependent on the sampling in λ^2, and is generally over-sampled to allow accurate characterisation of the signals present. It is also subject to problems of missing scales: small λ^2 values (high frequencies) are sensitive to features that are broad in Faraday depth that are depolarised at low frequencies, while low frequency (or more accurately, broad λ^2 coverage out to large λ^2) is needed to provide a high resolution in the Faraday depth domain. Deconvolution algorithms, such as RM-CLEAN (Heald et al., 2009), can be used to further process the Faraday spectra.

RM-synthesis has the advantage of being able to process data with low signal-to-noise per channel, as the Fourier transform has the effect of concentrating faint broadband signals into narrower, higher signal-to-noise signals in the Faraday depth

[13] For historical reasons, the term RM is generally associated with linear fitting and discrete sources while Faraday depth is associated with RM synthesis and diffuse polarised emission; a full exploration of the subtle definitional differences between these two terms is outside the scope of this work.

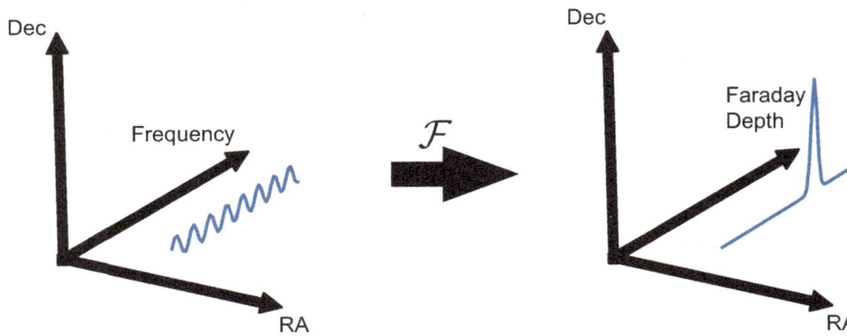

Fig. 9.3 Schematic of input and output data for RM synthesis: position-position-frequency cubes (for Stokes Q and U), with faint frequency-dependent signals, are Fourier transformed along the frequency axis to produce position-position-Faraday depth cubes which concentrates the signal while leaving the noise dispersed

domain. It also has the advantage of being a non-parametric algorithm, with no assumed model or fitting parameters, but these come with the disadvantages of not handling well large gaps in the frequency coverage (which introduce large side-lobes in the point spread function) and sometimes being sensitive to the details of the deconvolution process (again analogous to problems that can arise with aperture synthesis image CLEANing).

9.2.2 *Computing*

From a computing perspective, RM synthesis tools are typically divided into either '1D' or '3D' versions, where the 1D tools operate on individual spectra (typically of unresolved radio sources) and the 3D tools operate on position-position-frequency(-Stokes) data cubes covering some region of the sky. The 1D tools tend to be computationally fairly light, by virtue of outsourcing the problem of spectrum extraction to the user—this often ends up being more intensive than the RM synthesis step. Three-D tools have a much heavier computational footprint due to working with entire data cubes together. This footprint is primarily on the disk read/write (getting the data cubes into memory, and writing the results) and on RAM use (holding the frequency-domain cubes, the Faraday depth-domain cubes, and point spread function cubes all in memory simultaneously). Unfortunately, RM synthesis also makes the data volume problem of polarimetry worse: the output data has the same dimensionality as the input data but with the frequency axis replaced by a Faraday depth axis, which will typically have the same order-of-magnitude as many samples if not slightly more. To manage both the RAM use and disk volume, it is common to break data into smaller tiles or pieces of sky area (as RM synthesis treats each pixel as independent); this does not reduce the total data volume but can reduce the file sizes to more convenient levels (GB rather than TB).

QU-fitting modifies the traditional polarisation angle linear regression analysis by moving the fit to the Stokes Q and U parameters, which are better behaved in the low signal-to-noise regime and also allow linear superposition of multiple signals. QU-fitting can use various models, from the simple idealised one-source model used in angle-fitting to different internal Faraday depolarisation models and more, as well as models that include multiple polarised emission sources. As these models can potentially have many parameters, including non-linear parameters, the actual fitting routines used tend to involve sophisticated algorithms such as nested sampling or Markov chain Monte Carlo methods.

Compared with RM synthesis, QU-fitting generally requires more CPU resources to explore the model parameter space, especially if multiple models are being tested. Most applications have focused on applying the method to small numbers of individual lines of sight (which may include multiple lines of sight through resolved sources), rather than to every pixel across some sky area.

The identification of polarised sources generally proceeds in one of two ways. The first way is to begin with a catalogue of Stokes I sources, such as those generated by the source finders discussed in Sect. 8.1, and to determine whether the polarisation properties at those locations qualify as significant detections. This method is simple and generally effective, but there are cases of (partially) resolved sources where the polarised signal can occur away from the Stokes I peak and be missed by such analysis. The alternative way is process the entire sky area of the observation in polarisation and perform source-finding on the polarised intensity or related quantities (e.g., Farnes et al., 2018). These methods tend to struggle (particularly when using source finders intended for Stokes I) due to non-Gaussian noise properties in polarised intensity.

Polarisation is associated with both persistent and variable radio sources (e.g., radio galaxies and AGN) as well as with transient sources (e.g., pulsars and FRBs). Transient sources are often found to be significantly polarised, so a full analysis of these sources must add the Stokes axis as an additional dimension to be explored and characterised, in additional to the other dimensions (position, frequency, time) relevant for these sources.

9.3 Time Domain

9.3.1 Pulsar Astronomy

Pulsars (abbreviated from Pulsating stars) are neutron stars showcasing a wide variety of unique properties. They possess a high rotation spin frequency (current record holder spins at 716 Hz; Hessels et al., 2006) and an enormous magnetic field strength (order of up to $\sim 10^{15}$ T; MRI machines in comparison produce 3 T).

Furthermore, a mass of $\sim$1–2 $M_\odot$ is compressed onto the size of a typical city (10–15 km) making it one of the densest arrangements of matter known in the universe ($\sim 10^{14}$ g cm^{-3}). Pulsars are also known as 'cosmic lighthouses' owing to the twin electromagnetic beams emitted from their magnetic poles. When the pulsar spin-axis and magnetic-axis are misaligned, periodic signals (with a periodicity determined by the rotation frequency) can be observed with telescopes (assuming that the beam crosses the line of sight of Earth). Pulsars have been observed across the entire electromagnetic spectrum (in radio, optical, X-ray as well as gamma-rays). However, the radio frequency band combined with advancements in instrumentation and search techniques have maximised the scientific throughput of understanding pulsars. This has been achieved via new discoveries as well as in-depth studies of individual sources.

Since the serendipitous discovery of radio pulsars in 1967 (Hewish et al., 1968), several surveys since have increased the population of pulsars to the current number[14] of over 3300 (Manchester et al., 2005) including few discoveries outside the Galaxy (e.g., Ridley et al., 2013). Some pulsars are recycled and spun up to millisecond spin periods from their companions (Alpar et al., 1982; Radhakrishnan and Srinivasan, 1982) and hence called millisecond pulsars. These pulsars are highly stable rotators (spin period changes at 10^{-20} s^{-1}) thus behaving as precise clocks even rivalling atomic clocks on long timescales (Hobbs et al., 2020). A collection of these millisecond pulsars have been used to construct multiple pulsar timing arrays (Foster and Backer, 1990) under the umbrella of the International Pulsar Timing Array[15] (IPTA) with the aim of detecting the stochastic gravitational wave background produced from coalescing supermassive black-hole binaries (e.g., Taylor et al., 2019). Recent results have shown that such a detection is plausible in the near future (Antoniadis et al., 2022). Pulsars can also provide stringent constraints on the neutron star equation of state derived from mass (Antoniadis et al., 2013; Cromartie et al., 2020) and radius (Miller et al., 2021) measurements. Binary pulsars have particularly been excellent laboratories for testing gravity (general relativity as well as alternate theories) in the strong-field regime (Kramer et al., 2021). Motivated by such fundamental open questions, radio pulsar searching continues to be a key scientific driver for modern-day radio telescopes like FAST and MeerKAT as well as the upcoming SKA.

From a technical perspective, the steady increase in pulsar discoveries can be attributed to customised changes in hardware and software over the years. Increased bandwidths of receivers have played a major role in improving sensitivity given that pulsars emit broadband signals. For example, modern surveys with FAST use more than 400 MHz of bandwidth (Han et al., 2021). Adding to this has been the development of receivers and back-ends operating at low (50–100 MHz; Sanidas et al., 2019) to high frequencies (up to 86 GHz; Liu et al., 2021b). Pulsars predominantly behave as steep spectrum sources (Lorimer et al., 2006) and thus

[14] https://www.atnf.csiro.au/research/pulsar/psrcat/.

[15] https://ipta4gw.org/.

observed more brightly at lower frequencies. However, propagation effects through the interstellar medium (ISM) have a significant impact on detection. In similar lines to the RM (discussed in Sect. 9.2), the integrated column density of electrons along the line of sight can be quantified by the dispersion measure (DM) given by:

$$DM = \int_0^d n_e dL \tag{9.3}$$

with d indicating the distance between the pulsar and Earth. This can be derived from modelling the interstellar medium as a frequency dependent refractive index (e.g. Lorimer and Kramer, 2004). This leads to a time-delay on Earth between a high-frequency and low-frequency pulse emitted at the same time from a pulsar (commonly known as dispersion) given by

$$\Delta t \simeq k\mathrm{DM} \times \left(f_1^{-2} - f_2^{-2}\right) \tag{9.4}$$

where $k = 4.15 \times 10^6$ ms with f_1 and f_2 representing the reference frequencies. This leads to a dispersion sweep and can smear out the pulsar signal when averaged across the observing frequency band. Its generally corrected by splitting the power across the frequency band into spectral channels (using pulsar spectrometers) and applying a time shift per channel to rectify the delay (using Eq. 9.4 per channel with f_1 fixed) before averaging across frequency. This technique is computationally cheap and is better suited for searches where the DM value of the potential discovery is unknown apriori and thus demands a large range of DM trials to be applied.

The signal power averaged in time across each spectral channel can help determine corrupted regions of the frequency band caused by radio frequency interference (RFI). RFI mitigation is thus crucial and techniques are applied before dedispersion and after (once the spectral information is lost). The time-series containing a potential periodic signal is then converted to Fourier power. Peak frequencies in the Fourier spectrum are observed as potential candidate signals. An additional enhancement is the stretching and stacking of the Fourier spectrum to boost the signal strength using the fundamental as well as higher order harmonics (termed as harmonic summing). This periodicity search method would work reliably if the pulsar is isolated. However, pulsars in binary systems are impacted by Doppler modulation leading to variations in the observed periodicity. Signals would thus arrive earlier/later than expected and show a spread in Fourier power. The most commonly used method to correct for this effect is the modelling of the pulsar velocity with a constant acceleration term (Johnston and Kulkarni, 1991). Such a model is effective only when the observation time is roughly a tenth of the binary orbital period. Such acceleration searches have been largely successful in finding compact binary systems including PSR J1757-1854 which is the most accelerated pulsar currently known (Cameron et al., 2018). The techniques have either applied a resampling scheme on the time-series (Camilo et al., 2000) or a matched-filtering technique post the Fourier transform (Ransom et al., 2002).

Once candidates are identified based on a range of DM and acceleration trials, the information is used to assess the strength of the signal in the original time and frequency resolved data. The data is averaged/stacked modulo the candidate spin period in observing frequency as well as time (post dedispersion and binary demodulation). Furthermore, an optimisation scheme is applied to better tune the trial search parameters to maximise the signal-to-noise ratio. This method of *folding* the data helps boost the detection significance of the candidate by averaging out noise and enhancing the signal. A clear signature of a pulsar detection is a persistent signal in time and frequency with an exponential drop in significance when moving away from the true DM and acceleration parameters. When these signal parameters (with a pulsar signature) do not match the parameters from the known pulsar catalogue and are redetected with similar signal parameters in a follow-up observation, the pulsar is declared as a new discovery. If the barycentred (reference frame brought to the solar system barycentre) spin frequency has changed between epochs, the pulsar is confirmed to be in a binary system.

Above mentioned procedures have formed the foundation of a majority of search schemes for targeted and blind surveys (e.g., Manchester et al., 2001; Cordes et al., 2006a; Keith et al., 2010; Eatough et al., 2013; Stappers and Kramer, 2016; Cameron et al., 2020). However, each of the steps have been upgraded by adapting to improved algorithms, technology and computing power with time. Dedispersion is now commonly implemented on GPUs leading to order of magnitudes boost in processing speed (Magro et al., 2011; Barsdell et al., 2012; Bassa et al., 2017). Apart from porting to GPUs, efficient dedispersion algorithms have also been explored (e.g., Zackay and Ofek, 2017). The application of the Fast Fourier Transform (FFT) algorithm (Cooley and Tukey, 1965) and harmonic summing have also been incorporated for GPUs in modern search pipelines (Morello et al., 2019). Acceleration search algorithms have been ported to GPUs too with extensive research being conducted on bench-marking against existing legacy software (e.g., Dimoudi et al., 2018; White et al., 2023). Surveys now commonly utilise multi-pixel feeds on single dish telescopes (e.g., Han et al., 2021) or hundreds of synthesised beams using interferometers for enhanced coverage (e.g., Stappers and Kramer, 2016; Sanidas et al., 2019; Padmanabh et al., 2023). The multi-beam information has also been pivotal in exploring spatial filtering techniques to discern RFI (Kocz et al., 2012). While the acceleration search has been successful, higher dimensionality searches to find compact binary systems have begun to take precedence. Firstly, searches also accounting for acceleration derivative (commonly called 'jerk') have been particularly used for searching globular cluster targets (Andersen and Ransom, 2018; Douglas et al., 2022). Secondly, a fully coherent binary pulsar search accounting for all Keplerian parameters has also found applications. The earliest such example is the Einstein@Home volunteer distributed computing project that uses spare CPU and GPU cycles from volunteer machines to search not only for gravitational waves but also binary radio pulsars in circular orbits (Allen et al., 2002; Knispel et al., 2013).

There is also scope for different perspectives to searching. Apart from FFT based methods, other algorithms have been explored/re-explored. One such approach is the 'folding' of a dedispersed time-series with different spin-period trials. This method is coherent unlike FFT methods where the power is lost to multiple harmonics. Although the algorithm was formulated just a few years after the first pulsar discovery (Staelin, 1969), a pipeline implementation has come to light only recently owing to its computational demand (Morello et al., 2020). This method has also demonstrated better sensitivity to slow spinning pulsars (order of seconds) than the FFT method (Cameron et al., 2017; Morello et al., 2020). The current Keplerian parameter searches can be extended from circular to eccentric binaries for targeted sources like globular clusters where searching in the DM dimension can be reduced (Balakrishnan et al., 2022). Searches have predominantly used total intensity information but detection of pulses using and linear and circular polarisation information in time is poorly explored.[16] An important assumption so far is the persistent periodicity of a pulse received from a pulsar. However, a large class of pulsars exist that show intermittent and sporadic emission with time (Lyne, 2009). Single pulse search techniques are crucial in such scenarios. Here post dedispersion, matched-filtering techniques are applied to detect significant pulses. A ground-breaking consequence of such methods has been the emergence of the young field of fast radio bursts (see Sect. 9.3.2).

The expectation of the SKA is ~30,000 new pulsar discoveries (Keane et al., 2015) with some pulsars expected to be found in the intergalactic medium. However, a more difficult problem to tackle is data management and development of real-time processing pipelines. Nearly 2000 beams on sky are expected (500 for SKA-Low and 1500 for SKA-Mid) (Levin et al., 2018; Keane, 2018) to be operating on a telescope which would monitor the sky 24×7. The massive data rates would thus not allow for raw data to be stored. We are already seeing this with the MPIfR-MeerKAT Galactic plane survey being conducted (Padmanabh et al., 2023). Processing pipelines would thus have to find effective ways of reducing compute time. This would involve improved memory management as well as reduced input/output operations. Moreover, billions of candidates with diagnostic plots (Fig. 9.4) are expected from large scale surveys. This calls for automated procedures without human intervention, thus encouraging the implementation of AI/ML based algorithms (see Sect. 10.3).

9.3.2 FRBs

Fast Radio Bursts (FRBs) are millisecond-long bursts of radio emission of extragalactic origin. At the time of writing, the nature or FRBs is still unknown. It is also not clear if all FRBs are representatives of the same source population, or whether multiple underlying populations exist. Recent reviews of this exciting research field

[16] Although polarised sources identified through imaging have been targeted for radio pulsations and yielded success (e.g., Sobey et al., 2022).

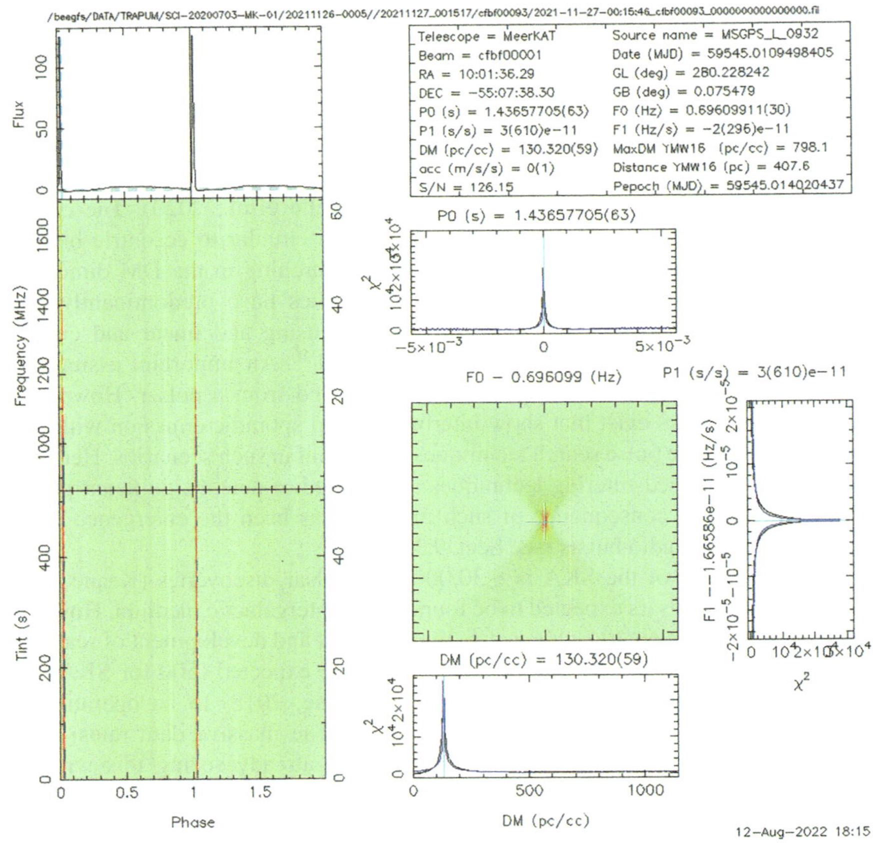

Fig. 9.4 Sample diagnostic plot of a known pulsar candidate in data taken with the MeerKAT telescope. Left panels show the signal in frequency-phase and time-phase. Other plots indicate the fine-tuning done using a reduced χ^2 metric to maximise the signal-to-noise ratio in DM, spin period and acceleration. The information on the top right corner is metadata about the candidate

were represented by Cordes and Chatterjee (2019), Petroff et al. (2019) or Caleb and Keane (2021). This summary refers to a recent status report given by Kramer (2022).

Observationally, there are currently (at least) two distinct classes of FRBs. On one hand, we have repeating FRBs ("repeaters") from which multiple bursts have been detected. On the other hand, for non-repeaters, only single bursts have been detected. In those cases, the monitoring has, of course, not been continuous, so that repeats could have been missed. In any case, for the repeaters, any successful model must not include a cataclysmic event in order to create the observed coherent radio emission. The majority of models currently discussed contains highly-magnetised neutron stars as the energy sources, so called "magnetars". For a recent review on magnetars see Kaspi and Beloborodov (2017).

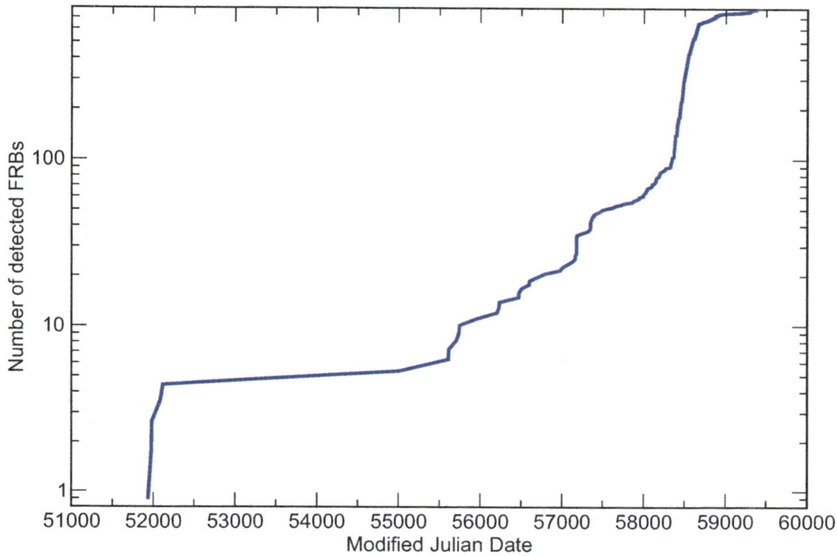

Fig. 9.5 Number of known FRBs as a function of time. Currently, more than 800 bursts from more than 600 FRBs are known

The first FRB was reported in 2007 by Lorimer et al. (2007). They had discovered a bright millisecond-long burst from a position on the sky near the Small Magellanic Cloud in archival pulsar search data. It was not until the work by Thornton et al. (2013) that it was realised that these radio bursts are common, appearing across the sky typically every 10 seconds or so,[17] and that they are of truly extragalactic origin. Thornton et al. (2013) also introduced the name "Fast Radio Bursts" for the first time, which has been used ever since.

Because FRBs have a cosmological origin, they can be used as a cosmological probes in order to examine the baryon content of the Universe, the nature of Dark Matter and Dark Energy, the ionisation history or the intergalactic magnetic field (Macquart et al., 2015; Keane et al., 2016; Macquart et al., 2020). Apart from attempts to understand their nature, such applications motivate the further discovery of many more FRBs.

Currently, at the time of writing, there are more than 800 FRBs known already (see Fig. 9.5). For cosmological applications, e.g. for studying Dark Matter or Dark Energy, a few thousands of FRBs may be needed (Macquart et al., 2015). New telescopes like the SKA promise to deliver such numbers relatively easily, even though finding FRBs is not trivial. The emission of FRBs is short in duration, typically a few or some tens of millisecond long, and their position on the sky is seemingly random. The most effective way to find FRBs is to scan large portions

[17] This statement obviously depends on the sensitivity of the observations.

of the sky simultaneously. Hence, telescopes with large fields-of-view are superior, such as the Canadian CHIME telescope. The field-of-view needs to pixelated and a time-series has to be obtained from each pixel, whereas the time resolution has to be sufficiently large to resolve the bursts against the background signals. Recording the data for offline processes may be possible, but an on-line analysis is clearly preferred, so that perhaps also other telescopes can be triggered. In such a case, but also for practical purposes, the analysis pipeline should be automated, and the human component should be replaced by an algorithmic process, which may also make use of machine learning and artificial intelligence. To make such algorithms successful in recognising an FRB, it is useful to exploit the unique characteristics of FRB emission, in order to distinguish it from other signals, including man-made radio interference signals.

The short duration of FRB emission suggests that the emission may be beamed. The large distances suggest a very energetic process. It is not uncommon for FRBs to be highly polarised and the brightness temperatures implied by the pulse structures are of the order of 10^{36} K or more (e.g., Cordes and Chatterjee, 2019). The cosmological origin of FRBs was established by Thornton et al. (2013), based on the large dispersion of the radio signal, i.e. the signal arrives first at high frequencies and then progressively later at lower frequencies (with a $\propto \nu^{-2}$ dependence). The amount of dispersion is given by the dispersion measure (DM), which is the integrated column density of the free electrons along our line-of-sight. In the Milky Way, it is dominated by the ionised interstellar medium and combined with electron density models, the DM provides a distance estimate. In contrast, the DM of FRBs is significantly larger than the amount of electron "available" in the Milky Way, confirming their extragalactic origin. Hence, the DM of FRBs is the sum of contributions from the Milky Way, the intergalactic medium, the host galaxy and potentially contributions local to the FRB source, such as a circum-object envelope.

Based on the CHIME FRB catalogue (Pleunis et al., 2021), one can identify four archetypical types of bursts: (a) Simple bursts comprised of one peak with a limited spectrum power-law, (b) narrow-band simple bursts with a Gaussian-like spectrum, (c) complex bursts with multiple peaks and similar (broad/narrow) frequency extent, and (d) complex bursts with multiple sub-bursts that show a downward drift in frequency as time progresses. The first two groups make up 90% of the whole population (i.e. 30% and 60%, respectively). The last two groups make up 5% each. Repeaters are typically part of the last group, and it is interesting that this is consistent with making up about 5% of the detected population in general. It is important to emphasise that none of the observed spectra (i.e. in neither group) is consistent with the radio spectrum of magnetars or pulsars. At the time of writing, there has been no published detection of FRBs below 100 MHz (e.g. with LOFAR or MWA), while the highest frequency of FRB detection is about 8 GHz with the GBT and Effelsberg.

It is notable that the emission properties of repeaters appear to be different from those of other FRBs. The downwards drift in frequency and a sub-dividing structure in the frequency-time plane is peculiar. The CHIME colleagues (Pleunis et al., 2021) also showed that bursts of repeaters are typically wider in time but narrower in bandwidth. These statistically significant differences may cast doubt on the possibility that all FRBs eventually repeat. Hence, based on the emission properties alone, the existence of (at least) two classes of FRBs appears plausible.

Apart from the burst shape and the peculiar frequency structure, also the polarisation properties seem to be different for repeating FRBs, though the picture is less clear here. There appears to be tendency for repeaters to be highly polarised with flat position angles and negligible circular polarisation, see e.g. FRB 20121102A (Michilli et al., 2018) or FRB 20201124A (Hilmarsson et al., 2021). However, other observations of FRB 20201124A paint a more complicated picture, as the polarisation properties appear to be a function of frequency (Kumar et al., 2021). In contrast, non-repeaters are often less polarised, but also sometimes with significant circular polarisation. Interestingly, the rotation measure is large, as for the Galactic centre magnetar (Desvignes et al., 2018). Not only for this reason, it is interesting to compare those properties with those of radio-loud magnetars.

One can use the described emission properties to not only distinguish between FRBs and other signals, but potentially also to search especially for repeaters and single-burst FRBs. It is obviously also very useful, to not only search in total intensity data, but to also take polarisation properties into account. The first survey that tried to automate detection and where also polarisation data were recorded and analysed was the SUPERB survey with the Parkes telescope (Keane et al., 2018), which was indeed successful in detecting FRBs (Petroff et al., 2015), though triggers were still vetted by humans before broadcasting. One should also note that exploiting a potential high degree of linear polarisation is made computationally challenging, since some FRBs show an extreme degree of Faraday rotation, which—when not accounted for—will smear out the polarisation and leaving a seemingly unpolarised signal behind. Pipelines therefore exists which do not only search in dispersion measure but also in the so-called rotation measure, describing the degree of Faraday rotation (Michilli et al., 2021).

The usage of artificial intelligence (AI) in this special subject of radio intensive radio astronomy is obviously well motivated. It is therefore no surprise that corresponding algorithms and software packages have been developed especially for FRBs. We refer the reader to recent examples (e.g., Agarwal et al., 2020; Bezuidenhout et al., 2022; Luo et al., 2023) and to the next chapter for a more general description of AI methods and applications.

9.3.3 Interstellar Scintillation

The emission from pulsars and FRBs must travel through space to reach the observer. Radio waves encounter many regions of plasma which can affect their signals, including the ionised interstellar medium (IISM), solar wind, and the Earth's ionosphere. As described in the previous sections, pulsars and FRBs experience 'dispersion', a delay proportional to the observing wavelength squared, which arises from the group velocity of radio waves in a cold plasma. This effect must be accounted for when searching for pulsed signals, and being proportional to the total electron column density along the line-of-sight, serves as a proxy of the distance. FRBs show column densities in excess of the contribution from the Milky Way, which helped distinguish them as extragalactic.

Variations in the electron density cause phase variations, and result in many scattered paths of light. As pulsars are point-like, their emission is spatially coherent and these scattered paths create a moving interference pattern at the observer. This effect is known as 'scintillation' (Fig. 9.6), and is analogous to the apparent twinkling of stars caused by distortions in the Earth's atmosphere. Pulsar scintillation has been known since 1968, shortly after the discovery of the first pulsars (Scheuer, 1968), which itself happened in the context of studying interplanetary scintillation (Hewish et al., 1968).

The main observable of scintillation is the 'dynamic spectrum', the intensity of the pulsed emission as a function of frequency and time. Contiguous patches of increased intensity in the dynamic spectrum are called 'scintles'. In strongly scattered sources, scintillation minima approach zero, while the brightest scintles can extend to many times the mean intensity. As such, scintillation can greatly aid or detract from detecting pulsars when the size of scintles is comparable to the observing band, if by chance the pulsar is observed during an interference maximum or minimum. As an example, this is seen dramatically in observations of globular clusters harbouring many pulsars; in many repeated observations of 47 Tucanae with Parkes at 20 cm, a total of 20 pulsars were detectable at different times, despite the mean flux of only 3 of the pulsars was above the detection threshold (Camilo et al., 2000). With new wide-band receivers this becomes less of an effect, as scintillation is more likely to average out over the band.

Scintillation was originally thought to be dominantly diffractive with the characteristic shapes of scintles being used to constrain models of interstellar turbulence (Rickett, 1970). However, periodic patterns were frequently observed, leading to the emphasis of refractive scintillation (Gupta et al., 1994), which can conveniently be reduced through the 2D power spectrum—'secondary spectrum'—to discrete interfering paths of propagation (Rickett et al., 1997). Secondary spectra often show parabolic scintillation arcs, which have been studied since 2001 (Stinebring et al., 2001). Their explanation as a collection of refracted images of the pulsar at points of stationary phase altered the picture of scintillation (Walker et al., 2004; Cordes et al., 2006b). Secondary spectra express the power in terms of the conjugate variables of frequency—related to the phase delay τ—and time—related to the Doppler shift f_{D} between scattered images. Often only a single thin scattering surface is visible, such that the shared dependence of f_{D} and τ on the angle of images θ results in a parabolic distribution of power constituting a 'scintillation arc'.

The curvature of these arcs encodes relative distances and velocities of the pulsar, scattering material, and the Earth. Annual variations can be fit to measure the distance and geometry of scattering screens, and in binary pulsars can be simultaneous fit with orbital variations to measure the full 3D geometry of pulsar orbits (Rickett et al., 2014; Reardon et al., 2020). When the scattering surface is

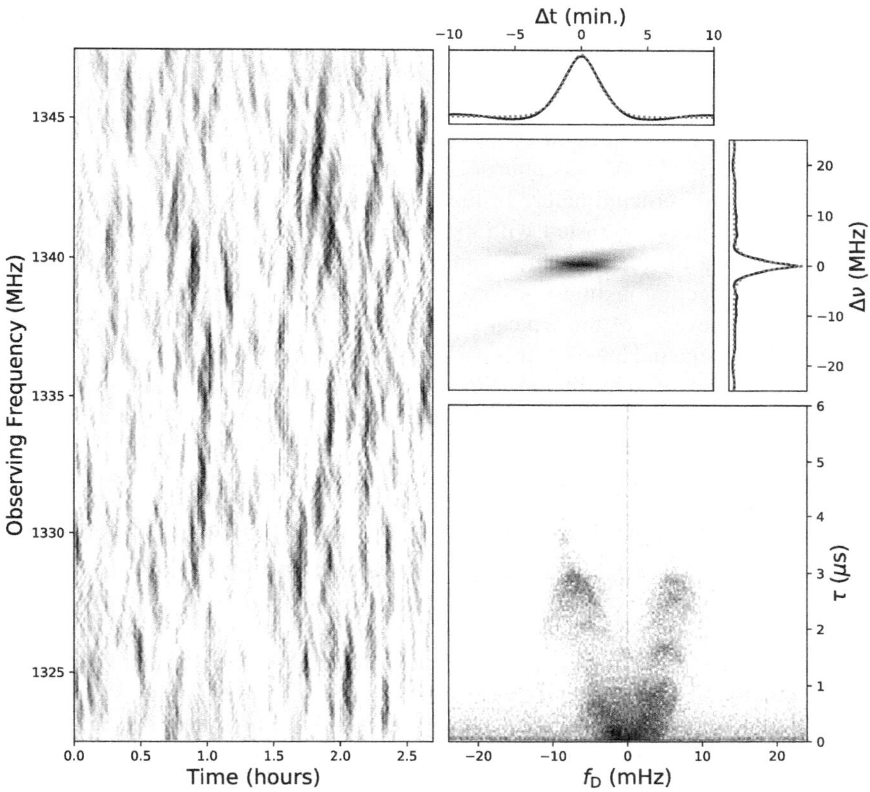

Fig. 9.6 Scintillation of the bright northern pulsar PSR B0355+54, observed at the Effelsberg 100 m telescope. *Left:* dynamic spectrum of scintillation, where the intensity scale extends from 0 to 7 times the mean intensity. Concentrations of high intensity are called 'scintles', extending to many times the mean, while the intensity does reach zero in many places—if the scintillation bandwidth is comparable to the observing bandwidth, this can easily aid, or detract in detection. *Top Right:* 2D auto-correlation function of the dynamic spectrum, and cuts through the $\Delta\nu = 0$ and $\Delta t = 0$ axis. Fitting these is a common way to derive the characteristic time and frequency scales of scintillation. *Bottom right:* the 2D power spectrum of scintillation, called the 'secondary spectrum', showing the power as a function of the phase delay τ and Doppler rate f_D between interfering paths. The parabolic shape is indicative of a thin scattering surface. The distinct features with inverted parabolic 'arclets' implies compact $\lesssim$0.1 AU structures distributed anisotropically. These discrete features lead to the highly ordered fine-scale fringing seen in the dynamic spectrum

known a priori (e.g. for pulsars embedded in supernova remnants), scintillation arcs can be used to measure the full 3D pulsar velocity (Yao et al., 2021).

Scattering screens are seen to be long lived, where for B1133+16 the same scattering screens are seen over 40 years (McKee et al., 2022, seen by re-reducing archival data), implying screen scales of $\gtrsim$5000 AU. In addition to these parabolic arcs, there often appears significant substructure, with compact scattering structures <0.1 AU which appear at fixed angular position over $\sim$month timescales (Hill et al., 2005; Sprenger et al., 2022). These structures can appear as inverted 'arclets' with the same curvature as the main parabola, and suggest highly anisotropic scattering—this was confirmed spectacularly using VLBI, which showed that the screen of PSR B0834+06 was almost one-dimensional on the sky (Brisken et al., 2010). The highly ordered nature of these scattering screens in the dynamic IISM was surprising, as it contrasted with the standard picture of diffraction caused by turbulent plasma fluctuations. From simple spherical arguments, compact regions capable of deflecting light to several milli-arcseconds seem to imply electron densities far in excess of the typical IISM values, which would be over-pressured and quickly dissipate. However, this can be aided through projection effects; recent theories to origin of scintillation screens involve corrugated plasma sheets seen at grazing incidence (Pen et al., 2014), or elongated 'noodles' of plasma (Gwinn, 2019), confined with magnetic fields.

Scintillation arcs appear to be prevalent, where at the time of writing they have been seen towards >100 pulsars (Stinebring et al., 2022; Main et al., 2023). However, the requirements to properly detect scintillation can be steep, requiring finely channelised data not typically used in pulsar observations, and high signal-to-noise as scattering features are seen down to intensities $<0.1\%$ of the pulsar's flux. In the case where scattering delays are of comparable to the pulse width, the pulse profile, and the scintillation pattern in frequency cannot be simultaneously resolved due to the Nyquist limit. A possible workaround is to have parallel datastreams with different observing parameters, which is currently being implemented in pulsar observations at the Effelsberg telescope. The method of 'cyclic spectroscopy' can go beyond the Nyquist limit, by taking advantage of the periodic nature of pulsars, forming the combination of the electric field between adjacent pulses within the scintillation timescale (Demorest, 2011). This method has been proven to work in at least one case (Walker et al., 2013), but is extremely computationally costly, scaling as N^3 with the number of channels.

One of the big promises of scintillation is the technique of 'scintillometry', using scattering screens to make precision measurements of the pulsar's emission and motion. If the interstellar response function is fully known, then these screens could be used as interferometers to study pulsars with $\lesssim$ nano-arcsecond precision (Pen et al., 2014). Dynamic spectra contain the total intensity (at present, often having $\sim 10^8$ points). Retrieving the underlying electric field falls under the well-known problem of 'phase retrieval'. This is a difficult problem without a unique solution, but has been achieved in some cases as scintillation is often highly ordered and sparse (Walker et al., 2008; Osłowski and Walker, 2023; Baker et al., 2022). The response function is much more readily measurable in baseband data, as phases are lost when forming intensity or averaging. The aforementioned cyclic spectroscopy must work on the baseband data, and in even simpler cases the impulse response function can be directly measured from bright, sharp pulses (Main et al., 2017). This is seldom done to date, due to the large computational and storage requirements of working with baseband data, although a backend for real-time cyclic spectroscopy is being developed for the Green Bank Telescope.

The required temporal and spectral resolution of a dynamic spectrum and, subsequently, the required data volume to be kept for analysis is determined by the size of the scintles, which are the smallest structures present. Thus, it varies strongly since the variation of possible geometrical lens configurations allows for any size of scintles. As a result, even filterbank data might require non-negligible amounts of data volume in some cases. Usually though, dynamic spectra can be reduced to a few gigabytes per hour of observation or much less, which is not expected to change in the future if scintles are already fully resolved.

Scintillation is also seen in FRBs, and can be used to constrain locations of the scattering material. In some cases, an FRB is seen with a temporal scattering tail, which is fully modulated by a second scintillation pattern consistent with the Milky Way. For this to occur, the two screens must appear 'point-like' to each other, meaning the angular extent of the second screen must be less than the resolution λ/D of the first screen. This forms an inequality, and requires the scattering screen to be located within the FRB's host galaxy (Masui et al., 2015). As FRBs are extragalactic, and sample the sky randomly, they will serve as useful probes of ionised structures in the Milky Way, helping to build improved galactic electron density models. When FRBs are localised to milli-arcsecond precision with VLBI, the angular broadening can be used to locate their scattering screens (Ocker et al., 2021). Although a rotational periodicity has not yet been discovered in any FRB, the techniques of scintillometry can be applied to repeating FRBs through the pairwise correlation of burst spectra (Main et al., 2022). For scintillation originating in the Milky Way, this provides a measure of local scattering screens, while for scintillation in the host galaxy, this could be a probe of the FRB's motion and source environment.

With next-generation telescopes, pulsars will be observed every day with much higher signal-to-noise. Increased sensitivity and wide-band receivers will allow the probing of more distant sources, as most studies of scintillation have focused on the nearby $\lesssim$1 kpc environment. Processing all of the data necessary to capture the wealth of information from scintillation will be a logistical challenge, requiring real-time methods to determine the interstellar response. AI techniques have not been applied in this field, but may need to be developed. Even with current facilities, there is a tremendous amount that can still be learned, as many bright pulsars have not been monitored with anywhere near the full available lever arm in time and frequency. Additionally, telescopes such as CHORD and HIRAX will have built-in VLBI capabilities sub-GHz frequencies, and will be able to image scattering screens daily, or with every FRB detection. Automatic VLBI pipelines will need to be developed, similar to the rapid $\sim$ arcsecond FRB localisation schemes from arrays such as VLA, ASKAP, MeerKAT and the DSA-100.

References

K.N. Abazajian, J.K. Adelman-McCarthy, M.A. Agüeros, S.S. Allam, C. Allende Prieto, D. An, K.S.J. Anderson, S.F. Anderson, J. Annis, N.A. Bahcall, C.A.L. Bailer-Jones, J.C. Barentine, B.A. Bassett, A.C. Becker, T.C. Beers, E.F. Bell, V. Belokurov, A.A. Berlind, E.F. Berman, M. Bernardi, S.J. Bickerton, D. Bizyaev, J.P. Blakeslee, M.R. Blanton, J.J. Bochanski, W.N. Boroski, H.J. Brewington, J. Brinchmann, J. Brinkmann, R.J. Brunner, T. Budavári, L.N. Carey, S. Carliles, M.A. Carr, F.J. Castander, D. Cinabro, A.J. Connolly, István Csabai, C.E. Cunha, P.C. Czarapata, J.R.A. Davenport, E. de Haas, B. Dilday, M. Doi, D.J. Eisenstein, M.L. Evans, N.W. Evans, X. Fan, S.D. Friedman, J.A. Frieman, M. Fukugita, B.T. Gänsicke, E. Gates, B. Gillespie, G. Gilmore, B. Gonzalez, C.F. Gonzalez, E.K. Grebel, J.E. Gunn, Z. Györy, P.B. Hall, P. Harding, F.H. Harris, M. Harvanek, S.L. Hawley, Jeffrey J.E. Hayes, T.M. Heckman, J.S. Hendry, G.S. Hennessy, R.B. Hindsley, J. Hoblitt, C.J. Hogan, D.W. Hogg, J.A. Holtzman, J.B. Hyde, S.-I. Ichikawa, T. Ichikawa, M. Im, Ž. Ivezić, S. Jester, L. Jiang, J.A. Johnson, A.M. Jorgensen, M. Jurić, S.M. Kent, R. Kessler, S.J. Kleinman, G.R. Knapp, K. Konishi, R.G. Kron, J. Krzesinski, N. Kuropatkin, H. Lampeitl, S. Lebedeva, M.G. Lee, Y.S. Lee, R. French Leger, Sébastien Lépine, N. Li, M. Lima, H. Lin, D.C. Long, C.P. Loomis, J. Loveday, R.H. Lupton, E. Magnier, O. Malanushenko, V. Malanushenko, R. Mandelbaum, B. Margon, J.P. Marriner, D. Martínez-Delgado, T. Matsubara, P.M. McGehee, T.A. McKay, A. Meiksin, H.L. Morrison, F. Mullally, J.A. Munn, T. Murphy, T. Nash, A. Nebot Jr., E.H. Neilsen, H.J. Newberg, P.R. Newman, R.C. Nichol, T. Nicinski, M. Nieto-Santisteban, A. Nitta, S. Okamura, D.J. Oravetz, J.P. Ostriker, R. Owen, N. Padmanabhan, K. Pan, C. Park, G. Pauls Jr., J. Peoples, W.J. Percival, J.R. Pier, A.C. Pope, D. Pourbaix, P.A. Price, N. Purger, T. Quinn, M.J. Raddick, P. Re Fiorentin, G.T. Richards, M.W. Richmond, A.G. Riess, H.-W. Rix, C.M. Rockosi, M. Sako, D.J. Schlegel, D.P. Schneider, R.-D. Scholz, M.R. Schreiber, A.D. Schwope, U. Seljak, B. Sesar, E. Sheldon, K. Shimasaku, V.C. Sibley, A.E. Simmons, T. Sivarani, J. Allyn Smith, M.C. Smith, V. Smolčić, S.A. Snedden, A. Stebbins, M. Steinmetz, C. Stoughton, M.A. Strauss, M. SubbaRao, Y. Suto, A.S. Szalay, I. Szapudi, P. Szkody, M. Tanaka, M. Tegmark, L.F.A. Teodoro, A.R. Thakar, C.A. Tremonti, D.L. Tucker, A. Uomoto, D.E. Vanden Berk, J. Vandenberg, S. Vidrih, M.S. Vogeley, W. Voges, N.P. Vogt, Y. Wadadekar, S. Watters, D.H. Weinberg, A.A. West, S.D.M. White, B.C. Wilhite, A.C. Wonders, B. Yanny, D.R. Yocum, D.G. York, I. Zehavi, S. Zibetti, D.B. Zucker, The seventh data release of the Sloan digital sky survey. Astrophys. J. Suppl. **182**(2), 543–558 (2009). https://doi.org/10.1088/0067-0049/182/2/543

D. Agarwal, K. Aggarwal, S. Burke-Spolaor, D.R. Lorimer, N. Garver-Daniels, FETCH: A deep-learning based classifier for fast transient classification. Mon. Not. R. Astron. Soc. **497**(2), 1661–1674 (2020). https://doi.org/10.1093/mnras/staa1856

B. Allen, M.A. Papa, B.F. Schutz, Optimal strategies for sinusoidal signal detection. Phys. Rev. D **66**(10), 102003 (2002). https://doi.org/10.1103/PhysRevD.66.102003

M.A. Alpar, A.F. Cheng, M.A. Ruderman, J. Shaham, A new class of radio pulsars. Nature **300**(5894), 728–730 (1982). https://doi.org/10.1038/300728a0

B.C. Andersen, S.M. Ransom, A Fourier D. "Jerk" search for binary pulsars. Astrophys. J. Lett. **863**(1), L13 (2018). https://doi.org/10.3847/2041-8213/aad59f

J. Antoniadis, P.C.C. Freire, N. Wex, T.M. Tauris, R.S. Lynch, M.H. van Kerkwijk, M. Kramer, C. Bassa, V.S. Dhillon, T. Driebe, J.W.T. Hessels, V.M. Kaspi, V.I. Kondratiev, N. Langer, T.R. Marsh, M.A. McLaughlin, T.T. Pennucci, S.M. Ransom, I.H. Stairs, J. van Leeuwen, J.P.W. Verbiest, D.G. Whelan, A massive pulsar in a compact relativistic binary. Science **340**(6131), 448 (2013). https://doi.org/10.1126/science.1233232

J. Antoniadis, Z. Arzoumanian, S. Babak, M. Bailes, A.S. Bak Nielsen, P.T. Baker, C.G. Bassa, B. Bécsy, A. Berthereau, M. Bonetti, A. Brazier, P.R. Brook, M. Burgay, S. Burke-Spolaor, R.N. Caballero, J.A. Casey-Clyde, A. Chalumeau, D.J. Champion, M. Charisi, S. Chatterjee, S. Chen, I. Cognard, J.M. Cordes, N.J. Cornish, F. Crawford, H.T. Cromartie, K. Crowter, S. Dai, M.E. DeCesar, P.B. Demorest, G. Desvignes, T. Dolch, B. Drachler, M. Falxa, E.C. Ferrara, W. Fiore, E. Fonseca, J.R. Gair, N. Garver-Daniels, B. Goncharov, D.C. Good,

E. Graikou, L. Guillemot, Y.J. Guo, J.S. Hazboun, G. Hobbs, H. Hu, K. Islo, G.H. Janssen, R.J. Jennings, A.D. Johnson, M.L. Jones, A.R. Kaiser, D.L. Kaplan, R. Karuppusamy, M.J. Keith, L.Z. Kelley, M. Kerr, J.S. Key, M. Kramer, M.T. Lam, W.G. Lamb, T.J.W. Lazio, K.J. Lee, L. Lentati, K. Liu, J. Luo, R.S. Lynch, A.G. Lyne, D.R. Madison, R.A. Main, R.N. Manchester, A. McEwen, J.W. McKee, M.A. McLaughlin, M.B. Mickaliger, C.M.F. Mingarelli, C. Ng, D.J. Nice, S. Osłowski, A. Parthasarathy, T.T. Pennucci, B.B.P. Perera, D. Perrodin, A. Petiteau, N.S. Pol, N.K. Porayko, A. Possenti, S.M. Ransom, P.S. Ray, D.J. Reardon, C.J. Russell, A. Samajdar, L.M. Sampson, S. Sanidas, J.M. Sarkissian, K. Schmitz, L. Schult, A. Sesana, G. Shaifullah, R.M. Shannon, B.J. Shapiro-Albert, X. Siemens, J. Simon, T.L. Smith, L. Speri, R. Spiewak, I.H. Stairs, B.W. Stappers, D.R. Stinebring, J.K. Swiggum, S.R. Taylor, G. Theureau, C. Tiburzi, M. Vallisneri, E. van der Wateren, A. Vecchio, J.P.W. Verbiest, S.J. Vigeland, H. Wahl, J.B. Wang, J. Wang, L. Wang, C.A. Witt, S. Zhang, X.J. Zhu, The International Pulsar Timing Array second data release: Search for an isotropic gravitational wave background. Mon. Not. R. Astron. Soc. **510**(4), 4873–4887 (2022). https://doi.org/10.1093/mnras/stab3418.

M. Aravena, R. Decarli, F. Walter, R. Bouwens, P.A. Oesch, C.L. Carilli, F.E. Bauer, E. Da Cunha, E. Daddi, J. Gónzalez-López, R.J. Ivison, D.A. Riechers, I. Smail, A.M. Swinbank, A. Weiss, T. Anguita, R. Bacon, E. Bell, F. Bertoldi, P. Cortes, P. Cox, J. Hodge, E. Ibar, H. Inami, L. Infante, A. Karim, B. Magnelli, K. Ota, G. Popping, P. van der Werf, J. Wagg, Y. Fudamoto, The ALMA spectroscopic survey in the Hubble ultra deep field: search for [CII] line and dust emission in 6. Astrophys. J. **833**(1), 71 (2016). https://doi.org/10.3847/1538-4357/833/1/71

D. Baker, W. Brisken, M.H. van Kerkwijk, R. Main, U.-L. Pen, T. Sprenger, O. Wucknitz, Interstellar interferometry: precise curvature measurement from pulsar secondary spectra. Mon. Not. R. Astron. Soc. **510**(3), 4573–4581 (2022). https://doi.org/10.1093/mnras/stab3599

V. Balakrishnan, D. Champion, E. Barr, M. Kramer, V. Venkatraman Krishnan, R.P. Eatough, R. Sengar, M. Bailes, Coherent search for binary pulsars across all Five Keplerian parameters in radio observations using the template-bank algorithm. Mon. Not. R. Astron. Soc. **511**(1), 1265–1284 (2022). https://doi.org/10.1093/mnras/stab3746

B.R. Barsdell, M. Bailes, D.G. Barnes, C.J. Fluke, Accelerating incoherent dedispersion. Mon. Not. R. Astron. Soc. **422**(1), 379–392 (2012). https://doi.org/10.1111/j.1365-2966.2012.20622.x

C.G. Bassa, Z. Pleunis, J.W.T. Hessels, Enabling pulsar and fast transient searches using coherent dedispersion. Astron. Comput. **18**, 40–46 (2017). https://doi.org/10.1016/j.ascom.2017.01.004

M. Béthermin, Y. Fudamoto, M. Ginolfi, F. Loiacono, Y. Khusanova, P.L. Capak, P. Cassata, A. Faisst, O. Le Fèvre, D. Schaerer, J.D. Silverman, L. Yan, R. Amorin, S. Bardelli, M. Boquien, A. Cimatti, I. Davidzon, M. Dessauges-Zavadsky, S. Fujimoto, C. Gruppioni, N.P. Hathi, E. Ibar, G.C. Jones, A.M. Koekemoer, G. Lagache, B.C. Lemaux, C. Moreau, P.A. Oesch, F. Pozzi, D.A. Riechers, M. Talia, S. Toft, L. Vallini, D. Vergani, G. Zamorani, E. Zucca, The ALPINE-ALMA [CII] survey: Data processing, catalogs, and statistical source properties. Astron. Astrophys. **643**, A2 (2020). https://doi.org/10.1051/0004-6361/202037649

M.C. Bezuidenhout, E. Barr, M. Caleb, L.N. Driessen, F. Jankowski, M. Kramer, M. Malenta, V. Morello, K. Rajwade, S. Sanidas, B.W. Stappers, M. Surnis, MeerTRAP: 12 Galactic fast transients detected in a real-time, commensal MeerKAT survey. Mon. Not. R. Astron. Soc. **512**(1), 1483–1498 (2022). https://doi.org/10.1093/mnras/stac579.

A.D. Bolatto, M. Wolfire, A.K. Leroy, The CO-to-H_2 Conversion Factor. Annu. Rev. Astron. Astrophys. **51**(1), 207–268 (2013). https://doi.org/10.1146/annurev-astro-082812-140944

C. Borys, S. Chapman, M. Halpern, D. Scott, The Hubble Deep Field North SCUBA Super-map - I. Submillimetre maps, sources and number counts. Mon. Not. R. Astron. Soc. **344**(2), 385–398 (2003). https://doi.org/10.1046/j.1365-8711.2003.06818.x

R.J. Bouwens, R. Smit, S. Schouws, M. Stefanon, R. Bowler, R. Endsley, V. Gonzalez, H. Inami, D. Stark, P. Oesch, J. Hodge, M. Aravena, E. da Cunha, P. Dayal, I. de Looze, A. Ferrara, Y. Fudamoto, L. Graziani, C. Li, T. Nanayakkara, A. Pallottini, R. Schneider, L. Sommovigo, M. Topping, P. van der Werf, H. Algera, L. Barrufet, A. Hygate, I. Labbé, D. Riechers, J. Witstok, Reionization era bright emission line survey: selection and characterization of

luminous interstellar medium reservoirs in the z > 6.5 universe. Astrophys. J. **931**(2), 160 (2022). https://doi.org/10.3847/1538-4357/ac5a4a
M.A. Brentjens, A.G. de Bruyn, Faraday rotation measure synthesis. Astron. Astrophys. **441**(3), 1217–1228 (2005). https://doi.org/10.1051/0004-6361:20052990
D. Brisbin, O. Miettinen, M. Aravena, V. Smolčić, I. Delvecchio, C. Jiang, B. Magnelli, M. Albrecht, A.M. Arancibia, H. Aussel, N. Baran, F. Bertoldi, M. Béthermin, P. Capak, C.M. Casey, F. Civano, C.C. Hayward, O. Ilbert, A. Karim, O. Le Fevre, S. Marchesi, H.J. McCracken, F. Navarrete, M. Novak, D. Riechers, N. Padilla, M. Salvato, K. Scott, E. Schinnerer, K. Sheth, L. Tasca, An ALMA survey of submillimeter galaxies in the COSMOS field: Multiwavelength counterparts and redshift distribution. Astron. Astrophys. **608**, A15 (2017). https://doi.org/10.1051/0004-6361/201730558
W.F. Brisken, J.P. Macquart, J.J. Gao, B.J. Rickett, W.A. Coles, A.T. Deller, S.J. Tingay, C.J. West, 100 μas resolution VLBI imaging of anisotropic interstellar scattering toward pulsar B0834+06. Astrophys. J. **708**(1), 232–243 (2010). https://doi.org/10.1088/0004-637X/708/1/232
R. Cañameras, C. Yang, N.P.H. Nesvadba, A. Beelen, R. Kneissl, S. Koenig, E. Le Floc'h, M. Limousin, S. Malhotra, A. Omont, D. Scott, Planck's dusty GEMS. VI. Multi-J CO excitation and interstellar medium conditions in dusty starburst galaxies at z = 2-4. Astron. Astrophys. **620**, A61 (2018). https://doi.org/10.1051/0004-6361/201833625
M. Caleb, E. Keane, A decade and a half of fast radio burst observations. Universe **7**(11), 453 (2021). https://doi.org/10.3390/universe7110453
A.D. Cameron, E.D. Barr, D.J. Champion, M. Kramer, W.W. Zhu, An investigation of pulsar searching techniques with the fast folding algorithm. Mon. Not. R. Astron. Soc. **468**(2), 1994–2010 (2017). https://doi.org/10.1093/mnras/stx589
A.D. Cameron, D.J. Champion, M. Kramer, M. Bailes, E.D. Barr, C.G. Bassa, S. Bhandari, N.D.R. Bhat, M. Burgay, S. Burke-Spolaor, R.P. Eatough, C.M.L. Flynn, P.C.C. Freire, A. Jameson, S. Johnston, R. Karuppusamy, M.J. Keith, L. Levin, D.R. Lorimer, A.G. Lyne, M.A. McLaughlin, C. Ng, E. Petroff, A. Possenti, A. Ridolfi, B.W. Stappers, W. van Straten, T.M. Tauris, C. Tiburzi, N. Wex, The high time resolution universe pulsar survey - XIII. PSR J1757-1854, the most accelerated binary pulsar. Mon. Not. R. Astron. Soc. **475**(1), L57–L61 (2018). https://doi.org/10.1093/mnrasl/sly003.
A.D. Cameron, D.J. Champion, M. Bailes, V. Balakrishnan, E.D. Barr, C.G. Bassa, S. Bates, S. Bhandari, N.D.R. Bhat, M. Burgay, S. Burke-Spolaor, C.M.L. Flynn, A. Jameson, S. Johnston, M.J. Keith, M. Kramer, L. Levin, A.G. Lyne, C. Ng, E. Petroff, A. Possenti, D.A. Smith, B.W. Stappers, W. van Straten, C. Tiburzi, J. Wu, The high time resolution universe pulsar survey - XVI. Discovery and timing of 40 pulsars from the southern Galactic plane. Mon. Not. R. Astron. Soc. **493**(1), 1063–1087 (2020). https://doi.org/10.1093/mnras/staa039
F. Camilo, D.R. Lorimer, P. Freire, A.G. Lyne, R.N. Manchester, Observations of 20 millisecond pulsars in 47 Tucanae at 20 centimeters Astrophys. J. **535**(2), 975–990 (2000). https://doi.org/10.1086/308859
C.L. Carilli, F. Walter, Cool gas in high-redshift galaxies. Annu. Rev. Astron. Astrophys. **51**(1), 105–161 (2013). https://doi.org/10.1146/annurev-astro-082812-140953
J. Caruana, L. Wisotzki, E.C. Herenz, J. Kerutt, T. Urrutia, K.B. Schmidt, R. Bouwens, J. Brinchmann, S. Cantalupo, M. Carollo, C. Diener, A. Drake, T. Garel, R.A. Marino, J. Richard, R. Saust, J. Schaye, A. Verhamme, The MUSE-Wide survey: a measurement of the L. α emitting fraction among z > 3 galaxies. Mon. Not. R. Astron. Soc. **473**(1), 30–37 (2018). https://doi.org/10.1093/mnras/stx2307
J. Cooley, J. Tukey, An algorithm for the machine calculation of complex fourier series. Math. Comput. **19**(90), 297–301 (1965)
J.M. Cordes, S. Chatterjee, Fast radio bursts: an extragalactic enigma. Ann. Rev. Astron. Astrophys. **57**, 417–465 (2019). https://doi.org/10.1146/annurev-astro-091918-104501
J.M. Cordes, P.C.C. Freire, D.R. Lorimer, F. Camilo, D.J. Champion, D.J. Nice, R. Ramachandran, J.W.T. Hessels, W. Vlemmings, J. van Leeuwen, S.M. Ransom, N.D.R. Bhat, Z. Arzoumanian, M.A. McLaughlin, V.M. Kaspi, L. Kasian, J.S. Deneva, B. Reid, S. Chatterjee, J.L. Han, D.C.

Backer, I.H. Stairs, A.A. Deshpande, C.A. Faucher-Giguère, Arecibo pulsar survey using ALFA. I. Survey strategy and first discoveries. Astrophys. J. **637**(1), 446–455 (2006a). https://doi.org/10.1086/498335

J.M. Cordes, B.J. Rickett, D.R. Stinebring, W.A. Coles, Theory of parabolic arcs in interstellar scintillation spectra. Astrophys. J. **637**(1), 346 (2006b)

H.T. Cromartie, E. Fonseca, S.M. Ransom, P.B. Demorest, Z. Arzoumanian, H. Blumer, P.R. Brook, M.E. DeCesar, T. Dolch, J.A. Ellis, R.D. Ferdman, E.C. Ferrara, N. Garver-Daniels, P.A. Gentile, M.L. Jones, M.T. Lam, D.R. Lorimer, R.S. Lynch, M.A. McLaughlin, C. Ng, D.J. Nice, T.T. Pennucci, R. Spiewak, I.H. Stairs, K. Stovall, J.K. Swiggum, W.W. Zhu, Relativistic Shapiro delay measurements of an extremely massive millisecond pulsar. Nat. Astron. **4**, 72–76 (2020). https://doi.org/10.1038/s41550-019-0880-2

I. De Looze, M. Baes, G.J. Bendo, L. Cortese, J. Fritz, The reliability of [C II] as an indicator of the star formation rate. Mon. Not. R. Astron. Soc. **416**(4), 2712–2724 (2011). https://doi.org/10.1111/j.1365-2966.2011.19223.x

R. Decarli, F. Walter, C. Carilli, D. Riechers, P. Cox, R. Neri, M. Aravena, E. Bell, F. Bertoldi, D. Colombo, E. Da Cunha, E. Daddi, M. Dickinson, D. Downes, R. Ellis, L. Lentati, R. Maiolino, K.M. Menten, H.W. Rix, M. Sargent, D. Stark, B. Weiner, A. Weiss, A molecular line scan in the Hubble Deep Field North. Astrophys. J. **782**(2), 78 (2014). https://doi.org/10.1088/0004-637X/782/2/78

R. Decarli, F. Walter, M. Aravena, C. Carilli, R. Bouwens, E. da Cunha, E. Daddi, D. Elbaz, D. Riechers, I. Smail, M. Swinbank, A. Weiss, R. Bacon, F. Bauer, E.F. Bell, F. Bertoldi, S. Chapman, L. Colina, P.C. Cortes, P. Cox, J. Gónzalez-López, H. Inami, R. Ivison, J. Hodge, A. Karim, B. Magnelli, K. Ota, Gergö Popping, H.-W. Rix, M. Sargent, A. van der Wel, P. van der Werf, The ALMA spectroscopic survey in the Hubble Ultra Deep Field: Molecular gas reservoirs in high-redshift galaxies. Astrophys. J. **833**(1), 70 (2016). https://doi.org/10.3847/1538-4357/833/1/70

R. Decarli, F. Walter, J. Gónzalez-López, M. Aravena, L. Boogaard, C. Carilli, P. Cox, E. Daddi, G. Popping, D. Riechers, B. Uzgil, A. Weiss, R.J. Assef, R. Bacon, F.E. Bauer, F. Bertoldi, R. Bouwens, T. Contini, P.C. Cortes, E. da Cunha, T. Díaz-Santos, D. Elbaz, H. Inami, J. Hodge, R. Ivison, O. Le Fèvre, B. Magnelli, M. Novak, P. Oesch, H.-W. Rix, M.T. Sargent, I. Smail, A.M. Swinbank, R.S. Somerville, P. van der Werf, J. Wagg, L. Wisotzki, The ALMA spectroscopic survey in the HUDF: CO luminosity functions and the molecular gas content of galaxies through cosmic history. Astrophys. J. **882**(2), 138 (2019). https://doi.org/10.3847/1538-4357/ab30fe

P.B. Demorest, Cyclic spectral analysis of radio pulsars. Mon. Not. R. Astron. Soc. **416**(4), 2821–2826 (2011). https://doi.org/10.1111/j.1365-2966.2011.19230.x

G. Desvignes, R.P. Eatough, U.L. Pen, K.J. Lee, S.A. Mao, R. Karuppusamy, D.H.F.M. Schnitzeler, H. Falcke, M. Kramer, O. Wucknitz, L.G. Spitler, P. Torne, K. Liu, G.C. Bower, I. Cognard, A.G. Lyne, B.W. Stappers, Large magneto-ionic variations toward the galactic center magnetar, PSR J1745-2900. Astrophys. J. Lett. **852**(1), L12 (2018). https://doi.org/10.3847/2041-8213/aaa2f8

S. Dimoudi, K. Adamek, P. Thiagaraj, S.M. Ransom, A. Karastergiou, W. Armour, A GPU implementation of the correlation technique for real-time Fourier domain pulsar acceleration searches. Astrophys. J. Suppl. **239**(2), 28 (2018). https://doi.org/10.3847/1538-4365/aabe88

W.J. Dirienzo, C. Brogan, R. Indebetouw, C.J. Chandler, R.K. Friesen, K.E. Devine, Physical conditions of the earliest phases of massive star formation: single-dish and interferometric observations of ammonia and CCS in infrared dark clouds. Astron. J. **150**(5), 159 (2015). https://doi.org/10.1088/0004-6256/150/5/159

A. Douglas, P.V. Padmanabh, S.M. Ransom, A. Ridolfi, P. Freire, V.V. Krishnan, E.D. Barr, C. Pallanca, M. Cadelano, A. Possenti, I. Stairs, J.W.T. Hessels, M.E. DeCesar, R.S. Lynch, M. Bailes, M. Burgay, D.J. Champion, R. Karuppusamy, M. Kramer, B. Stappers, L. Vleeschower, Two new black widow millisecond pulsars in M28. Astrophys. J. **927**(1), 126 (2022). https://doi.org/10.3847/1538-4357/ac4744

L. Dunne, S.J. Maddox, P.P. Papadopoulos, R.J. Ivison, H.L. Gomez, Dust, CO and [C I]: Cross-calibration of molecular gas mass tracers in metal-rich galaxies across cosmic time. Mon. Not. R. Astron. Soc. (2022). https://doi.org/10.1093/mnras/stac2098

R.P. Eatough, M. Kramer, A.G. Lyne, M.J. Keith, A coherent acceleration search of the Parkes multibeam pulsar survey - techniques and the discovery and timing of 16 pulsars. Mon. Not. R. Astron. Soc. **431**(1), 292–307 (2013). https://doi.org/10.1093/mnras/stt161

D. Elbaz, M. Dickinson, H.S. Hwang, T. Díaz-Santos, G. Magdis, B. Magnelli, D. Le Borgne, F. Galliano, M. Pannella, P. Chanial, L. Armus, V. Charmandaris, E. Daddi, H. Aussel, P. Popesso, J. Kartaltepe, B. Altieri, I. Valtchanov, D. Coia, H. Dannerbauer, K. Dasyra, R. Leiton, J. Mazzarella, D.M. Alexander, V. Buat, D. Burgarella, R.R. Chary, R. Gilli, R.J. Ivison, S. Juneau, E. Le Floc'h, D. Lutz, G.E. Morrison, J.R. Mullaney, E. Murphy, A. Pope, D. Scott, M. Brodwin, D. Calzetti, C. Cesarsky, S. Charlot, H. Dole, P. Eisenhardt, H.C. Ferguson, N. Förster Schreiber, D. Frayer, M. Giavalisco, M. Huynh, A.M. Koekemoer, C. Papovich, N. Reddy, C. Surace, H. Teplitz, M.S. Yun, G. Wilson, GOODS-Herschel: an infrared main sequence for star-forming galaxies. Astron. Astrophys. **533**, A119 (2011). https://doi.org/10.1051/0004-6361/201117239

J.S. Farnes, G. Heald, H. Junklewitz, D.D. Mulcahy, M. Haverkorn, C.L. Van Eck, C.J. Riseley, M. Brentjens, C. Horellou, V. Vacca, D.I. Jones, A. Horneffer, R. Paladino, Source finding in linear polarization for LOFAR, and SKA predecessor surveys, using Faraday moments. Mon. Not. R. Astron. Soc. **474**(3), 3280–3296 (2018). https://doi.org/10.1093/mnras/stx2915

K. Ferrière, J.L. West, T.R. Jaffe, The correct sense of Faraday rotation. Mon. Not. R. Astron. Soc. **507**(4), 4968–4982 (2021). https://doi.org/10.1093/mnras/stab1641

N.M. Förster Schreiber, S. Wuyts, Star-forming galaxies at cosmic noon. Annu. Rev. Astron. Astrophys. **58**, 661–725 (2020). https://doi.org/10.1146/annurev-astro-032620-021910

R.S. Foster, D.C. Backer, Constructing a pulsar timing array. Astrophys. J. **361**, 300 (1990). https://doi.org/10.1086/169195

B. Garilli, O. Le Fèvre, L. Guzzo, D. Maccagni, V. Le Brun, S. de la Torre, B. Meneux, L. Tresse, P. Franzetti, G. Zamorani, A. Zanichelli, L. Gregorini, D. Vergani, D. Bottini, R. Scaramella, M. Scodeggio, G. Vettolani, C. Adami, S. Arnouts, S. Bardelli, M. Bolzonella, A. Cappi, S. Charlot, P. Ciliegi, T. Contini, S. Foucaud, I. Gavignaud, O. Ilbert, A. Iovino, F. Lamareille, H.J. McCracken, B. Marano, C. Marinoni, A. Mazure, R. Merighi, S. Paltani, R. Pellò, A. Pollo, L. Pozzetti, M. Radovich, E. Zucca, J. Blaizot, A. Bongiorno, O. Cucciati, Y. Mellier, C. Moreau, L. Paioro, The Vimos VLT deep survey. Global properties of 20,000 galaxies in the $I_{AB} < 22.5$ WIDE survey. Astron. Astrophys. **486**(3), 683–695 (2008). https://doi.org/10.1051/0004-6361:20078878

B. Garilli, L. Guzzo, M. Scodeggio, M. Bolzonella, U. Abbas, C. Adami, S. Arnouts, J. Bel, D. Bottini, E. Branchini, A. Cappi, J. Coupon, O. Cucciati, I. Davidzon, G. De Lucia, S. de la Torre, P. Franzetti, A. Fritz, M. Fumana, B.R. Granett, O. Ilbert, A. Iovino, J. Krywult, V. Le Brun, O. Le Fèvre, D. Maccagni, K. Małek, F. Marulli, H.J. McCracken, L. Paioro, M. Polletta, A. Pollo, H. Schlagenhaufer, L.A.M. Tasca, R. Tojeiro, D. Vergani, G. Zamorani, A. Zanichelli, A. Burden, C. Di Porto, A. Marchetti, C. Marinoni, Y. Mellier, L. Moscardini, R.C. Nichol, J.A. Peacock, W.J. Percival, S. Phleps, M. Wolk, The VIMOS public extragalactic survey (VIPERS). First data release of 57 204 spectroscopic measurements. Astron. Astrophys. **562**, A23 (2014). https://doi.org/10.1051/0004-6361/201322790

J. González-López, R. Decarli, R. Pavesi, F. Walter, M. Aravena, C. Carilli, L. Boogaard, G. Popping, A. Weiss, R.J. Assef, F.E. Bauer, F. Bertoldi, R. Bouwens, T. Contini, P.C. Cortes, P. Cox, E. da Cunha, E. Daddi, T. Díaz-Santos, H. Inami, J. Hodge, R. Ivison, O. Le Fèvre, B. Magnelli, P. Oesch, D. Riechers, H.-W. Rix, I. Smail, A.M. Swinbank, R.S. Somerville, B. Uzgil, P. van der Werf, The Atacama large millimeter/submillimeter array spectroscopic survey in the Hubble Ultra Deep Field: CO emission lines and 3 mm continuum sources. Astrophys. J. **882**(2), 139 (2019). https://doi.org/10.3847/1538-4357/ab3105

Y. Gupta, B.J. Rickett, A.G. Lyne, Refractive interstellar scintillation in pulsar dynamic spectra. Mon. Not. R. Astron. Soc. **269**, 1035–1068 (1994). https://doi.org/10.1093/mnras/269.4.1035

L. Guzzo, M. Scodeggio, B. Garilli, B.R. Granett, A. Fritz, U. Abbas, C. Adami, S. Arnouts, J. Bel, M. Bolzonella, D. Bottini, E. Branchini, A. Cappi, J. Coupon, O. Cucciati, I. Davidzon, G. De Lucia, S. de la Torre, P. Franzetti, M. Fumana, P. Hudelot, O. Ilbert, A. Iovino, J. Krywult, V. Le Brun, O. Le Fèvre, D. Maccagni, K. Małek, F. Marulli, H.J. McCracken, L. Paioro, J.A. Peacock, M. Polletta, A. Pollo, H. Schlagenhaufer, L.A.M. Tasca, R. Tojeiro, D. Vergani, G. Zamorani, A. Zanichelli, A. Burden, C. Di Porto, A. Marchetti, C. Marinoni, Y. Mellier, L. Moscardini, R.C. Nichol, W.J. Percival, S. Phleps, M. Wolk, The VIMOS Public Extragalactic Redshift Survey (VIPERS). An unprecedented view of galaxies and large-scale structure at $0.5 < z < 1.2$. Astron. Astrophys. **566**, A108 (2014). https://doi.org/10.1051/0004-6361/201321489

C.R. Gwinn, Noodle model for scintillation arcs. Mon. Not. R. Astron. Soc. **486**(2), 2809–2826 (2019). https://doi.org/10.1093/mnras/stz894

J.L. Han, C. Wang, P.F. Wang, T. Wang, D.J. Zhou, J.-H. Sun, Y. Yan, W.-Q. Su, W.-C. Jing, X. Chen, X.Y. Gao, L.-G. Hou, J. Xu, K.J. Lee, N. Wang, P. Jiang, R.-X. Xu, J. Yan, H.-Q. Gan, X. Guan, W.-J. Huang, J.-C. Jiang, H. Li, Y.-P. Men, C. Sun, B.-J. Wang, H.G. Wang, S.-Q. Wang, J.-T. Xie, H. Xu, R. Yao, X.-P. You, D.J. Yu, J.-P. Yuan, R. Yuen, C.-F. Zhang, Y. Zhu, The FAST galactic plane pulsar snapshot survey: I. Project design and pulsar discoveries. Res. Astron. Astrophys. **21**(5), 107 (2021). https://doi.org/10.1088/1674-4527/21/5/107

K.C. Harrington, A. Weiss, M.S. Yun, B. Magnelli, C.E. Sharon, T.K.D. Leung, A. Vishwas, Q.D. Wang, D.T. Frayer, E.F. Jiménez-Andrade, D. Liu, P. García, E. Romano-Díaz, B.L. Frye, S. Jarugula, T. Bădescu, D. Berman, H. Dannerbauer, A. Díaz-Sánchez, L. Grassitelli, P. Kamieneski, W.J. Kim, A. Kirkpatrick, J.D. Lowenthal, H. Messias, J. Puschnig, G.J. Stacey, P. Torne, F. Bertoldi, Turbulent gas in lensed Planck-selected Starbursts at $z \sim 1$-3.5. Astrophys. J. **908**(1), 95 (2021). https://doi.org/10.3847/1538-4357/abcc01

G. Heald, R. Braun, R. Edmonds, The Westerbork SINGS survey. II Polarization, Faraday rotation, and magnetic fields. Astron. Astrophys. **503**(2), 409–435 (2009). https://doi.org/10.1051/0004-6361/200912240

E.C. Herenz, L. Wisotzki, LSDCat: Detection and cataloguing of emission-line sources in integral-field spectroscopy datacubes. Astron. Astrophys. **602**, A111 (2017). https://doi.org/10.1051/0004-6361/201629507

E.C. Herenz, T. Urrutia, L. Wisotzki, J. Kerutt, R. Saust, M. Werhahn, K.B. Schmidt, J. Caruana, C. Diener, R. Bacon, J. Brinchmann, J. Schaye, M. Maseda, P.M. Weilbacher, The MUSE-Wide survey: A first catalogue of 831 emission line galaxies. Astron. Astrophys. **606**, A12 (2017). https://doi.org/10.1051/0004-6361/201731055

J.W.T. Hessels, S.M. Ransom, I.H. Stairs, P.C.C. Freire, V.M. Kaspi, F. Camilo, A radio pulsar spinning at 716 Hz. Science **311**(5769), 1901–1904 (2006). https://doi.org/10.1126/science.1123430

A. Hewish, S.J. Bell, J.D.H. Pilkington, P.F. Scott, R.A. Collins, Observation of a rapidly pulsating radio source. Nature **217**(5130), 709–713 (1968). https://doi.org/10.1038/217709a0

A.S. Hill, D.R. Stinebring, C.T. Asplund, D.E. Berwick, W.B. Everett, N.R. Hinkel, Deflection of pulsar signal reveals compact structures in the galaxy. Astrophys. J. Lett. **619**(2), L171–L174 (2005). https://doi.org/10.1086/428347

G.H. Hilmarsson, L.G. Spitler, R.A. Main, D.Z. Li, Polarization properties of FRB 20201124A from detections with the Effelsberg 100-m radio telescope. Mon. Not. R. Astron. Soc. **508**(4), 5354–5361 (2021). https://doi.org/10.1093/mnras/stab2936

G. Hobbs, L. Guo, R.N. Caballero, W. Coles, K.J. Lee, R.N. Manchester, D.J. Reardon, D. Matsakis, M.L. Tong, Z. Arzoumanian, M. Bailes, C.G. Bassa, N.D.R. Bhat, A. Brazier, S. Burke-Spolaor, D.J. Champion, S. Chatterjee, I. Cognard, S. Dai, G. Desvignes, T. Dolch, R.D. Ferdman, E. Graikou, L. Guillemot, G.H. Janssen, M.J. Keith, M. Kerr, M. Kramer, M.T. Lam, K. Liu, A. Lyne, T.J.W. Lazio, R. Lynch, J.W. McKee, M.A. McLaughlin, C.M.F. Mingarelli, D.J. Nice, S. Osłowski, T.T. Pennucci, B.B.P. Perera, D. Perrodin, A. Possenti, C.J. Russell, S. Sanidas, A. Sesana, G. Shaifullah, R.M. Shannon, J. Simon, R. Spiewak, I.H. Stairs, B.W. Stappers, J.K. Swiggum, S.R. Taylor, G. Theureau, L. Toomey, R. van Haasteren, J.B. Wang, Y. Wang, X.J. Zhu, A pulsar-based time-scale from the International Pulsar Timing

Array. Mon. Not. R. Astron. Soc. **491**(4), 5951–5965 (2020). https://doi.org/10.1093/mnras/stz3071

J.A. Hodge, I. Smail, F. Walter, E. da Cunha, A.M. Swinbank, M. Rybak, B. Venemans, W.N. Brandt, G. Calistro Rivera, S.C. Chapman, C.-C. Chen, P. Cox, H. Dannerbauer, R. Decarli, T.R. Greve, K.K. Knudsen, K.M. Menten, E. Schinnerer, J.M. Simpson, P. van der Werf, J.L. Wardlow, A. Weiss, ALMA reveals potential evidence for spiral arms, bars, and rings in high-redshift submillimeter galaxies. Astrophys. J. **876**(2), 130 (2019). https://doi.org/10.3847/1538-4357/ab1846

D.H. Hughes, S. Serjeant, J. Dunlop, M. Rowan-Robinson, A. Blain, R.G. Mann, R. Ivison, J. Peacock, A. Efstathiou, W. Gear, S. Oliver, A. Lawrence, M. Longair, P. Goldschmidt, T. Jenness, High-redshift star formation in the Hubble Deep Field revealed by a submillimetre-wavelength survey. Nature **394**(6690), 241–247 (1998). https://doi.org/10.1038/28328

S. Jin, E. Daddi, D. Liu, V. Smolčić, E. Schinnerer, A. Calabrò, Q. Gu, J. Delhaize, I. Delvecchio, Yu Gao, M. Salvato, A. Puglisi, M. Dickinson, F. Bertoldi, M. Sargent, M. Novak, G. Magdis, I. Aretxaga, G.W. Wilson, P. Capak, "Super-deblended" dust emission in galaxies. II. Far-IR to (Sub)millimeter photometry and high-redshift galaxy candidates in the full COSMOS field. Astrophys. J. **864**(1), 56 (2018). https://doi.org/10.3847/1538-4357/aad4af

S. Jin, E. Daddi, G.E. Magdis, D. Liu, E. Schinnerer, P.P. Papadopoulos, Q. Gu, Y. Gao, A. Calabrò, Discovery of four apparently cold dusty galaxies at z = 3.62-5.85 in the COSMOS field: direct evidence of cosmic microwave background impact on high-redshift galaxy observables. Astrophys. J. **887**(2), 144 (2019). https://doi.org/10.3847/1538-4357/ab55d6

S. Jin, H. Dannerbauer, B. Emonts, P. Serra, C.D.P. Lagos, A.P. Thomson, L. Bassini, M. Lehnert, J.R. Allison, J.B. Champagne, B. Indermühle, R.P. Norris, N. Seymour, R. Shimakawa, C.M. Casey, C. De Breuck, G. Drouart, N. Hatch, T. Kodama, Y. Koyama, P. Macgregor, G. Miley, R. Overzier, J.M. Pérez-Martínez, J.M. Rodríguez-Espinosa, H. Röttgering, M. Sánchez Portal, B. Ziegler, COALAS. I. ATCA CO(1-0) survey and luminosity function in the Spiderweb protocluster at z = 2.16. Astron. Astrophys. **652**, A11 (2021). https://doi.org/10.1051/0004-6361/202040232

S. Jin, E. Daddi, G.E. Magdis, D. Liu, J.R. Weaver, Q. Tan, F. Valentino, Y. Gao, E. Schinnerer, A. Calabrò, Q. Gu, D.B. Sese, Diagnosing deceivingly cold dusty galaxies at 3.5 < z < 6: A substantial population of compact starbursts with high infrared optical depths. Astron. Astrophys. **665**, A3 (2022). https://doi.org/10.1051/0004-6361/202243341

H.M. Johnston, S.R. Kulkarni, On the detectability of pulsars in close binary systems. Astrophys. J. **368**, 504 (1991). https://doi.org/10.1086/169715

P. Kamphuis, G.I.G. Józsa, S.H. Oh, K. Spekkens, N. Urbancic, P. Serra, B.S. Koribalski, R.J. Dettmar, Automated kinematic modelling of warped galaxy discs in large H I surveys: 3D tilted-ring fitting of H I emission cubes. Mon. Not. R. Astron. Soc. **452**(3), 3139–3158 (2015). https://doi.org/10.1093/mnras/stv1480

V.M. Kaspi, A.M. Beloborodov, Magnetars. Annu. Rev. Astron. Astrophys. **55**(1), 261–301 (2017). https://doi.org/10.1146/annurev-astro-081915-023329

E.F. Keane, Pulsar science with the SKA, in *Pulsar Astrophysics the Next Fifty Years*, ed. by P. Weltevrede, B.B.P. Perera, L.L. Preston, S. Sanidas, vol. 337 (2018), pp. 158–164. https://doi.org/10.1017/S1743921317009188

E. Keane, B. Bhattacharyya, M. Kramer, B. Stappers, E.F. Keane, B. Bhattacharyya, M. Kramer, B.W. Stappers, S.D. Bates, M. Burgay, S. Chatterjee, D.J. Champion, R.P. Eatough, J.W.T. Hessels, G. Janssen, K.J. Lee, J. van Leeuwen, J. Margueron, M. Oertel, A. Possenti, S. Ransom, G. Theureau, P. Torne, A cosmic census of radio pulsars with the SKA, in *Advancing Astrophysics with the Square Kilometre Array (AASKA14)* (2015), p. 40. https://doi.org/10.22323/1.215.0040

E.F. Keane, S. Johnston, S. Bhandari, E. Barr, N.D.R. Bhat, M. Burgay, M. Caleb, C. Flynn, A. Jameson, M. Kramer, E. Petroff, A. Possenti, W. van Straten, M. Bailes, S. Burke-Spolaor, R.P. Eatough, B.W. Stappers, T. Totani, M. Honma, H. Furusawa, T. Hattori, T. Morokuma, Y. Niino, H. Sugai, T. Terai, N. Tominaga, S. Yamasaki, N. Yasuda, R. Allen, J. Cooke, J. Jencson, M.M. Kasliwal, D.L. Kaplan, S.J. Tingay, A. Williams, R. Wayth, P. Chandra,

D. Perrodin, M. Berezina, M. Mickaliger, C. Bassa, The host galaxy of a fast radio burst. Nature **530**, 453–456 (2016). https://doi.org/10.1038/nature17140

E.F. Keane, E.D. Barr, A. Jameson, V. Morello, M. Caleb, S. Bhandari, E. Petroff, A. Possenti, M. Burgay, C. Tiburzi, M. Bailes, N.D.R. Bhat, S. Burke-Spolaor, R.P. Eatough, C. Flynn, F. Jankowski, S. Johnston, M. Kramer, L. Levin, C. Ng, W. van Straten, V.V. Krishnan, The survey for pulsars and extragalactic radio bursts - I. Survey description and overview. Mon. Not. R. Astron. Soc. **473**(1), 116–135 (2018). https://doi.org/10.1093/mnras/stx2126

M.J. Keith, A. Jameson, W. van Straten, M. Bailes, S. Johnston, M. Kramer, A. Possenti, S.D. Bates, N.D.R. Bhat, M. Burgay, S. Burke-Spolaor, N. D'Amico, L. Levin, P.L. McMahon, S. Milia, B.W. Stappers, The high time resolution universe pulsar survey - I. System configuration and initial discoveries. Mon. Not. R. Astron. Soc. **409**(2), 619–627 (2010). https://doi.org/10.1111/j.1365-2966.2010.17325.x

A. Klitsch, L. Christensen, F. Valentino, N. Kanekar, P. Møller, M.A. Zwaan, J.P.U. Fynbo, M. Neeleman, J.X. Prochaska, CO excitation and line energy distributions in gas-selected galaxies. Mon. Not. R. Astron. Soc. **514**(2), 2346–2355 (2022). https://doi.org/10.1093/mnras/stac1190

B. Knispel, R.P. Eatough, H. Kim, E.F. Keane, B. Allen, D. Anderson, C. Aulbert, O. Bock, F. Crawford, H.B. Eggenstein, H. Fehrmann, D. Hammer, M. Kramer, A.G. Lyne, B. Machenschalk, R.B. Miller, M.A. Papa, D. Rastawicki, J. Sarkissian, X. Siemens, B.W. Stappers, Einstein@Home discovery of 24 pulsars in the Parkes multi-beam pulsar survey. Astrophys. J. **774**(2), 93 (2013). https://doi.org/10.1088/0004-637X/774/2/93

J. Kocz, M. Bailes, D. Barnes, S. Burke-Spolaor, L. Levin, Enhanced pulsar and single pulse detection via automated radio frequency interference detection in multipixel feeds. Mon. Not. R. Astron. Soc. **420**(1), 271–278 (2012). https://doi.org/10.1111/j.1365-2966.2011.20029.x

J. Koda, T. Sawada, M.C.H. Wright, P. Teuben, S.A. Corder, J. Patience, N. Scoville, J. Donovan Meyer, F. Egusa, CO(J = 1-0) Imaging of M51 with CARMA and the Nobeyama 45 m telescope. Astrophys. J. Suppl. **193**(1), 19 (2011). https://doi.org/10.1088/0067-0049/193/1/19

J. Koda, P. Teuben, T. Sawada, A. Plunkett, E. Fomalont, Total power map to visibilities (TP2VIS): joint deconvolution of ALMA 12m, 7m, and total power array data. Publ. Astron. Soc. Pac. **131**(999), 054505 (2019). https://doi.org/10.1088/1538-3873/ab047e

B.S. Koribalski, L. Staveley-Smith, T. Westmeier, P. Serra, K. Spekkens, O.I. Wong, K. Lee-Waddell, C.D.P. Lagos, D. Obreschkow, E.V. Ryan-Weber, M. Zwaan, V. Kilborn, G. Bekiaris, K. Bekki, F. Bigiel, A. Boselli, A. Bosma, B. Catinella, G. Chauhan, M.E. Cluver, M. Colless, H.M. Courtois, R.A. Crain, W.J.G. de Blok, H. Dénes, A.R. Duffy, A. Elagali, C.J. Fluke, B.Q. For, G. Heald, P.A. Henning, K.M. Hess, B.W. Holwerda, C. Howlett, T. Jarrett, D.H. Jones, M.G. Jones, G.I.G. Józsa, R. Jurek, E. Jütte, P. Kamphuis, I. Karachentsev, J. Kerp, D. Kleiner, R.C. Kraan-Korteweg, Á.R. López-Sánchez, J. Madrid, M. Meyer, J. Mould, C. Murugeshan, R.P. Norris, S.H. Oh, T.A. Oosterloo, A. Popping, M. Putman, T.N. Reynolds, J. Rhee, A.S.G. Robotham, S. Ryder, A.C. Schröder, Li Shao, A.R.H. Stevens, E.N. Taylor, J.M. vanÂ der Hulst, L. Verdes-Montenegro, B.P. Wakker, J. Wang, M. Whiting, B. Winkel, C. Wolf, WALLABY - an SKA Pathfinder HI survey. Astrophys. Space Sci. **365**(7), 118 (2020). https://doi.org/10.1007/s10509-020-03831-4

M. Kramer, Fast radio bursts: neutron stars, magnetars or something else? in *IAU Symposium 363: Neutron Star Astrophysics at the Crossroads: Magnetars and the Multimessenger Revolution*, vol. 10 (2022)

M. Kramer, I.H. Stairs, R.N. Manchester, N. Wex, A.T. Deller, W.A. Coles, M. Ali, M. Burgay, F. Camilo, I. Cognard, T. Damour, G. Desvignes, R.D. Ferdman, P.C.C. Freire, S. Grondin, L. Guillemot, G.B. Hobbs, G. Janssen, R. Karuppusamy, D.R. Lorimer, A.G. Lyne, J.W. McKee, M. McLaughlin, L.E. Münch, B.B.P. Perera, N. Pol, A. Possenti, J. Sarkissian, B.W. Stappers, G. Theureau, Strong-field gravity tests with the double pulsar. Phys. Rev. X **11**(4), 041050 (2021). https://doi.org/10.1103/PhysRevX.11.041050

P. Kumar, R.M. Shannon, M.E. Lower, S. Bhandari, A.T. Deller, C. Flynn, E.F. Keane, Circularly polarized radio emission from the repeating fast radio burst source FRB 20201124A. arXiv e-prints, art. arXiv:2109.11535 (2021)

O. Le Fèvre, P. Cassata, O. Cucciati, B. Garilli, O. Ilbert, V. Le Brun, D. Maccagni, C. Moreau, M. Scodeggio, L. Tresse, G. Zamorani, C. Adami, S. Arnouts, S. Bardelli, M. Bolzonella, M. Bondi, A. Bongiorno, D. Bottini, A. Cappi, S. Charlot, P. Ciliegi, T. Contini, S. de la Torre, S. Foucaud, P. Franzetti, I. Gavignaud, L. Guzzo, A. Iovino, B. Lemaux, C. López-Sanjuan, H.J. McCracken, B. Marano, C. Marinoni, A. Mazure, Y. Mellier, R. Merighi, P. Merluzzi, S. Paltani, R. Pellò, A. Pollo, L. Pozzetti, R. Scaramella, L. Tasca, D. Vergani, G. Vettolani, A. Zanichelli, E. Zucca, The VIMOS VLT deep survey final data release: a spectroscopic sample of 35 016 galaxies and AGN out to z ~6.7 selected with $17.5 \leq i_{AB} \leq 24.75$. Astron. Astrophys. **559**, A14 (2013). https://doi.org/10.1051/0004-6361/201322179

O. Le Fèvre, L.A.M. Tasca, P. Cassata, B. Garilli, V. Le Brun, D. Maccagni, L. Pentericci, R. Thomas, E. Vanzella, G. Zamorani, E. Zucca, R. Amorin, S. Bardelli, P. Capak, L. Cassarà, M. Castellano, A. Cimatti, J.G. Cuby, O. Cucciati, S. de la Torre, A. Durkalec, A. Fontana, M. Giavalisco, A. Grazian, N.P. Hathi, O. Ilbert, B.C. Lemaux, C. Moreau, S. Paltani, B. Ribeiro, M. Salvato, D. Schaerer, M. Scodeggio, V. Sommariva, M. Talia, Y. Taniguchi, L. Tresse, D. Vergani, P.W. Wang, S. Charlot, T. Contini, S. Fotopoulou, C. López-Sanjuan, Y. Mellier, N. Scoville, The VIMOS ultra-deep survey: ~10 000 galaxies with spectroscopic redshifts to study galaxy assembly at early epochs $2 < z < 6$. Astron. Astrophys. **576**, A79 (2015). https://doi.org/10.1051/0004-6361/201423829

O. Le Fèvre, M. Béthermin, A. Faisst, G.C. Jones, P. Capak, P. Cassata, J.D. Silverman, D. Schaerer, L. Yan, R. Amorin, S. Bardelli, M. Boquien, A. Cimatti, M. Dessauges-Zavadsky, M. Giavalisco, N.P. Hathi, Y. Fudamoto, S. Fujimoto, M. Ginolfi, C. Gruppioni, S. Hemmati, E. Ibar, A. Koekemoer, Y. Khusanova, G. Lagache, B.C. Lemaux, F. Loiacono, R. Maiolino, C. Mancini, D. Narayanan, L. Morselli, H. Méndez-Hernàndez, P.A. Oesch, F. Pozzi, M. Romano, D. Riechers, N. Scoville, M. Talia, L.A.M. Tasca, R. Thomas, S. Toft, L. Vallini, D. Vergani, F. Walter, G. Zamorani, E. Zucca, The ALPINE-ALMA [CII] survey. Survey strategy, observations, and sample properties of 118 star-forming galaxies at $4 < z < 6$. Astron. Astrophys. **643**, A1 (2020). https://doi.org/10.1051/0004-6361/201936965

F. Lelli, E.M. Di Teodoro, F. Fraternali, A.W.S. Man, Z.-Y. Zhang, C. De Breuck, T.A. Davis, R. Maiolino, A massive stellar bulge in a regularly rotating galaxy 1.2 billion years after the Big Bang. Science **371**(6530), 713–716 (2021). https://doi.org/10.1126/science.abc1893

A.K. Leroy, A. Hughes, D. Liu, J. Pety, E. Rosolowsky, T. Saito, E. Schinnerer, A. Schruba, A. Usero, C.M. Faesi, C.N. Herrera, M. Chevance, A.P.S. Hygate, A.A. Kepley, E.W. Koch, M. Querejeta, K. Sliwa, D. Will, C.D. Wilson, G.S. Anand, A. Barnes, F. Belfiore, I. Bešlić, F. Bigiel, G.A. Blanc, A.D. Bolatto, M. Boquien, Y. Cao, R. Chandar, J. Chastenet, I.-D. Chiang, E. Congiu, D.A. Dale, S. Deger, J.S. den Brok, C. Eibensteiner, E. Emsellem, A. García-Rodríguez, S.C.O. Glover, K. Grasha, B. Groves, J.D. Henshaw, M.J. Jiménez Donaire, J. Kim, R.S. Klessen, K. Kreckel, J.M.D. Kruijssen, K.L. Larson, J.C. Lee, N. Mayker, R. McElroy, S.E. Meidt, A. Mok, H.-A. Pan, J. Puschnig, A. Razza, P. Sánchez-Bl'azquez, K.M. Sandstrom, F. Santoro, A. Sardone, F. Scheuermann, J. Sun, D.A. Thilker, J.A. Turner, L. Ubeda, D. Utomo, E.J. Watkins, T.G. Williams, PHANGS-ALMA data processing and pipeline. Astrophys. J. Suppl. **255**(1), 19 (2021a). https://doi.org/10.3847/1538-4365/abec80

A.K. Leroy, E. Schinnerer, A. Hughes, E. Rosolowsky, J. Pety, A. Schruba, A. Usero, G.A. Blanc, M. Chevance, E. Emsellem, C.M. Faesi, C.N. Herrera, D. Liu, S.E. Meidt, M. Querejeta, T. Saito, K.M. Sandstrom, J. Sun, T.G. Williams, G.S. Anand, A.T. Barnes, E.A. Behrens, F. Belfiore, S.M. Benincasa, I. Bešlić, F. Bigiel, A.D. Bolatto, J.S. den Brok, Y. Cao, R. Chandar, J. Chastenet, I.D. Chiang, E. Congiu, D.A. Dale, S. Deger, C. Eibensteiner, O.V. Egorov, A. García-Rodríguez, S.C.O. Glover, K. Grasha, J.D. Henshaw, I.T. Ho, A.A. Kepley, J. Kim, R.S. Klessen, K. Kreckel, E.W. Koch, J.M.D. Kruijssen, K.L. Larson, J.C. Lee, L.A. Lopez, J. Machado, N. Mayker, R. McElroy, E.J. Murphy, E.C. Ostriker, H.-A. Pan, I. Pessa, J. Puschnig, A. Razza, P. Sánchez-Blázquez, F. Santoro, A. Sardone, F. Scheuermann, K. Sliwa, M.C. Sormani, S.K. Stuber, D.A. Thilker, J.A. Turner, D. Utomo, E.J. Watkins, B. Whitmore,

PHANGS-ALMA: Arcsecond CO(2-1) imaging of nearby star-forming galaxies. Astrophys. J. Suppl. **257**(2), 43 (2021b). https://doi.org/10.3847/1538-4365/ac17f3

L. Levin, W. Armour, C. Baffa, E. Barr, S. Cooper, R. Eatough, A. Ensor, E. Giani, A. Karastergiou, R. Karuppusamy, M. Keith, M. Kramer, R. Lyon, M. Mackintosh, M. Mickaliger, R. van Nieuwpoort, M. Pearson, T. Prabu, J. Roy, O. Sinnen, L. Spitler, H. Spreeuw, B.W. Stappers, W. van Straten, C. Williams, H. Wang, K. Wiesner, SKA TDT Team, Pulsar searches with the SKA, in *Pulsar Astrophysics the Next Fifty Years*, ed. by P. Weltevrede, B.B.P. Perera, L.L. Preston, S. Sanidas, vol. 337 (2018), pp. 171–174. https://doi.org/10.1017/S1743921317009528

S.J. Lilly, O. Le Fèvre, A. Renzini, G. Zamorani, M. Scodeggio, T. Contini, C.M. Carollo, G. Hasinger, J.P. Kneib, A. Iovino, V. Le Brun, C. Maier, V. Mainieri, M. Mignoli, J. Silverman, L.A.M. Tasca, M. Bolzonella, A. Bongiorno, D. Bottini, P. Capak, K. Caputi, A. Cimatti, O. Cucciati, E. Daddi, R. Feldmann, P. Franzetti, B. Garilli, L. Guzzo, O. Ilbert, P. Kampczyk, K. Kovac, F. Lamareille, A. Leauthaud, J.F. Le Borgne, H.J. McCracken, C. Marinoni, R. Pello, E. Ricciardelli, C. Scarlata, D. Vergani, D.B. Sanders, E. Schinnerer, N. Scoville, Y. Taniguchi, S. Arnouts, H. Aussel, S. Bardelli, M. Brusa, A. Cappi, P. Ciliegi, A. Finoguenov, S. Foucaud, A. Franceschini, C. Halliday, C. Impey, C. Knobel, A. Koekemoer, J. Kurk, D. Maccagni, S. Maddox, B. Marano, G. Marconi, B. Meneux, B. Mobasher, C. Moreau, J.A. Peacock, C. Porciani, L. Pozzetti, R. Scaramella, D. Schiminovich, P. Shopbell, I. Smail, D. Thompson, L. Tresse, G. Vettolani, A. Zanichelli, E. Zucca, zCOSMOS: A large VLT/VIMOS redshift survey covering $0 < z < 3$ in the COSMOS field. Astrophys. J. Suppl. **172**(1), 70–85 (2007). https://doi.org/10.1086/516589

S.J. Lilly, V. Le Brun, C. Maier, V. Mainieri, M. Mignoli, M. Scodeggio, G. Zamorani, M. Carollo, T. Contini, J.-P. Kneib, O. Le Fèvre, A. Renzini, S. Bardelli, M. Bolzonella, A. Bongiorno, K. Caputi, G. Coppa, O. Cucciati, S. de la Torre, L. de Ravel, P. Franzetti, B. Garilli, A. Iovino, P. Kampczyk, K. Kovac, C. Knobel, F. Lamareille, J.-F. Le Borgne, R. Pello, Y. Peng, E. Pérez-Montero, E. Ricciardelli, J.D. Silverman, M. Tanaka, L. Tasca, L. Tresse, D. Vergani, E. Zucca, O. Ilbert, M. Salvato, P. Oesch, U. Abbas, D. Bottini, P. Capak, A. Cappi, P. Cassata, A. Cimatti, M. Elvis, M. Fumana, L. Guzzo, G. Hasinger, A. Koekemoer, A. Leauthaud, D. Maccagni, C. Marinoni, H. McCracken, P. Memeo, B. Meneux, C. Porciani, L. Pozzetti, D. Sanders, R. Scaramella, C. Scarlata, N. Scoville, P. Shopbell, Y. Taniguchi, The zCOSMOS 10k-bright spectroscopic sample. Astrophys. J. Suppl. **184**(2), 218–229 (2009). https://doi.org/10.1088/0067-0049/184/2/218

D. Liu, Y. Gao, K. Isaak, E. Daddi, C. Yang, N. Lu, P. van der Werf, High-J CO versus far-infrared relations in normal and starburst galaxies. Astrophys. J. Lett. **810**(2), L14 (2015). https://doi.org/10.1088/2041-8205/810/2/L14

D. Liu, E. Daddi, M. Dickinson, F. Owen, M. Pannella, M. Sargent, M. Béthermin, G. Magdis, Y. Gao, X. Shu, T. Wang, S. Jin, H. Inami, "Super-deblended" Dust emission in galaxies. I. The GOODS-North catalog and the cosmic star formation rate density out to redshift 6. Astrophys. J. **853**(2), 172 (2018). https://doi.org/10.3847/1538-4357/aaa600

D. Liu, E. Daddi, E. Schinnerer, T. Saito, A. Leroy, J.D. Silverman, F. Valentino, G.E. Magdis, Y. Gao, S. Jin, A. Puglisi, B. Groves, CO excitation, molecular gas density, and interstellar radiation field in local and high-redshift galaxies. Astrophys. J. **909**(1), 56 (2021a). https://doi.org/10.3847/1538-4357/abd801

K. Liu, G. Desvignes, R.P. Eatough, R. Karuppusamy, M. Kramer, P. Torne, R. Wharton, S. Chatterjee, J.M. Cordes, G.B. Crew, C. Goddi, S.M. Ransom, H. Rottmann, F. Abbate, G.C. Bower, C.D. Brinkerink, H. Falcke, A. Noutsos, A. Hernández-Gómez, W. Jiang, M.D. Johnson, R.-S. Lu, Y. Pidopryhora, L. Rezzolla, L. Shao, Z. Shen, N. Wex, An 86 GHz search for pulsars in the galactic center with the atacama large millimeter/submillimeter array. Astrophys. J. **914**(1), 30 (2021b). https://doi.org/10.3847/1538-4357/abf9a2

D.R. Lorimer, M. Kramer, *Handbook of Pulsar Astronomy*, vol. 4 (2004)

D.R. Lorimer, A.J. Faulkner, A.G. Lyne, R.N. Manchester, M. Kramer, M.A. McLaughlin, G. Hobbs, A. Possenti, I.H. Stairs, F. Camilo, M. Burgay, N. D'Amico, A. Corongiu, F. Crawford, The Parkes Multibeam Pulsar Survey - VI. Discovery and timing of 142 pulsars and a Galactic population analysis. Mon. Not. R. Astron. Soc. **372**(2), 777–800 (2006). https://doi.org/10.1111/j.1365-2966.2006.10887.x

D.R. Lorimer, M. Bailes, M.A. McLaughlin, D.J. Narkevic, F. Crawford, A bright millisecond radio burst of extragalactic origin. Science **318**, 777 (2007)

J.-W. Luo, J.-M. Zhu-Ge, B. Zhang, Machine learning classification of CHIME fast radio bursts - I. Supervised methods. Mon. Not. R. Astron. Soc. **518**(2), 1629–1641 (2023). https://doi.org/10.1093/mnras/stac3206

A.G. Lyne, Intermittent pulsars, in *Astrophysics and Space Science Library*, ed. by W. Becker, volume 357 of Astrophysics and Space Science Library (2009), p. 67. https://doi.org/10.1007/978-3-540-76965-1_4

J.P. Macquart, E. Keane, K. Grainge, M. McQuinn, R. Fender, J. Hessels, A. Deller, R. Bhat, R. Breton, S. Chatterjee, C. Law, D. Lorimer, E.O. Ofek, M. Pietka, L. Spitler, B. Stappers, C. Trott, Fast transients at cosmological distances with the SKA, in *Advancing Astrophysics with the Square Kilometre Array (AASKA14)* (2015), p. 55

J.P. Macquart, J.X. Prochaska, M. McQuinn, K.W. Bannister, S. Bhandari, C.K. Day, A.T. Deller, R.D. Ekers, C.W. James, L. Marnoch, S. Osłowski, C. Phillips, S.D. Ryder, D.R. Scott, R.M. Shannon, N. Tejos, A census of baryons in the Universe from localized fast radio bursts. Nature **581**(7809), 391–395 (2020). https://doi.org/10.1038/s41586-020-2300-2

S.C. Madden, D. Cormier, S. Hony, V. Lebouteiller, N. Abel, M. Galametz, I. De Looze, M. Chevance, F.L. Polles, M.Y. Lee, F. Galliano, A. Lambert-Huyghe, D. Hu, L. Ramambason, Tracing the total molecular gas in galaxies: [CII] and the CO-dark gas. Astron. Astrophys. **643**, A141 (2020). https://doi.org/10.1051/0004-6361/202038860

A. Magro, A. Karastergiou, S. Salvini, B. Mort, F. Dulwich, K. Zarb Adami, Real-time, fast radio transient searches with GPU de-dispersion. Mon. Not. R. Astron. Soc. **417**(4), 2642–2650 (2011). https://doi.org/10.1111/j.1365-2966.2011.19426.x

R. Main, M. van Kerkwijk, U.-L. Pen, N. Mahajan, K. Vanderlinde, Descattering of giant pulses in PSR B1957+20. Astrophys. J. Lett. **840**(2), L15 (2017). https://doi.org/10.3847/2041-8213/aa6f03

R.A. Main, G.H. Hilmarsson, V.R. Marthi, L.G. Spitler, R.S. Wharton, S. Bethapudi, D.Z. Li, H.H. Lin, Scintillation time-scale measurement of the highly active FRB20201124A. Mon. Not. R. Astron. Soc. **509**(3), 3172–3180 (2022). https://doi.org/10.1093/mnras/stab3218

R.A. Main, A. Parthasarathy, S. Johnston, A. Karastergiou, A. Basu, A.D. Cameron, M.J. Keith, L.S. Oswald, B. Posselt, D.J. Reardon, X. Song, P. Weltevrede, The Thousand Pulsar Array programme on MeerKAT - X. Scintillation arcs of 107 pulsars. Mon. Not. R. Astron. Soc. **518**(1), 1086–1097 (2023). https://doi.org/10.1093/mnras/stac3149

R.N. Manchester, A.G. Lyne, F. Camilo, J.F. Bell, V.M. Kaspi, N. D'Amico, N.P.F. McKay, F. Crawford, I.H. Stairs, A. Possenti, M. Kramer, D.C. Sheppard, The Parkes multi-beam pulsar survey - I. Observing and data analysis systems, discovery and timing of 100 pulsars. Mon. Not. R. Astron. Soc. **328**(1), 17–35 (2001). https://doi.org/10.1046/j.1365-8711.2001.04751.x

R.N. Manchester, G.B. Hobbs, A. Teoh, M. Hobbs, The Australia telescope national facility pulsar catalogue. Astron. J. **129**(4), 1993–2006 (2005). https://doi.org/10.1086/428488

K. Masui, H.-H. Lin, J. Sievers, C.J. Anderson, T.-C. Chang, X. Chen, A. Ganguly, M. Jarvis, C.-Y. Kuo, Y.-C. Li, Y.-W. Liao, M. McLaughlin, U.-L. Pen, J.B. Peterson, A. Roman, P.T. Timbie, T. Voytek, J.K. Yadav, Dense magnetized plasma associated with a fast radio burst. Nature **528**, 523–525 (2015). https://doi.org/10.1038/nature15769

J.W. McKee, H. Zhu, D.R. Stinebring, J.M. Cordes, Probing the local interstellar medium with scintillometry of the bright pulsar B1133 + 16. Astrophys. J. **927**(1), 99 (2022). https://doi.org/10.3847/1538-4357/ac460b

D. Michilli, A. Seymour, J.W.T. Hessels, L.G. Spitler, V. Gajjar, A.M. Archibald, G.C. Bower, S. Chatterjee, J.M. Cordes, K. Gourdji, G.H. Heald, V.M. Kaspi, C.J. Law, C. Sobey, E.A.K. Adams, C.G. Bassa, S. Bogdanov, C. Brinkman, P. Demorest, F. Fernandez, G. Hellbourg, T.J.W. Lazio, R.S. Lynch, N. Maddox, B. Marcote, M.A. McLaughlin, Z. Paragi, S.M. Ransom, P. Scholz, A.P.V. Siemion, S.P. Tendulkar, P. van Rooy, R.S. Wharton, D. Whitlow, An extreme magneto-ionic environment associated with the fast radio burst source FRB 121102. Nature **553**(7687), 182–185 (2018). https://doi.org/10.1038/nature25149

D. Michilli, K.W. Masui, R. Mckinven, D. Cubranic, M. Bruneault, C. Brar, C. Patel, P.J. Boyle, I.H. Stairs, A. Renard, K. Bandura, S. Berger, D. Breitman, T. Cassanelli, M. Dobbs, V.M. Kaspi, C. Leung, J. Mena-Parra, Z. Pleunis, L. Russell, P. Scholz, S.R. Siegel, S.P. Tendulkar, K. Vanderlinde, An analysis pipeline for CHIME/FRB full-array baseband data. Astrophys. J. **910**(2), 147 (2021). https://doi.org/10.3847/1538-4357/abe626

M.C. Miller, F.K. Lamb, A.J. Dittmann, S. Bogdanov, Z. Arzoumanian, K.C. Gendreau, S. Guillot, W.C.G. Ho, J.M. Lattimer, M. Loewenstein, S.M. Morsink, P.S. Ray, M.T. Wolff, C.L. Baker, T. Cazeau, S. Manthripragada, C.B. Markwardt, T. Okajima, S. Pollard, I. Cognard, H.T. Cromartie, E. Fonseca, L. Guillemot, M. Kerr, A. Parthasarathy, T.T. Pennucci, S. Ransom, I. Stairs, The radius of PSR J0740+6620 from NICER and XMM-Newton data. Astrophys. J. Lett. **918**(2), L28 (2021). https://doi.org/10.3847/2041-8213/ac089b

V. Morello, E.D. Barr, S. Cooper, M. Bailes, S. Bates, N.D.R. Bhat, M. Burgay, S. Burke-Spolaor, A.D. Cameron, D.J. Champion, R.P. Eatough, C.M.L. Flynn, A. Jameson, S. Johnston, M.J. Keith, E.F. Keane, M. Kramer, L. Levin, C. Ng, E. Petroff, A. Possenti, B.W. Stappers, W. van Straten, C. Tiburzi, The high time resolution universe survey - XIV. Discovery of 23 pulsars through GPU-accelerated reprocessing. Mon. Not. R. Astron. Soc. **483**(3), 3673–3685 (2019). https://doi.org/10.1093/mnras/sty3328

V. Morello, E.D. Barr, B.W. Stappers, E.F. Keane, A.G. Lyne, Optimal periodicity searching: revisiting the fast folding algorithm for large-scale pulsar surveys. Mon. Not. R. Astron. Soc. **497**(4), 4654–4671 (2020). https://doi.org/10.1093/mnras/staa2291

R. Neri, P. Cox, A. Omont, A. Beelen, S. Berta, T. Bakx, M. Lehnert, A.J. Baker, V. Buat, A. Cooray, H. Dannerbauer, L. Dunne, S. Dye, S. Eales, R. Gavazzi, A.I. Harris, C.N. Herrera, D. Hughes, R. Ivison, S. Jin, M. Krips, G. Lagache, L. Marchetti, H. Messias, M. Negrello, I. Perez-Fournon, D.A. Riechers, S. Serjeant, S. Urquhart, C. Vlahakis, A. Weiß, P. van der Werf, C. Yang, A.J. Young, NOEMA redshift measurements of bright Herschel galaxies. Astron. Astrophys. **635**, A7 (2020). https://doi.org/10.1051/0004-6361/201936988

S.K. Ocker, J.M. Cordes, S. Chatterjee, Constraining galaxy halos from the dispersion and scattering of fast radio bursts and pulsars. Astrophys. J. **911**(2), 102 (2021). https://doi.org/10.3847/1538-4357/abeb6e

S.J. Oliver, L. Wang, A.J. Smith, B. Altieri, A. Amblard, V. Arumugam, R. Auld, H. Aussel, T. Babbedge, A. Blain, J. Bock, A. Boselli, V. Buat, D. Burgarella, N. Castro-Rodríguez, A. Cava, P. Chanial, D.L. Clements, A. Conley, L. Conversi, A. Cooray, C.D. Dowell, E. Dwek, S. Eales, D. Elbaz, M. Fox, A. Franceschini, W. Gear, J. Glenn, M. Griffin, M. Halpern, E. Hatziminaoglou, E. Ibar, K. Isaak, R.J. Ivison, G. Lagache, L. Levenson, N. Lu, S. Madden, B. Maffei, G. Mainetti, L. Marchetti, K. Mitchell-Wynne, A.M.J. Mortier, H.T. Nguyen, B. O'Halloran, A. Omont, M.J. Page, P. Panuzzo, A. Papageorgiou, C.P. Pearson, I. Pérez-Fournon, M. Pohlen, J.I. Rawlings, G. Raymond, D. Rigopoulou, D. Rizzo, I.G. Roseboom, M. Rowan-Robinson, M. Sánchez Portal, R. Savage, B. Schulz, D. Scott, N. Seymour, D.L. Shupe, J.A. Stevens, M. Symeonidis, M. Trichas, K.E. Tugwell, M. Vaccari, E. Valiante, I. Valtchanov, J.D. Vieira, L. Vigroux, R. Ward, G. Wright, C.K. Xu, M. Zemcov, HerMES: SPIRE galaxy number counts at 250, 350, and 500 μm. Astron. Astrophys. **518**, L21 (2010). https://doi.org/10.1051/0004-6361/201014697

S. Osłowski, M.A. Walker, H-FISTA: a hierarchical algorithm for phase retrieval with application to pulsar dynamic spectra. Mon. Not. R. Astron. Soc. **519**(1), 1261–1276 (2023). https://doi.org/10.1093/mnras/stac3412

P.V. Padmanabh, E.D. Barr, S.S. Sridhar, M.R. Rugel, A. Damas-Segovia, A.M. Jacob, V. Balakrishnan, M. Berezina, M.C.I. Bernadich, A. Brunthaler, D.J. Champion, P.C.C. Freire, S. Khan, H.R. Klöckner, M. Kramer, Y.K. Ma, S.A. Mao, Y.P. Men, K.M. Menten, S. Sengupta, V. Venkatraman Krishnan, O. Wucknitz, F. Wyrowski, M.C. Bezuidenhout, S. Buchner, M. Burgay, W. Chen, C.J. Clark, L. Künkel, L. Nieder, B. Stappers, L.S. Legodi, M.M. Nyamai, The MPIfR-MeerKAT Galactic Plane survey I – System setup and early results. arXiv e-prints, art. arXiv:2303.09231 (2023). https://doi.org/10.48550/arXiv.2303.09231

R. Pavesi, C.E. Sharon, D.A. Riechers, J.A. Hodge, R. Decarli, F. Walter, C.L. Carilli, E. Daddi, I. Smail, M. Dickinson, R.J. Ivison, M. Sargent, E. da Cunha, M. Aravena, J. Darling, V. Smolčić, N.Z. Scoville, P.L. Capak, J. Wagg, The CO luminosity density at high-z (COLDz) survey: a sensitive, large-area blind search for low-J CO emission from cold gas in the early universe with the Karl G. Jansky very large array. Astrophys. J. **864**(1), 49 (2018). https://doi.org/10.3847/1538-4357/aacb79

U.L. Pen, J.P. Macquart, A.T. Deller, W. Brisken, 50 picoarcsec astrometry of pulsar emission. Mon. Not. R. Astron. Soc. **440**, L36–L40 (2014). https://doi.org/10.1093/mnrasl/slu010

K. Penner, A. Pope, E.L. Chapin, T.R. Greve, F. Bertoldi, M. Brodwin, R.-R. Chary, C.J. Conselice, K. Coppin, M. Giavalisco, D.H. Hughes, R.J. Ivison, T. Perera, D. Scott, K. Scott, G. Wilson, Origins of the extragalactic background at 1 mm from a combined analysis of the AzTEC and MAMBO data in GOODS-N. Mon. Not. R. Astron. Soc. **410**(4), 2749–2759 (2011). https://doi.org/10.1111/j.1365-2966.2010.17650.x

T.A. Perera, E.L. Chapin, J.E. Austermann, K.S. Scott, G.W. Wilson, M. Halpern, A. Pope, D. Scott, M.S. Yun, J.D. Lowenthal, G. Morrison, I. Aretxaga, J.J. Bock, K. Coppin, M. Crowe, L. Frey, D.H. Hughes, Y. Kang, S. Kim, P.D. Mauskopf, An AzTEC 1.1mm survey of the GOODS-N field - I. Maps, catalogue and source statistics. Mon. Not. R. Astron. Soc. **391**(3), 1227–1238 (2008). https://doi.org/10.1111/j.1365-2966.2008.13902.x

E. Petroff, M. Bailes, E.D. Barr, B.R. Barsdell, N.D.R. Bhat, F. Bian, S. Burke-Spolaor, M. Caleb, D. Champion, P. Chandra, G. Da Costa, C. Delvaux, C. Flynn, N. Gehrels, J. Greiner, A. Jameson, S. Johnston, M.M. Kasliwal, E.F. Keane, S. Keller, J. Kocz, M. Kramer, G. Leloudas, D. Malesani, J.S. Mulchaey, C. Ng, E.O. Ofek, D.A. Perley, A. Possenti, B.P. Schmidt, Y. Shen, B. Stappers, P. Tisserand, W. van Straten, C. Wolf, A real-time fast radio burst: polarization detection and multiwavelength follow-up. Mon. Not. R. Astron. Soc. **447**(1), 246–255 (2015). https://doi.org/10.1093/mnras/stu2419

E. Petroff, J.W.T. Hessels, D.R. Lorimer, Fast radio bursts. Astron. Astrophys. Rev. **27**(1), 4 (2019). https://doi.org/10.1007/s00159-019-0116-6

J. Pety, E. Schinnerer, A.K. Leroy, A. Hughes, S.E. Meidt, D. Colombo, G. Dumas, S. García-Burillo, K.F. Schuster, C. Kramer, C.L. Dobbs, T.A. Thompson, The Plateau de Bure + 30 m arcsecond whirlpool survey reveals a thick disk of diffuse molecular gas in the M51 galaxy. Astrophys. J. **779**(1), 43 (2013). https://doi.org/10.1088/0004-637X/779/1/43

Z. Pleunis, D.C. Good, V.M. Kaspi, R. Mckinven, S.M. Ransom, P. Scholz, K. Bandura, M. Bhardwaj, P.J. Boyle, C. Brar, T. Cassanelli, P. Chawla, Fengqiu, Dong, E. Fonseca, B.M. Gaensler, A. Josephy, J.F. Kaczmarek, C. Leung, H.-H. Lin, K.W. Masui, J. Mena-Parra, D. Michilli, C. Ng, C. Patel, M. Rafiei-Ravandi, M. Rahman, P. Sanghavi, K. Shin, K.M. Smith, I.H. Stairs, S.P. Tendulkar, Fast radio burst morphology in the first CHIME/FRB catalog. arXiv e-prints, art. arXiv:2106.04356 (2021)

V. Radhakrishnan, G. Srinivasan, On the origin of the recently discovered ultra-rapid pulsar. Curr. Sci. **51**, 1096–1099 (1982)

S.M. Ransom, S.S. Eikenberry, J. Middleditch, Fourier techniques for very long astrophysical time-series analysis. Astron. J. **124**(3), 1788–1809 (2002). https://doi.org/10.1086/342285

D.J. Reardon, W.A. Coles, M. Bailes, N.D.R. Bhat, S. Dai, G.B. Hobbs, M. Kerr, R.N. Manchester, S. Osłowski, A. Parthasarathy, C.J. Russell, R.M. Shannon, R. Spiewak, L. Toomey, A.V. Tuntsov, W. van Straten, M.A. Walker, J. Wang, L. Zhang, X.-J. Zhu, Precision orbital dynamics from interstellar scintillation arcs for PSR J0437-4715. Astrophys. J. **904**(2), 104 (2020). https://doi.org/10.3847/1538-4357/abbd40

C. Reuter, J.D. Vieira, J.S. Spilker, A. Weiss, M. Aravena, M. Archipley, M. Béthermin, S.C. Chapman, C. De Breuck, C. Dong, W.B. Everett, J. Fu, T.R. Greve, C.C. Hayward, R. Hill, Y. Hezaveh, S. Jarugula, K. Litke, M. Malkan, D.P. Marrone, D. Narayanan, K.A. Phadke, A.A. Stark, M.L. Strandet, The complete redshift distribution of dusty star-forming galaxies from the SPT-SZ survey. Astrophys. J. **902**(1), 78 (2020). https://doi.org/10.3847/1538-4357/abb599

B.J. Rickett, Interstellar scintillation and pulsar intensity variations. Mon. Not. R. Astron. Soc. **150**, 67 (1970). https://doi.org/10.1093/mnras/150.1.67

B.J. Rickett, A.G. Lyne, Y. Gupta, Interstellar fringes from pulsar B0834+06. Mon. Not. R. Astron. Soc. **287**(4), 739–752 (1997). https://doi.org/10.1093/mnras/287.4.739

B.J. Rickett, W.A. Coles, C.F. Nava, M.A. McLaughlin, S.M. Ransom, F. Camilo, R.D. Ferdman, P.C.C. Freire, M. Kramer, A.G. Lyne, I.H. Stairs, Interstellar scintillation of the double pulsar J0737-3039. Astrophys. J. **787**(2), 161 (2014). https://doi.org/10.1088/0004-637X/787/2/161

J.P. Ridley, F. Crawford, D.R. Lorimer, S.R. Bailey, J.H. Madden, R. Anella, J. Chennamangalam, Eight new radio pulsars in the large magellanic cloud. Mon. Not. R. Astron. Soc. **433**(1), 138–146 (2013). https://doi.org/10.1093/mnras/stt709

D.A. Riechers, C.M. Bradford, D.L. Clements, C.D. Dowell, I. Pérez-Fournon, R.J. Ivison, C. Bridge, A. Conley, H. Fu, J.D. Vieira, J. Wardlow, J. Calanog, A. Cooray, P. Hurley, R. Neri, J. Kamenetzky, J.E. Aguirre, B. Altieri, V. Arumugam, D.J. Benford, M. Béthermin, J. Bock, D. Burgarella, A. Cabrera-Lavers, S.C. Chapman, P. Cox, J.S. Dunlop, L. Earle, D. Farrah, P. Ferrero, A. Franceschini, R. Gavazzi, J. Glenn, E.A.G. Solares, M.A. Gurwell, M. Halpern, E. Hatziminaoglou, A. Hyde, E. Ibar, A. Kovács, M. Krips, R.E. Lupu, P.R. Maloney, P. Martinez-Navajas, H. Matsuhara, E.J. Murphy, B.J. Naylor, H.T. Nguyen, S.J. Oliver, A. Omont, M.J. Page, G. Petitpas, N. Rangwala, I.G. Roseboom, D. Scott, A.J. Smith, J.G. Staguhn, A. Streblyanska, A.P. Thomson, I. Valtchanov, M. Viero, L. Wang, M. Zemcov, J. Zmuidzinas, A dust-obscured massive maximum-starburst galaxy at a redshift of 6.34. Nature **496**(7445), 329–333 (2013). https://doi.org/10.1038/nature12050

D.A. Riechers, R. Pavesi, C.E. Sharon, J.A. Hodge, R. Decarli, F. Walter, C.L. Carilli, M. Aravena, E. da Cunha, E. Daddi, M. Dickinson, I. Smail, P.L. Capak, R.J. Ivison, M. Sargent, N.Z. Scoville, J. Wagg, COLDz: Shape of the CO luminosity function at high redshift and the cold gas history of the universe. Astrophys. J. **872**(1), 7 (2019). https://doi.org/10.3847/1538-4357/aafc27

D.A. Riechers, L.A. Boogaard, R. Decarli, J. González-López, I. Smail, F. Walter, M. Aravena, C.L. Carilli, P.C. Cortes, P. Cox, T. Díaz-Santos, J.A. Hodge, H. Inami, R.J. Ivison, M. Kaasinen, J. Wagg, A. Weiß, P. van der Werf, VLA-ALMA spectroscopic survey in the hubble ultra deep field (VLASPECS): total cold gas masses and CO line ratios for z = 2-3 main-sequence galaxies. Astrophys. J. Lett. **896**(2), L21 (2020). https://doi.org/10.3847/2041-8213/ab9595

D.A. Riechers, A. Weiss, F. Walter, C.L. Carilli, P. Cox, R. Decarli, R. Neri, Microwave background temperature at a redshift of 6.34 from H_2O absorption. Nature **602**(7895), 58–62 (2022). https://doi.org/10.1038/s41586-021-04294-5

A. Saintonge, G. Kauffmann, C. Kramer, L.J. Tacconi, C. Buchbender, B. Catinella, S. Fabello, J. Graciá-Carpio, J. Wang, L. Cortese, J. Fu, R. Genzel, R. Giovanelli, Q. Guo, M.P. Haynes, T.M. Heckman, M.R. Krumholz, J. Lemonias, C. Li, S. Moran, N. Rodriguez-Fernandez, D. Schiminovich, K. Schuster, A. Sievers, COLD GASS, an IRAM legacy survey of molecular gas in massive galaxies - I. Relations between H_2, H I, stellar content and structural properties. Mon. Not. R. Astron. Soc. **415**(1), 32–60 (2011). https://doi.org/10.1111/j.1365-2966.2011.18677.x

A. Saintonge, B. Catinella, L.J. Tacconi, G. Kauffmann, R. Genzel, L. Cortese, R. Davé, T.J. Fletcher, J. Graciá-Carpio, C. Kramer, T.M. Heckman, S. Janowiecki, K. Lutz, D. Rosario, D. Schiminovich, K. Schuster, J. Wang, S. Wuyts, S. Borthakur, I. Lamperti, G.W. Roberts-Borsani, xCOLD GASS: The complete IRAM 30 m legacy survey of molecular gas for galaxy evolution studies. Astrophys. J. Suppl. **233**(2), 22 (2017). https://doi.org/10.3847/1538-4365/aa97e0

A. Saintonge, C.D. Wilson, T. Xiao, L. Lin, H.S. Hwang, T. Tosaki, M. Bureau, P.J. Cigan, C.J.R. Clark, D.L. Clements, I. De Looze, T. Dharmawardena, Y. Gao, W.K. Gear, J. Greenslade, I. Lamperti, J.C. Lee, C. Li, M.J. Michałowski, A. Mok, H.-A. Pan, A.E. Sansom, M. Sargent, M.W.L. Smith, T. Williams, C. Yang, M. Zhu, G. Accurso, P. Barmby, E. Brinks, N. Bourne, T. Brown, A. Chung, E.J. Chung, A. Cibinel, K. Coppin, J. Davies, T.A. Davis, S. Eales, L. Fanciullo, T. Fang, Y. Gao, D.H.W. Glass, H.L. Gomez, T. Greve, J. He, L.C. Ho, F. Huang, H. Jeong, X. Jiang, Q. Jiao, F. Kemper, J.H. Kim, M. Kim, T. Kim, J. Ko, X. Kong, K. Lacaille, C.G. Lacey, B. Lee, J.H. Lee, W.-K. Lee, K. Masters, S.-H. Oh, P. Papadopoulos, C. Park,

S.-J. Park, H. Parsons, K. Rowlands, P. Scicluna, J.M. Scudder, R. Sethuram, S. Serjeant, Y. Shao, Y.-K. Sheen, Y. Shi, H. Shim, C.M.A. Smith, K. Spekkens, A.-L. Tsai, A. Verma, S. Urquhart, G. Violino, S. Viti, D. Wake, J. Wang, J. Wouterloot, Y. Yang, K. Yim, F. Yuan, Z. Zheng, JINGLE, a JCMT legacy survey of dust and gas for galaxy evolution studies - I. Survey overview and first results. Mon. Not. R. Astron. Soc. **481**(3), 3497–3519 (2018). https://doi.org/10.1093/mnras/sty2499

S. Sanidas, S. Cooper, C.G. Bassa, J.W.T. Hessels, V.I. Kondratiev, D. Michilli, B.W. Stappers, C.M. Tan, J. van Leeuwen, L. Cerrigone, R.A. Fallows, M. Iacobelli, E. Orrú, R.F. Pizzo, A. Shulevski, M.C. Toribio, S. ter Veen, P. Zucca, L. Bondonneau, J.M. Grießmeier, A. Karastergiou, M. Kramer, C. Sobey, The LOFAR tied-array all-sky survey (LOTAAS): Survey overview and initial pulsar discoveries. Astron. Astrophys. **626**, A104 (2019). https://doi.org/10.1051/0004-6361/201935609

M.T. Sargent, E. Daddi, M. Béthermin, H. Aussel, G. Magdis, H.S. Hwang, S. Juneau, D. Elbaz, E. da Cunha, Regularity underlying complexity: a redshift-independent description of the continuous variation of galaxy-scale molecular gas properties in the mass-star formation rate plane. Astrophys. J. **793**(1), 19 (2014). https://doi.org/10.1088/0004-637X/793/1/19

P.A.G. Scheuer, Amplitude variations in pulsed radio sources. Nature **218**(5145), 920–922 (1968). https://doi.org/10.1038/218920a0

E. Schinnerer, S.E. Meidt, J. Pety, A. Hughes, D. Colombo, S. García-Burillo, K.F. Schuster, G. Dumas, C.L. Dobbs, A.K. Leroy, C. Kramer, T.A. Thompson, M.W. Regan, The PdBI arcsecond whirlpool survey (PAWS). I. A cloud-scale/multi-wavelength view of the interstellar medium in a grand-design spiral galaxy. Astrophys. J. **779**(1), 42 (2013). https://doi.org/10.1088/0004-637X/779/1/42

M. Scodeggio, L. Guzzo, B. Garilli, B.R. Granett, M. Bolzonella, S. de la Torre, U. Abbas, C. Adami, S. Arnouts, D. Bottini, A. Cappi, J. Coupon, O. Cucciati, I. Davidzon, P. Franzetti, A. Fritz, A. Iovino, J. Krywult, V. Le Brun, O. Le Fèvre, D. Maccagni, K. Małek, A. Marchetti, F. Marulli, M. Polletta, A. Pollo, L.A.M. Tasca, R. Tojeiro, D. Vergani, A. Zanichelli, J. Bel, E. Branchini, G. De Lucia, O. Ilbert, H.J. McCracken, T. Moutard, J.A. Peacock, G. Zamorani, A. Burden, M. Fumana, E. Jullo, C. Marinoni, Y. Mellier, L. Moscardini, W.J. Percival, The VIMOS public extragalactic redshift survey (VIPERS). Full spectroscopic data and auxiliary information release (PDR-2). Astron. Astrophys. **609**, A84 (2018). https://doi.org/10.1051/0004-6361/201630114

P. Serra, B. Koribalski, V. Kilborn, J.R. Allison, S.W. Amy, L. Ball, K. Bannister, M.E. Bell, D.C.J. Bock, R. Bolton, M. Bowen, B. Boyle, S. Broadhurst, D. Brodrick, M. Brothers, J.D. Bunton, J. Chapman, W. Cheng, A.P. Chippendale, Y. Chung, F. Cooray, T. Cornwell, D. DeBoer, P. Diamond, R. Forsyth, R. Gough, N. Gupta, G.A. Hampson, L. Harvey-Smith, S. Hay, D.B. Hayman, I. Heywood, A.W. Hotan, S. Hoyle, B. Humphreys, B. Indermuehle, C. Jacka, C.A. Jackson, S. Jackson, K. Jeganathan, S. Johnston, J. Joseph, P. Kamphuis, M. Leach, E. Lenc, E. Lensson, S. Mackay, M. Marquarding, J. Marvil, N. McClure-Griffiths, D. McConnell, M. Meyer, P. Mirtschin, S. Neuhold, A. Ng, R.P. Norris, J. O'Sullivan, J. Pathikulangara, S. Pearce, C. Phillips, A. Popping, R.Y. Qiao, J.E. Reynolds, P. Roberts, R.J. Sault, A.E.T. Schinckel, R. Shaw, T.W. Shimwell, L. Staveley-Smith, M. Storey, A.W. Sweetnam, E. Troup, A. Tzioumis, M.A. Voronkov, T. Westmeier, M. Whiting, C. Wilson, O.I. Wong, X. Wu, ASKAP H I imaging of the galaxy group IC 1459. Mon. Not. R. Astron. Soc. **452**(3), 2680–2691 (2015a). https://doi.org/10.1093/mnras/stv1326

P. Serra, T. Westmeier, N. Giese, R. Jurek, L. Flöer, A. Popping, B. Winkel, T. van der Hulst, M. Meyer, B.S. Koribalski, L. Staveley-Smith, H. Courtois, SoFiA: a flexible source finder for 3D spectral line data. Mon. Not. R. Astron. Soc. **448**(2), 1922 (2015b). ISSN 0035-8711. https://doi.org/10.1093/mnras/stv079

M.W.L. Smith, E. Ibar, S.J. Maddox, E. Valiante, L. Dunne, S. Eales, S. Dye, C. Furlanetto, N. Bourne, P. Cigan, R.J. Ivison, H. Gomez, D.J.B. Smith, S. Viaene, The Herschel-ATLAS data release 2, Paper I. Submillimeter and far-infrared images of the south and north galactic poles: the largest Herschel survey of the extragalactic sky. Astrophys. J. Suppl. **233**(2), 26 (2017). https://doi.org/10.3847/1538-4365/aa9b35

C. Sobey, C.G. Bassa, S.P. O'Sullivan, J.R. Callingham, C.M. Tan, J.W.T. Hessels, V.I. Kondratiev, B.W. Stappers, C. Tiburzi, G. Heald, T. Shimwell, R.P. Breton, M. Kirwan, H.K. Vedantham, E. Carretti, J.M. Grießmeier, M. Haverkorn, A. Karastergiou, Searching for pulsars associated with polarised point sources using LOFAR: Initial discoveries from the TULIPP project. Astron. Astrophys. **661**, A87 (2022). https://doi.org/10.1051/0004-6361/202142636

T. Sprenger, R. Main, O. Wucknitz, G. Mall, J. Wu, Double-lens scintillometry: the variable scintillation of pulsar B1508 + 55. Mon. Not. R. Astron. Soc. **515**(4), 6198–6216 (2022). https://doi.org/10.1093/mnras/stac2160

D.H. Staelin, Fast folding algorithm for detection of periodic pulse trains. IEEE Proc. **57**, 724–725 (1969). https://doi.org/10.1109/PROC.1969.7051

B. Stappers, M. Kramer, An update on TRAPUM, in *MeerKAT Science: On the Pathway to the SKA* (2016), p. 9. https://doi.org/10.22323/1.277.0009

D.R. Stinebring, M.A. McLaughlin, J.M. Cordes, K.M. Becker, J.E. Espinoza Goodman, M.A. Kramer, J.L. Sheckard, C.T. Smith, Faint scattering around pulsars: probing the interstellar medium on solar system size scales. Astrophys. J. Lett. **549**(1), L97 (2001)

D.R. Stinebring, B.J. Rickett, A.H. Minter, A.S. Hill, A.P. Jussila, L. Mathis, M.A. McLaughlin, S.K. Ocker, S.M. Ransom, A scintillation arc survey of 22 pulsars with low to moderate dispersion measures. arXiv e-prints, art. arXiv:2207.08756 (2022)

K. Tadaki, D. Iono, M.S. Yun, I. Aretxaga, B. Hatsukade, D.H. Hughes, S. Ikarashi, T. Izumi, R. Kawabe, K. Kohno, M. Lee, Y. Matsuda, K. Nakanishi, T. Saito, Y. Tamura, J. Ueda, H. Umehata, G.W. Wilson, T. Michiyama, M. Ando, P. Kamieneski, The gravitationally unstable gas disk of a starburst galaxy 12 billion years ago. Nature **560**(7720), 613–616 (2018). https://doi.org/10.1038/s41586-018-0443-1

S. Taylor, S. Burke-Spolaor, P.T. Baker, M. Charisi, K. Islo, L.Z. Kelley, D.R. Madison, J. Simon, S. Vigeland, Nanograv Collaboration, Supermassive black-hole demographics &environments with pulsar timing arrays. Bull. Am. Astron. Soc. **51**(3), 336 (2019)

D. Thornton, B. Stappers, M. Bailes, B. Barsdell, S. Bates, N.D.R. Bhat, M. Burgay, S. Burke-Spolaor, D.J. Champion, P. Coster, N. D'Amico, A. Jameson, S. Johnston, M. Keith, M. Kramer, L. Levin, S. Milia, C. Ng, A. Possenti, W. van Straten, A population of fast radio bursts at cosmological distances. Science **341**, 53–56 (2013). https://doi.org/10.1126/science.1236789

T. Urrutia, L. Wisotzki, J. Kerutt, K.B. Schmidt, E.C. Herenz, J. Klar, R. Saust, M. Werhahn, C. Diener, J. Caruana, D. Krajnović, R. Bacon, L. Boogaard, J. Brinchmann, H. Enke, M. Maseda, T. Nanayakkara, J. Richard, M. Steinmetz, P.M. Weilbacher, The MUSE-Wide Survey: survey description and first data release. Astron. Astrophys. **624**, A141 (2019). https://doi.org/10.1051/0004-6361/201834656

E. Valiante, M.W.L. Smith, S. Eales, S.J. Maddox, E. Ibar, R. Hopwood, L. Dunne, P.J. Cigan, S. Dye, E. Pascale, E.E. Rigby, N. Bourne, C. Furlanetto, R.J. Ivison, The Herschel-ATLAS data release 1 - I. Maps, catalogues and number counts. Mon. Not. R. Astron. Soc. **462**(3), 3146–3179 (2016). https://doi.org/10.1093/mnras/stw1806

M.A. Walker, D.B. Melrose, D.R. Stinebring, C.M. Zhang, Interpretation of parabolic arcs in pulsar secondary spectra. Mon. Not. R. Astron. Soc. **354**(1), 43–54 (2004)

M.A. Walker, L.V.E. Koopmans, D.R. Stinebring, W. van Straten, Interstellar holography. Mon. Not. R. Astron. Soc. **388**(3), 1214–1222 (2008). https://doi.org/10.1111/j.1365-2966.2008.13452.x

M.A. Walker, P.B. Demorest, W. van Straten, Cyclic spectroscopy of the millisecond pulsar, B1937+21. Astrophys. J. **779**(2), 99 (2013). https://doi.org/10.1088/0004-637X/779/2/99

F. Walter, R. Decarli, M. Sargent, C. Carilli, M. Dickinson, D. Riechers, R. Ellis, D. Stark, B. Weiner, M. Aravena, E. Bell, F. Bertoldi, P. Cox, E. Da Cunha, E. Daddi, D. Downes, L. Lentati, R. Maiolino, K.M. Menten, R. Neri, H.W. Rix, A. Weiss, A molecular line scan in the Hubble Deep Field North: constraints on the CO luminosity function and the cosmic H_2 density. Astrophys. J. **782**(2), 79 (2014). https://doi.org/10.1088/0004-637X/782/2/79

F. Walter, R. Decarli, M. Aravena, C. Carilli, R. Bouwens, E. da Cunha, E. Daddi, R.J. Ivison, D. Riechers, I. Smail, M. Swinbank, A. Weiss, T. Anguita, R. Assef, R. Bacon, F. Bauer, E.F. Bell,

F. Bertoldi, S. Chapman, L. Colina, P.C. Cortes, P. Cox, M. Dickinson, D. Elbaz, J. Gónzalez-López, E. Ibar, H. Inami, L. Infante, J. Hodge, A. Karim, O. Le Fevre, B. Magnelli, R. Neri, P. Oesch, K. Ota, G. Popping, H.-W. Rix, M. Sargent, K. Sheth, A. van der Wel, P. van der Werf, J. Wagg, ALMA spectroscopic survey in the Hubble Ultra Deep Field: survey description. Astrophys. J. **833**(1), 67 (2016). https://doi.org/10.3847/1538-4357/833/1/67

F. Walter, M. Neeleman, R. Decarli, B. Venemans, R. Meyer, A. Weiss, E. Bañados, S.E.I. Bosman, C. Carilli, X. Fan, D. Riechers, H.-W. Rix, T.A. Thompson, ALMA 200 pc Imaging of a z 7 quasar reveals a compact, disk-like host galaxy. Astrophys. J. **927**(1), 21 (2022). https://doi.org/10.3847/1538-4357/ac49e8

J. White, K. Adámek, J. Roy, S. Dimoudi, S.M. Ransom, W. Armour, Bits missing: finding exotic pulsars using bfloat16 on NVIDIA GPUs. Astrophys. J. Suppl. **265**(1), 13 (2023). https://doi.org/10.3847/1538-4365/acb351

J.P. Williams, E.J. de Geus, L. Blitz, Clumpfind: determining structure in molecular clouds. Astrophysics Source Code Library, record ascl:1107.014, July (2011)

C. Yang, R. Gavazzi, A. Beelen, P. Cox, A. Omont, M.D. Lehnert, Y. Gao, R.J. Ivison, A.M. Swinbank, L. Barcos-Muñoz, R. Neri, A. Cooray, S. Dye, S. Eales, H. Fu, E. González-Alfonso, E. Ibar, M.J. Michałowski, H. Nayyeri, M. Negrello, J. Nightingale, I. Pérez-Fournon, D.A. Riechers, I. Smail, P. van der Werf, CO, H_2O, H_2O^+ line and dust emission in a z = 3.63 strongly lensed starburst merger at sub-kiloparsec scales. Astron. Astrophys. **624**, A138 (2019). https://doi.org/10.1051/0004-6361/201833876

J. Yao, W. Zhu, R.N. Manchester, W.A. Coles, D. Li, N. Wang, M. Kramer, D.R. Stinebring, Y. Feng, W. Yan, C. Miao, M. Yuan, P. Wang, J. Lu, Evidence for three-dimensional spin-velocity alignment in a pulsar. Nat. Astron. **5**, 788–795 (2021). https://doi.org/10.1038/s41550-021-01360-w

D.G. York, J. Adelman, J.E. Anderson Jr., S.F. Anderson, J. Annis, N.A. Bahcall, J.A. Bakken, R. Barkhouser, S. Bastian, E. Berman, W.N. Boroski, S. Bracker, C. Briegel, J.W. Briggs, J. Brinkmann, R. Brunner, S. Burles, L. Carey, M.A. Carr, F.J. Castander, B. Chen, P.L. Colestock, A.J. Connolly, J.H. Crocker, I. Csabai, P.C. Czarapata, J.E. Davis, M. Doi, T. Dombeck, D. Eisenstein, N. Ellman, B.R. Elms, M.L. Evans, X. Fan, G.R. Federwitz, L. Fiscelli, S. Friedman, J.A. Frieman, M. Fukugita, B. Gillespie, J.E. Gunn, V.K. Gurbani, E. de Haas, M. Haldeman, F.H. Harris, J. Hayes, T.M. Heckman, G.S. Hennessy, R.B. Hindsley, S. Holm, D.J. Holmgren, C.-H. Huang, C. Hull, D. Husby, S.-I. Ichikawa, T. Ichikawa, Ž. Ivezić, S. Kent, R.S.J. Kim, E. Kinney, M. Klaene, A.N. Kleinman, S. Kleinman, G.R. Knapp, J. Korienek, R.G. Kron, P.Z. Kunszt, D.Q. Lamb, B. Lee, R.F. Leger, S. Limmongkol, C. Lindenmeyer, D.C. Long, C. Loomis, J. Loveday, R. Lucinio, R.H. Lupton, B. MacKinnon, E.J. Mannery, P.M. Mantsch, B. Margon, P. McGehee, T.A. McKay, A. Meiksin, A. Merelli, D.G. Monet, J.A. Munn, V.K. Narayanan, T. Nash, E. Neilsen, R. Neswold, H.J. Newberg, R.C. Nichol, T. Nicinski, M. Nonino, N. Okada, S. Okamura, J.P. Ostriker, R. Owen, A.G. Pauls, J. Peoples, R.L. Peterson, D. Petravick, J.R. Pier, A. Pope, R. Pordes, A. Prosapio, R. Rechenmacher, T.R. Quinn, G.T. Richards, M.W. Richmond, C.H. Rivetta, C.M. Rockosi, K. Ruthmansdorfer, D. Sandford, D.J. Schlegel, D.P. Schneider, M. Sekiguchi, G. Sergey, K. Shimasaku, W.A. Siegmund, S. Smee, J.A. Smith, S. Snedden, R. Stone, C. Stoughton, M.A. Strauss, C. Stubbs, M. SubbaRao, A.S. Szalay, I. Szapudi, G.P. Szokoly, A.R. Thakar, C. Tremonti, D.L. Tucker, A. Uomoto, D. Vanden Berk, M.S. Vogeley, P. Waddell, S.-I. Wang, M. Watanabe, D.H. Weinberg, B. Yanny, N. Yasuda, SDSS Collaboration, The Sloan digital sky survey: technical summary. Astron. J. **120**(3), 1579–1587 (2000). https://doi.org/10.1086/301513

B. Zackay, E.O. Ofek, An accurate and efficient algorithm for detection of radio bursts with an unknown dispersion measure, for single-dish telescopes and interferometers. Astrophys. J. **835**(1), 11 (2017). https://doi.org/10.3847/1538-4357/835/1/11

Y. Zhao, N. Lu, C.K. Xu, Y. Gao, S.D. Lord, V. Charmandaris, T. Diaz-Santos, A. Evans, J. Howell, A.O. Petric, P.P. van der Werf, D.B. Sanders, The [NII] 205 μm emission in local luminous infrared galaxies. Astrophys. J. **819**(1), 69 (2016). https://doi.org/10.3847/0004-637X/819/1/69

Y. Zheng, J. Wang, J. Irwin, J. English, Q. Ma, R. Wang, K. Wang, Q.D. Wang, M. Krause, T.H. Randriamampandry, J. Li, R. Beck, CHANG-ES XXV: H I imaging of nearby edge-on galaxies - Data Release 4. Mon. Not. R. Astron. Soc. **513**(1), 1329–1353 (2022). https://doi.org/10.1093/mnras/stac760

Chapter 10
Using AI for Radio (Big) Data

Caroline Heneka, Julia Niebling, Hongming Tang, Vishnu Balakrishnan, Jakob Gawlikowski, Gregor Kasieczka, Gary Segal, Hyoyin Gan, and Sireesha Chamarthi

10.1 Motivation

The advent of both large-scale as well as higher sensitivity and high-resolution radio astronomical surveys warrant the need for efficient data reduction and automation. This includes the ability to extract more and less biased information while mining

C. Heneka (✉)
Institut für Theoretische Physik, Universität Heidelberg, Heidelberg, Germany
e-mail: heneka@thphys.uni-heidelberg.de

J. Niebling · J. Gawlikowski · S. Chamarthi
Institute of Data Science, German Aerospace Center (DLR), Jena, Germany
e-mail: julia.niebling@dlr.de; jakob.gawlikowski@dlr.de; sireesha.Chamarthi@dlr.de

H. Tang
Department of Astronomy, Tsinghua University, Beijing, China
e-mail: hongmingt@mail.tsinghua.edu.cn

V. Balakrishnan
Max-Planck-Institut für Radioastronomie, Bonn, Germany
e-mail: vishnu@mpifr-bonn.mpg.de

G. Kasieczka
Institut für Experimentalphysik, Universität Hamburg, Hamburg, Germany
e-mail: gregor.kasieczka@uni-hamburg.de

G. Segal
School of Mathematics and Physics, University of Queensland, Brisbane, QLD, Australia

CSIRO Space and Astronomy, Epping, NSW, Australia
e-mail: g.segal@uq.edu.au

H. Gan
Department of Physics and Astronomy, The University of British Columbia, Vancouver, BC, Canada
e-mail: hgan@phas.ubc.ca

E. Vardoulaki et al. (eds.), *Data-Intensive Radio Astronomy*, Astrophysics and Space Science Library 472, https://doi.org/10.1007/978-3-031-58468-8_10

through increasing amounts of data. For radio astronomical data this prominently concerns the SKA (Braun et al. 2015) and its precursor instruments such as LOFAR, MeerKAT, HERA and MWA (Shimwell et al. 2017; Jonas and MeerKAT Team 2016; DeBoer et al. 2017; Lonsdale et al. 2009). With current radio interferometers generating data rates at 10s to 100s of gigabits per second (about 100 times faster than current glass fibre internet connections), with a couple of terabyte per processed observation, and storing dozens of Petabyte of data (closing in on the data content of all written works produced by humanity Sabater et al. 2017), the SKA is set to outshine this with projected data rates of terabits per second to store hundreds of petabytes of data per year (Davis et al. 2016). AI driven by the use of deep neural networks is suitable for this feat as it can improve performance with increasing amounts of data (better) as compared to traditional data analysis based on statistical and conventional machine learning techniques. Another reason for the power and versatility of AI and most prominently DL and further advanced ML methods lies in their ability to learn representations and flexibly adapt models to additional incoming data (Bengio et al. 2012; Walmsley et al. 2022).

AI has become indispensable in industry applications such as self-driving cars, robotic manufacturing, natural language processing, visual recognition or health-care. At the same time, its application to, and further development for, different fields of science such as astronomy accelerates (Djorgovski et al. 2022). Examples of underlying tasks performed by networks are object detection and classification (Girshick 2015; He et al. 2017), image processing and generation (Kingma and Dhariwal 2018; Ardizzone et al. 2019), or segmentation (Ronneberger et al. 2015; Hu et al. 2017), all tasks that we as radio astronomers are interested in. Thus as well for radio astronomy applications beyond 'traditional' ML gain traction, as they do in astronomy at wavelengths other than radio, for example for source detection and deblending (González et al. 2018; Burke et al. 2019; Boucaud et al. 2020; Yi et al. 2022), inference tasks (Alsing et al. 2019; Jeffrey et al. 2021; Ho et al. 2022) or map reconstruction and denoising (Shen et al. 2017; Caldeira et al. 2019; Zhang et al. 2019). We note though, that some of the scientific use-cases for AI methods go beyond what industry previously required, for example when a robust and reliable uncertainty quantification is of utmost importance for scientific inference tasks; see Sect. 10.6.2.

10.1.1 A Wealth of AI Applications

In this chapter we introduce and discuss crucial developments of AI methods for radio astronomy, in relation to radio continuum and radio intensity mappings as well as radio source studies. Suitable tasks include denoising such as the cleaning of (high levels of) multi-component noise, foreground removal and removal of radio beam artefacts. Successful applications of deep learning models (for example U-Net, AE and ResNet network architectures) to specific radio astronomical tasks include RFI mitigation (Akeret et al. 2017; Kerrigan et al. 2019; Vafaei Sadr et al. 2020), mitigation of primary beam effects in HI intensity maps (Ni et al. 2022), treatment of instrumental noise and spurious artefacts (Gheller and Vazza 2022), and 21 cm foreground removal (Makinen et al. 2021). Further tasks are the reconstruction of uv-coverage and radio map-making; so have super-resolution inspired models been used for phase and amplitude reconstruction in visibility space (Schmidt et al. 2022), and U-Net as well as further network variants been shown to perform better than CLEAN (see Chap. 9) on simulated MeerKAT data for radio-interferometric image deconvolution (Yu et al. 2019; Laine et al. 2019; Nammour et al. 2022). Finally, networks enable reaction in real time, e.g. for data quality assurance and pipeline hyperparameter calibration (Mesarcik et al. 2022; Yatawatta and Avruch 2021). Given the many interesting applications of DL and other modern ML methods in radio astronomy, we will focus here on specific science tasks. In the following, we highlight multiple-component source finding and characterisation in Sect. 10.2 (see also Chap. 8 for detection algorithms not based on DL), detection of pulsars in Sect. 10.3, radio galaxy morphology identification and classification in Sect. 10.4, and characterisation of properties from both single sources and full maps in Sect. 10.5. Additionally, we will showcase how AI can significantly contribute to better our knowledge of uncertainties as well as the hunt for rare radio objects and anomalies, see Sect. 10.6. Finally, we present in Sect. 10.8 methodologies used outside radio astronomy (e.g. particle physics), from which data-intensive radio astronomy can benefit.

10.1.2 What Are ML, DL and AI for Radio Astronomy

To understand the potential of ML, DL and AI and its use-case for radio astronomy, let us briefly introduce the concepts, algorithms and architectures of importance that current applications in radio astronomy are building upon. Machine and statistical learning have been around for many years and are widely used in astronomical research, laying the ground for modern ML (Ball and Brunner 2010; James et al. 2013; Feigelson and Babu 2013; Hobson et al. 2014; Baron 2019; Feigelson et al. 2021). A wealth of methods has proven useful in our analysis toolkit, for example mixture models, sampling methods, and clustering algorithms, also meta-heuristics such as genetic algorithms and graphs, maps as well as decision trees (Borne 2009; Geach 2012; Lee et al. 2012; Heinis et al. 2016; Baron and Poznanski 2017; Blei et al. 2017; Kuhn and Feigelson 2017). Recently, as discussed in Sect. 10.1, modern machine learning methods based on deep learning with neural networks have started to revolutionise data analysis, creating so-called AI models (Fluke and Jacobs 2020). We will focus on deep learning, complemented by graph, map and tree-based ML models, in the following, as these will likely be important in the years to come.

These ML methods can be deployed in a supervised manner, where the model is adjusted based on a (usually large, where large depends on the application and distribution to be represented) set of training data, via back propagation set to optimise a cost or loss function that matches the requested task. For example, for segmentation of radio galaxies we would typically train on a set of images with known ground-truth segmentation maps to minimise the L2 loss (typically the mean-squared error) between labelled ground truth and output image presented by the network. Supervised methods are extremely powerful when sufficient (in amount and variety, again, to represent the target distribution) training data either from simulations and/or observations is available. They are not restricted to DL-based network models, but also entail modern ML methods such as random forests (Breiman 2001, see also Sect. 10.4 for their application). When only a small labelled data-set is available, as is the case for example in radio galaxy classification, semi-supervised learning leverages unlabelled data for example by pseudo-labelling or training in combination with auxiliary tasks where labels are more widely available (Lin et al. 2022; Slijepcevic et al. 2022a). Also when no labelled data is available, self-supervised learning using for example contrastive learning with auto-encoders or AE (Hayat et al. 2021; Slijepcevic et al. 2022b), or fully unsupervised learning such as a wide range of clustering algorithms, or map-based algorithms based on a variety of distance measures between sub-sets of the data-sets are able to group and detect different classes and characteristics (Ralph et al. 2019; Yu and Hou 2022).

Far from being able to give a full introduction to deep learning models as an example of modern ML, we briefly introduce some key notions here, see for example Erdmann et al. (2021) for a pedagogical introduction. To build a successful deep learning model, both the architecture, i.e. how layers are set up and connected and choice of layers, including their hyperparameters such as number of neurons or filters are key to properly match the datatype and task faced. Most often architectural building blocks are based on dense layers of fully connected neurons, suitable for regression, emulation and classification from a set or vector of features, and convolutional layers comprised of a set of filters that convolve with kernels of fixed sizes, suitable to treat 1D series, 2D imaging and higher-dimensional data to learn representations. Layers are then connected to build an architecture, which can range from simple linear setups to blocks of layers that are connected in various ways, for example when downsampling and upsampling layers (e.g. U-Net), different linear blocks (e.g. ResNet) or blocks that run in parallel (as for region proposal networks) are concatenated. The architecture is then optimised with regards to hyperparameters such as kernel filter sizes, depth or number of layers, or activation functions, as well as the tuning of measures against overfitting such as dropout. In supervised methods models are trained to minimise a loss function that is chosen to match our scientific task and their performance is finally tested on data not used during training.

This brings us to the crucial point of training data, at least in the case of algorithms that are not fully unsupervised. For supervised training, a critical assumption is that the distribution of our training data represents the data we want the model apply to sufficiently well, i.e. it is independent but identically distributed. If this is not the case, we risk through sample selection bias a covariate shift between source and target domain, which manifests itself through lack of generalisation in reduced model accuracy or performance. Luckily, there are methods to saveguard against covariate shift such as batch normalisation, and advanced ML methods are often able to generalise and transfer learn between learning a task on a source domain and improved performance of learning a target task on a target domain (where target $\neq$ source) (Sugiyama et al. 2007; Ioffe and Szegedy 2015; Tan et al. 2018; Schneider et al. 2020). A simple example is to learn a task on simulations to then transfer to observed data. When we transfer between distributions, but for the same task, we also call this domain adaptation (Wang and Deng 2018). As often in (radio) astronomy, labelling of data is an expensive (and impossible concerning the data volumes we face) task that requires expert knowledge and comes with its own biases, both transfer learning and active learning are interesting applications, together with semi-supervised and self-supervised learning already mentioned. Here active learning algorithms iteratively learn in a setting where from a majority unlabelled training set labels are requested by the learner, for example based on a confidence score, reducing the amount of labelled data needed and helping against covariate shift (Lochner and Bassett 2021; Leoni et al. 2022).

As mentioned in the previous introductory chapter, a wealth of AI (meaning foremost DL and further modern ML algorithms) applications to radio astronomy awaits to be explored, and we will highlight in the following chapters important applications for the golden years in radio astronomy to come.

10.2 Source Detection

Source finding, or object detection and identification, that uses AI and particularly DL entails a range of similar tasks, from binary classification (if a target source is present in a portion of data taken), region proposal (designating a bounding box for the region where a source is expected), to full segmentation of the object (identifying the pixel, voxel, or generally data portion that represent the object). These detection tasks can both be performed for narrowly pre-defined types of objects (e.g. class of radio sources) and for objects just sharing for example certain properties (e.g. any radio galaxy), or even to detect objects deviating from a pre-defined norm (see Sect. 10.6.2). This is discussed in Chap. 8 for radio source finding based on algorithms that do not fall under AI methods. Depending on the task, a variety of well-tailored deep learning architectures and further modern machine learning methods have proven successful at source detection in astronomy (González et al. 2018; Panes et al. 2021; Chatterjee et al. 2021; Doorenbos et al. 2022), as recently also shown for radio astronomy (Barkai et al. 2022; Delli Veneri et al. 2023). For example source detection with variants of region-based convolutional networks, R-CNNs (Girshick 2015), and networks that frame detection as a regression problem such as you-only-look-once, YOLO (Redmon et al. 2015), have proven their ability to outperform other methods during the second SKA data challenge when detecting radio sources in a large (1TB) data cube of neutral hydrogen emission (e.g., Hartley 2023). We give more details on the SKA data challenge below. AI and modern ML methods with the more with increasing data volumes become an integral part of data analysis pipelines of (radio) astronomical surveys in the coming years.

Depending on the scientific goal and available data, we can choose between supervised networks trained on labelled ground truth data for detecting a certain type of objects, and semi-supervised and unsupervised complementary methods. Regarding input data types, source finding in radio astronomy ranges from 3D cubes (tomography), over 2D imaging (often in the spatial plane on-sky), to 1D time series or spectroscopic measurements. Especially for the SKA, 3D tomographic cubes of the brightness temperature of the 21 cm forbidden spin-flip transition of neutral hydrogen will be mapped out across cosmic time and thereby reach data volumes at peta- and exabyte scales. A typical example in radio astronomy is the detection of radio-bright sources such as galaxies, for example SFG and AGN. Here CNN (LeCun et al. 1998; Fukushima 2004) have been applied to detect not only compact radio-bright sources, but created for LOFAR continuum maps new binary detection masks of extended sources such as clusters of galaxies and filaments (Gheller et al. 2018). This source finding task thus often is entangled with a classification and/or subsequent source characterisation task depending on our scientific question. Prominent subsequent tasks are the classification of radio sources (see Sect. 10.4), photometry or flux measurements of radio galaxies, as well as geometric properties such as shape, size, position and inclination angles (see Sect. 10.5). In particular for radio interferometers, artefacts and systematics stemming for example from RFI mitigation, residual beam side-lobes and foregrounds can prove challenging when striving to avoid falsely detected sources in automated detection. Furthermore, algorithms able to detect both single- and multi-component sources, such as region proposal networks (RPNs; Ren et al. 2015), avoid over-prediction of objects, mitigating false positives. Besides the ability to treat large 3D and 2D data volumes, CNN-based networks have also been shown to keep up with a large data stream in real-time to detect radio transient candidates in LOFAR and expected SKA time series data (Ruhe et al. 2022). These networks successfully employed to detect astronomical sources are usually based on convolutional layers as core building blocks.

10.2.1 Radio Source Detection with (Convolutional) Networks

Starting from basic CNN architectures, a variety of additions and modifications improves performance and enlarges the scope of source detection tasks:

- A basic CNN architecture coupled with dense regressive layers: This can mark portions of data, of the network input size, as containing a detection; see Lukic et al. (2019) for an application to radio source detection. Additional characterisation tasks are possible and usually perform well.
- Region proposal networks: They can identify several candidate bounding boxes of varying size and for multiple classes per network input image. This method bypasses the problem that for unknown size and number of objects a standard CNN would have to select and classify a huge number of regions by extracting

CNN feature maps and classifying proposed regions. This is especially advantageous if we do not know how many and what sized radio sources are to be expected. This algorithm has been refined until fast enough for real-time object detection, going from the first R-CNN (Girshick et al. 2013), to Fast R-CNN (Girshick 2015) and Faster R-CNN (Ren et al. 2015). Faster R-CNN have also been extended from radio source classification to detection for the first SKA science data challenge (Wu et al. 2019; Bonaldi et al. 2021).
- YOLO (you-only-look-once) networks (Redmon et al. 2015; Redmon and Farhadi 2016): This network is an order of magnitude faster than previous algorithms and turns classification of proposed regions into regression by creating class probability maps and bounding boxes not localised but for the entire data input at once; this is an important advantage when facing large data volumes. Especially the ability of class probability mappings is advantageous when striving to detect multi-component sources that often appear in radio imaging.
- Transformers (Vaswani et al. 2017): DEtection TRansformer (DETR; Carion et al. 2020) uses CNN features to encode-decode for feed-forward networks to derive class and bounding box predictions. This attention-based approach allows to focus on and refine different regions of the input simultaneously.

For overlapping radio sources or components the source detection task is entangled with deblending. For example for deep photometric surveys CNN-based architectures such as U-Net have been shown to be able to deblend galaxies an order of magnitude more precise and with considerably less bias than a 'traditional' source detection method (Boucaud et al. 2020); for other deblending methods see Sect. 8.2

10.2.2 Network-Based Segmentation

For the segmentation of sources, i.e. the creation of maps that assign to every data element a class or class probability, important network models in use are:

- U-Net (Ronneberger et al. 2015) which were designed for bio-medical image segmentation and have already proven useful to detect and segment blended galaxy images (Boucaud et al. 2020). This unique architecture is characterised by the ability to connect both small and large scale information present in the data by concatenating each downsampling step, i.e. convolution and pooling, with the upsampling step of the same dimension.
- Auto-Encoders (AE; see also Sect. 10.6.2) consist of an encoder with downsampling layers, that after the latent space bottleneck of minimal dimensionality is followed by the decoder with upsampling layers. As such, they are sequential and feature unconnected encoder-decoder layers as compared to the U-Net above. Like U-Net, AE perform tasks such as source segmentation and emulation (Lucie-Smith et al. 2022; Rothschild et al. 2022). Here the segmented image can be reconstructed from an input image, or directly via trained decoder from latent

space variables of the bottleneck. Variational Auto-Encoder (VAE) often show improved performance, as their latent space is regularised via imposed latent space distributions for variables.
- Non-binary probabilistic mappings that assign a probability for each data point to belong to an object type can be created both by using ensembles of network segmentation models, via appropriate choice of output layer activation (for example cross-entropy; Bretonnière et al. 2021). These probabilistic mappings can then be used as an uncertainty map and as a prior, for example for the fitting of galaxy brightness profiles.

Besides source detection, these methods are equally able to segment, or flag, systematics and high-noise regions in the data. Especially important for radio astronomy is the flagging of RFI, where AE models have shown good performance on LOFAR and HERA data (Mesarcik et al. 2022). Generally, high fidelity segmentation maps lead to an improved performance in characterisation and inference tasks. Finally, besides supervised models described above, some models are suitable for semi-supervised and unsupervised detection and segmentation, for example VAE for decoder generation trained on auxiliary tasks, contrastive learning, or in the case of generative models for data augmentation; see also Sect. 10.6 for a discussion of the use of VAE for uncertainty estimation. Also clustering algorithms such as DBSCAN (Density-Based Spatial Clustering of Applications with Noise; Ester et al. 1996), and map algorithms such as self-organising maps (SOM; Kohonen 1982) can support source detection.

10.2.3 Example: The Second SKA Data Challenge (SDC2)

The Square Kilometre Array Observatory (SKAO) will map the radio sky over so far largely unexplored cosmic volumes. SKAO data products will thus be unprecedently large and complex for radio astronomy with data rates of the order of TB/s, requiring advanced analysis techniques (Davis et al. 2016). To this end, the Science Data Challenge 2 was designed to find and characterise neutral hydrogen (HI) sources in a 1TB data cube simulated for 2000h SKA-Mid observations, see Fig. 10.1. Before, the first SKA Science data challenge had aimed for the analysis of 2D radio imaging at three different frequencies (Bonaldi et al. 2021). The winning SDC2 analysis combined two network-based techniques, a modified YOLO architecture and a traditional detection algorithm coupled with CNN for classification of true versus false positive detection and for inference of properties (Hartley 2023). For large (radio) astronomical data this ensemble network approach has thus proven to be very useful. Remarkably, 3D convolutional architectures based on U-Net allowed for a high fidelity 3D reconstruction in both spatial and frequency space, that in turn allowed for accurate and precise estimates of radio source properties such as integrated HI flux and HI size, and even for low S/N radio sources. Besides DL-based approaches also source detection based on 'traditional' signal processing

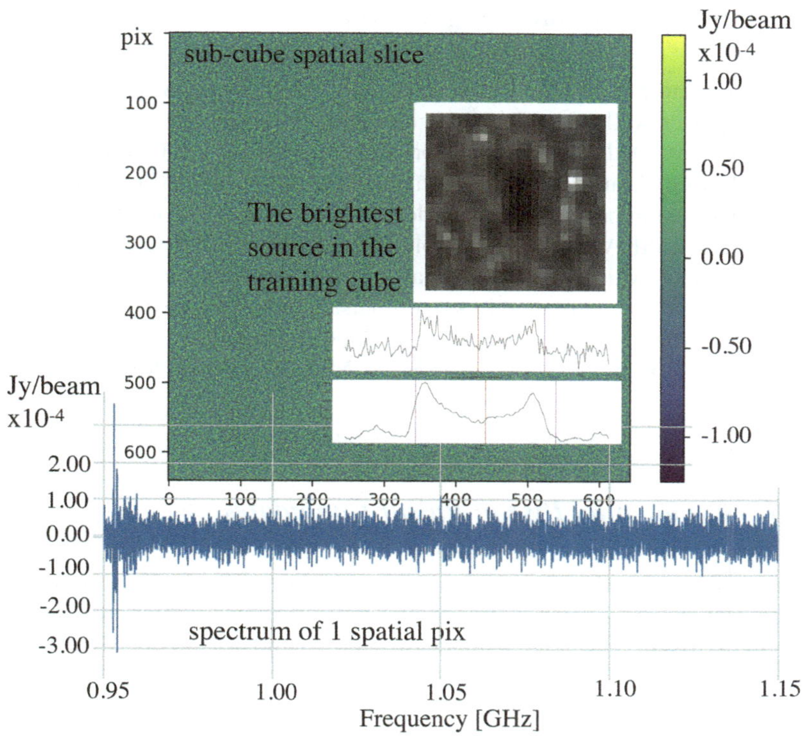

Fig. 10.1 A glimpse of the SKA Science Data Challenge 2 data: Spatial slice of a sub-cube of the full (25,714 × 25,714 × 6667) voxel cube (top back, blue to yellow flux in Jy/beam), cutout both spatially and in frequency of the brightest source in the training cube (front, black), and full spectrum of one spatial pixel as Jy/beam as a function of frequency in GHz (bottom, blue)

techniques such as SoFiA and modern ML such as 3D wavelet filtering methods (Starck et al. 2007; Vonesch et al. 2007) performed well. Generally, ML-based methods were able to detect fainter (lower integrated line flux) and lower S/N sources at higher completeness. At the same time some of both ML and traditional algorithms over-predict the occurrence of rare bright sources, with a tendency of traditional signal processing for example with SoFiA performing better at the bright end of the HI source distribution. For supervised DL methods this might be due to the rareness of these brightest sources in the training data cube. Regarding the characterisation of properties, the ML and DL-based methods on average could derive source properties with smaller fractional error and across a larger parameter range (especially when it comes to the smallest, faintest, or lowest S/N objects), somewhat mirroring the results for the detection task.

10.3 Pulsar Detection Using ML/AI

Discovering new radio pulsars at scale in modern pulsar surveys requires computational power well beyond what is available from a single desktop or laptop workstation unit. Dedicated high-performance computing (HPC) clusters hosting several CPUs, GPUs and Field Programmable Gate Arrays FPGAs have been a norm in radio pulsar astronomy over the past decade and will continue to be indispensable in the upcoming SKA era. Massive volunteer-distributed computing projects like Einstein@Home[1] which currently has an average computing capacity of 11732.1 TFLOPS[2] have also contributed to several pulsar discoveries in the Pulsar Arecibo L-band Feed Array (PALFA; Cordes et al. 2006; Knispel 2011; Allen et al. 2002) and Parkes Multibeam Pulsar Survey (PMPS; Manchester et al. 2001; Knispel et al. 2013).

10.3.1 Exponential Rise of Pulsar Candidates

Pulsar candidates produced from folding (as explained in detail in Sect. 9.3) are four-dimensional data cubes consisting of time, frequency, rotational phase and power of a signal. The total number of such candidates produced from pulsar surveys has been steadily increasing over the past few decades. For example—the second 2nd Molonglo Survey in 1978 (Manchester et al. 1978) produced 2500 total pulsar candidates. A decade later in 1988, the Parkes-20 cm survey increased that number to a total of 150,000 candidates (Johnston et al. 1992). The early 2000s saw the emergence of arguably the most successful pulsar survey to date—PMPS (Manchester et al. 2001) which continues to produce discoveries to date. The initial processing run created a total of 8 million pulsar candidates which then expanded to 55 million candidates for the next generation pulsar survey in the 2010s—the High Time Resolution Universe (HTRU; Keith et al. 2010).

Traditionally, these candidates were visually inspected by plotting 1D projections (pulse profile, dispersion measure DM curve) and 2D projections (frequency-phase and time-phase) in order to identify potentially promising signals which were confirmed with dedicated telescope follow-up time. Clearly, manual inspection is no longer feasible. For example, assuming a highly optimistic time of one second to decide on a pulsar candidate, working 16 hours a day, it would take a person 955 days or around 2.6 years to go through the entire data-set. This is precisely the reason why machine learning algorithms have become extremely popular in pulsar searching. The reasons for this increase have largely been due to improvements in instrumentation which has led to finer frequency channels and faster sampling

[1] https://einsteinathome.org/.

[2] https://einsteinathome.org/server_status.php.

rates, computation and algorithm advancements which have led to expansions in the search space to find the most exciting compact binary pulsars (Allen et al. 2002; Balakrishnan et al. 2022). We refer the interested readers to section two of Lyon et al. (2016) for a more in-depth discussion of the increase in pulsar candidate yield over time and a model to quantify the increase based on the survey specifications.

10.3.2 ML in Radio Pulsar Astronomy

The first use of ML in radio pulsar astronomy was done by Eatough et al. (2010) using a multi-layer perceptron (MLP) to classify between pulsars and noise/interference signals in the PMPS survey. Similar MLP approaches were also used by Bates et al. (2012), Morello et al. (2014) to identify pulsar signals in the mid-latitude ($-120^\circ \leq l \leq 30^\circ$ and $|b| \leq 15^\circ$) portion of the HTRU survey. These initial ML approaches relied on domain knowledge for feature extraction as a first step. Examples of features used include the signal-to-noise (S/N) ratio of the pulse profile, dispersion measure value detected by the pulsar-search pipeline, and metrics to calculate the persistence of signal in the time and frequency domain. Values from these features were then passed onto a neural network for training and final classification. This feature extraction step was necessary as the training sets back then were small (70–100 pulsar detections), which therefore required the use of relatively shallow (fewer hidden layers) neural networks for regularisation and to avoid the curse of dimensionality. The first deep learning solution—Pulsar Image Classification System (PICS) where raw pixel values were used to identify pulsar signals was done by Zhu et al. (2014). This pipeline used a combination of CNNs and support vector machines (SVMs) for 2D projections of the pulsar candidate datacube and SVMs and MLPs for the 1D projections. Scores from each of these algorithms were then passed on to a logistic regression (lr) model with an L2 penalty. This algorithm then outputs one final score, i.e probability of the signal being a pulsar. This approach of combining the output of various classifiers is referred to in the literature as an ensemble ML pipeline. Wang et al. (2019) improved on this further by replacing the CNN architecture in PICS with a ResNet which was used to identify pulsars in the FAST Drift-Scan Survey (CRAFTS; Li et al. 2018). Inspired by the success of boosting algorithms in other fields—for example, eXtreme Gradient Boosting (XGBoost; Chen and Guestrin 2016) in the HEP meets ML Kaggle challenge (Chen and He 2015)—and photometric redshift estimation using Adaptive Boosting (Adaboost; Hoyle et al. 2015), similar architectures have also successfully been demonstrated to aid in pulsar candidate identification (Bethapudi and Desai 2018; Zhao et al. 2022a).

While all the above-mentioned approaches led to several pulsar discoveries, a drawback of this is that the training of the neural network needs to be done offline before deployment. The RFI environment near a telescope changes over time. Additionally, the amount of data produced by pulsar surveys are rising at an exponential rate. For example, the recently concluded MPIfR MeerKAT-L band

Galactic Plane survey (MMGPS-L; Padmanabh et al. 2023) produced 350 TB of data per week which needed to be analysed in quasi-real time as the raw data is deleted after analysis. Future pulsar surveys using SKA will increase the size of the datasets further. Therefore, the use of ML algorithms that are trained online and better suited for large-scale data streams to solve the pulsar candidate classification problem will likely play an important role going forward. Lyon et al. (2016) used a Gaussian Hellinger Very Fast Decision Tree (GH-VFDT) which is capable of classifying one million pulsar candidates every 15 seconds. This approach led to the discovery of several new pulsars in the LOFAR Tied-Array All-Sky Survey (LOTAAS; Sanidas et al. 2019). The gain in speed for training and classification for GH-VFDT comes at the cost of lower overall precision and recall rates compared to algorithms trained offline.

One of the challenges of pulsar candidate classification is the class-imbalance problem. In a typical training set, there are hundreds to a few thousand pulsar detections compared to several millions of noise and RFI detections. Supervised methods require that the training set be equally sampled between the different classification categories. Different approaches have been attempted to address this problem. The canonical approach has been to undersample the noise and RFI samples (Eatough et al. 2010; Bates et al. 2012; Morello et al. 2014; Zhu et al. 2014). Alternatively, oversampling the pulsar samples using Synthetic Minority Oversampling Technique (SMOTE) has also been used successfully (e.g., Bethapudi and Desai 2018). Since 2019, Generative Adversarial networks (GANs; Goodfellow et al. 2014) have also been used to augment the training set (Guo et al. 2019; Mahmoud and Guo 2021) including using GANs to build a semi-supervised classifier (Balakrishnan et al. 2021). Unsupervised ML techniques like Gaussian mixture models have found applications in other areas of pulsar astronomy including identifying clusters in the P-$\dot{P}$ parameter space (Igoshev and Popov 2013) and population synthesis of radio pulsars (Igoshev and Popov 2013). The use of neural networks in dedispersing and denoising of pulsar data before passing to traditional techniques like the Fast Fourier Transform (FFT) and Fourier Fringe Analysis (FFA) has also been recently explored (Künkel et al. 2021) and although this is not yet practical in terms of processing time and memory demands, it acts as a proof of concept for various ways in which ML algorithms can replace traditional techniques in pulsar searching. Looking ahead, it is clear that ML algorithms will play a vital role in our ability to discover pulsars which in turn will be a gateway for new and exciting science. For the interested reader(s), we have compiled a list of standard training data-sets that can be readily used to build your own ML classifier.

List of Standard Labelled Pulsar Data-Sets

- HTRU Medlat Training Data set[3] (Morello et al. 2014).
- HTRU2 Dataset[4] (Lyon et al. 2016).

[3] http://astronomy.swin.edu.au/~vmorello/.

[4] https://archive.ics.uci.edu/ml/datasets/HTRU2.

- FAST pulsar dataset[5] (Wang et al. 2019).
- HTRU lowlat Training dataset[6] (Balakrishnan et al. 2021).

10.4 Radio Galaxy Morphology Classification

Though some of the ML source finders also support radio galaxy classification functionalities (i.e., Wu et al. 2019; Lao et al. 2021; Zhang et al. 2022, see Sect. 10.2 for more examples of ML source finders), most radio galaxy classification algorithms are developed assuming the source awaited for classification is real. Classifying radio sources themselves are of interests as each type of radio galaxy might share similar physical properties or evolution histories, and thus algorithm developers usually concentrate on those classes they are interested in when training their models, rather than taking all possible source morphological tags into account (see Rudnick 2021, for a summary of tags).

When talking about ML classification algorithms, one may either hunt for objects with complex, unusual or even peculiar morphologies, or classify radio sources to specific given types. We here summarise several representative technical routines of both targets:

1. *Hunting the unexpected*: Unsupervised learning approaches are usually implemented to facilitate the work of finding unusual, peculiar or unexpected objects (also see Sects. 10.6.2 and 10.6.3). SOM based methods (i.e. PINK; Polsterer et al. 2015) are perhaps the most popular ones (i.e, Ralph et al. 2019; Galvin et al. 2020; Mostert et al. 2021; Gupta et al. 2022). Project scientists usually manually examine these dimensionality-reduced results and induce sample object classes themselves. Alternatively, a recent approach made use of *complexity* (see Sects. 8.3 and 10.6.3), so as to help the hunt of anomalous objects (Segal et al. 2023).
2. *Classify source with pre-defined classes*
    - *Image-based direct classification*: Thanks to the extraordinary pattern recognition ability of CNNs, a significant number of CNN-based approaches have been developed using various model architectures and training strategies (i.e., Aniyan and Thorat 2017; Lukic et al. 2018; Becker et al. 2021; Bowles et al. 2021b,a; Scaife and Porter 2021; Alegre et al. 2022; Kummer et al. 2022; Mohan et al. 2022; Tang et al. 2022; Slijepcevic et al. 2022a,b). For an example, see Fig. 10.2.
    - *Derived feature based indirect classification*: Instead of asking a model to extract morphological features from images and use them for classification, this routine requires developers to first extract parametric features themselves,

[5] https://github.com/dzuwhf/FAST_label_data.

[6] https://github.com/vishnubk/sgan.

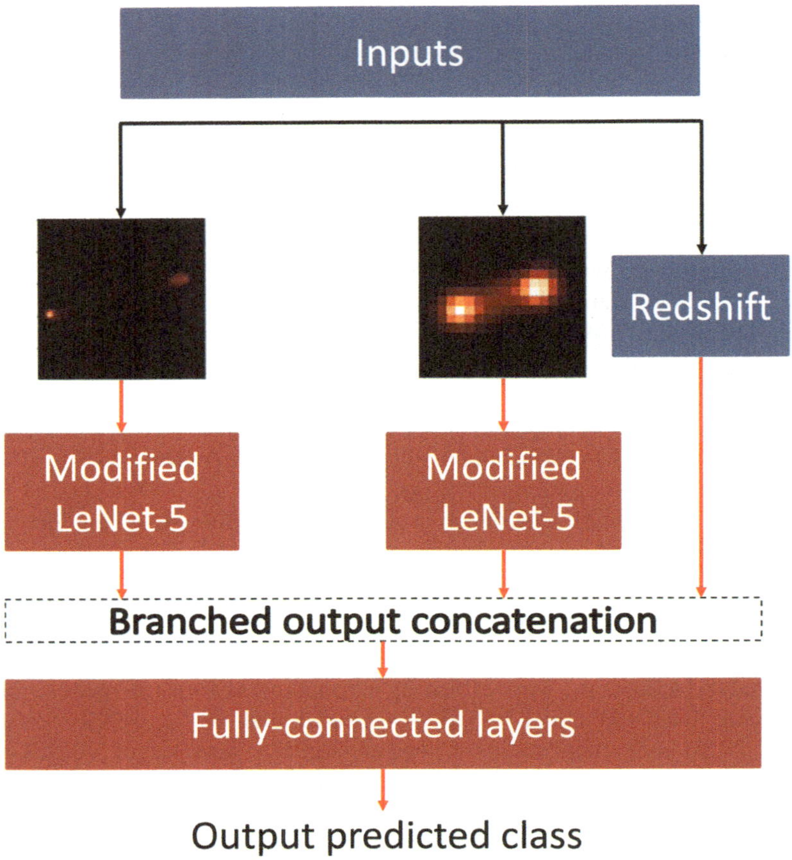

Fig. 10.2 Schematic of a multi-domain CNN classification algorithm to identify whether a radio galaxy is a Giant Radio Galaxy or not (Tang et al. 2022). Blue blocks and sample input images represent input data of a sample source await for classification; red blocks represent the machine learning components used to build the network. The red arrows on the diagram explain how we forward propagate the input data throughout the network and perform concatenation operations. Finally, the white dashed block accounts for the concatenation operation that connects network outputs from previous machine learning blocks and the input source redshift, which are then propagated to the following full-connected layers

and then feed these features as inputs to classification algorithms. This can be feeding sample image moments to SVM classification algorithm (Sadeghi et al. 2021); having radio component pair properties fed into a Random Forest (RF) model (Proctor 2016); or deriving Haralick features before using them to train an HDBSCAN clustering algorithm (Ntwaetsile and Geach 2021).

Even not considering the desire to hunt the unexpected, unsupervised learning approaches are valued in the last years not only because of the lack of large-scale well-labelled data, but also due to the worries of the transferability or reusability of current supervised learning approaches. Supervised (i.e., Aniyan and Thorat 2017) or semi-supervised (i.e., Slijepcevic et al. 2022a) learning models need to prove they are able to perform well on variant surveys with cost-effective modification or fine-tuning. In terms of unsupervised learning methods, on the other hand, the potential work loan of manually identifying objects of interest need to be evaluated and treated carefully.

While it is true that the generalisation ability of supervised learning approaches is largely constrained by the sample size of well-labelled sources available for model training and validation, it is also true these approaches can be superior when we have sufficient data to feed them. Many of these applications can perform morphological class prediction in human-comparable accuracy with given sample images. For instance, the very early application in the field was able to identify edge-darkened sources (FR I, see the lower right panel of Fig. 8.2 for example), edge-brightened sources (FR II, see the lower left panel of Fig. 8.2 for example) and bent-tailed sources with model general accuracy of $\sim$90% (Aniyan and Thorat 2017).

However, data quality and sample size shall constrain these approaches' potential. Efforts then were made to enlarge the sample of well-labelled object data. By far, three main approaches are trying to tackle this data barrier:

1. Simulation: Simulated data intuitively equipped with user-defined object classes, with little difficulty in generating large data samples for model training, i.e. Bonaldi et al. (2019) for SKA SDC1 (Bonaldi et al. 2021).
2. Generative models: Such approaches could generate almost infinite 'fake objects' that are similar but different from training samples, and thus enlarge the available sample size for training other models (i.e. through VAE method; Bastien et al. 2021).
3. Citizen Science: By asking volunteers to help identify the morphological type of realistic radio sources from given image cutouts (i.e., Fig. 10.3), a well-labelled large-scale catalogue of real radio objects can be received if a project receives popularity. These samples then could be used for model training with no doubt (Banfield et al. 2015).

The first two approaches for sure shall benefit model training and enable people to develop sub-optimal cataloguing and data-mining methods towards the next generation sky surveys. The first two SKA data challenges are obvious examples of that, with tens of developer teams around the world attended the competition and produced dozens of ML based or non-ML based data post-processing pipelines able to perform source finding, characterisation and classification at once.

Though simulated or generated data can relieve the data stress, the performance drop caused by the so-called simulation-to-real gap is still non-negligible. After all, it is necessary to have representative data. One of the promising approaches, perhaps, is citizen science (Liang et al. 2022).

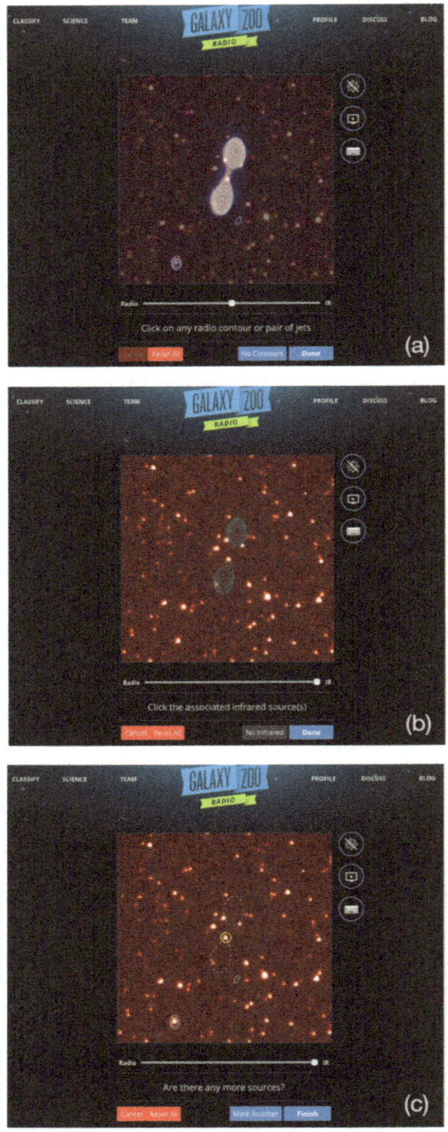

Fig. 10.3 An interface illustration of RGZ classification workflow, which is consistent with Figure 3 of Banfield et al. (2015). Each 'subject' refers to a $3' \times 3'$ field-of-view. A typical classification can be done in three steps: (**a**) step 1: mark all radio components that belong to the same parent radio source. (**b**) Step 2: select the corresponding infrared host galaxy. (**c**) Step 3: continue classification if there are other radio sources in the image, otherwise move on to the next subject

As mentioned in Sect. 8.3, RGZ managed to build one of the biggest human-identified catalogue of well-extended radio sources. The resulting radio galaxy catalogue (RGZ Data Release 1) with considerable consensus level has enabled project scientists to develop multiple classification algorithms (Alger et al. 2018; Lukic et al. 2018; Galvin et al. 2019; Ralph et al. 2019; Slijepcevic et al. 2022a; Tang et al. 2022).

However, its lack of scientifically meaningful radio morphology labels has significantly constrained the project team to develop algorithms to classify source meaningful morphology (i.e. FR I, FR II). If similar offshoot project Radio Galaxy Zoo (LOFAR) and Radio Galaxy Zoo (EMU) want to solve this issue in the future, data quality and potential biases should be treated carefully, and at best considered such issues before releasing such a project to the public (Rudnick 2021; Bowles et al. 2022).

10.5 Characterisation of Properties

Characterisation of properties entails both inference of global properties and model parameters from full (continuum) or spectral line maps, such as global properties that determine fluctuations of 21 cm hydrogen intensity maps (see Sect. 10.5.1), as well as characterisation of properties of single objects such as flux density or ellipticity measurements of radio sources, (see Sect. 10.5.2). The ability of advanced machine learning methods to learn representations and beyond-Gaussian information comes in handy here, and even enables likelihood-free inference (Alsing et al. 2018; Ardizzone et al. 2018). Also, methods sometimes perform almost surprisingly well when transfer learning for example between different levels of noise and foregrounds (Neutsch et al. 2022). In addition, the ability to process large amounts of data fast (easily seconds for tens of thousands of inference tasks for supervised methods after training) is almost unique to modern ML. Of course, for every ML model the interpretability of results on the way to explainable AI (see Sect. 10.7), e.g. feature visualisation, class activation (Zhou et al. 2015), saliency (Simonyan et al. 2013), layer-wise relevance propagation (Montavon et al. 2019), and uncertainty estimation with probabilistic Bayesian or ensemble methods (see Sect. 10.6.2) are important to validate results.

10.5.1 (Direct) Inference from Maps

Radio continuum and spectral line mappings at radio wavelengths, prominent example is 21 cm intensity mapping of neutral hydrogen, have to deal with high levels of foregrounds and noise (Choudhury et al. 2020; Spinelli et al. 2022). Furthermore, the cosmological fluctuation signal we aim to measure is intrinsically non-Gaussian, meaning, for example power spectra as summary statistics do not encode all physical potential interesting information present in the signal. Other summary statistics and measures such as bispectra (Shimabukuro et al. 2016; Hutter et al. 2020) or cross-correlation signals (Heneka et al. 2017; Hutter et al. 2018; Heneka and Mesinger 2020) can partly account for this, but lack the flexibility for example of DL models. Equally modern ML methods offer the possibility to flexibly extract and discard highly non-Gaussian noise and foreground contributions for map cleaning purposes (Wang et al. 2021; Makinen et al. 2021; Ni et al. 2022). As often radio interferometric measurements offer the possibility of a quite finely gridded frequency resolution, these signals offer 1D spectra in frequency space, 2D imaging and for example for the upcoming SKAO even 3D radio data cubes, as well as multi-channel inputs such as radio continuum and 21 cm HI maps. Here for example deep networks offer a lot of flexibility to adjust to the data input form, e.g. using 1D, 2D or even 3D convolutional layers as well as multiple input channels.

Scientific inference goals for radio continuum and spectral line mapping are astrophysical such as galaxy evolution determined by typical luminosities and star formation rates, and cosmological striving to measure the large-scale structure of our Universe and thus derive cosmological parameters such as the matter density parameter. Besides mapping the more local Universe that was already targeted by galaxy surveys the new frontier in astronomy and astrophysics is the mapping of cosmic dawn (CD) at redshifts beyond 10 and the epoch of reionisation (EoR) at redshifts $z > 5$–6 (Mesinger 2016; Trott and Pober 2019; Liu and Shaw 2020). Here the 21 cm fluctuations of neutral hydrogen measure the heating signal of the very first stars and galaxies that formed during CD, and our Universe's last big phase transition where regions of before neutral hydrogen start to be ionised until at the and of the EoR the Universe is fully ionised again. This 21 cm signal is targeted by the SKAO in imaging (finely gridded in frequency) and by the instruments preceding it such as LOFAR, HERA and MWA as a statistical detection of the signal power.

Tools and architectures that are very much suited to infer global properties from this type of interferometric 1D, 2D, and 3D data are again often models based on convolutions. These learn filters to take into account correlations across voxels and reduce dimensionality until the feature space is fed into a dense network part for the actual regression task (other regressors are also possible). Besides learned filters in CNN-based architectures, for example wavelet scattering transforms (WST) (Bruna and Mallat 2012) offer the possibility of physically motivated filters that need not be trained. Both CNN-based architectures and WST have already been successfully employed for inference of reionisation era astrophysics (Gillet et al. 2019; Prelogović et al. 2022; Greig et al. 2022), where also relatively shallow and fast-to-train 3D architectures perform extremely well on SKAO mocks (see Fig. 10.4). These architectures can be improved further upon to provide full parameter likelihoods via likelihood-free inference by turning them into BNN or via simulation-based inference that estimates the likelihood shape for example with normalising flows (Zhao et al. 2022a,b, see also Sect. 10.6.2). Other interesting applications of DL for large-scale radio fluctuation maps are the use of networks as emulators (to for example derive power spectra from maps; Schmit and Pritchard 2018; Jennings et al. 2019), simulators (to create large samples of mocks or change observable;

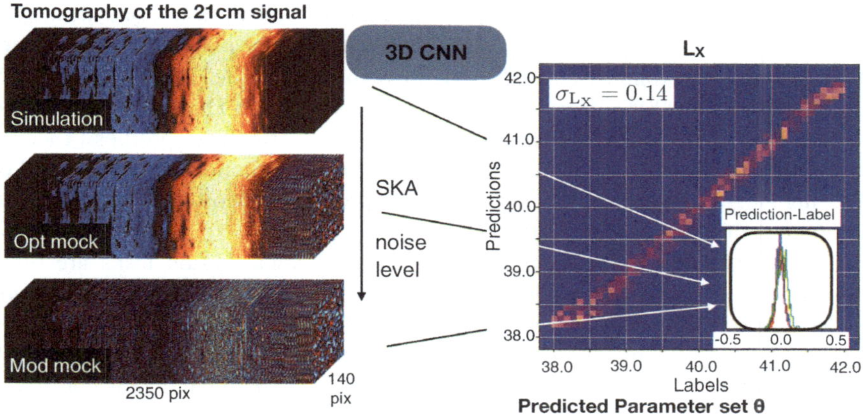

Fig. 10.4 Scheme of joint regression of parameter set θ from 21 cm lightcones. Left: Simulated lightcones (top) generated with 21cmFAST (Mesinger et al. 2011; Murray et al. 2020), including optimistic (middle) and moderate (bottom) noise levels as expected for SKA-Low (Pober et al. 2013; Heneka and Cooray 2021). Right: Example scatter plot of predictions against labels for the regressed X-ray luminosity of ionising sources L_X (Neutsch et al. 2022); note in the inlet histogram of (prediction—label) the insensitivity to noise levels

Rafieferantsoa et al. 2018) and for the purpose of foreground and noise cleaning directly on maps; so have networks been and certainly will be employed to clean artefacts in reconstructed radio images due to incomplete sampling of the sky in Fourier space (Schmidt et al. 2022; Dabbech et al. 2022; Terris et al. 2023).

Further DL advantages are the possibility to explore very high-dimensional parameter spaces fast, the fact that inferred parameter values are not biased by the chosen summary statistics ('let the network choose'), and the complementarity to more conventional methods. An important final word of warning, this DL-based inference is only then able to outperform other methods, when appropriate pre-processing is applied (usually found empirically in combination with domain knowledge), and in the case of supervised learners when the training data-set is chosen appropriately. This can for example mean to train on a wide range of fluctuation fields without biasing the result by excessive noise present.

10.5.2 *Object Characterisation*

In connection to detection of sources at radio wavelengths (see Chaps. 8 and 9 and Sect. 10.2), we often want to also characterise sources detected with regard to their properties of scientific interest. For example measuring the integrated flux density of HI sources can be used to construct a luminosity function and study the radiative output of the sources of interest; line widths measured inform about the dynamic state of the gaseous media within radio sources as well as accretion; measuring size, axis ratio and angles on-sky and relative to the sky plane yields important information for morphological and formation/evolution studies (Rezaei et al. 2022; Delli Veneri et al. 2023; Hartley 2023). Furthermore, with DL we are flexible to strive for fitting procedures such as the derivation of brightness profiles (Tuccillo et al. 2018), or potentially even symbolic regression of functional forms instead of deriving single properties or a combination of properties themselves (Udrescu and Tegmark 2020; Delgado et al. 2022).

Typical models that perform well for object characterisation tasks are often similar in their basic building blocks to models suitable for inference from maps as described in the previous Sect. 10.5.1. Main differences lie in the target output trained for and the typically less important larger scale correlations, or in network terms, a sufficient smaller size of convolutional kernels the input data is convolved with. Of course, this is only true for radio astronomical data in the case where most sources measured are relatively small in extension as compared to the interferometrically measured full map. As already mentioned in Sect. 10.2, CNN-based architectures were able to for example derive from 3D HI source images both accurate and precise properties of HI sources such as their source size and integrated flux, which has also been shown for other wavelengths than radio (Boucaud et al. 2020; Hartley 2023). Also, object characterisation or regression of object properties can be closely related to classification tasks (see Sect. 10.4), as in DL we can move by adjusting training labels, output dimension as well as activation function from an inference task to a more or less finely binned classification task (and vice versa).

Object characterisation with DL methods itself profits from a before applied pre-cleaning or segmentation treatment of the radio objects of interest, where high-fidelity reconstruction significantly improves performance of inference tasks. This pre-processing can both successfully be performed with more conventional methods and with DL-based methods as discussed in Sect. 10.2. In Fig. 10.5 an example for such a successful pre-processing is shown, where network-based HI source reconstruction in 3D was used to both precisely and accurately provide unbiased

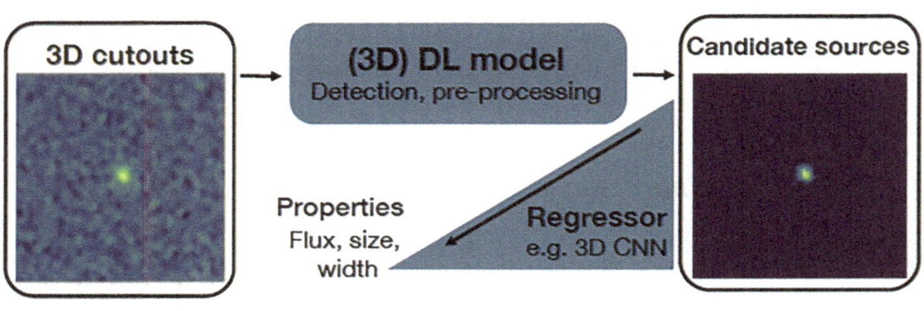

Fig. 10.5 A pre-cleaned HI source image from the SKA Data Challenge 2 (left) and a network-based reconstruction of this HI source (right) that was used for high fidelity inference of source properties; both images are 2D spatial slices from a full 3D data cube

estimates of HI source properties for the SKA Data Challenge 2. We need to be careful though that no crucial part of information encoded in the data gets lost or becomes biased, where for example a bias towards smaller HI source sizes could be caused as faint edges get cut away, via careful comparison with other algorithms and methods of explainable AI (see Sect. 10.7).

10.6 Quality Control

10.6.1 Dimensionality Reduction

Radio interferometers produce massive datasets in high dimensions. In machine learning and pattern recognition, an input dataset is described by a set of measurable characteristics, namely, features. A number of features form a feature vector. Adding more features increases the dimension of the feature vector. While increasing the number of features improves the modelling of the input data to a certain extent, having too many features slows down the training process, introduces noise and ends up overfitting the data.

The curse of dimensionality (Bellman and Kalaba 1959) addresses difficulties that arise especially when analysing and organising high-dimensional data. The curse of dimensionality happens because the data density decreases exponentially when the dimensionality increases. The sparse data often lead to overfitting.

One of the most common approaches to overcoming the curse of dimensionality is dimensionality reduction. Dimensionality reduction refers to the process of reducing the dimension of a dataset while preserving the important characteristics of the data. It is especially useful for classifying and processing massive radio astronomical data.

Dimensionality reduction is often performed as a preprocessing step before training a model in machine learning. There are a number of advantages:

- Reduced data save computing time and storage space.
- Dimensionality reduction helps remove redundant data and noise and helps combine highly correlated features.
- Lower dimensional data are easier for visualisation and interpretation.
- It also helps remove noise and redundant features, which leads to a better model when the data are used for machine learning.

A disadvantage is that original data may be lost, typically corresponding to 1-15% of the original data.

Dimensionality reduction has two main types: feature selection and feature extraction. Feature selection finds a subset from the original set of features by removing redundant features. Hence it does not modify the original data values. Filters, wrappers and embedded methods are three common ways to perform feature selection.

Filter methods evaluate the correlation between the input and output data using statistical measures and select the most relevant features based on the correlation scores. The Chi-square test, Fisher's score, and correlation coefficient are commonly used statistical measures.

Wrapper methods evaluate the performance of possible subsets of features when a machine learning algorithm is given to fit the input data. Wrappers often produce more accurate predictions compared to filters, however, wrappers are computationally more expensive. Forward selection and backward elimination (Marill and Green 1963) are commonly used techniques.

Embedded methods are techniques that have built-in feature selection as part of their model construction. They combine the benefits of the filter and wrapper methods. In each iteration of training, the most relevant features are extracted. Random forest (Breiman 2001) and LASSO (Least Absolute Shrinkage and Selection Operator) are commonly used techniques.

Feature extraction, on the other hand, finds a combination of new features by transforming a high-dimensional feature space into a lower-dimensional one. Hence it involves modification of the original data. Principal Component Analysis (PCA; Hotelling 1933; Pearson 1901), Self Organising Map (SOM; Kohonen 1982, 1990) and t-distributed Stochastic Neighbour Embedding (t-SNE; Van der Maaten and Hinton 2008) are widely used techniques. These methods become more popular in astronomical data analysis.

PCA projects data by finding a number of orthogonal eigenvectors that maximise the variance of the projected data, the eigenvectors are known as principal components (PCs). The number of PCs is typically lower than the dimension of the original data. PCA can mitigate overfitting and it is computationally efficient. However, the performance worsens on nonlinear data and PCs are often difficult to interpret. PCA has been applied to various problems in astronomy, for example, stellar spectrum classification and analysis (Deeming 1964; Bailer-Jones et al. 1998; Re Fiorentin, P. et al. 2007) and morphological and spectral classification of galaxies (Lahav et al. 1996; Kelly and McKay 2004; Peth et al. 2016).

A **SOM** or **Kohonen map** consists of two layers, an input layer and an output layer. Figure 10.6 shows a typical structure of a SOM. The output layer is typically a two-dimensional grid of nodes or neurons, inspired by neural networks. As a rule of thumb, the maximal grid size is suggested by $5\sqrt{N}$ where N is the sample size (Estévez et al. 2013; Shalaginov and Franke 2015). Each node in the output layer is fully connected to the input layer by a weight vector. In each iteration, the distances between a sample input vector and all neurons are calculated to find the best matching unit (BMU) that has the shortest distance from the input. The weights of neurons are updated to resemble the BMU. The resulting weight vector maps successfully the input data into the two-dimensional grid. SOMs have a few drawbacks: they are often computationally expensive and the resulting maps could be inconsistent due to the random initialisation of neurons at the beginning. Figures 10.6 and 10.7 show a SOM application on LOFAR-EoR (Low Frequency Array Epoch of Reionisation) data. The LOFAR-EoR data have a complex structure with high dimensions. To efficiently explore various morphological features in the

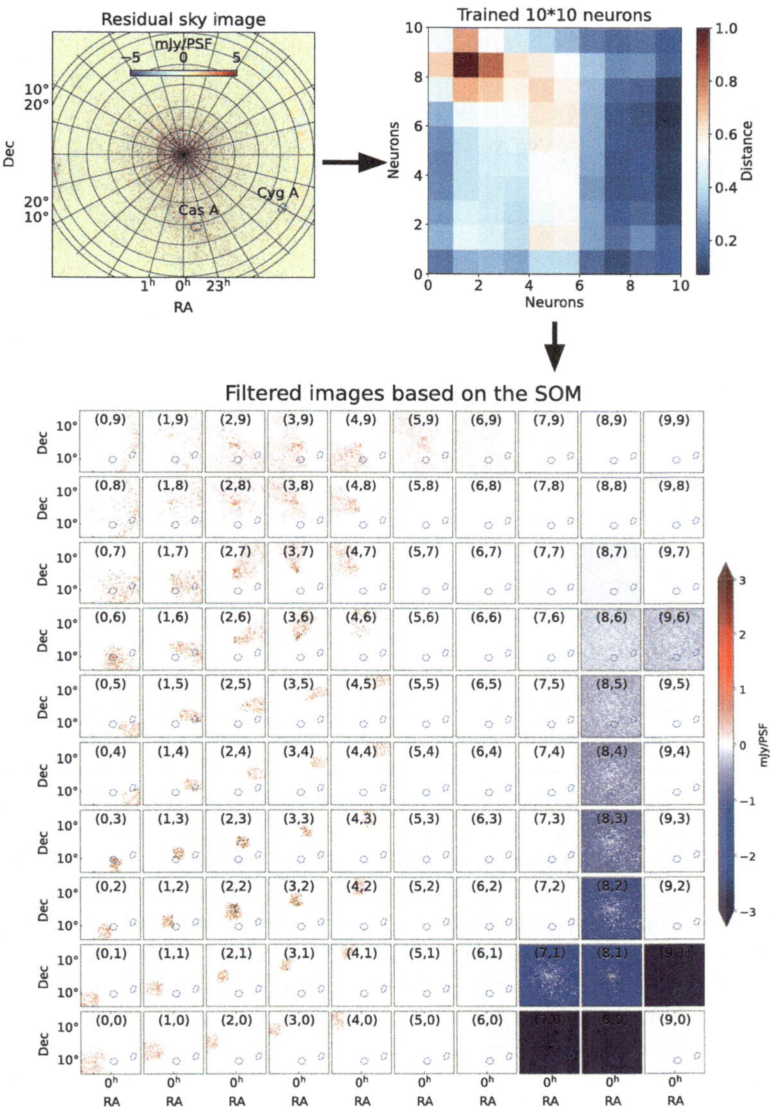

Fig. 10.6 A self-organising map (SOM) application on a residual image from LOFAR-EoR data (top left; Gan et al. 2022). To detect morphological features in the image, vector attributes that describe different shapes are calculated from the image. The obtained vector attributes are used as features and are fed to the SOM algorithm. The original image is filtered from the trained 10 by 10 SOM (top right). Each neuron is sensitive to different features in the original image (bottom)

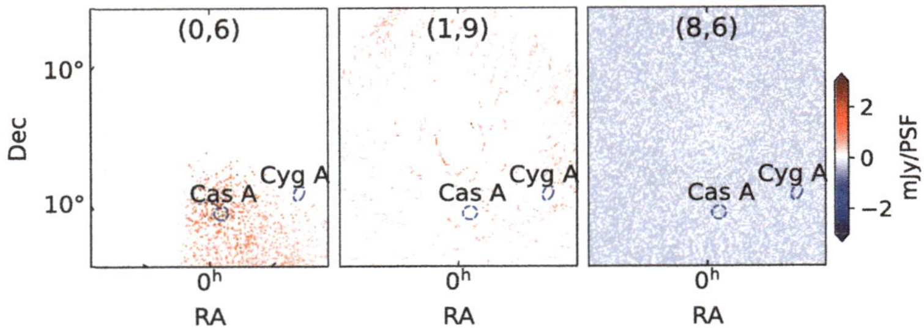

Fig. 10.7 Zoomed in filtered sky images of three neurons (0,6), (1,9) and (8,6) from the trained SOM on the bottom panel of Fig. 10.6. The position of very bright sources, Cassiopeia A and Cygnus A, are marked with blue circles. The three neurons are sensitive to distinct features in the residual sky image: neuron (0,6) is sensitive to the residual power coming from the direction of Cassiopeia A; neuron (1,9) is sensitive to the LOFAR beam characterised by a ring shape and neuron (8,6) is sensitive to negative residuals

LOFAR-EoR residual sky images (Gan et al. 2022), one can use the concept of vector attributes from computer vision and the SOM. Vector attributes describe the shapes and sizes of morphological characteristics of connected components in images and they can be used to filter specific features in images. However, filtering often requires the manual thresholding of vector attributes by experts. In this example, the calculated vector attributes are fed directly to a 10 by 10 SOM and we give the SOM freedom to explore features in the sky image. Each neuron can be used as an image filter and the filtered images are shown on the bottom panel of Fig. 10.6. Figure 10.7 shows zoomed filtered images based on three neurons of the trained SOM. In this example, the three neurons are sensitive to different features in the residual sky image: neuron (0,6) is sensitive to the residual power coming from the direction of Cassiopeia A; neuron (1,9) is sensitive to the LOFAR beam characterised by a ring shape and neuron (8,6) is sensitive to negative residuals. Another astronomical application of SOMs is PINK (Parallelized rotation/flipping INvariant Kohonen maps; Polsterer et al. 2015). PINK is a framework which uses a modified SOM to perform automatic galaxy classification. Applications of SOMs are also found in photometric redshift calibration (Wright et al. 2020).

t-SNE first converts pairwise distances between data points into similarities, using the t-distribution. The goal is to minimise the difference between the similarities of the original data and its corresponding low-dimensional data in the embedding. Unlike the traditional SNE, which uses the Gaussian distribution, the t-distribution has longer tails that can distinguish intra-cluster points better. t-SNE can handle nonlinear data and is good at preserving local and global data structures. A disadvantage is that t-SNE is computationally expensive. Astronomical applications of t-SNE include, for example, the classification and diagnostics of spectral information from the Galah (GALactic Archaeology with Hermes) survey (Traven et al. 2017) and quantification of the similarity between planetary systems (Alibert 2019).

10.6.2 Uncertainty Quantifications and Anomaly Detection in Deep Learning

For the quantification of uncertainty in a neural network's prediction, one has to differentiate between two different main sources of uncertainties, the data and the model (Hüllermeier and Waegeman 2021; Kendall and Gal 2017). In general one differentiates between uncertainty which is caused by the nature of the data (data or aleatoric uncertainty) and uncertainty stemming from the lack of knowledge of the model (model or epistemic).

The former one covers uncertainties as measurement noise, label noise or ambiguities in the data representations. Data uncertainty is considered as irreducible, since the uncertainty is already in the data and hence there is no chance for a model to become more certain. In contrast to this, the model uncertainty covers causes as insufficient model capacities, errors in the training process or insufficient coverage of the data distribution by the training data. These are factors that are changeable by providing a model the additional knowledge, for example in the form of improved training strategies, additional training data or a more advanced model structure and additional parameters. Hence, the model uncertainty is seen as reducible uncertainty. Since the data distribution can never be fully covered by a training set, a third type of uncertainty, the distributional uncertainty, is sometimes considered separately to the model uncertainty. It covers uncertainties which are caused by a shift in the data distribution between the training data and the testing data (Malinin and Gales 2018).

In Fig. 10.8 the different types of uncertainty described above are sketched for regression and classification tasks.

Calibration of Neural Networks A neural network is called calibrated, the predicted uncertainty represents the reality, i.e., if it matches the actual error-rate (Guo et al. 2017). For example for classification tasks, a calibrated neural network would be false in 10% of the cases where the predicted probability, also called confidence, of a class is 90%. For classification tasks the calibration can be checked straight forward, since the softmax output already delivers class probabilities. The calibration can be either measured for example with the expected calibration error, or visually represented by reliability diagrams (See Fig. 10.9). If the confidence of a neural network is higher than the actual predictive performance, the network is called to be over-confident. For the case that the confidence is below the actual predictive performance it's called under-confident. It is important to note, that basic neural network have been found to be often over-confident, especially when the non-linearities are given by the rectified linear unit (ReLu) activation function.

The calibration is an important measure to quantify the quality of neural network uncertainty quantification methods. Besides approaches for modelling and quantifying uncertainties in neural networks, which are described in the next part of this section, post-processing approaches that are applied after the training process are the most used approaches to calibrate a neural network. Especially the approach of temperature scaling is popular and easy to apply for classification tasks (Guo et al.

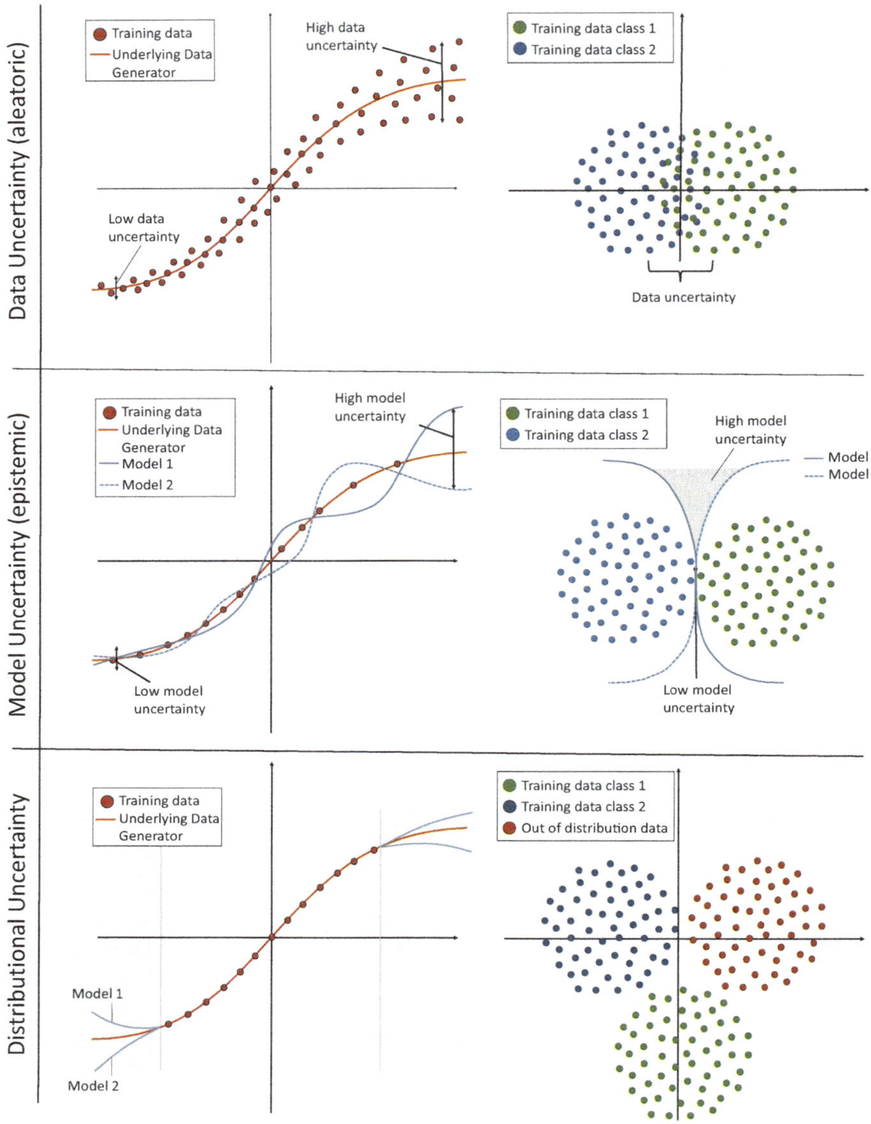

Fig. 10.8 A sketch of the different types of uncertainty for the tasks of classification (right) and for regression (left). While the data uncertainty (top row) is caused by the nature of the data, the model uncertainty (centre row) is caused by lack-of-knowledge of the model. Distributional uncertainty (bottom row) is a special case of model uncertainty and is caused by shifts from the training data distribution on which a model was trained

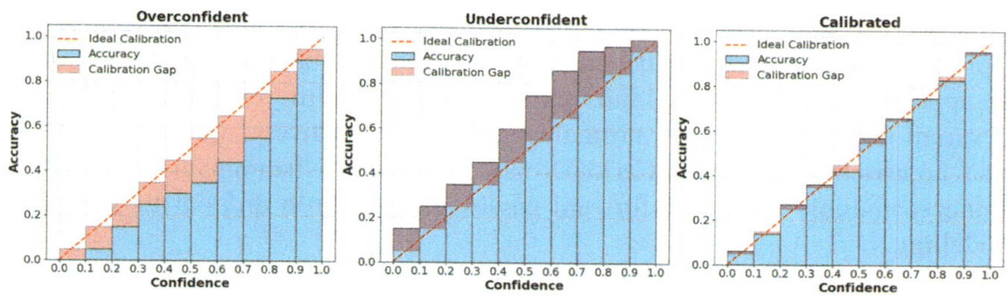

Fig. 10.9 Reliability diagrams of three different classifiers: An overconfident predictor (left), an underconfident predictor (centre) and well calibrated predictor (right)

2017). For temperature-scaling a scalar parameter (the temperature) is trained on a hold-out set to maximise the log-likelihood function, where the temperature equal to one is equivalent to the normal softmax output, a temperature larger than one reduces the confidence, and a temperature between 0 and 1 increase the confidence.

10.6.2.1 Approaches for Modelling and Quantifying Uncertainty in Neural Networks

Bayesian Neural Networks

Bayesian Neural networks (BNN; Denker and LeCun 1991; MacKay 1992) model the epistemic uncertainty by inferring a probability distribution over the network parameters θ. Given a set of training data samples, $\mathcal{D}$ and a prior distribution $p(\theta)$ over the model parameters, the posterior distribution over the model parameters can be derived by applying Bayes' theorem,

$$p\,(\theta|\mathcal{D}) = \frac{p(y|x,\theta)p(\theta)}{p(y|x)} \propto p(y|x,\theta)p(\theta)\,.$$

For a derived posterior distribution $p(\theta|\mathcal{D})$ and for a a test-time sample-target pair (x^*, y^*), the prediction can be modelled based on the law of total probability,

$$p(y^*|x^*,\mathcal{D}) = \int p(y^*|x^*,\theta)p(\theta|\mathcal{D})d\theta\,.$$

Since the computation of the predictive uncertainty $p(y^*|x^*,\mathcal{D})$ is in general intractable, common Bayesian networks work with approximation schemes, as the sampling based Monte Carlo Approximation.

The main difference between existing Bayesian neural networks is given in the way how the (in general intractable) posterior distribution is derived. Here, the variational inference and the Laplace approximation are two different ways that are widely used.

Variational inference (Hinton and Van Camp 1993) is a common technique to approximate the posterior distribution by tractable distributions. There exist different approaches for variational inference, ranging from different direct applications of variational inference, over normalising flows, to relatively simple to apply and train approaches as Monte Carlo Dropout (Gal and Ghahramani 2016), which connects multiple stochastic forward passes under neuron dropout to a Bayesian modelling.

Laplace approximation (Denker et al. 1987; Tishby et al. 1989) derives the posterior distribution in a post-hoc manner, after the training of a neural network, by locally approximating the neural network with a quadratic function around the mean prediction and interpreting the Hessian as the information matrix, i.e. the inverse of the covariance matrix of a Gaussian distribution. While the main idea stays the same, the approximation of the Hessian is the key for the Laplace approximation and different versions have been introduced (Ritter et al. 2018; Lee et al. 2020).

For a more detailed derivation and description of Bayesian neural networks and for a more detailed description of the different approaches, we refer to works as Gawlikowski et al. (2021) or Jospin et al. (2022).

Ensembles of Neural Networks

In contrast to the Bayesian Neural Networks which model a probability distribution over the model parameters, ensembles of neural network model the epistemic uncertainty based on multiple networks with different (locally optimal) parameters, the so called ensemble members (Lakshminarayanan et al. 2017). Due to the highly non-linear loss landscape on which a neural network is trained and due to the stochasticity in the training process, multiple training runs (in general) lead to multiple different (locally optimal) model parameters. For an ensemble of $M \in \mathbb{N}_+$ neural networks f_{θ_i} with parameters θ_i, $i \in \{1, ..., M\}$, a prediction $\hat{y}$ for an input x^* is then simply given as the mean prediction of the ensemble members,

$$\hat{y} = \frac{1}{M} \sum_{i=1}^{M} f_{\theta_i}(x^*) \, .$$

Ensembles have been shown to be efficient in receiving calibrated uncertainty predictions as well as improving the test performance compared to single neural networks. In the literature also different approaches to make ensembles of neural networks more efficient in training time (Huang et al. 2017) and memory and computational effort (Mariet et al. 2020).

10.6.2.2 Machine Learning Based Anomaly Detection

What Is an Anomaly?

Finding anomalies or outliers is a prevalent task in nearly every case where a running system is observed and defects or dis-functionalities should be prevented. Defining an anomaly is often harder than the detection itself and it highly depends on the special use case as well. Usually, we understand anomalies as something which differs from the normal state in a certain sense. Hawkins (1980) defines it as "an observation which deviates so much from other observation as to arouse suspicions that it was generated by a different mechanism".

If an anomaly can be described in a very clear mathematical way, e.g. by certain rules, an anomaly detection algorithm would simply check for those well-known rules. This procedure is also know as *rule-based anomaly detection*. However, those rules are usually not known. However, another approach is to derive rules from a sufficiently large dataset itself.

Finding anomalies often depends on the data in which the anomaly should be found. For example, finding unusual behaviour in time series may be more complicated than finding defects or irregularities in images, because of the time dependency. The approaches described in this chapter are applicable for both kind of data if not further indicated. Thus, the time dependency might be neglected in some methods or it is included, for example, by applying a sliding window approach on the time series.

Further, anomaly detection can be done in a supervised or unsupervised manner. For supervised methods, labels are needed that already indicate which data points are anomalous. Often labelling data is time consuming or not possible without any expert knowledge. Therefore, we want to focus on unsupervised anomaly detection, i.e. finding the anomalies without the explicit pre-knowledge about the anomalies at training time. This is also known as novelty detection. In this case, machine learning models learn the normal behaviour from the given data with the assumption that most of the data is normal. An anomaly is then detected when the model predicts that this data point differs from the learned normal behaviour. There are different types of anomalies which have significantly different properties. For example, a slow shift in the signal is completely different to single outliers or even the occurrence of unknown patterns in the data.

Anomaly Detection Approaches

While there are multiple approaches for anomaly detection, they cover a wide range of ideas and backgrounds, mainly based on statistical properties. Recently, also neural network including approaches have been proposed in various articles (e.g., Dai and Chen 2022; Hinton and Salakhutdinov 2006; Niebling et al. 2020; Tuli et al. 2022).

The possibly most classical way of finding anomalies is to assume that the given data follows a certain distribution. Based on that assumption, the parameters for this distribution, for example, mean and variance of a Gaussian distribution, can be estimated by common statistical estimators. One popular method is the maximum likelihood estimation. A new data point is then tested whether it belongs to the estimated distribution.

Another method to find anomalies are Isolation Forests (Liu et al. 2008). This approach is based on the Random Forest algorithm. The latter one uses decision trees to find features which are helpful to differ between certain classes in which the data has to be divided. The Isolation Forest algorithm is based on the intuitive assumption that outliers are simpler to separate from the normal data. Thus, if it needs only a few features to separate a new data point from the given data, this new point is more likely an anomaly.

SVM or support vector machines are used for finding hyperplanes which separate two classes such that the margin from the hyperplane to each class is maximised (Vapnik 1999). In general to use SVMs for data which cannot be separated with a linear function, the data is transformed to a higher dimensional space. A so-called one-class SVM is the basis for anomaly detection with SVMs. This one-class SVM is trained in such a way that the given data is separated from everything else, i.e. a decision boundary around the given data is computed (Schölkopf et al. 1999). Thus, new data points which are outside of this decision boundary are more likely to be an anomaly than data points which are located inside the decision boundary.

Many other approaches are based on metrics to score data. Based on these anomaly scores, anomalous data can be identified. One of such score is the local outlier factor (LOF; Breunig et al. 2000). This factor measures "how isolated the object is with respect to the surrounding neighbourhood" (Breunig et al. 2000). It takes the local density around a certain data point as well as the global density into account. In particular, a data point which is in a dense surrounding and easily reachable is rather not an anomaly compared with a data point which is in a more sparse neighbourhood and harder to reach.

Those described approaches are classical machine learning approaches not using any kind of neural network structure. However, deep learning, i.e. the usage of artificial neural networks (see also Sect. 10.1), offers some advantage over classical methods: Deep learning can handle high dimensional and multivariate data as well as it is capable of finding complex patterns or non-linear relations within the data. A typical neural network is relatively easy to deploy as a developer does not need that much pre-knowledge about the certain use case. The hyperparameters of a neural network to be chosen are, for example, the type and number of neurons, layers, and some hyperparameters for the training process like regularisation parameters or parameters of the optimisation algorithm. With that, one can already obtain a relatively good working model.

A typical neural network based approach for finding anomalies are auto-encoder (AE; Hinton and Salakhutdinov 2006). AE are special neural networks which are trained to learn the identity map between inputs and outputs. Therefore, the training is unsupervised as the label of each data point is the given data itself.

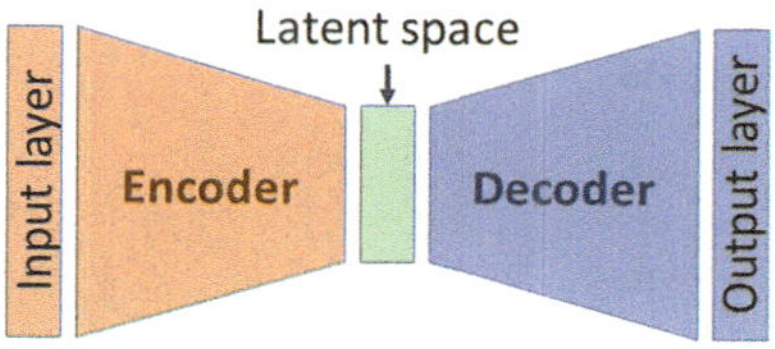

Fig. 10.10 Schematic structure of an auto-encoder (AE)

A schematic visualisation of a typical AE can be found in Fig. 10.10. Each AE consists of an encoder and a decoder. The encoder is responsible to embed the input to a lower dimensional latent space while the decoder maps data from the latent space back to the input dimensions. The choice of suitable layers of the encoder and decoder of an AE depends on the application. The basic way is to use dense layers. For analysing images, convolutional layers are often a better choice, because filters, which are learned, take the spatial structure into account. Often, the structure of the decoder is nearly the reversed structure of the encoder. To include time-dependency in particular, certain layers suitable for sequential data can be used. Typical examples for that type of architecture are recurrent neural networks (RNN), gated recurrent units (GRU) or long short-term memories (LSTM). Auto-encoders not only learn to reduce dimensionality (see also Sect. 10.6.1). They further are able to find outliers in the data. One possibility is to use the reconstruction error between the input and predicted output. If it is above a special threshold, for example defined by a value which takes the training or validation loss into account, the data point is probably an anomaly. The other possibility is to check whether the encoded data point in the lower dimensional latent space is an outlier compared to the trained data. That can be done, for example, with the above mentioned traditional anomaly detection methods which work well on less dimensions.

10.6.2.3 Applications of Anomaly Detection in Time Domain Astronomy

Anomaly Detection in Transit Observations

As mentioned in Sect. 9.3, Time domain astronomy aims for exploring the changes in a wide range of cosmic objects like exoplanets, novae, supernovae, active galactic nuclei, gamma ray bursts, pulsars and binary stars. A light curve is used to study the changes in the celestial objects observed by the transit observations. There has been an exponential growth in the volumes of data generated in time-domain astronomy over the last two decades. Many transit surveys like *Sloan Digital Sky Survey* (SDSS; York et al. 2000), *Kepler* (Borucki et al. 2008) and *TESS* (Ricker et al. 2015) space telescopes have contributed to the progress of time domain (transit) observations. The results from the future advanced surveys like *Legacy Survey of Space and Time* (LSST; Ivezic et al. 2009) and *Square Kilometre Array* (SKA; Dewdney et al. 2009) might push the data volumes to a much higher orders of magnitude in the range of zettabytes per year. With the increase in the data volumes, automated

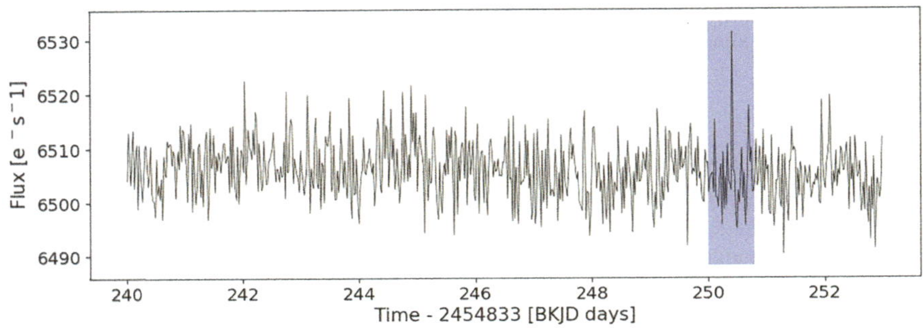

Fig. 10.11 Cosmic ray detection in the light curve observed in Kepler KIC 2436365 plotted using Lightkurve package (Lightkurve Collaboration et al. 2018). The cosmic ray event is highlighted in blue

detection techniques especially anomaly detection has become more relevant in transient analysis. Even though the discussions in the following section deals with optical surveys, the approach of anomaly detection can be applied to all wavelength domains of time-series observations notably radio astronomy surveys which are data intensive. One of the areas where transit surveys are extensively used is exoplanet detection.

The transit detection of exoplanets have been revolutionised with the advent of *telescope, Kepler* (Borucki et al. 2008) and *TESS* (Ricker et al. 2015) space telescopes. The light curve data from these telescopes would require expert data processing for the detection of exoplanets and other relevant information. Most of the survey telescopes have a dedicated data analysis pipeline for the noise removal and light curve extractions (Jenkins et al. 2010). In order to handle the huge volumes of data, Transit model-fitting and multiple planet search algorithms are implemented to analyse the final data from the pipeline (Li et al. 2019). For the detection of the planets in a precise way, the primary step of the pre-processing pipeline is to remove the outliers and then correct for the systematics in the data. Outliers occur in various forms like cosmic rays, bad pixels and argabrightening events (Jenkins et al. 2010). Whereas systematic errors occur due to the optical distortions, thermal transients and sudden pixel sensitivity dropouts. Each outlier and systematic errors has to be addressed separately. For instance, one of the most common outlier is cosmic rays which is detected as the very high fluctuations in the flux observed over a short timescales. Most of the data processing pipelines remove the cosmic ray hits. In the case of Kepler data the Presearch Data Conditioning (PDC) pipeline (Twicken et al. 2010) removes most of the high intensity cosmic ray hits, but in some cases the outliers might not be caught during the data reduction. The cosmic ray shown in Fig. 10.11 is one such anomaly, as also discussed in *lightkurve* package documentation.[7]

[7] https://docs.lightkurve.org/index.html.

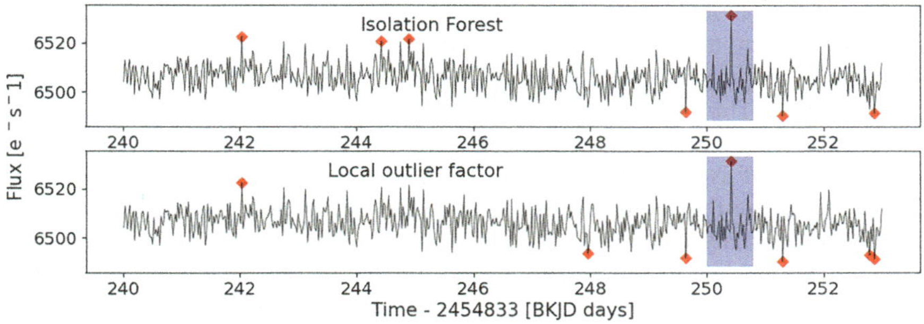

Fig. 10.12 Anomalies detected using isolation forest method (top) and local outlier factor (bottom). The anomalies detected in the data are highlighted by red dots. The blue highlighted region is the cosmic ray events in the data. Both the algorithms detected that event as an anomaly with a high score. It can also be noted that the two approaches highlight different anomalies

An example is presented in Fig. 10.12, where two anomaly detection algorithms *Isolation Forest* and *Local outlier Factor* were used to evaluate if the cosmic ray outliers can be predicted. For this purpose, the models present in ML libraries like `scikit learn` (Pedregosa et al. 2012) and `pyod` (Zhao et al. 2019) were utilised. The outlier is detected and also the highest anomaly score is assigned to the cosmic ray event among the other anomalies.

New Approaches in Anomaly Detection for Transit Observations

As seen in Fig. 10.12, the anomalies identified by the two methods are different as the algorithms might consider different points as anomalous based on the implementation and the logic behind it. There is no one single state-of-the-art algorithm which predicts all anomalies present in the data. Also, many of the anomalies detected might not be of interest to all end users. For instance, artefacts of the instrument can be of interest for the instrumentation engineer, whereas astrophysicists could find the astronomical transient outliers interesting. Apart from this, the data quality control team would be interested in the artefacts in the data altogether. Hence the relevance in the anomaly detection is dependent to the interest of the user. Any typical machine learning algorithm cannot differentiate between relevant and irrelevant anomalies. This need resulted in the development of a classification algorithm to search for the anomaly of interest, resulting in the development of the *Astronomaly* package (Lochner and Bassett 2021). This framework fuses active learning with anomaly detection on various kinds of data including images, spectra and other time series data. The package has a python backend that handles the data pre-processing, feature extraction and various anomaly detection algorithms. Whereas the frontend is used for user interaction for scoring the objects of interest/relevance for the anomaly detection. This novel approach incorporates the user relevance score with the anomaly detection scores for identifying anomalies of interest. This approach was demonstrated on Galaxy Zoo dataset (Lintott et al. 2008). Along with images, *Astronomaly* package was also further applied on light curves for anomaly detection (Webb et al. 2020; Lochner and Bassett 2021). This resulted in identifying previously undiscovered stellar flares, classical flares and other objects of interest.

The future of astronomical surveys demands the ability to investigate the light curves to explore the anomalies or objects of interest. There is a huge potential in combining anomaly detection with stellar object classification techniques to find unidentified transient variables on big (transit) data. Efforts have also been made to apply anomaly detection techniques on transit data of astronomy for large transient surveys (Martínez-Galarza et al. 2021). In particular, *Astronomaly* framework was built to handle large data volumes from telescopes such as the SKA (Lochner and Bassett 2021).

10.6.3 Applications of Anomaly Detection in Radio Surveys Using a Complexity Based Approach

Detecting the Unexpected in Large-Scale Radio Surveys

Next-generation (NxGen) large-scale radio surveys such as the Evolutionary Map of the Universe (EMU; Norris et al. 2011), LOFAR Two-metre Sky Survey (LoTSS; Shimwell et al. 2022, 2017) and the Square Kilometre Array (Dewdney et al. 2009, SKA) will expand and explore new regions of the Observable Parameter Space, presenting the opportunity for new and unexpected discoveries.

There is a risk that source extraction algorithms and observer expectations may overlook unexpected new discoveries that do not align with current models, past experience or planned science goals (Robinson 1987; Norris 2017). A potential solution is to explicitly search for the unexpected in data before it is discarded. Due to the large volume of data coming from large-scale radio surveys it is not feasible for scientists to look for unexpected sources by eye (e.g., Norris 2017).

The Coarse-grained complexity (Segal et al. 2019, 2023) is an intrinsic property of an observation, meaning it is a quantity that can be determined for an observation of interest without needing to determine the complexity of other observations or the distances between them. This is distinct from extrinsic properties such as being an outlier or anomaly which depend on relations to other observations.

This makes complexity a useful quantity in two complimentary ways:

1. it can be compared with other complexity values, obtained at the time of analysis or gradually over time, to determine if the observation is highly complex or an anomaly, and
2. it is the property of a single observation and can be determined without reference to a sample. This is distinct from the traditional unsupervised anomaly detection which requires comparisons with a sample.

Furthermore the running time for computing the Coarse-grained complexity scales linearly with increasing sample size.

These attributes make Coarse-grained complexity a useful tool for finding anomalies in new and large data that extends coverage of the Observable Parameter Space. Unsupervised learning algorithms that rely on 'between member' operations within a sample (such as the classic k nearest neighbours and Self-organising Map Algorithms) will often have running times that scale quadratically $O(N^2)$ with the increasing sample size. Supervised learning approaches run into issues as they rely on learning from the features of past observations which may differ significantly from the features of unexpected future discoveries.

To avoid expectation bias, complexity does not require existing catalogue data and does not learn specific features from past observations. Rather it has general parameters that extend across different samples and surveys (Segal et al. 2019, 2023).

The Coarse-Grained Complexity

The Coarse-grained complexity, $C(f(x))$, is calculated by applying a lossless data compression algorithm C to the description of an observation x after applying a smoothing function f. The issue with using the compressed file size directly as measure of complexity is that it is maximised by random information. The Coarse-grained complexity of a random sequence becomes small with smoothing, as the smoothed information content becomes homogeneous at the coarse-grained resolution. The measure will return larger values for data with structure and regularities and small values where the data is either simple or random.

The lossless data compression provides an upper bound approximation of the uncomputable Kolmogorov complexity of x, that is the length of the shortest binary program $l(p)$, for the reference universal prefix Turing machine U, that outputs x; denoted as $K(x)$:

$$K(x) = min_p\{l(p) : U(p) = x\} .$$

The Kolmogorov complexity has the advantage can be defined for a particular description of an observation, such as a radio image of a galaxy, and can serve as a proxy for the Shannon entropy which is defined in terms of the unknown distribution P for possible states of the system (Aaronson et al. 2014):

$$H(P) = -\sum_{x \in X} P(x) \log P(x) .$$

The approach is based on the implementation of *apparent complexity* (Aaronson et al. 2014) and the notion of *effective complexity* (Gell-Mann 1994; Gell-Mann and Lloyd 1996), which can be defined as the information required to describe a system's regularities.

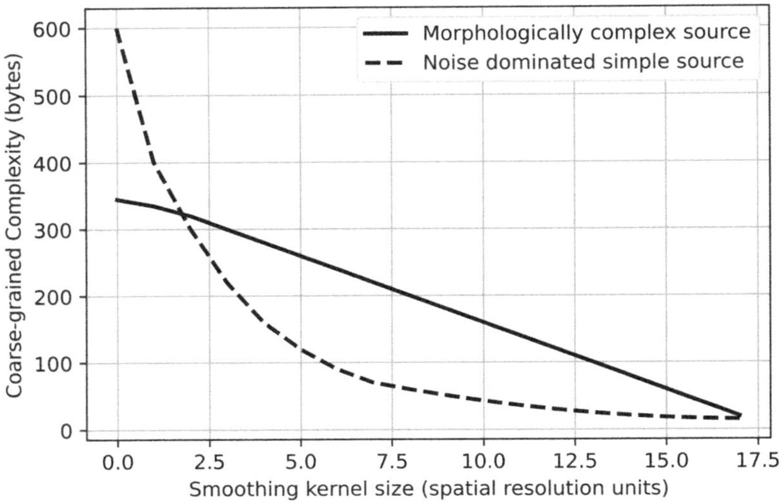

Fig. 10.13 A comparison of the Coarse-grained complexity between morphologically complex sources (e.g. multiple components, tails, etc.) and noise dominated simple sources (i.e. unresolved) across changes in the size of the smoothing kernel. The complexity measurements for morphologically complex sources decrease more gradually as the size of the smoothing kernel increases, due to the preservation of information at a coarser resolution as a result of macroscopic structure

A key additional step in the implementation of the Coarse-grained complexity, is the calibration of the size of the smoothing kernel to produce maximum separation between the complexity values of morphologically complex observations (e.g. multiple components, tails, etc) and noise dominated simple observations (i.e. unresolved sources), where complex sources have higher values, as illustrated in Fig. 10.13. As the size of the smoothing kernel increases, the Coarse-grained complexity of morphologically complex and noise dominated simple sources decrease at different rates. The complexity measurements for morphologically complex sources decrease more gradually as the size of the smoothing kernel increases, due to the preservation of information at a coarser resolution as a result of macroscopic structure.

The Coarse-grained complexity can be calibrated to align with expert distinctions between meaningful structure and noise by adjusting the measurement resolution through smoothing. The measure does not rely on the presence of any particular structures or structural elements, it rather makes only explicit assumptions regarding the choice of coarse-grained level with respect to the scale of the image. Previous data is therefore used only to calibrate the coarse-grained level (i.e. the appropriate measurement resolution).

After being calibrated on the ATLAS DR1 CDFS (Chandra Deep Field-South Survey) sample of only approximately 708 observations (Segal et al. 2019), the approach has been shown to generalise well, identifying complex and anomalous observations in the larger ATLAS DR3 CDFS (n=2861) and ELAIS (n=1964) (Segal et al. 2019) and from scanning the EMU Pilot Survey (EMU-PS) data (365,000 sampled frames containing at least the ~220,000 *Selavy* catalogue sources; Segal et al. 2023). The effectiveness of the EMU-PS complexity scan was evaluated using crowd-sourced labels generated via the Zooniverse.org platform demonstrating that the high-value tail of the complexity distribution comprises many unusual and complex objects (Segal et al. 2023). Extended sources featured heavily in the tail above the 95th percentile, with objects of a more anomalous appearance, such as odd radio circles (Norris et al. 2021), apparent from above the 99th percentile (Segal et al. 2023).

10.7 Interpretability of Machine Learning

While reproducibility and reusability have been emphasised when developing ML based data analysis pipelines in radio astronomy[8] that are essential for AI reliability, so as dimensionality reduction, uncertainty quantification and anomaly detection (see also Sects. 10.6.2 and 10.6.3); AI interpretability (Doshi-Velez and Kim 2017) as another factor of trustworthy machine learning, received less attention. In the context of machine learning, interpretability has following definition:

10.7.1 Interpretability

The ability to explain or to present in understandable terms to a human (Doshi-Velez and Kim 2017).

Some parametric machine learning techniques, such as gradient boosted trees, are fairly interpretable by themselves (Alger et al. 2021). That said, one can understand how a model makes predictions with little difficulty if the model is highly interpretable. If one wants to further investigate input feature importance of a typical model, eXplainable Artificial Intelligence (XAI) methods developed by academia and industry together, such as SHapley Additive exPlanation (SHAP; Lundberg and Lee 2017) pronounced by Su-In Lee's lab and Microsoft research department, can also be used to quantitatively compute how these features contribute to typical model prediction outcomes (Alegre et al. 2022).

Non-parametric algorithms, especially image based DL algorithms, however, usually require more effort to interpret, and are thought to be "black boxes". Recent progress to overcome this issue can be categorised into two general approaches:

1. increase model interpretability by changing model architecture.
2. apply post-hoc model explanation method to interpret model prediction results.

---

[8] https://sdc2.astronomers.skatelescope.org.

One example of changing model architecture is a recent radio galaxy classifier algorithm which introduced attention-gating technique to the model, enabling model user to explain model's Region of Interest (RoI) when predicting radio galaxy morphology type with a corresponding sample attention map (Bowles et al. 2021b).

Alternatively, post-hoc model explanation methods such as saliency mapping was implemented to enhance image data feature for pulsar and FRB hunting (Zhang et al. 2020). Considering applications in broader astronomy scheme, some attempts have also used XAI methods like Grad-CAM for model mis-classification diagnostics (Bhambra et al. 2022). After all, model interpretation methods are still under development, and awaited for standardisation.

10.8 Techniques Outside Radio Astronomy

Apart from the obvious difference that one domain (radio astronomy) studies the universe at the largest possible scales, while the other (particle physics) is concerned with the structure of elementary particles (particle physics), there are two key practical differences in how data is analysed in particle physics collider experiments and astronomy respectively.

First, particle colliders such as the Large Hadron Collider (LHC; Evans and Bryant 2008) produce discrete *events* that can be considered to be drawn independently from an underlying physical distribution corresponding to the standard model of particle physics (and potential physical theories beyond the standard model ΛCDM). In practice, events are the result of collisions between bunches of particles as recorded by complex detectors such as ALICE (ALICE Collaboration et al. 2008), ATLAS (ATLAS Collaboration et al. 2008), CMS (CMS Collaboration et al. 2008), or LHCb (ALICE Collaboration et al. 2008).

Each of these detectors possesses a large numbers of read-out channels (O(100M)) and subsystems sensitive to different types of particles that were produced in these collisions. For example, around one billion particle collisions take place in the CMS experiment per second, leading to a raw datarate of O(1 PB/second). Multi-stage selection lossy selection algorithms (triggers) reduce this rate down to Gb/s by only selecting potentially interesting events for storage and further processing. The first of these stages (termed the Level 1 trigger) needs to operate with microsecond latency. Among rule based selection algorithms, ML methods implemented on Field Programmable Gate Array (FPGA) hardware are employed (Duarte et al. 2018).

Second, highly sophisticated and problem specific chains of simulation tools (e.g. `Pythia`; Sjöstrand et al. (2015), `Herwig`; Bähr et al. (2008), `Sherpa`; Gleisberg et al. (2009) for the hard process and `Geant4`; Agostinelli et al. (2003) for subsequent interactions of particles with detectors) exist that allow producing realistic events either according to the standard model or alternate theories of physics beyond the standard model. Events produced by such simulations mimic *real* events—i.e. collider-data—to such a degree that the same subsequent reconstruction and processing steps can be applied to extract high-level physically meaningful features from low-level detector readouts. The availability of these simulations shapes how data analysis in collider-based physics is carried out as precise predictions of different processes can be compared to recorded data.

We can roughly distinguish between two broad scientific goals in collider experiments: direct searches for new particles—such as dark matter candidate particles—and precision measurements of standard model parameters (of course, deviations from theory-predictions in precision measurements also act as indirect evidence for new physics). Machine learning, and especially modern deep learning, plays an increasingly large role in both activities (see e.g. Karagiorgi et al. (2022) for a review of machine learning in searches and Carleo et al. (2019) for a general overview of machine learning in the physical sciences).

A prototypical machine learning task is identifying (*tagging*) the presence of specific particles (e.g. top quarks) in an event (see Kasieczka et al. 2019, for a community study of such top-taggers). Knowing that top quarks were present in an event can be used as a selection criterion to, for example, search for hypothetical new particles that subsequently decayed into top quarks. Such classifiers can be trained in a fully supervised fashion using simulated examples of top quarks and typical background particles and then applied to analyse collider data for which this truth information is a-priori unknown. Despite the high quality of simulation, there typically remains a simulation gap that can be addressed with additional calibration factors derived from data (Sirunyan et al. 2020).

How to best represent the symmetries of physical data in the architecture of deep learning networks is a widely studied topic and many approaches using from high-level observables to representing the low-level detector read-outs as images, lists, sets, or graphs have been considered (Kasieczka et al. 2019). Constructing graphs where individual nodes correspond to individual measured particles inside an event, and connections are built according to the geometrical distribution quickly emerged as a powerful solution to this problem (Qu and Gouskos 2020). For a general review of graph networks in particle physics see Shlomi et al. (2020). Recently, combining transformer networks (Qu et al. 2022) based on the attention mechanism (Vaswani et al. 2017) with graphs has shown additional improvements. Note that while the discussion focused on top quarks as an exemplary use-case, similar results also exist for other types of particles.

While searches for new physics signals were traditionally carried out in a model-dependent fashion (e.g. by training a supervised classifier for a specific hypothetical particle), machine-learning based anomaly detection (see also Sects. 10.6.2 and 10.6.3) offers an alternative strategy (Kasieczka et al. 2021; Aarrestad et al. 2022; D'Agnolo and Wulzer 2019).

One such strategy is to search for new physical particles that are resonant (i.e. produce a peak over a smoothly falling background when considering an invariant mass distribution over a set of events). First, a slice of the invariant mass spectrum is chosen to test for new physics (termed signal region) while treating the others as background dominated (termed sideband). Unsupervised learning algorithms (e.g. auto-encoders; Heimel et al. 2019; Farina et al. 2020; ATLAS Collaboration 2022a) can then be trained in the sideband and evaluated in the signal region, with a change in loss indicating the presence of a signal. Alternatively, a classifier between signal region and sideband can be used to identify new physics present in the signal region (Collins et al. 2018, 2019; Aad et al. 2020).

In both cases, the machine learning algorithms are trained using additional features of the data as input and problems arise if these features are correlated with the mass. For example the signal region/sideband classifier could learn to distinguish these regions based on the correlation profile of features for the background, and not due to the presence of a signal. See Nachman and Shih (2020), Hallin et al. (2022a,b), Raine et al. (2022), for recent results mitigating this issue by using a generative model such as a conditional normalising flow (Papamakarios et al. 2019)—i.e. a trainable diffeomorphism between data and a well-defined latent space—to produce an in-situ estimate of the background in the signal region learned from data. Note that similar techniques also already found application outside particle physics for the measurement of galactic dark matter (Buckley et al. 2023) or finding stellar streams (Shih et al. 2022).

Beyond their use in searches for new particles, generative models in particle physics are also widely considered as a possible strategy to augment the classical simulation tools discussed above. See Butter et al. (2022), Alanazi et al. (2021) and Adelmann et al. (2022) for recent overviews of such generative models for hard-interactions and complex detectors, respectively and ATLAS Collaboration (2022b) for a concrete application by the ATLAS collaboration. Finally, Butter et al. (2021), Bieringer et al. (2022) investigate the statistical properties of generative models applied to physical data.

While the above discussion focused on a few specific applications and limited examples of possible interest, machine learning in particle physics is a much broader endeavour and also covers topics such as statistical inference methods (Brehmer et al. 2019), reconstructing events from detector readouts (Pata et al. 2023), or monitoring detector operations (Pol et al. 2018). The *Living Review*[9] provides a well-organised entry to the literature.

As an example for synergies, for large astronomical surveys to be useful for astrophysical studies, sources need to be classified. Radio surveys of more than one million sources are completely changing the analysis requirements. Development of reliable classification algorithms is an active field of research with the goal to develop classifiers capable of probabilistic class membership assignment for very large datasets. Examples are image-based classification with gradient boosting, decision trees and basic CNN architectures (Euclid Collaboration et al. 2022; Alegre et al. 2022). Modern ML-methods can be adapted from the LHC environment to the specific requirements of radio astronomical data, defining an efficient path to cutting-edge ML-classification there and generating synergies through the feedback from astrophysics to LHC.

---

[9] https://iml-wg.github.io/HEPML-LivingReview/.

Of course, machine learning developments from particle physics will, in most cases, not be directly applicable to challenges in modern radio astronomy. However, there are enough common issues such as the unsupervised discovery of interesting properties in data; the efficient use of scarce experimental observations; the well-motivated inclusion of symmetry properties; the proper accounting of uncertainties and other sources of noise; and finally the great similarity in fundamental physical questions to motivate a close exchange between these two domains.

References

G. Aad, B. Abbott, D.C. Abbott, A. Abed Abud, K. Abeling, D.K. Abhayasinghe, S.H. Abidi, O.S. Abouzeid, N.L. Abraham, H. Abramowicz, H. Abreu, Y. Abulaiti, B.S. Acharya, B. Achkar, L. Adam, C. Adam Bourdarios, L. Adamczyk, L. Adamek, J. Adelman, M. Adersberger, A. Adiguzel, S. Adorni, T. Adye, A.A. Affolder, Y. Afik, C. Agapopoulou, M.N. Agaras, A. Aggarwal, C. Agheorghiesei, J.A. Aguilar-Saavedra, A. Ahmad, F. Ahmadov, W.S. Ahmed, X. Ai, G. Aielli, S. Akatsuka, M. Akbiyik, T.P.A. Åkesson, E. Akilli, A.V. Akimov, K. Al Khoury, G.L. Alberghi, J. Albert, M.J. Alconada Verzini, S. Alderweireldt, M. Aleksa, I.N. Aleksandrov, C. Alexa, T. Alexopoulos, A. Alfonsi, F. Alfonsi, M. Alhroob, B. Ali, S. Ali, M. Aliev, G. Alimonti, C. Allaire, B.M.M. Allbrooke, B.W. Allen, P.P. Allport, A. Aloisio, F. Alonso, C. Alpigiani, E. Alunno Camelia, M. Alvarez Estevez, M.G. Alviggi, Y. Amaral Coutinho, A. Ambler, L. Ambroz, C. Amelung, D. Amidei, S.P. Amor Dos Santos, S. Amoroso, C.S. Amrouche, F. An, C. Anastopoulos, N. Andari, T. Andeen, J.K. Anders, S.Y. Andrean, A. Andreazza, V. Andrei, C.R. Anelli, S. Angelidakis, A. Angerami, A.V. Anisenkov, A. Annovi, C. Antel, M.T. Anthony, E. Antipov, M. Antonelli, D.J.A. Antrim, F. Anulli, M. Aoki, J.A. Aparisi Pozo, M.A. Aparo, L. Aperio Bella, N. Aranzabal Barrio, V. Araujo Ferraz, R. Araujo Pereira, C. Arcangeletti, A.T.H. Arce, F.A. Arduh, J.F. Arguin, S. Argyropoulos, J.H. Arling, A.J. Armbruster, A. Armstrong, O. Arnaez, H. Arnold, Z.P. Arrubarrena Tame, G. Artoni, K. Asai, S. Asai, T. Asawatavonvanich, N. Asbah, E.M. Asimakopoulou, L. Asquith, J. Assahsah, K. Assamagan, R. Astalos, R.J. Atkin, M. Atkinson, N.B. Atlay, H. Atmani, K. Augsten, V.A. Austrup, G. Avolio, M.K. Ayoub, G. Azuelos, H. Bachacou, K. Bachas, M. Backes, F. Backman, P. Bagnaia, M. Bahmani, H. Bahrasemani, A.J. Bailey, V.R. Bailey, J.T. Baines, C. Bakalis, O.K. Baker, P.J. Bakker, E. Bakos, D. Bakshi Gupta, S. Balaji, E.M. Baldin, P. Balek, F. Balli, W.K. Balunas, J. Balz, E. Banas, M. Bandieramonte, A. Bandyopadhyay, Sw. Banerjee, L. Barak, W.M. Barbe, E.L. Barberio, D. Barberis, M. Barbero, G. Barbour, T. Barillari, M.S. Barisits, J. Barkeloo, T. Barklow, R. Barnea, B.M. Barnett, R.M. Barnett, Z. Barnovska-Blenessy, A. Baroncelli, G. Barone, A.J. Barr, L. Barranco Navarro, F. Barreiro, J. Barreiro Guimarães da Costa, U. Barron, S. Barsov, F. Bartels, R. Bartoldus, G. Bartolini, A.E. Barton, P. Bartos, A. Basalaev, A. Basan, A. Bassalat, M.J. Basso, R.L. Bates, S. Batlamous, J.R. Batley, B. Batool, M. Battaglia, M. Bauce, F. Bauer, K.T. Bauer, P. Bauer, H.S. Bawa, A. Bayirli, J.B. Beacham, T. Beau, P.H. Beauchemin, F. Becherer, P. Bechtle, H.C. Beck, H.P. Beck, K. Becker, C. Becot, A. Beddall, A.J. Beddall, V.A. Bednyakov, M. Bedognetti, C.P. Bee, T.A. Beermann, M. Begalli, M. Begel, A. Behera, J.K. Behr, F. Beisiegel, M. Belfkir, A.S. Bell, G. Bella, L. Bellagamba, A. Bellerive, P. Bellos, K. Beloborodov, K. Belotskiy, N.L. Belyaev, D. Benchekroun, N. Benekos, Y. Benhammou, D.P. Benjamin, M. Benoit, J.R. Bensinger, S. Bentvelsen, L. Beresford, M. Beretta, D. Berge, E. Bergeaas Kuutmann, N. Berger, B. Bergmann, L.J. Bergsten, J. Beringer, S. Berlendis, G. Bernardi, C. Bernius, F.U. Bernlochner, T. Berry, P. Berta, C. Bertella, A. Berthold, I.A. Bertram, O. Bessidskaia Bylund, N. Besson, A. Bethani, S. Bethke, A. Betti, A.J. Bevan, J. Beyer, D.S. Bhattacharya, P. Bhattarai, V.S. Bhopatkar, R. Bi, R.M. Bianchi, O. Biebel,

D. Biedermann, R. Bielski, K. Bierwagen, N.V. Biesuz, M. Biglietti, T.R.V. Billoud, M. Bindi, A. Bingul, C. Bini, S. Biondi, C.J. Birch-Sykes, M. Birman, T. Bisanz, J.P. Biswal, D. Biswas, A. Bitadze, C. Bittrich, K. Bjørke, T. Blazek, I. Bloch, C. Blocker, A. Blue, U. Blumenschein, G.J. Bobbink, V.S. Bobrovnikov, S.S. Bocchetta, D. Boerner, D. Bogavac, A.G. Bogdanchikov, C. Bohm, V. Boisvert, P. Bokan, T. Bold, A.E. Bolz, M. Bomben, M. Bona, J.S. Bonilla, M. Boonekamp, C.D. Booth, A.G. Borbély, H.M. Borecka-Bielska, L.S. Borgna, A. Borisov, G. Borissov, D. Bortoletto, D. Boscherini, M. Bosman, J.D. Bossio Sola, K. Bouaouda, J. Boudreau, E.V. Bouhova-Thacker, D. Boumediene, S.K. Boutle, A. Boveia, J. Boyd, D. Boye, I.R. Boyko, A.J. Bozson, J. Bracinik, N. Brahimi, G. Brandt, O. Brandt, F. Braren, B. Brau, J.E. Brau, W.D. Breaden Madden, K. Brendlinger, R. Brener, L. Brenner, R. Brenner, S. Bressler, B. Brickwedde, D.L. Briglin, D. Britton, D. Britzger, I. Brock, R. Brock, G. Brooijmans, W.K. Brooks, E. Brost, P.A. Bruckman de Renstrom, B. Brüers, D. Bruncko, A. Bruni, G. Bruni, L.S. Bruni, S. Bruno, M. Bruschi, N. Bruscino, L. Bryngemark, T. Buanes, Q. Buat, P. Buchholz, A.G. Buckley, I.A. Budagov, M.K. Bugge, F. Bührer, O. Bulekov, B.A. Bullard, T.J. Burch, S. Burdin, C.D. Burgard, A.M. Burger, B. Burghgrave, J.T.P. Burr, C.D. Burton, J.C. Burzynski, V. Büscher, E. Buschmann, P.J. Bussey, J.M. Butler, C.M. Buttar, J.M. Butterworth, P. Butti, W. Buttinger, C.J. Buxo Vazquez, A. Buzatu, A.R. Buzykaev, G. Cabras, S. Cabrera Urbán, D. Caforio, H. Cai, V.M.M. Cairo, O. Cakir, N. Calace, P. Calafiura, G. Calderini, P. Calfayan, G. Callea, L.P. Caloba, A. Caltabiano, S. Calvente Lopez, D. Calvet, S. Calvet, T.P. Calvet, M. Calvetti, R. Camacho Toro, S. Camarda, D. Camarero Munoz, P. Camarri, M.T. Camerlingo, D. Cameron, C. Camincher, S. Campana, M. Campanelli, A. Camplani, V. Canale, A. Canesse, M. Cano Bret, J. Cantero, T. Cao, Y. Cao, M.D.M. Capeans Garrido, M. Capua, R. Cardarelli, F. Cardillo, G. Carducci, I. Carli, T. Carli, G. Carlino, B.T. Carlson, E.M. Carlson, L. Carminati, R.M.D. Carney, S. Caron, E. Carquin, S. Carrá, G. Carratta, J.W.S. Carter, T.M. Carter, M.P. Casado, A.F. Casha, F.L. Castillo, L. Castillo Garcia, V. Castillo Gimenez, N.F. Castro, A. Catinaccio, J.R. Catmore, A. Cattai, V. Cavaliere, V. Cavasinni, E. Celebi, F. Celli, K. Cerny, A.S. Cerqueira, A. Cerri, L. Cerrito, F. Cerutti, A. Cervelli, S.A. Cetin, Z. Chadi, D. Chakraborty, J. Chan, W.S. Chan, W.Y. Chan, J.D. Chapman, B. Chargeishvili, D.G. Charlton, T.P. Charman, C.C. Chau, S. Che, S. Chekanov, S.V. Chekulaev, G.A. Chelkov, B. Chen, C. Chen, C.H. Chen, H. Chen, J. Chen, J. Chen, J. Chen, S. Chen, S.J. Chen, X. Chen, Y. Chen, Y.H. Chen, H.C. Cheng, H.J. Cheng, A. Cheplakov, E. Cheremushkina, R. Cherkaoui El Moursli, E. Cheu, K. Cheung, T.J.A. Chevalérias, L. Chevalier, V. Chiarella, G. Chiarelli, G. Chiodini, A.S. Chisholm, A. Chitan, I. Chiu, Y.H. Chiu, M.V. Chizhov, K. Choi, A.R. Chomont, Y.S. Chow, L.D. Christopher, M.C. Chu, X. Chu, J. Chudoba, J.J. Chwastowski, L. Chytka, D. Cieri, K.M. Ciesla, D. Cinca, V. Cindro, I.A. Cioară, A. Ciocio, F. Cirotto, Z.H. Citron, M. Citterio, D.A. Ciubotaru, B.M. Ciungu, A. Clark, M.R. Clark, P.J. Clark, S.E. Clawson, C. Clement, Y. Coadou, M. Cobal, A. Coccaro, J. Cochran, R. Coelho Lopes de Sa, H. Cohen, A.E.C. Coimbra, B. Cole, A.P. Colijn, J. Collot, P. Conde Muiño, S.H. Connell, I.A. Connelly, S. Constantinescu, F. Conventi, A.M. Cooper-Sarkar, F. Cormier, K.J.R. Cormier, L.D. Corpe, M. Corradi, E.E. Corrigan, F. Corriveau, M.J. Costa, F. Costanza, D. Costanzo, G. Cowan, J.W. Cowley, J. Crane, K. Cranmer, R.A. Creager, S. Crépé-Renaudin, F. Crescioli, M. Cristinziani, V. Croft, G. Crosetti, A. Cueto, T. Cuhadar Donszelmann, H. Cui, A.R. Cukierman, W.R. Cunningham, S. Czekierda, P. Czodrowski, M.M. Czurylo, M.J. da Cunha Sargedas de Sousa, J.V. da Fonseca Pinto, C. da Via, W. Dabrowski, F. Dachs, T. Dado, S. Dahbi, T. Dai, C. Dallapiccola, M. Dam, G. D'Amen, V. D'Amico, J. Damp, J.R. Dandoy, M.F. Daneri, M. Danninger, V. Dao, G. Darbo, O. Dartsi, A. Dattagupta, T. Daubney, S. D'Auria, C. David, T. Davidek, D.R. Davis, I. Dawson, K. de, R. de Asmundis, M. de Beurs, S. de Castro, N. de Groot, P. de Jong, H. de la Torre, A. de Maria, D. de Pedis, A. de Salvo, U. de Sanctis, M. de Santis, A. de Santo, J.B. de Vivie de Regie, C. Debenedetti, D.V. Dedovich, A.M. Deiana, J. Del Peso, Y. Delabat Diaz, D. Delgove, F. Deliot, C.M. Delitzsch, M. Della Pietra, D. Della Volpe, A. Dell'Acqua, L. Dell'Asta, M. Delmastro, C. Delporte, P.A. Delsart, D.A. Demarco, S. Demers, M. Demichev, G. Demontigny, S.P. Denisov, L. D'Eramo, D. Derendarz, J.E. Derkaoui, F. Derue, P. Dervan, K. Desch, K. Dette, C. Deutsch, M.R. Devesa, P.O. Deviveiros, F.A. di Bello, A. di Ciaccio, L. di Ciaccio, W.K. di Clemente, C. di

Donato, A. di Girolamo, G. di Gregorio, B. di Micco, R. di Nardo, K.F. di Petrillo, R. di Sipio, C. Diaconu, F.A. Dias, T. Dias Do Vale, M.A. Diaz, F.G. Diaz Capriles, J. Dickinson, M. Didenko, E.B. Diehl, J. Dietrich, S. Díez Cornell, C. Diez Pardos, A. Dimitrievska, W. Ding, J. Dingfelder, S.J. Dittmeier, F. Dittus, F. Djama, T. Djobava, J.I. Djuvsland, M.A.B. Do Vale, M. Dobre, D. Dodsworth, C. Doglioni, J. Dolejsi, Z. Dolezal, M. Donadelli, B. Dong, J. Donini, A. D'Onofrio, M. D'Onofrio, J. Dopke, A. Doria, M.T. Dova, A.T. Doyle, E. Drechsler, E. Dreyer, T. Dreyer, A.S. Drobac, D. Du, T.A. Du Pree, Y. Duan, F. Dubinin, M. Dubovsky, A. Dubreuil, E. Duchovni, G. Duckeck, O.A. Ducu, D. Duda, A. Dudarev, A.C. Dudder, E.M. Duffield, M. D'Uffizi, L. Duflot, M. Dührssen, C. Dülsen, M. Dumancic, A.E. Dumitriu, M. Dunford, A. Duperrin, H. Duran Yildiz, M. Düren, A. Durglishvili, D. Duschinger, B. Dutta, D. Duvnjak, G.I. Dyckes, M. Dyndal, S. Dysch, B.S. Dziedzic, M.G. Eggleston, T. Eifert, G. Eigen, K. Einsweiler, T. Ekelof, H. El Jarrari, V. Ellajosyula, M. Ellert, F. Ellinghaus, A.A. Elliot, N. Ellis, J. Elmsheuser, M. Elsing, D. Emeliyanov, A. Emerman, Y. Enari, M.B. Epland, J. Erdmann, A. Ereditato, P.A. Erland, M. Errenst, M. Escalier, C. Escobar, O. Estrada Pastor, E. Etzion, H. Evans, M.O. Evans, A. Ezhilov, F. Fabbri, L. Fabbri, V. Fabiani, G. Facini, R.M. Fakhrutdinov, S. Falciano, P.J. Falke, S. Falke, J. Faltova, Y. Fang, Y. Fang, G. Fanourakis, M. Fanti, M. Faraj, A. Farbin, A. Farilla, E.M. Farina, T. Farooque, S.M. Farrington, P. Farthouat, F. Fassi, P. Fassnacht, D. Fassouliotis, M. Faucci Giannelli, W.J. Fawcett, L. Fayard, O.L. Fedin, W. Fedorko, A. Fehr, M. Feickert, L. Feligioni, A. Fell, C. Feng, M. Feng, M.J. Fenton, A.B. Fenyuk, S.W. Ferguson, J. Ferrando, A. Ferrante, A. Ferrari, P. Ferrari, R. Ferrari, D.E. Ferreira de Lima, A. Ferrer, D. Ferrere, C. Ferretti, F. Fiedler, A. Filipčič, F. Filthaut, K.D. Finelli, M.C.N. Fiolhais, L. Fiorini, F. Fischer, J. Fischer, W.C. Fisher, T. Fitschen, I. Fleck, P. Fleischmann, T. Flick, B.M. Flierl, L. Flores, L.R. Flores Castillo, F.M. Follega, N. Fomin, J.H. Foo, G.T. Forcolin, B.C. Forland, A. Formica, F.A. Förster, A.C. Forti, E. Fortin, M.G. Foti, D. Fournier, H. Fox, P. Francavilla, S. Francescato, M. Franchini, S. Franchino, D. Francis, L. Franco, L. Franconi, M. Franklin, G. Frattari, A.N. Fray, P.M. Freeman, B. Freund, W.S. Freund, E.M. Freundlich, D.C. Frizzell, D. Froidevaux, J.A. Frost, M. Fujimoto, C. Fukunaga, E. Fullana Torregrosa, T. Fusayasu, J. Fuster, A. Gabrielli, A. Gabrielli, S. Gadatsch, P. Gadow, G. Gagliardi, L.G. Gagnon, G.E. Gallardo, E.J. Gallas, B.J. Gallop, R. Gamboa Goni, K.K. Gan, S. Ganguly, J. Gao, Y. Gao, Y.S. Gao, F.M. Garay Walls, C. García, J.E. García Navarro, J.A. García Pascual, C. Garcia-Argos, M. Garcia-Sciveres, R.W. Gardner, N. Garelli, S. Gargiulo, C.A. Garner, V. Garonne, S.J. Gasiorowski, P. Gaspar, A. Gaudiello, G. Gaudio, I.L. Gavrilenko, A. Gavrilyuk, C. Gay, G. Gaycken, E.N. Gazis, A.A. Geanta, C.M. Gee, C.N.P. Gee, J. Geisen, M. Geisen, C. Gemme, M.H. Genest, C. Geng, S. Gentile, S. George, T. Geralis, L.O. Gerlach, P. Gessinger-Befurt, G. Gessner, S. Ghasemi, M. Ghasemi Bostanabad, M. Ghneimat, A. Ghosh, A. Ghosh, B. Giacobbe, S. Giagu, N. Giangiacomi, P. Giannetti, A. Giannini, G. Giannini, S.M. Gibson, M. Gignac, D.T. Gil, B.J. Gilbert, D. Gillberg, G. Gilles, D.M. Gingrich, M.P. Giordani, P.F. Giraud, G. Giugliarelli, D. Giugni, F. Giuli, S. Gkaitatzis, I. Gkialas, E.L. Gkougkousis, P. Gkountoumis, L.K. Gladilin, C. Glasman, J. Glatzer, P.C.F. Glaysher, A. Glazov, G.R. Gledhill, I. Gnesi, M. Goblirsch-Kolb, D. Godin, S. Goldfarb, T. Golling, D. Golubkov, A. Gomes, R. Goncalves Gama, R. Gonçalo, G. Gonella, L. Gonella, A. Gongadze, F. Gonnella, J.L. Gonski, S. González de La Hoz, S. Gonzalez Fernandez, R. Gonzalez Lopez, C. Gonzalez Renteria, R. Gonzalez Suarez, S. Gonzalez-Sevilla, G.R. Gonzalvo Rodriguez, L. Goossens, N.A. Gorasia, P.A. Gorbounov, H.A. Gordon, B. Gorini, E. Gorini, A. Gorišek, A.T. Goshaw, M.I. Gostkin, C.A. Gottardo, M. Gouighri, A.G. Goussiou, N. Govender, C. Goy, I. Grabowska-Bold, E.C. Graham, J. Gramling, E. Gramstad, S. Grancagnolo, M. Grandi, V. Gratchev, P.M. Gravila, F.G. Gravili, C. Gray, H.M. Gray, C. Grefe, K. Gregersen, I.M. Gregor, P. Grenier, K. Grevtsov, C. Grieco, N.A. Grieser, A.A. Grillo, K. Grimm, S. Grinstein, J.F. Grivaz, S. Groh, E. Gross, J. Grosse-Knetter, Z.J. Grout, C. Grud, A. Grummer, J.C. Grundy, L. Guan, W. Guan, C. Gubbels, J. Guenther, A. Guerguichon, J.G.R. Guerrero Rojas, F. Guescini, D. Guest, R. Gugel, A. Guida, T. Guillemin, S. Guindon, U. Gul, J. Guo, W. Guo, Y. Guo, Z. Guo, R. Gupta, S. Gurbuz, G. Gustavino, M. Guth, P. Gutierrez, C. Gutschow, C. Guyot, C. Gwenlan, C.B. Gwilliam, E.S. Haaland, A. Haas, C. Haber, H.K. Hadavand, A. Hadef, M. Haleem, J. Haley, J.J. Hall,

G. Halladjian, G.D. Hallewell, K. Hamano, H. Hamdaoui, M. Hamer, G.N. Hamity, K. Han, L. Han, S. Han, Y.F. Han, K. Hanagaki, M. Hance, D.M. Handl, M.D. Hank, R. Hankache, E. Hansen, J.B. Hansen, J.D. Hansen, M.C. Hansen, P.H. Hansen, E.C. Hanson, K. Hara, T. Harenberg, S. Harkusha, P.F. Harrison, N.M. Hartman, N.M. Hartmann, Y. Hasegawa, A. Hasib, S. Hassani, S. Haug, R. Hauser, L.B. Havener, M. Havranek, C.M. Hawkes, R.J. Hawkings, S. Hayashida, D. Hayden, C. Hayes, R.L. Hayes, C.P. Hays, J.M. Hays, H.S. Hayward, S.J. Haywood, F. He, Y. He, M.P. Heath, V. Hedberg, S. Heer, A.L. Heggelund, C. Heidegger, K.K. Heidegger, W.D. Heidorn, J. Heilman, S. Heim, T. Heim, B. Heinemann, J.G. Heinlein, J.J. Heinrich, L. Heinrich, J. Hejbal, L. Helary, A. Held, S. Hellesund, C.M. Helling, S. Hellman, C. Helsens, R.C.W. Henderson, Y. Heng, L. Henkelmann, A.M. Henriques Correia, H. Herde, Y. Hernández Jiménez, H. Herr, M.G. Herrmann, T. Herrmann, G. Herten, R. Hertenberger, L. Hervas, T.C. Herwig, G.G. Hesketh, N.P. Hessey, H. Hibi, A. Higashida, S. Higashino, E. Higón-Rodriguez, K. Hildebrand, J.C. Hill, K.K. Hill, K.H. Hiller, S.J. Hillier, M. Hils, I. Hinchliffe, F. Hinterkeuser, M. Hirose, S. Hirose, D. Hirschbuehl, B. Hiti, O. Hladik, D.R. Hlaluku, J. Hobbs, N. Hod, M.C. Hodgkinson, A. Hoecker, D. Hohn, D. Hohov, T. Holm, T.R. Holmes, M. Holzbock, L.B.A.H. Hommels, T.M. Hong, J.C. Honig, A. Hönle, B.H. Hooberman, W.H. Hopkins, Y. Horii, P. Horn, L.A. Horyn, S. Hou, A. Hoummada, J. Howarth, J. Hoya, M. Hrabovsky, J. Hrdinka, J. Hrivnac, A. Hrynevich, T. Hryn'ova, P.J. Hsu, S.C. Hsu, Q. Hu, S. Hu, Y.F. Hu, D.P. Huang, Y. Huang, Y. Huang, Z. Hubacek, F. Hubaut, M. Huebner, F. Huegging, T.B. Huffman, M. Huhtinen, R. Hulsken, R.F.H. Hunter, P. Huo, N. Huseynov, J. Huston, J. Huth, R. Hyneman, S. Hyrych, G. Iacobucci, G. Iakovidis, I. Ibragimov, L. Iconomidou-Fayard, P. Iengo, R. Ignazzi, O. Igonkina, R. Iguchi, T. Iizawa, Y. Ikegami, M. Ikeno, N. Ilic, F. Iltzsche, H. Imam, G. Introzzi, M. Iodice, K. Iordanidou, V. Ippolito, M.F. Isacson, M. Ishino, W. Islam, C. Issever, S. Istin, F. Ito, J.M. Iturbe Ponce, R. Iuppa, A. Ivina, H. Iwasaki, J.M. Izen, V. Izzo, P. Jacka, P. Jackson, R.M. Jacobs, B.P. Jaeger, V. Jain, G. Jäkel, K.B. Jakobi, K. Jakobs, T. Jakoubek, J. Jamieson, K.W. Janas, R. Jansky, M. Janus, P.A. Janus, G. Jarlskog, A.E. Jaspan, N. Javadov, T. Javå¯Rek, M. Javurkova, F. Jeanneau, L. Jeanty, J. Jejelava, P. Jenni, N. Jeong, S. Jézéquel, H. Ji, J. Jia, H. Jiang, Y. Jiang, Z. Jiang, S. Jiggins, F.A. Jimenez Morales, J. Jimenez Pena, S. Jin, A. Jinaru, O. Jinnouchi, H. Jivan, P. Johansson, K.A. Johns, C.A. Johnson, R.W.L. Jones, S.D. Jones, T.J. Jones, J. Jongmanns, J. Jovicevic, X. Ju, J.J. Junggeburth, A. Juste Rozas, A. Kaczmarska, M. Kado, H. Kagan, M. Kagan, A. Kahn, C. Kahra, T. Kaji, E. Kajomovitz, C.W. Kalderon, A. Kaluza, A. Kamenshchikov, M. Kaneda, N.J. Kang, S. Kang, Y. Kano, J. Kanzaki, L.S. Kaplan, D. Kar, K. Karava, M.J. Kareem, I. Karkanias, S.N. Karpov, Z.M. Karpova, V. Kartvelishvili, A.N. Karyukhin, E. Kasimi, A. Kastanas, C. Kato, J. Katzy, K. Kawade, K. Kawagoe, T. Kawaguchi, T. Kawamoto, G. Kawamura, E.F. Kay, S. Kazakos, V.F. Kazanin, R. Keeler, R. Kehoe, J.S. Keller, E. Kellermann, D. Kelsey, J.J. Kempster, J. Kendrick, K.E. Kennedy, O. Kepka, S. Kersten, B.P. Kerševan, S. Ketabchi Haghighat, M. Khader, F. Khalil-Zada, M. Khandoga, A. Khanov, A.G. Kharlamov, T. Kharlamova, E.E. Khoda, A. Khodinov, T.J. Khoo, G. Khoriauli, E. Khramov, J. Khubua, S. Kido, M. Kiehn, C.R. Kilby, E. Kim, Y.K. Kim, N. Kimura, A. Kirchhoff, D. Kirchmeier, J. Kirk, A.E. Kiryunin, T. Kishimoto, D.P. Kisliuk, V. Kitali, C. Kitsaki, O. Kivernyk, T. Klapdor-Kleingrothaus, M. Klassen, C. Klein, M.H. Klein, M. Klein, U. Klein, K. Kleinknecht, P. Klimek, A. Klimentov, T. Klingl, T. Klioutchnikova, F.F. Klitzner, P. Kluit, S. Kluth, E. Kneringer, E.B.F.G. Knoops, A. Knue, D. Kobayashi, M. Kobel, M. Kocian, T. Kodama, P. Kodys, D.M. Koeck, P.T. Koenig, T. Koffas, N.M. Köhler, M. Kolb, I. Koletsou, T. Komarek, T. Kondo, K. Köneke, A.X.Y. Kong, A.C. König, T. Kono, V. Konstantinides, N. Konstantinidis, B. Konya, R. Kopeliansky, S. Koperny, K. Korcyl, K. Kordas, G. Koren, A. Korn, I. Korolkov, E.V. Korolkova, N. Korotkova, O. Kortner, S. Kortner, V.V. Kostyukhin, A. Kotsokechagia, A. Kotwal, A. Koulouris, A. Kourkoumeli-Charalampidi, C. Kourkoumelis, E. Kourlitis, V. Kouskoura, R. Kowalewski, W. Kozanecki, A.S. Kozhin, V.A. Kramarenko, G. Kramberger, D. Krasnopevtsev, M.W. Krasny, A. Krasznahorkay, D. Krauss, J.A. Kremer, J. Kretzschmar, P. Krieger, F. Krieter, A. Krishnan, M. Krivos, K. Krizka, K. Kroeninger, H. Kroha, J. Kroll, J. Kroll, K.S. Krowpman, U. Kruchonak, H. Krüger, N. Krumnack, M.C. Kruse, J.A. Krzysiak, A. Kubota, O. Kuchinskaia, S. Kuday,

J.T. Kuechler, S. Kuehn, T. Kuhl, V. Kukhtin, Y. Kulchitsky, S. Kuleshov, Y.P. Kulinich, M. Kuna, T. Kunigo, A. Kupco, T. Kupfer, O. Kuprash, H. Kurashige, L.L. Kurchaninov, Y.A. Kurochkin, A. Kurova, M.G. Kurth, E.S. Kuwertz, M. Kuze, A.K. Kvam, J. Kvita, T. Kwan, F. La Ruffa, C. Lacasta, F. Lacava, D.P.J. Lack, H. Lacker, D. Lacour, E. Ladygin, R. Lafaye, B. Laforge, T. Lagouri, S. Lai, I.K. Lakomiec, J.E. Lambert, S. Lammers, W. Lampl, C. Lampoudis, E. Lançon, U. Landgraf, M.P.J. Landon, M.C. Lanfermann, V.S. Lang, J.C. Lange, R.J. Langenberg, A.J. Lankford, F. Lanni, K. Lantzsch, A. Lanza, A. Lapertosa, J.F. Laporte, T. Lari, F. Lasagni Manghi, M. Lassnig, T.S. Lau, A. Laudrain, A. Laurier, M. Lavorgna, S.D. Lawlor, M. Lazzaroni, B. Le, E. Le Guirriec, A. Lebedev, M. Leblanc, T. Lecompte, F. Ledroit-Guillon, A.C.A. Lee, C.A. Lee, G.R. Lee, L. Lee, S.C. Lee, S. Lee, B. Lefebvre, H.P. Lefebvre, M. Lefebvre, C. Leggett, K. Lehmann, N. Lehmann, G. Lehmann Miotto, W.A. Leight, A. Leisos, M.A.L. Leite, C.E. Leitgeb, R. Leitner, D. Lellouch, K.J.C. Leney, T. Lenz, S. Leone, C. Leonidopoulos, A. Leopold, C. Leroy, R. Les, C.G. Lester, M. Levchenko, J. Levêque, D. Levin, L.J. Levinson, D.J. Lewis, B. Li, B. Li, C.Q. Li, F. Li, H. Li, H. Li, J. Li, K. Li, L. Li, M. Li, Q. Li, Q.Y. Li, S. Li, X. Li, Y. Li, Z. Li, Z. Li, Z. Li, Z. Liang, M. Liberatore, B. Liberti, A. Liblong, K. Lie, S. Lim, C.Y. Lin, K. Lin, R.A. Linck, R.E. Lindley, J.H. Lindon, A. Linss, A.L. Lionti, E. Lipeles, A. Lipniacka, T.M. Liss, A. Lister, J.D. Little, B. Liu, B.L. Liu, H.B. Liu, J.B. Liu, J.K.K. Liu, K. Liu, M. Liu, P. Liu, X. Liu, Y. Liu, Y. Liu, Y.L. Liu, Y.W. Liu, M. Livan, A. Lleres, J. Llorente Merino, S.L. Lloyd, C.Y. Lo, E.M. Lobodzinska, P. Loch, S. Loffredo, T. Lohse, K. Lohwasser, M. Lokajicek, J.D. Long, R.E. Long, I. Longarini, L. Longo, K.A. Looper, I. Lopez Paz, A. Lopez Solis, J. Lorenz, N. Lorenzo Martinez, A.M. Lory, P.J. Lösel, A. Lösle, X. Lou, X. Lou, A. Lounis, J. Love, P.A. Love, J.J. Lozano Bahilo, M. Lu, Y.J. Lu, H.J. Lubatti, C. Luci, F.L. Lucio Alves, A. Lucotte, F. Luehring, I. Luise, L. Luminari, B. Lund-Jensen, M.S. Lutz, D. Lynn, H. Lyons, R. Lysak, E. Lytken, F. Lyu, V. Lyubushkin, T. Lyubushkina, H. Ma, L.L. Ma, Y. Ma, D.M. Mac Donell, G. Maccarrone, A. Macchiolo, C.M. MacDonald, J.C. MacDonald, J. Machado Miguens, D. Madaffari, R. Madar, W.F. Mader, M. Madugoda Ralalage Don, N. Madysa, J. Maeda, T. Maeno, M. Maerker, V. Magerl, N. Magini, J. Magro, D.J. Mahon, C. Maidantchik, T. Maier, A. Maio, K. Maj, O. Majersky, S. Majewski, Y. Makida, N. Makovec, B. Malaescu, Pa. Malecki, V.P. Maleev, F. Malek, D. Malito, U. Mallik, D. Malon, C. Malone, S. Maltezos, S. Malyukov, J. Mamuzic, G. Mancini, I. Mandić, L. Manhaes de Andrade Filho, I.M. Maniatis, J. Manjarres Ramos, K.H. Mankinen, A. Mann, A. Manousos, B. Mansoulie, I. Manthos, S. Manzoni, A. Marantis, G. Marceca, L. Marchese, G. Marchiori, M. Marcisovsky, L. Marcoccia, C. Marcon, C.A. Marin Tobon, M. Marjanovic, Z. Marshall, M.U.F. Martensson, S. Marti-Garcia, C.B. Martin, T.A. Martin, V.J. Martin, B. Martin Dit Latour, L. Martinelli, M. Martinez, P. Martinez Agullo, V.I. Martinez Outschoorn, S. Martin-Haugh, V.S. Martoiu, A.C. Martyniuk, A. Marzin, S.R. Maschek, L. Masetti, T. Mashimo, R. Mashinistov, J. Masik, A.L. Maslennikov, L. Massa, P. Massarotti, P. Mastrandrea, A. Mastroberardino, T. Masubuchi, D. Matakias, A. Matic, N. Matsuzawa, P. Mättig, J. Maurer, B. Maček, D.A. Maximov, R. Mazini, I. Maznas, S.M. Mazza, J.P. Mc Gowan, S.P. Mc Kee, T.G. McCarthy, W.P. McCormack, E.F. McDonald, J.A. McFayden, G. McHedlidze, M.A. McKay, K.D. McLean, S.J. McMahon, P.C. McNamara, C.J. McNicol, R.A. McPherson, J.E. Mdhluli, Z.A. Meadows, S. Meehan, T. Megy, S. Mehlhase, A. Mehta, B. Meirose, D. Melini, B.R. Mellado Garcia, J.D. Mellenthin, M. Melo, F. Meloni, A. Melzer, E.D. Mendes Gouveia, L. Meng, X.T. Meng, S. Menke, E. Meoni, S. Mergelmeyer, S.A.M. Merkt, C. Merlassino, P. Mermod, L. Merola, C. Meroni, G. Merz, O. Meshkov, J.K.R. Meshreki, J. Metcalfe, A.S. Mete, C. Meyer, J.P. Meyer, M. Michetti, R.P. Middleton, L. Mijović, G. Mikenberg, M. Mikestikova, M. Mikuž, H. Mildner, A. Milic, C.D. Milke, D.W. Miller, A. Milov, D.A. Milstead, R.A. Mina, A.A. Minaenko, I.A. Minashvili, A.I. Mincer, B. Mindur, M. Mineev, Y. Minegishi, L.M. Mir, M. Mironova, A. Mirto, K.P. Mistry, T. Mitani, J. Mitrevski, V.A. Mitsou, M. Mittal, O. Miu, A. Miucci, P.S. Miyagawa, A. Mizukami, J.U. Mjörnmark, T. Mkrtchyan, M. Mlynarikova, T. Moa, S. Mobius, K. Mochizuki, P. Mogg, S. Mohapatra, R. Moles-Valls, K. Mönig, E. Monnier, A. Montalbano, J. Montejo Berlingen, M. Montella, F. Monticelli, S. Monzani, N. Morange, A.L. Moreira de Carvalho, D. Moreno, M. Moreno Llácer, C. Moreno Martinez, P. Morettini, M. Morgenstern, S. Morgenstern,

D. Mori, M. Morii, M. Morinaga, V. Morisbak, A.K. Morley, G. Mornacchi, A.P. Morris, L. Morvaj, P. Moschovakos, B. Moser, M. Mosidze, T. Moskalets, J. Moss, E.J.W. Moyse, S. Muanza, J. Mueller, R.S.P. Mueller, D. Muenstermann, G.A. Mullier, D.P. Mungo, J.L. Munoz Martinez, F.J. Munoz Sanchez, P. Murin, W.J. Murray, A. Murrone, J.M. Muse, M. Muškinja, C. Mwewa, A.G. Myagkov, A.A. Myers, G. Myers, J. Myers, M. Myska, B.P. Nachman, O. Nackenhorst, A. Nag Nag, K. Nagai, K. Nagano, Y. Nagasaka, J.L. Nagle, E. Nagy, A.M. Nairz, Y. Nakahama, K. Nakamura, T. Nakamura, H. Nanjo, F. Napolitano, R.F. Naranjo Garcia, R. Narayan, I. Naryshkin, T. Naumann, G. Navarro, P.Y. Nechaeva, F. Nechansky, T.J. Neep, A. Negri, M. Negrini, C. Nellist, C. Nelson, M.E. Nelson, S. Nemecek, M. Nessi, M.S. Neubauer, F. Neuhaus, M. Neumann, R. Newhouse, P.R. Newman, C.W. Ng, Y.S. Ng, Y.W.Y. Ng, B. Ngair, H.D.N. Nguyen, T. Nguyen Manh, E. Nibigira, R.B. Nickerson, R. Nicolaidou, D.S. Nielsen, J. Nielsen, M. Niemeyer, N. Nikiforou, V. Nikolaenko, I. Nikolic-Audit, K. Nikolopoulos, P. Nilsson, H.R. Nindhito, Y. Ninomiya, A. Nisati, N. Nishu, R. Nisius, I. Nitsche, T. Nitta, T. Nobe, D.L. Noel, Y. Noguchi, I. Nomidis, M.A. Nomura, M. Nordberg, J. Novak, T. Novak, O. Novgorodova, R. Novotny, L. Nozka, K. Ntekas, E. Nurse, F.G. Oakham, H. Oberlack, J. Ocariz, A. Ochi, I. Ochoa, J.P. Ochoa-Ricoux, K. O'Connor, S. Oda, S. Odaka, S. Oerdek, A. Ogrodnik, A. Oh, C.C. Ohm, H. Oide, M.L. Ojeda, H. Okawa, Y. Okazaki, M.W. O'Keefe, Y. Okumura, T. Okuyama, A. Olariu, L.F. Oleiro Seabra, S.A. Olivares Pino, D. Oliveira Damazio, J.L. Oliver, M.J.R. Olsson, A. Olszewski, J. Olszowska, Ö. O. Öncel, D.C. O'Neil, A.P. O'Neill, A. Onofre, P.U.E. Onyisi, H. Oppen, R.G. Oreamuno Madriz, M.J. Oreglia, G.E. Orellana, D. Orestano, N. Orlando, R.S. Orr, V. O'Shea, R. Ospanov, G. Otero Y Garzon, H. Otono, P.S. Ott, G.J. Ottino, M. Ouchrif, J. Ouellette, F. Ould-Saada, A. Ouraou, Q. Ouyang, M. Owen, R.E. Owen, V.E. Ozcan, N. Ozturk, J. Pacalt, H.A. Pacey, K. Pachal, A. Pacheco Pages, C. Padilla Aranda, S. Pagan Griso, G. Palacino, S. Palazzo, S. Palestini, M. Palka, P. Palni, C.E. Pandini, J.G. Panduro Vazquez, P. Pani, G. Panizzo, L. Paolozzi, C. Papadatos, K. Papageorgiou, S. Parajuli, A. Paramonov, C. Paraskevopoulos, D. Paredes Hernandez, S.R. Paredes Saenz, B. Parida, T.H. Park, A.J. Parker, M.A. Parker, F. Parodi, E.W. Parrish, J.A. Parsons, U. Parzefall, L. Pascual Dominguez, V.R. Pascuzzi, J.M.P. Pasner, F. Pasquali, E. Pasqualucci, S. Passaggio, F. Pastore, P. Pasuwan, S. Pataraia, J.R. Pater, A. Pathak, J. Patton, T. Pauly, J. Pearkes, B. Pearson, M. Pedersen, L. Pedraza Diaz, R. Pedro, T. Peiffer, S.V. Peleganchuk, O. Penc, H. Peng, B.S. Peralva, M.M. Perego, A.P. Pereira Peixoto, L. Pereira Sanchez, D.V. Perepelitsa, E. Perez Codina, F. Peri, L. Perini, H. Pernegger, S. Perrella, A. Perrevoort, K. Peters, R.F.Y. Peters, B.A. Petersen, T.C. Petersen, E. Petit, V. Petousis, A. Petridis, C. Petridou, P. Petroff, F. Petrucci, M. Pettee, N.E. Pettersson, K. Petukhova, A. Peyaud, R. Pezoa, L. Pezzotti, T. Pham, F.H. Phillips, P.W. Phillips, M.W. Phipps, G. Piacquadio, E. Pianori, A. Picazio, R.H. Pickles, R. Piegaia, D. Pietreanu, J.E. Pilcher, A.D. Pilkington, M. Pinamonti, J.L. Pinfold, C. Pitman Donaldson, M. Pitt, L. Pizzimento, A. Pizzini, M.A. Pleier, V. Plesanovs, V. Pleskot, E. Plotnikova, P. Podberezko, R. Poettgen, R. Poggi, L. Poggioli, I. Pogrebnyak, D. Pohl, I. Pokharel, G. Polesello, A. Poley, A. Policicchio, R. Polifka, A. Polini, C.S. Pollard, V. Polychronakos, D. Ponomarenko, L. Pontecorvo, S. Popa, G.A. Popeneciu, L. Portales, D.M. Portillo Quintero, S. Pospisil, K. Potamianos, I.N. Potrap, C.J. Potter, H. Potti, T. Poulsen, J. Poveda, T.D. Powell, G. Pownall, M.E. Pozo Astigarraga, P. Pralavorio, S. Prell, D. Price, M. Primavera, M.L. Proffitt, N. Proklova, K. Prokofiev, F. Prokoshin, S. Protopopescu, J. Proudfoot, M. Przybycien, D. Pudzha, A. Puri, P. Puzo, D. Pyatiizbyantseva, J. Qian, Y. Qin, A. Quadt, M. Queitsch-Maitland, M. Racko, F. Ragusa, G. Rahal, J.A. Raine, S. Rajagopalan, A. Ramirez Morales, K. Ran, D.M. Rauch, F. Rauscher, S. Rave, B. Ravina, I. Ravinovich, J.H. Rawling, M. Raymond, A.L. Read, N.P. Readioff, M. Reale, D.M. Rebuzzi, G. Redlinger, K. Reeves, J. Reichert, D. Reikher, A. Reiss, A. Rej, C. Rembser, A. Renardi, M. Renda, M.B. Rendel, A.G. Rennie, S. Resconi, E.D. Resseguie, S. Rettie, B. Reynolds, E. Reynolds, O.L. Rezanova, P. Reznicek, E. Ricci, R. Richter, S. Richter, E. Richter-Was, M. Ridel, P. Rieck, O. Rifki, M. Rijssenbeek, A. Rimoldi, M. Rimoldi, L. Rinaldi, T.T. Rinn, G. Ripellino, I. Riu, P. Rivadeneira, J.C. Rivera Vergara, F. Rizatdinova, E. Rizvi, C. Rizzi, S.H. Robertson, M. Robin, D. Robinson, C.M. Robles Gajardo, M. Robles Manzano, A. Robson, A. Rocchi,

E. Rocco, C. Roda, S. Rodriguez Bosca, A.M. Rodríguez Vera, S. Roe, J. Roggel, O. Røhne, R. Röhrig, R.A. Rojas, B. Roland, C.P.A. Roland, J. Roloff, A. Romaniouk, M. Romano, N. Rompotis, M. Ronzani, L. Roos, S. Rosati, G. Rosin, B.J. Rosser, E. Rossi, E. Rossi, E. Rossi, L.P. Rossi, L. Rossini, R. Rosten, M. Rotaru, B. Rottler, D. Rousseau, G. Rovelli, A. Roy, D. Roy, A. Rozanov, Y. Rozen, X. Ruan, T.A. Ruggeri, F. Rühr, A. Ruiz-Martinez, A. Rummler, Z. Rurikova, N.A. Rusakovich, H.L. Russell, L. Rustige, J.P. Rutherfoord, E.M. Rüttinger, M. Rybar, G. Rybkin, E.B. Rye, A. Ryzhov, J.A. Sabater Iglesias, P. Sabatini, L. Sabetta, S. Sacerdoti, H.F.W. Sadrozinski, R. Sadykov, F. Safai Tehrani, B. Safarzadeh Samani, M. Safdari, P. Saha, S. Saha, M. Sahinsoy, A. Sahu, M. Saimpert, M. Saito, T. Saito, H. Sakamoto, D. Salamani, G. Salamanna, A. Salnikov, J. Salt, A. Salvador Salas, D. Salvatore, F. Salvatore, A. Salvucci, A. Salzburger, J. Samarati, D. Sammel, D. Sampsonidis, D. Sampsonidou, J. Sánchez, A. Sanchez Pineda, H. Sandaker, C.O. Sander, I.G. Sanderswood, M. Sandhoff, C. Sandoval, D.P.C. Sankey, M. Sannino, Y. Sano, A. Sansoni, C. Santoni, H. Santos, S.N. Santpur, A. Santra, K.A. Saoucha, A. Sapronov, J.G. Saraiva, O. Sasaki, K. Sato, F. Sauerburger, E. Sauvan, P. Savard, R. Sawada, C. Sawyer, L. Sawyer, I. Sayago Galvan, C. Sbarra, A. Sbrizzi, T. Scanlon, J. Schaarschmidt, P. Schacht, D. Schaefer, L. Schaefer, S. Schaepe, U. Schäfer, A.C. Schaffer, D. Schaile, R.D. Schamberger, E. Schanet, C. Scharf, N. Scharmberg, V.A. Schegelsky, D. Scheirich, F. Schenck, M. Schernau, C. Schiavi, L.K. Schildgen, Z.M. Schillaci, E.J. Schioppa, M. Schioppa, K.E. Schleicher, S. Schlenker, K.R. Schmidt-Sommerfeld, K. Schmieden, C. Schmitt, S. Schmitt, J.C. Schmoeckel, L. Schoeffel, A. Schoening, P.G. Scholer, E. Schopf, M. Schott, J.F.P. Schouwenberg, J. Schovancova, S. Schramm, F. Schroeder, A. Schulte, H.C. Schultz-Coulon, M. Schumacher, B.A. Schumm, Ph. Schune, A. Schwartzman, T.A. Schwarz, Ph. Schwemling, R. Schwienhorst, A. Sciandra, G. Sciolla, M. Scornajenghi, F. Scuri, F. Scutti, L.M. Scyboz, C.D. Sebastiani, P. Seema, S.C. Seidel, A. Seiden, B.D. Seidlitz, T. Seiss, C. Seitz, J.M. Seixas, G. Sekhniaidze, S.J. Sekula, N. Semprini-Cesari, S. Sen, C. Serfon, L. Serin, L. Serkin, M. Sessa, H. Severini, S. Sevova, F. Sforza, A. Sfyrla, E. Shabalina, J.D. Shahinian, N.W. Shaikh, D. Shaked Renous, L.Y. Shan, M. Shapiro, A. Sharma, A.S. Sharma, P.B. Shatalov, K. Shaw, S.M. Shaw, M. Shehade, Y. Shen, A.D. Sherman, P. Sherwood, L. Shi, S. Shimizu, C.O. Shimmin, Y. Shimogama, M. Shimojima, I.P.J. Shipsey, S. Shirabe, M. Shiyakova, J. Shlomi, A. Shmeleva, M.J. Shochet, J. Shojaii, D.R. Shope, S. Shrestha, E.M. Shrif, E. Shulga, P. Sicho, A.M. Sickles, E. Sideras Haddad, O. Sidiropoulou, A. Sidoti, F. Siegert, Dj. Sijacki, Jr. Silva, Manuel, M.V. Silva Oliveira, S.B. Silverstein, S. Simion, R. Simoniello, C.J. Simpson-Allsop, S. Simsek, P. Sinervo, V. Sinetckii, S. Singh, M. Sioli, I. Siral, S.Yu. Sivoklokov, J. Sjölin, A. Skaf, E. Skorda, P. Skubic, M. Slawinska, K. Sliwa, R. Slovak, V. Smakhtin, B.H. Smart, J. Smiesko, N. Smirnov, S.Yu. Smirnov, Y. Smirnov, L.N. Smirnova, O. Smirnova, E.A. Smith, H.A. Smith, M. Smizanska, K. Smolek, A. Smykiewicz, A.A. Snesarev, H.L. Snoek, I.M. Snyder, S. Snyder, R. Sobie, A. Soffer, A. Søgaard, F. Sohns, C.A. Solans Sanchez, E.Yu. Soldatov, U. Soldevila, A.A. Solodkov, A. Soloshenko, O.V. Solovyanov, V. Solovyev, P. Sommer, H. Son, W. Song, W.Y. Song, A. Sopczak, A.L. Sopio, F. Sopkova, S. Sottocornola, R. Soualah, A.M. Soukharev, D. South, S. Spagnolo, M. Spalla, M. Spangenberg, F. Spanò, D. Sperlich, T.M. Spieker, G. Spigo, M. Spina, D.P. Spiteri, M. Spousta, A. Stabile, B.L. Stamas, R. Stamen, M. Stamenkovic, E. Stanecka, B. Stanislaus, M.M. Stanitzki, M. Stankaityte, B. Stapf, E.A. Starchenko, G.H. Stark, J. Stark, P. Staroba, P. Starovoitov, S. Stärz, R. Staszewski, G. Stavropoulos, M. Stegler, P. Steinberg, A.L. Steinhebel, B. Stelzer, H.J. Stelzer, O. Stelzer-Chilton, H. Stenzel, T.J. Stevenson, G.A. Stewart, M.C. Stockton, G. Stoicea, M. Stolarski, S. Stonjek, A. Straessner, J. Strandberg, S. Strandberg, M. Strauss, T. Strebler, P. Strizenec, R. Ströhmer, D.M. Strom, R. Stroynowski, A. Strubig, S.A. Stucci, B. Stugu, J. Stupak, N.A. Styles, D. Su, W. Su, X. Su, V.V. Sulin, M.J. Sullivan, D.M.S. Sultan, S. Sultansoy, T. Sumida, S. Sun, X. Sun, K. Suruliz, C.J.E. Suster, M.R. Sutton, S. Suzuki, M. Svatos, M. Swiatlowski, S.P. Swift, T. Swirski, A. Sydorenko, I. Sykora, M. Sykora, T. Sykora, D. Ta, K. Tackmann, J. Taenzer, A. Taffard, R. Tafirout, E. Tagiev, R. Takashima, K. Takeda, T. Takeshita, E.P. Takeva, Y. Takubo, M. Talby, A.A. Talyshev, K.C. Tam, N.M. Tamir, J. Tanaka, R. Tanaka, S. Tapia Araya, S. Tapprogge, A. Tarek Abouelfadl Mohamed, S. Tarem, K. Tariq, G. Tarna,

G.F. Tartarelli, P. Tas, M. Tasevsky, T. Tashiro, E. Tassi, A. Tavares Delgado, Y. Tayalati, A.J. Taylor, G.N. Taylor, W. Taylor, H. Teagle, A.S. Tee, R. Teixeira de Lima, P. Teixeira-Dias, H. Ten Kate, J.J. Teoh, S. Terada, K. Terashi, J. Terron, S. Terzo, M. Testa, R.J. Teuscher, S.J. Thais, N. Themistokleous, T. Theveneaux-Pelzer, F. Thiele, D.W. Thomas, J.O. Thomas, J.P. Thomas, E.A. Thompson, P.D. Thompson, E. Thomson, E.J. Thorpe, R.E. Ticse Torres, V.O. Tikhomirov, Yu.A. Tikhonov, S. Timoshenko, P. Tipton, S. Tisserant, K. Todome, S. Todorova-Nova, S. Todt, J. Tojo, S. Tokár, K. Tokushuku, E. Tolley, R. Tombs, K.G. Tomiwa, M. Tomoto, L. Tompkins, P. Tornambe, E. Torrence, H. Torres, E. Torró Pastor, C. Tosciri, J. Toth, D.R. Tovey, A. Traeet, C.J. Treado, T. Trefzger, F. Tresoldi, A. Tricoli, I.M. Trigger, S. Trincaz-Duvoid, D.A. Trischuk, W. Trischuk, B. Trocmé, A. Trofymov, C. Troncon, F. Trovato, L. Truong, M. Trzebinski, A. Trzupek, F. Tsai, J.C.L. Tseng, P.V. Tsiareshka, A. Tsirigotis, V. Tsiskaridze, E.G. Tskhadadze, M. Tsopoulou, I.I. Tsukerman, V. Tsulaia, S. Tsuno, D. Tsybychev, Y. Tu, A. Tudorache, V. Tudorache, T.T. Tulbure, A.N. Tuna, S. Turchikhin, D. Turgeman, I. Turk Cakir, R.J. Turner, R. Turra, P.M. Tuts, S. Tzamarias, E. Tzovara, K. Uchida, F. Ukegawa, G. Unal, M. Unal, A. Undrus, G. Unel, F.C. Ungaro, Y. Unno, K. Uno, J. Urban, P. Urquijo, G. Usai, Z. Uysal, V. Vacek, B. Vachon, K.O.H. Vadla, T. Vafeiadis, A. Vaidya, C. Valderanis, E. Valdes Santurio, M. Valente, S. Valentinetti, A. Valero, L. Valéry, R.A. Vallance, A. Vallier, J.A. Valls Ferrer, T.R. van Daalen, P. van Gemmeren, S. van Stroud, I. van Vulpen, M. Vanadia, W. Vandelli, M. Vandenbroucke, E.R. Vandewall, A. Vaniachine, D. Vannicola, R. Vari, E.W. Varnes, C. Varni, T. Varol, D. Varouchas, K.E. Varvell, M.E. Vasile, G.A. Vasquez, F. Vazeille, D. Vazquez Furelos, T. Vazquez Schroeder, J. Veatch, V. Vecchio, M.J. Veen, L.M. Veloce, F. Veloso, S. Veneziano, A. Ventura, A. Verbytskyi, V. Vercesi, M. Verducci, C.M. Vergel Infante, C. Vergis, W. Verkerke, A.T. Vermeulen, J.C. Vermeulen, C. Vernieri, M.C. Vetterli, N. Viaux Maira, T. Vickey, O.E. Vickey Boeriu, G.H.A. Viehhauser, L. Vigani, M. Villa, M. Villaplana Perez, E.M. Villhauer, E. Vilucchi, M.G. Vincter, G.S. Virdee, A. Vishwakarma, C. Vittori, I. Vivarelli, M. Vogel, P. Vokac, S.E. von Buddenbrock, E. von Toerne, V. Vorobel, K. Vorobev, M. Vos, J.H. Vossebeld, M. Vozak, N. Vranjes, M. Vranjes Milosavljevic, V. Vrba, M. Vreeswijk, R. Vuillermet, I. Vukotic, S. Wada, P. Wagner, W. Wagner, J. Wagner-Kuhr, S. Wahdan, H. Wahlberg, R. Wakasa, V.M. Walbrecht, J. Walder, R. Walker, S.D. Walker, W. Walkowiak, V. Wallangen, A.M. Wang, A.Z. Wang, C. Wang, C. Wang, F. Wang, H. Wang, H. Wang, J. Wang, P. Wang, Q. Wang, R.J. Wang, R. Wang, R. Wang, S.M. Wang, W.T. Wang, W. Wang, W.X. Wang, Y. Wang, Z. Wang, C. Wanotayaroj, A. Warburton, C.P. Ward, D.R. Wardrope, N. Warrack, A.T. Watson, M.F. Watson, G. Watts, B.M. Waugh, A.F. Webb, C. Weber, M.S. Weber, S.A. Weber, S.M. Weber, A.R. Weidberg, J. Weingarten, M. Weirich, C. Weiser, P.S. Wells, T. Wenaus, B. Wendland, T. Wengler, S. Wenig, N. Wermes, M. Wessels, T.D. Weston, K. Whalen, A.M. Wharton, A.S. White, A. White, M.J. White, D. Whiteson, B.W. Whitmore, W. Wiedenmann, C. Wiel, M. Wielers, N. Wieseotte, C. Wiglesworth, L.A.M. Wiik-Fuchs, H.G. Wilkens, L.J. Wilkins, H.H. Williams, S. Williams, S. Willocq, P.J. Windischhofer, I. Wingerter-Seez, E. Winkels, F. Winklmeier, B.T. Winter, M. Wittgen, M. Wobisch, A. Wolf, R. Wölker, J. Wollrath, M.W. Wolter, H. Wolters, V.W.S. Wong, N.L. Woods, S.D. Worm, B.K. Wosiek, K.W. Woźniak, K. Wraight, S.L. Wu, X. Wu, Y. Wu, J. Wuerzinger, T.R. Wyatt, B.M. Wynne, S. Xella, L. Xia, J. Xiang, X. Xiao, X. Xie, I. Xiotidis, D. Xu, H. Xu, H. Xu, L. Xu, T. Xu, W. Xu, Z. Xu, Z. Xu, B. Yabsley, S. Yacoob, K. Yajima, D.P. Yallup, N. Yamaguchi, Y. Yamaguchi, A. Yamamoto, M. Yamatani, T. Yamazaki, Y. Yamazaki, J. Yan, Z. Yan, H.J. Yang, H.T. Yang, S. Yang, T. Yang, X. Yang, Y. Yang, Z. Yang, W.M. Yao, Y.C. Yap, Y. Yasu, E. Yatsenko, H. Ye, J. Ye, S. Ye, I. Yeletskikh, M.R. Yexley, E. Yigitbasi, P. Yin, K. Yorita, K. Yoshihara, C.J.S. Young, C. Young, J. Yu, R. Yuan, X. Yue, M. Zaazoua, B. Zabinski, G. Zacharis, E. Zaffaroni, J. Zahreddine, A.M. Zaitsev, T. Zakareishvili, N. Zakharchuk, S. Zambito, D. Zanzi, D.R. Zaripovas, S.V. Zeißner, C. Zeitnitz, G. Zemaityte, J.C. Zeng, O. Zenin, T. Ženiš, D. Zerwas, M. Zgubič, B. Zhang, D.F. Zhang, G. Zhang, J. Zhang, Kaili Zhang, L. Zhang, L. Zhang, M. Zhang, R. Zhang, S. Zhang, X. Zhang, X. Zhang, Y. Zhang, Z. Zhang, Z. Zhang, P. Zhao, Z. Zhao, A. Zhemchugov, Z. Zheng, D. Zhong, B. Zhou, C. Zhou, H. Zhou, M.S. Zhou, M. Zhou, N. Zhou, Y. Zhou, C.G. Zhu, C. Zhu, H.L. Zhu, H. Zhu, J. Zhu,

Y. Zhu, X. Zhuang, K. Zhukov, V. Zhulanov, D. Zieminska, N.I. Zimine, S. Zimmermann, Z. Zinonos, M. Ziolkowski, L. Živković, G. Zobernig, A. Zoccoli, K. Zoch, T.G. Zorbas, R. Zou, L. Zwalinski, and Atlas Collaboration, Dijet resonance search with weak supervision using $\sqrt{\{s\}} = 13$ TeV p p collisions in the ATLAS detector. Phys. Rev. Lett. **125**(13), 131801 (2020). https://doi.org/10.1103/PhysRevLett.125.131801

S. Aaronson, S.M. Carroll, L. Ouellette, Quantifying the rise and fall of complexity in closed systems: the coffee automaton. arXiv e-prints, art. arXiv:1405.6903 (May 2014)

T. Aarrestad, M. van Beekveld, M. Bona, A. Boveia, S. Caron, J. Davies, A. de Simone, C. Doglioni, J. Duarte, A. Farbin, H. Gupta, L. Hendriks, L.A. Heinrich, J. Howarth, P. Jawahar, A. Jueid, J. Lastow, A. Leinweber, J. Mamuzic, E. Merényi, A. Morandini, P. Moskvitina, C. Nellist, J. Ngadiuba, B. Ostdiek, M. Pierini, B. Ravina, R. Ruiz de Austri, S. Sekmen, M. Touranakou, M. Vaškeviciute, R. Vilalta, J.-R. Vlimant, R. Verheyen, M. White, E. Wulff, E. Wallin, K.A. Wozniak, Z. Zhang, The dark machines anomaly score challenge: benchmark data and model Independent event classification for the Large Hadron Collider. SciPost Phys. **12**(1), 043 (2022). https://doi.org/10.21468/SciPostPhys.12.1.043

A. Adelmann, W. Hopkins, E. Kourlitis, M. Kagan, G. Kasieczka, C. Krause, D. Shih, V. Mikuni, B. Nachman, K. Pedro, D. Winklehner, New directions for surrogate models and differentiable programming for High Energy Physics detector simulation. arXiv e-prints, art. arXiv:2203.08806 (March 2022). https://doi.org/10.48550/arXiv.2203.08806

S. Agostinelli, J. Allison, K. Amako, J. Apostolakis, H. Araujo, P. Arce, M. Asai, D. Axen, S. Banerjee, G. Barrand, F. Behner, L. Bellagamba, J. Boudreau, L. Broglia, A. Brunengo, H. Burkhardt, S. Chauvie, J. Chuma, R. Chytracek, G. Cooperman, G. Cosmo, P. Degtyarenko, A. Dell'Acqua, G. Depaola, D. Dietrich, R. Enami, A. Feliciello, C. Ferguson, H. Fesefeldt, G. Folger, F. Foppiano, A. Forti, S. Garelli, S. Giani, R. Giannitrapani, D. Gibin, J.J. Gómez Cadenas, I. González, G. Gracia Abril, G. Greeniaus, W. Greiner, V. Grichine, A. Grossheim, S. Guatelli, P. Gumplinger, R. Hamatsu, K. Hashimoto, H. Hasui, A. Heikkinen, A. Howard, V. Ivanchenko, A. Johnson, F.W. Jones, J. Kallenbach, N. Kanaya, M. Kawabata, Y. Kawabata, M. Kawaguti, S. Kelner, P. Kent, A. Kimura, T. Kodama, R. Kokoulin, M. Kossov, H. Kurashige, E. Lamanna, T. Lampén, V. Lara, V. Lefebure, F. Lei, M. Liendl, W. Lockman, F. Longo, S. Magni, M. Maire, E. Medernach, K. Minamimoto, P. Mora de Freitas, Y. Morita, K. Murakami, M. Nagamatu, R. Nartallo, P. Nieminen, T. Nishimura, K. Ohtsubo, M. Okamura, S. O'Neale, Y. Oohata, K. Paech, J. Perl, A. Pfeiffer, M.G. Pia, F. Ranjard, A. Rybin, S. Sadilov, E. Di Salvo, G. Santin, T. Sasaki, N. Savvas, Y. Sawada, S. Scherer, S. Sei, V. Sirotenko, D. Smith, N. Starkov, H. Stoecker, J. Sulkimo, M. Takahata, S. Tanaka, E. Tcherniaev, E. Safai Tehrani, M. Tropeano, P. Truscott, H. Uno, L. Urban, P. Urban, M. Verderi, A. Walkden, W. Wander, H. Weber, J.P. Wellisch, T. Wenaus, D.C. Williams, D. Wright, T. Yamada, H. Yoshida, D. Zschiesche, G EANT4 Collaboration, G EANT4—a simulation toolkit. Nucl. Instrum. Methods Phys. Res. A **506**(3), 250–303 (2003). https://doi.org/10.1016/S0168-9002(03)01368-8

J. Akeret, C. Chang, A. Lucchi, A. Refregier, Radio frequency interference mitigation using deep convolutional neural networks. Astron. Comput. **18**, 35–39 (2017). https://doi.org/10.1016/j.ascom.2017.01.002

Y. Alanazi, N. Sato, P. Ambrozewicz, A.N. Hiller Blin, W. Melnitchouk, M. Battaglieri, T. Liu, Y. Li, A survey of machine learning-based physics event generation. arXiv e-prints, art. arXiv:2106.00643 (June 2021). https://doi.org/10.48550/arXiv.2106.00643

L. Alegre, J. Sabater, P. Best, R.I.J. Mostert, W.L. Williams, G. Gürkan, M.J. Hardcastle, R. Kondapally, T.W. Shimwell, D.J.B. Smith, A machine learning classifier for LOFAR radio galaxy cross-matching techniques. arXiv e-prints, art. arXiv:2207.01645 (July 2022)

M.J. Alger, J.K. Banfield, C.S. Ong, L. Rudnick, O.I. Wong, C. Wolf, H. Andernach, R.P. Norris, S.S. Shabala, Radio Galaxy Zoo: machine learning for radio source host galaxy cross-identification. Mon. Not. R. Astron. Soc. **478**(4), 5547–5563 (2018). https://doi.org/10.1093/mnras/sty1308

M.J. Alger, J.D. Livingston, N.M. McClure-Griffiths, J.L. Nabaglo, O.I. Wong, C.S. Ong, Interpretable Faraday complexity classification. Publ. Astron. Soc. Aust. **38**, e022 (2021). https://doi.org/10.1017/pasa.2021.10

Y. Alibert, New metric to quantify the similarity between planetary systems: application to dimensionality reduction using t-sne. Astron. & Astrophys. **624**, A45 (2019). https://doi.org/10.1051/0004-6361/201834592

ALICE Collaboration, K. Aamodt, A. Abrahantes Quintana, R. Achenbach, S. Acounis, D. Adamová, C. Adler, M. Aggarwal, F. Agnese, G. Aglieri Rinella, Z. Ahammed, A. Ahmad, N. Ahmad, S. Ahmad, A. Akindinov, P. Akishin, D. Aleksandrov, B. Alessandro, R. Alfaro, G. Alfarone, A. Alici, J. Alme, T. Alt, S. Altinpinar, W. Amend, C. Andrei, Y. Andres, A. Andronic, G. Anelli, M. Anfreville, V. Angelov, A. Anzo, C. Anson, T. Anticić, V. Antonenko, D. Antonczyk, F. Antinori, S. Antinori, P. Antonioli, L. Aphecetche, H. Appelshäuser, V. Aprodu, M. Arba, S. Arcelli, A. Argentieri, N. Armesto, R. Arnaldi, A. Arefiev, I. Arsene, A. Asryan, A. Augustinus, T.C. Awes, J. Äysto, M. Danish Azmi, S. Bablock, A. Badalà, S.K. Badyal, J. Baechler, S. Bagnasco, R. Bailhache, R. Bala, A. Baldisseri, A. Baldit, J. Bán, R. Barbera, P.L. Barberis, J.M. Barbet, G. Barnäfoldi, V. Barret, J. Bartke, D. Bartos, M. Basile, V. Basmanov, N. Bastid, G. Batigne, B. Batyunya, J. Baudot, C. Baumann, I. Bearden, B. Becker, J. Belikov, R. Bellwied, E. Belmont-Moreno, A. Belogianni, S. Belyaev, A. Benato, J.L. Beney, L. Benhabib, F. Benotto, S. Beolé, I. Berceanu, A. Bercuci, E. Berdermann, Y. Berdnikov, C. Bernard, R. Berny, J.D. Berst, H. Bertelsen, L. Betev, A. Bhasin, P. Baskar, A. Bhati, N. Bianchi, J. Bielčik, J. Bielčiková, L. Bimbot, G. Blanchard, F. Blanco, F. Blanco, D. Blau, C. Blume, S. Blyth, M. Boccioli, A. Bogdanov, H. Bøggild, M. Bogolyubsky, L. Boldizsár, M. Bombara, C. Bombonati, M. Bondila, D. Bonnet, V. Bonvicini, H. Borel, F. Borotto, V. Borshchov, Y. Bortoli, O. Borysov, S. Bose, L. Bosisio, M. Botje, S. Böttger, G. Bourdaud, O. Bourrion, S. Bouvier, A. Braem, M. Braun, P. Braun-Munzinger, L. Bravina, M. Bregant, G. Bruckner, R. Brun, E. Bruna, O. Brunasso, G.E. Bruno, D. Bucher, V. Budilov, D. Budnikov, H. Buesching, P. Buncic, M. Burns, S. Burachas, O. Busch, J. Bushop, X. Cai, H. Caines, F. Calaon, M. Caldogno, I. Cali, P. Camerini, R. Campagnolo, M. Campbell, X. Cao, G.P. Capitani, G. Cara Romeo, M. Cardenas-Montes, H. Carduner, F. Carena, W. Carena, P. Cariola, F. Carminati, J. Casado, A. Casanova Diaz, M. Caselle, J. Castillo Castellanos, J. Castor, V. Catanescu, E. Cattaruzza, D. Cavazza, P. Cerello, S. Ceresa, V. Černý, V. Chambert, S. Chapeland, A. Charpy, D. Charrier, M. Chartoire, J.L. Charvet, S. Chattopadhyay, S. Chattopadhyay, V. Chepurnov, S. Chernenko, M. Cherney, C. Cheshkov, B. Cheynis, P. Chochula, E. Chiavassa, V. Chibante Barroso, J. Choi, P. Christakoglou, P. Christiansen, C. Christensen, O.A. Chykalov, C. Cicalo, L. Cifarelli-Strolin, M. Ciobanu, F. Cindolo, C. Cirstoiu, O. Clausse, J. Cleymans, O. Cobanoglu, J.P. Coffin, S. Coli, A. Colla, C. Colledani, C. Combaret, M. Combet, M. Comets, G. Conesa Balbastre, Z. Conesa del Valle, G. Contin, J. Contreras, T. Cormier, F. Corsi, P. Cortese, F. Costa, E. Crescio, P. Crochet, E. Cuautle, J. Cussonneau, M. Dahlinger, A. Dainese, H.H. Dalsgaard, L. Daniel, I. Das, T. Das, A. Dash, R. Da Silva, M. Davenport, H. Daues, A. DeCaro, G. de Cataldo, J. DeCuveland, A. DeFalco, M. de Gaspari, P. de Girolamo, J. de Groot, D. DeGruttola, A. DeHaas, N. DeMarco, S. DePasquale, P. DeRemigis, D. de Vaux, G. Decock, H. Delagrange, M. DelFranco, G. Dellacasa, C. Dell'Olio, D. Dell'Olio, A. Deloff, V. Demanov, E. Dénes, G. D'Erasmo, D. Derkach, A. Devaux, D. Di Bari, A. Di Bartelomen, C. Di Giglio, S. Di Liberto, A. Di Mauro, P. Di Nezza, M. Dialinas, L. Diaz, R. Díaz Valdes, T. Dietel, R. Dima, H. Ding, C. Dinca, R. Divià, V. Dobretsov, A. Dobrin, B. Doenigus, T. Dobrowolski, I. Domínguez, M. Dorn, S. Drouet, A.E. Dubey, L. Ducroux, F. Dumitrache, E. Dumonteil, P. Dupieux, V. Duta, A. Dutta Majumdar, M. Dutta Majumdar, Th Dyhre, L. Efimov, A. Efremov, D. Elia, D. Emschermann, C. Engster, A. Enokizono, B. Espagnon, M. Estienne, A. Evangelista, D. Evans, S. Evrard, C.W. Fabjan, D. Fabris, J. Faivre, D. Falchieri, A. Fantoni, R. Farano, R. Fearick, O. Fedorov, V. Fekete, D. Felea, G. Feofilov, A. Férnandez Téllez, A. Ferretti, F. Fichera, S. Filchagin, E. Filoni, C. Finck, R. Fini, E.M. Fiore, D. Flierl, M. Floris, Z. Fodor, Y. Foka, S. Fokin, P. Force, F. Formenti, E. Fragiacomo, M. Fragkiadakis, D. Fraissard, A. Franco, M. Franco, U. Frankenfeld, U. Fratino, S. Fresneau, A. Frolov, U. Fuchs, J. Fujita,

C. Furget, M. Furini, M. Fusco Girard, J.J. Gaardhøje, A. Gabrielli, S. Gadrat, M. Gagliardi, A. Gago, L. Gaido, A. Gallas Torreira, M. Gallio, E. Gandolfi, P. Ganoti, M. Ganti, J. Garabatos, A. Garcia Lopez, L. Garizzo, L. Gaudichet, R. Gemme, M. Germain, A. Gheata, M. Gheata, B. Ghidini, P. Ghosh, G. Giolu, G. Giraudo, P. Giubellino, R. Glasow, P. Glässel, E.G. Ferreiro, C. Gonzalez Gutierrez, L.H. Gonzales-Trueba, S. Gorbunov, Y. Gorbunov, H. Gos, J. Gosset, S. Gotovac, H. Gottschlag, D. Gottschalk, V. Grabski, T. Grassi, H. Gray, O. Grebenyuk, K. Grebieszkow, C. Gregory, C. Grigoras, N. Grion, V. Grigoriev, A. Grigoryan, C. Grigoryan, S. Grigoryan, Y. Grishuk, P. Gros, J. Grosse-Oetringhaus, J.Y. Grossiord, R. Grosso, B. Grynyov, C. Guarnaccia, F. Guber, F. Guerin, R. Guernane, M. Guerzoni, A. Guichard, M. Guida, G. Guilloux, H. Gulkanyan, K. Gulbrandsen, T. Gunji, A. Gupta, V. Gupta, H.A. Gustafsson, H. Gutbrod, C. Hadjidakis, M. Haiduc, G. Hamar, H. Hamagaki, J. Hamblen, J.C. Hansen, P. Hardy, D. Hatzifotiadou, J.W. Harris, M. Hartig, A. Harutyunyan, A. Hayrapetyan, D. Hasch, D. Hasegan, J. Hehner, N. Heine, M. Heinz, H. Helstrup, A. Herghelegiu, S. Herlant, G. Herrera Corral, N. Herrmann, K. Hetland, P. Hille, H. Hinke, B. Hippolyte, M. Hoch, H. Hoebbel, H. Hoedlmoser, T. Horaguchi, M. Horner, P. Hristov, I. Hřivnáčová, S. Hu, C. Hu Guo, T. Humanic, A. Hurtado, D.S. Hwang, J.C. Ianigro, M. Idzik, S. Igolkin, R. Ilkaev, I. Ilkiv, M. Imhoff, P.G. Innocenti, E. Ionescu, M. Ippolitov, M. Irfan, C. Insa, M. Inuzuka, C. Ivan, A. Ivanov, M. Ivanov, V. Ivanov, P. Jacobs, A. Jacholkowski, L. Jančurová, R. Janik, M. Jasper, C. Jena, L. Jirden, D.P. Johnson, G.T. Jones, C. Jorgensen, F. Jouve, P. Jovanović, A. Junique, A. Jusko, H. Jung, W. Jung, K. Kadija, A. Kamal, R. Kamermans, S. Kapusta, A. Kaidalov, V. Kakoyan, S. Kalcher, E. Kang, J. Kapitan, V. Kaplin, K. Karadzhev, O. Karavichev, T. Karavicheva, E. Karpechev, K. Karpio, A. Kazantsev, U. Kebschull, R. Keidel, M. Mohsin Khan, A. Khanzadeev, Y. Kharlov, D. Kikola, B. Kileng, D. Kim, D.S. Kim, D.W. Kim, H.N. Kim, J.S. Kim, S. Kim, J.B. Kinson, S.K. Kiprich, I. Kisel, S. Kiselev, A. Kisiel, T. Kiss, V. Kiworra, J. Klay, C. Klein Bösing, M. Kliemant, A. Klimov, A. Klovning, A. Kluge, R. Kluit, S. Kniege, R. Kolevatov, T. Kollegger, A. Kolojvari, V. Kondratiev, E. Kornas, E. Koshurnikov, I. Kotov, R. Kour, M. Kowalski, S. Kox, K. Kozlov, I. Králik, F. Kramer, I. Kraus, A. Kravčáková, T. Krawutschke, M. Krivda, E. Kryshen, Y. Kucheriaev, A. Kugler, C. Kuhn, P. Kuijer, L. Kumar, N. Kumar, P. Kumpumaeki, A. Kurepin, A.N. Kurepin, S. Kushpil, V. Kushpil, M. Kutovsky, H. Kvaerno, M. Kweon, J.C. Labbé, F. Lackner, P. Ladron de Guevara, V. Lafage, P. La Rocca, M. Lamont, C. Lara, D.T. Larsen, G. Laurenti, C. Lazzeroni, Y. LeBornec, N. LeBris, C. LeGailliard, V. Lebedev, J. Lecoq, K.S. Lee, S.C. Lee, F. Lefévre, I. Legrand, T. Lehmann, L. Leistam, P. Lenoir, V. Lenti, H. Leon, I. Leon Monzon, P. Lévai, Q. Li, X. Li, F. Librizzi, R. Lietava, N. Lindegaard, V. Lindenstruth, C. Lippmann, M. Lisa, O.M. Listratenko, F. Littel, Y. Liu, J. Lo, V. Lobanov, V. Loginov, M. López Noriega, R. López-Ramírez, E. López Torres, P.M. Lorenzo, G. Løvhøiden, S. Lu, W. Ludolphs, M. Lunardon, L. Luquin, S. Lusso, J.R. Lutz, M. Luvisetto, V. Lyapin, A. Maevskaya, C. Magureanu, A. Mahajan, S. Majahan, T. Mahmoud, A. Mairani, D. Mahapatra, A. Makarov, I. Makhlyueva, M. Malek, T. Malkiewicz, D. Mal'Kevich, P. Malzacher, A. Mamonov, C. Manea, L.K. Mangotra, D. Maniero, V. Manko, F. Manso, V. Manzari, Y. Mao, A. Marcel, S. Marchini, J. Mareš, G.V. Margagliotti, A. Margotti, A. Marin, J.C. Marin, D. Marras, P. Martinengo, M.I. Martínez, A. Martinez-Davalos, G. Martínez Garcia, S. Martini, A. Marzari Chiesa, C. Marzocca, S. Masciocchi, M. Masera, M. Masetti, N.I. Maslov, A. Masoni, F. Massera, M. Mast, A. Mastroserio, Z.L. Matthews, B. Mayer, G. Mazza, M.D. Mazzaro, A. Mazzoni, F. Meddi, E. Meleshko, A. Menchaca-Rocha, S. Meneghini, M. Meoni, J. Mercado Perez, P. Mereu, O. Meunier, Y. Miake, A. Michalon, R. Michinelli, N. Miftakhov, M. Mignone, K. Mikhailov, J. Milosevic, Y. Minaev, F. Minafra, A. Mischke, D. Miśkowiec, V. Mitsyn, C. Mitu, B. Mohanty, D. Moisa, L. Molnar, M. Mondal, N. Mondal, L. Montaño Zetina, M. Monteno, M. Morando, M. Morel, S. Moretto, Th Morhardt, A. Morsch, T. Moukhanova, M. Mucchi, V. Muccifora, E. Mudnic, H. Müller, W. Müller, J. Munoz, D. Mura, L. Musa, J.F. Muraz, A. Musso, R. Nania, B. Nandi, E. Nappi, F. Navach, S. Navin, T. Nayak, S. Nazarenko, G. Nazarov, L. Nellen, F. Nendaz, A. Nianine, M. Nicassio, B.S. Nielsen, S. Nikolaev, V. Nikolic, S. Nikulin, V. Nikulin, B. Nilsen, M. Nitti, F. Noferini, P. Nomokonov, G. Nooren, F. Noto, D. Nouais, A. Nyiri, J. Nystrand, G. Odyniec, H. Oeschler, M. Oinonen,

M. Oldenburg, I. Oleks, E.K. Olsen, V. Onuchin, C. Oppedisano, F. Orsini, A. Ortiz-Velázquez, C. Oskamp, A. Oskarsson, F. Osmic, L. Österman, I. Otterlund, G. Ovrebekk, K. Oyama, M. Pachr, P. Pagano, G. Paić, C. Pajares, S. Pal, S. Pal, G. Pálla, A. Palmeri, G. Pancaldi, R. Panse, A. Pantaleo, G.S. Pappalardo, B. Pastirčák, C. Pastore, O. Patarakin, V. Paticchio, G. Patimo, A. Pavlinov, T. Pawlak, T. Peitzmann, Y. Pénichot, A. Pepato, H. Pereira, D. Peresunko, C. Perez, J. Perez Griffo, D. Perini, D. Perrino, W. Peryt, A. Pesci, V. Peskov, Y. Pestov, A.J. Peters, V. Petráček, A. Petridis, M. Petris, V. Petrov, V. Petrov, M. Petrovici, J. Peyré, S. Piano, A. Piccotti, P. Pichot, C. Piemonte, M. Pikna, R. Pilastrini, P. Pillot, O. Pinazza, B. Pini, L. Pinsky, V. Pinto Morais, V. Pismennaya, F. Piuz, R. Platt, M. Ploskon, S. Plumeri, J. Pluta, T. Pocheptsov, P. Podesta, F. Poggio, M. Poghosyan, T. Poghosyan, K. Polák, B. Polichtchouk, P. Polozov, V. Polyakov, B. Pommeresch, F. Pompei, A. Pop, S. Popescu, F. Posa, V. Pospíšil, B. Potukuchi, J. Pouthas, S. Prasad, R. Preghenella, F. Prino, L. Prodan, G. Prono, M.A. Protsenko, C.A. Pruneau, A. Przybyla, I. Pshenichnov, G. Puddu, P. Pujahari, A. Pulvirenti, A. Punin, V. Punin, J. Putschke, J. Quartieri, E. Quercigh, I. Rachevskaya, A. Rachevski, A. Rademakers, S. Radomski, A. Radu, J. Rak, L. Ramello, R. Raniwala, S. Raniwala, O.B. Rasmussen, J. Rasson, V. Razin, K. Read, J. Real, K. Redlich, C. Reichling, C. Renard, G. Renault, R. Renfordt, A.R. Reolon, A. Reshetin, J.P. Revol, K. Reygers, H. Ricaud, L. Riccati, R.A. Ricci, M. Richter, P. Riedler, L.M. Rigalleau, F. Riggi, W. Riegler, E. Rindel, J. Riso, A. Rivetti, M. Rizzi, V. Rizzi, M. Rodriguez Cahuantzi, K. Røed, D. Röhrich, S. Román-López, M. Romanato, R. Romita, F. Ronchetti, P. Rosinsky, P. Rosnet, S. Rossegger, A. Rossi, V. Rostchin, F. Rotondo, F. Roukoutakis, S. Rousseau, C. Roy, D. Roy, P. Roy, L. Royer, G. Rubin, A. Rubio, R. Rui, I. Rusanov, G. Russo, V. Ruuskanen, E. Ryabinkin, A. Rybicki, S. Sadovsky, K. Šafařík, R. Sahoo, J. Saini, P. Saiz, S. Salur, S. Sambyal, V. Samsonov, L. Šándor, A. Sandoval, H. Sann, J.C. Santiard, R. Santo, R. Santoro, G. Sargsyan, P. Saturnini, E. Scapparone, F. Scarlassara, B. Schackert, C. Schiaua, R. Schicker, T. Schioler, J.D. Schippers, C. Schmidt, H. Schmidt, R. Schneider, K. Schossmaier, J. Schukraft, Y. Schutz, K. Schwarz, K. Schweda, E. Schyns, G. Scioli, E. Scomparin, H. Snow, S. Sedykh, G. Segato, S. Sellitto, F. Semeria, S. Senyukov, H. Seppänen, S. Serci, L. Serkin, S. Serra, T. Sesselmann, A. Sevcenco, I. Sgura, G. Shabratova, R. Shahoyan, E. Sharkov, S. Sharma, K. Shigaki, K. Shileev, P. Shukla, A. Shurygin, M. Shurygina, Y. Sibiriak, E. Siddi, T. Siemiarczuk, M.H. Sigward, A. Silenzi, D. Silvermyr, R. Silvestri, E. Simili, V. Simion, R. Simon, L. Simonetti, R. Singaraju, V. Singhal, B. Sinha, T. Sinha, M. Siska, B. Sitár, M. Sitta, B. Skaali, P. Skowronski, M. Slodkowski, N. Smirnov, L. Smykov, R. Snellings, W. Snoeys, C. Soegaard, J. Soerensen, O. Sokolov, A. Soldatov, A. Soloviev, H. Soltveit, R. Soltz, W. Sommer, C. Soos, F. Soramel, S. Sorensen, D. Soyk, M. Spyropoulou-Stassinaki, J. Stachel, F. Staley, I. Stan, A. Stavinskiy, J. Steckert, G. Stefanini, G. Stefanek, T. Steinbeck, H. Stelzer, E. Stenlund, D. Stocco, M. Stockmeier, G. Stoicea, P. Stolpovsky, P. Strmeň, J.S. Stutzmann, G. Su, T. Sugitate, M. Šumbera, C. Suire, T. Susa, K. Sushil Kumar, D. Swoboda, J. Symons, I. Szarka, A. Szostak, M. Szuba, P. Szymanski, M. Tadel, C. Tagridis, L. Tan, D. Tapia Takaki, H. Taureg, A. Tauro, M. Tavlet, G. Tejeda Munoz, J. Thäder, R. Tieulent, P. Timmer, T. Tolyhy, N. Topilskaya, C. Torcato de Matos, H. Torii, L. Toscano, F. Tosello, A. Tournaire, T. Traczyk, G. Tröger, W. Tromeur, D. Truesdale, W. Trzaska, G. Tsiledakis, E. Tsilis, A. Tsvetkov, M. Turcato, R. Turrisi, M. Tuveri, T. Tveter, H. Tydesjo, L. Tykarski, K. Tywoniuk, E. Ugolini, K. Ullaland, J. Urbán, G.M. Urciuoli, G.L. Usai, M. Usseglio, A. Vacchi, M. Vala, F. Valiev, P. Vande Vyvre, A. Van Den Brink, N. Van Eijndhoven, N. Van Der Kolk, M. van Leeuwen, L. Vannucci, S. Vanzetto, J.P. Vanuxem, M.A. Vargas, R. Varma, A. Vascotto, A. Vasiliev, M. Vassiliou, P. Vasta, V. Vechernin, M. Venaruzzo, E. Vercellin, S. Vergara, W. Verhoeven, F. Veronese, I. Vetlitskiy, R. Vernet, V. Victorov, L. Vidak, G. Viesti, O. Vikhlyantsev, Z. Vilakazi, O. Villalobos Baillie, A. Vinogradov, L. Vinogradov, Y. Vinogradov, T. Virgili, Y. Viyogi, A. Vodopianov, G. Volpe, D. Vranic, J. Vrláková, B. Vulpescu, C. Wabnitz, V. Wagner, L. Wallet, R. Wan, Y. Wang, Y. Wang, R. Wheadon, R. Weis, Q. Wen, J. Wessels, J. Westergaard, J. Wiechula, A. Wiesenaecker, J. Wikne, A. Wilk, G. Wilk, C. Williams, N. Willis, B. Windelband, R. Witt, H. Woehri, K. Wyllie, C. Xu, C. Yang, H. Yang, F. Yermia, Z. Yin, Z. Yin, B. Yun Ky, I. Yushmanov,

B. Yuting, E. Zabrodin, S. Zagato, B. Zagreev, P. Zaharia, A. Zalite, G. Zampa, C. Zampolli, Y. Zanevskiy, A. Zarochentsev, O. Zaudtke, P. Závada, H. Zbroszczyk, A. Zepeda, V. Zeter, I. Zgura, M. Zhalov, D. Zhou, S. Zhou, G. Zhu, A. Zichichi, A. Zinchenko, G. Zinovjev, Y. Zoccarato, A. Zubarev, A. Zucchini, M. Zuffa, The ALICE experiment at the CERN LHC. J. Instrum. **3**(8), S08002 (2008). https://doi.org/10.1088/1748-0221/3/08/S08002

B. Allen, M. Alessandra Papa, B.F. Schutz, Optimal strategies for sinusoidal signal detection. Phys. Rev. D **66**(10), 102003 (2002). https://doi.org/10.1103/PhysRevD.66.102003

J. Alsing, B. Wandelt, S. Feeney, Massive optimal data compression and density estimation for scalable, likelihood-free inference in cosmology. Mon. Not. R. Astron. Soc. **477**(3), 2874–2885 (2018). https://doi.org/10.1093/mnras/sty819

J. Alsing, T. Charnock, S. Feeney, B. Wandelt, Fast likelihood-free cosmology with neural density estimators and active learning. Mon. Not. R. Astron. Soc. **488**(3), 4440–4458 (2019). https://doi.org/10.1093/mnras/stz1960

A.K. Aniyan, K. Thorat, Classifying radio galaxies with the convolutional neural network. Astrophy. J. Suppl. Ser. **230**(2), 20 (2017). https://doi.org/10.3847/1538-4365/aa7333

L. Ardizzone, J. Kruse, S. Wirkert, D. Rahner, E.W. Pellegrini, R.S. Klessen, L. Maier-Hein, C. Rother, U. Köthe, Analyzing inverse problems with invertible neural networks. arXiv e-prints, art. arXiv:1808.04730 (August 2018)

L. Ardizzone, C. Lüth, J. Kruse, C. Rother, U. Köthe, Guided image generation with conditional invertible neural networks. arXiv e-prints, art. arXiv:1907.02392 (July 2019)

ATLAS Collaboration, Anomaly detection search for new resonances decaying into a Higgs boson and a generic new particle X in hadronic final states using $\sqrt{s} = 13\,\text{TeV}$ pp collisions with the ATLAS detector. Technical report, CERN, Geneva (2022a). https://cds.cern.ch/record/2816323. All figures including auxiliary figures are available at https://atlas.web.cern.ch/Atlas/GROUPS/PHYSICS/CONFNOTES/ATLAS-CONF-2022-045

ATLAS Collaboration, Deep generative models for fast photon shower simulation in ATLAS. arXiv e-prints, art. arXiv:2210.06204 (October 2022b). https://doi.org/10.48550/arXiv.2210.06204

ATLAS Collaboration, G. Aad, E. Abat, J. Abdallah, A.A. Abdelalim, A. Abdesselam, O. Abdinov, B.A. Abi, M. Abolins, H. Abramowicz, E. Acerbi, B.S. Acharya, R. Achenbach, M. Ackers, D.L. Adams, F. Adamyan, T.N. Addy, M. Aderholz, C. Adorisio, P. Adragna, M. Aharrouche, S.P. Ahlen, F. Ahles, A. Ahmad, H. Ahmed, G. Aielli, P.F. Åkesson, T.P.A. Åkesson, A.V. Akimov, S.M. Alam, J. Albert, S. Albrand, M. Aleksa, I.N. Aleksandrov, M. Aleppo, F. Alessandria, C. Alexa, G. Alexander, T. Alexopoulos, G. Alimonti, M. Aliyev, P.P. Allport, S.E. Allwood-Spiers, A. Aloisio, J. Alonso, R. Alves, M.G. Alviggi, K. Amako, P. Amaral, S.P. Amaral, G. Ambrosini, G. Ambrosio, C. Amelung, V.V. Ammosov, A. Amorim, N. Amram, C. Anastopoulos, B. Anderson, K.J. Anderson, E.C. Anderssen, A. Andreazza, V. Andrei, L. Andricek, M.L. Andrieux, X.S. Anduaga, F. Anghinolfi, A. Antonaki, M. Antonelli, S. Antonelli, R. Apsimon, G. Arabidze, I. Aracena, Y. Arai, A.T.H. Arce, J.P. Archambault, J.F. Arguin, E. Arik, M. Arik, K.E. Arms, S.R. Armstrong, M. Arnaud, C. Arnault, A. Artamonov, S. Asai, S. Ask, B. Åsman, D. Asner, L. Asquith, K. Assamagan, A. Astbury, B. Athar, T. Atkinson, B. Aubert, B. Auerbach, E. Auge, K. Augsten, V.M. Aulchenko, N. Austin, G. Avolio, R. Avramidou, A. Axen, C. Ay, G. Azuelos, G. Baccaglioni, C. Bacci, H. Bachacou, K. Bachas, G. Bachy, E. Badescu, P. Bagnaia, D.C. Bailey, J.T. Baines, O.K. Baker, F. Ballester, F. Baltasar Dos Santos Pedrosa, E. Banas, D. Banfi, A. Bangert, V. Bansal, S.P. Baranov, S. Baranov, A. Barashkou, E.L. Barberio, D. Barberis, G. Barbier, P. Barclay, D.Y. Bardin, P. Bargassa, T. Barillari, M. Barisonzi, B.M. Barnett, R.M. Barnett, S. Baron, A. Baroncelli, M. Barone, A.J. Barr, F. Barreiro, J. Barreiro Guimarães da Costa, P. Barrillon, A. Barriuso Poy, N. Barros, V. Bartheld, H. Bartko, R. Bartoldus, S. Basiladze, J. Bastos, L.E. Batchelor, R.L. Bates, J.R. Batley, S. Batraneanu, M. Battistin, G. Battistoni, V. Batusov, F. Bauer, B. Bauss, D.E. Baynham, M. Bazalova, A. Bazan, P.H. Beauchemin, B. Beaugiraud, R.B. Beccherle, G.A. Beck, H.P. Beck, K.H. Becks, I. Bedajanek, A.J. Beddall, A. Beddall, P. Bednár, V.A. Bednyakov, C. Bee, S. Behar Harpaz, G.A.N. Belanger, C. Belanger-Champagne, B. Belhorma, P.J. Bell, W.H. Bell, G. Bella, F. Bellachia, L. Bellagamba, F. Bellina, G. Bellomo, M. Bellomo,

O. Beltramello, A. Belymam, S. Ben Ami, M. Ben Moshe, O. Benary, D. Benchekroun, C. Benchouk, M. Bendel, B.H. Benedict, N. Benekos, J. Benes, Y. Benhammou, G.P. Benincasa, D.P. Benjamin, J.R. Bensinger, K. Benslama, S. Bentvelsen, M. Beretta, D. Berge, E. Bergeaas, N. Berger, F. Berghaus, S. Berglund, F. Bergsma, J. Beringer, J. Bernabéu, K. Bernardet, C. Berriaud, T. Berry, H. Bertelsen, A. Bertin, F. Bertinelli, S. Bertolucci, N. Besson, A. Beteille, S. Bethke, W. Bialas, R.M. Bianchi, M. Bianco, O. Biebel, M. Bieri, M. Biglietti, H. Bilokon, M. Binder, S. Binet, N. Bingefors, A. Bingul, C. Bini, C. Biscarat, R. Bischof, M. Bischofberger, A. Bitadze, J.P. Bizzell, K.M. Black, R.E. Blair, J.J. Blaising, O. Blanch, G. Blanchot, C. Blocker, J. Blocki, A. Blondel, W. Blum, U. Blumenschein, C. Boaretto, G.J. Bobbink, A. Bocci, D. Bocian, R. Bock, M. Boehm, J. Boek, J.A. Bogaerts, A. Bogouch, C. Bohm, J. Bohm, V. Boisvert, T. Bold, V. Boldea, V.G. Bondarenko, R. Bonino, J. Bonis, W. Bonivento, P. Bonneau, M. Boonekamp, G. Boorman, M. Boosten, C.N. Booth, P.S.L. Booth, P. Booth, J.R.A. Booth, K. Borer, A. Borisov, I. Borjanovic, K. Bos, D. Boscherini, F. Bosi, M. Bosman, M. Bosteels, B. Botchev, H. Boterenbrood, D. Botterill, J. Boudreau, E.V. Bouhova-Thacker, C. Boulahouache, C. Bourdarios, M. Boutemeur, K. Bouzakis, G.R. Boyd, J. Boyd, B.H. Boyer, I.R. Boyko, N.I. Bozhko, S. Braccini, A. Braem, P. Branchini, G.W. Brandenburg, A. Brandt, O. Brandt, U. Bratzler, H.M. Braun, S. Bravo, I.P. Brawn, B. Brelier, J. Bremer, R. Brenner, S. Bressler, D. Breton, N.D. Brett, P. Breugnon, P.G. Bright-Thomas, F.M. Brochu, I. Brock, R. Brock, T.J. Brodbeck, E. Brodet, F. Broggi, Z. Broklova, C. Bromberg, G. Brooijmans, G. Brouwer, J. Broz, E. Brubaker, P.A. Bruckman de Renstrom, D. Bruncko, A. Bruni, G. Bruni, M. Bruschi, T. Buanes, N.J. Buchanan, P. Buchholz, I.A. Budagov, V. Büscher, L. Bugge, D. Buira-Clark, E.J. Buis, F. Bujor, T. Buran, H. Burckhart, D. Burckhart-Chromek, S. Burdin, R. Burns, E. Busato, J.J.F. Buskop, K.P. Buszello, F. Butin, J.M. Butler, C.M. Buttar, J. Butterworth, J.M. Butterworth, T. Byatt, S. Cabrera Urbán, E. Cabruja Casas, M. Caccia, D. Caforio, O. Cakir, P. Calafiura, G. Calderini, D. Calderón Terol, J. Callahan, L.P. Caloba, R. Caloi, D. Calvet, A. Camard, F. Camarena, P. Camarri, M. Cambiaghi, D. Cameron, J. Cammin, F. Campabadal Segura, S. Campana, V. Canale, J. Cantero, M.D.M. Capeans Garrido, I. Caprini, M. Caprini, M. Caprio, D. Caracinha, C. Caramarcu, Y. Carcagno, R. Cardarelli, C. Cardeira, L. Cardiel Sas, A. Cardini, T. Carli, G. Carlino, L. Carminati, B. Caron, S. Caron, C. Carpentieri, F.S. Carr, A.A. Carter, J.R. Carter, J. Carvalho, D. Casadei, M.P. Casado, M. Cascella, C. Caso, J. Castelo, V. Castillo Gimenez, N. Castro, F. Castrovillari, G. Cataldi, F. Cataneo, A. Catinaccio, J.R. Catmore, A. Cattai, S. Caughron, D. Cauz, A. Cavallari, P. Cavalleri, D. Cavalli, M. Cavalli-Sforza, V. Cavasinni, F. Ceradini, C. Cerna, C. Cernoch, A.S. Cerqueira, A. Cerri, F. Cerutti, M. Cervetto, S.A. Cetin, F. Cevenini, M. Chalifour, M. Chamizo llatas, A. Chan, J.W. Chapman, D.G. Charlton, S. Charron, S.V. Chekulaev, G.A. Chelkov, H. Chen, L. Chen, T. Chen, X. Chen, S. Cheng, T.L. Cheng, A. Cheplakov, V.F. Chepurnov, R. Cherkaoui El Moursli, D. Chesneanu, E. Cheu, L. Chevalier, J.L. Chevalley, F. Chevallier, V. Chiarella, G. Chiefari, L. Chikovani, A. Chilingarov, G. Chiodini, S. Chouridou, D. Chren, T. Christiansen, I.A. Christidi, A. Christov, M.L. Chu, J. Chudoba, A.G. Chuguev, G. Ciapetti, E. Cicalini, A.K. Ciftci, V. Cindro, M.D. Ciobotaru, A. Ciocio, M. Cirilli, M. Citterio, M. Ciubancan, J.V. Civera, A. Clark, W. Cleland, J.C. Clemens, B.C. Clement, C. Clément, D. Clements, R.W. Clifft, M. Cobal, A. Coccaro, J. Cochran, R. Coco, P. Coe, S. Coelli, E. Cogneras, C.D. Cojocaru, J. Colas, A.P. Colijn, C. Collard, C. Collins-Tooth, J. Collot, R. Coluccia, G. Comune, P. Conde Muiño, E. Coniavitis, M. Consonni, S. Constantinescu, C. Conta, F.A. Conventi, J. Cook, M. Cooke, N.J. Cooper-Smith, T. Cornelissen, M. Corradi, S. Correard, A. Corso-Radu, J. Coss, G. Costa, M.J. Costa, D. Costanzo, T. Costin, R. Coura Torres, L. Courneyea, C. Couyoumtzelis, G. Cowan, B.E. Cox, J. Cox, D.A. Cragg, K. Cranmer, J. Cranshaw, M. Cristinziani, G. Crosetti, C. Cuenca Almenar, S. Cuneo, A. Cunha, M. Curatolo, C.J. Curtis, P. Cwetanski, Z. Czyczula, S. D'Auria, M. D'Onofrio, A. Da Rocha Gesualdi Mello, P.V.M. Da Silva, R. Da Silva, W. Dabrowski, A. Dael, A. Dahlhoff, T. Dai, C. Dallapiccola, S.J. Dallison, J. Dalmau, C.H. Daly, M. Dam, D. Damazio, M. Dameri, K.M. Danielsen, H.O. Danielsson, R. Dankers, D. Dannheim, G. Darbo, P. Dargent, C. Daum, J.P. Dauvergne, M. David, T. Davidek, N. Davidson, R. Davidson, I. Dawson, J.W. Dawson, R.K. Daya, K. De,

R. de Asmundis, R. de Boer, S. DeCastro, N. DeGroot, P. de Jong, X. de La Broise, E. DeLa Cruz-Burelo, C. DeLa Taille, B. DeLotto, M. DeOliveira Branco, D. DePedis, P. de Saintignon, A. DeSalvo, U. DeSanctis, A. DeSanto, J.B. DeVivie DeRegie, G. DeZorzi, S. Dean, G. Dedes, D.V. Dedovich, P.O. Defay, R. Degele, M. Dehchar, M. Deile, C. DelPapa, J. DelPeso, T. DelPrete, E. Delagnes, P. Delebecque, A. Dell'Acqua, M. Della Pietra, D. della Volpe, M. Delmastro, P. Delpierre, N. Delruelle, P.A. Delsart, C. Deluca Silberberg, S. Demers, M. Demichev, P. Demierre, B. Demirköz, W. Deng, S.P. Denisov, C. Dennis, C.J. Densham, M. Dentan, J.E. Derkaoui, F. Derue, P. Dervan, K.K. Desch, A. Dewhurst, A. Di Ciaccio, L. Di Ciaccio, A. Di Domenico, A. Di Girolamo, B. Di Girolamo, S. Di Luise, A. Di Mattia, A. Di Simone, M.M. Diaz Gomez, E.B. Diehl, H. Dietl, J. Dietrich, W. Dietsche, S. Diglio, M. Dima, K. Dindar, B. Dinkespiler, C. Dionisi, R. Dipanjan, P. Dita, S. Dita, F. Dittus, S.D. Dixon, F. Djama, R. Djilkibaev, T. Djobava, M.A.B. do Vale, M. Dobbs, R. Dobinson, D. Dobos, E. Dobson, M. Dobson, J. Dodd, O.B. Dogan, T. Doherty, Y. Doi, J. Dolejsi, I. Dolenc, Z. Dolezal, B.A. Dolgoshein, E. Domingo, M. Donega, J. Dopke, D.E. Dorfan, O. Dorholt, A. Doria, A. Dos Anjos, M. Dosil, A. Dotti, M.T. Dova, J.D. Dowell, A.T. Doyle, G. Drake, D. Drakoulakos, Z. Drasal, J. Drees, N. Dressnandt, H. Drevermann, C. Driouichi, M. Dris, J.G. Drohan, J. Dubbert, T. Dubbs, E. Duchovni, G. Duckeck, A. Dudarev, M. Dührssen, H. Dür, I.P. Duerdoth, S. Duffin, L. Duflot, M.A. Dufour, N. Dumont Dayot, H. Duran Yildiz, D. Durand, A. Dushkin, R. Duxfield, M. Dwuznik, F. Dydak, D. Dzahini, S. Díez Cornell, M. Düren, W.L. Ebenstein, S. Eckert, S. Eckweiler, P. Eerola, I. Efthymiopoulos, U. Egede, K. Egorov, W. Ehrenfeld, T. Eifert, G. Eigen, K. Einsweiler, E. Eisenhandler, T. Ekelof, L.M. Eklund, M. El Kacimi, M. Ellert, S. Elles, N. Ellis, J. Elmsheuser, M. Elsing, R. Ely, D. Emeliyanov, R. Engelmann, M. Engström, P. Ennes, B. Epp, A. Eppig, V.S. Epshteyn, A. Ereditato, V. Eremin, D. Eriksson, I. Ermoline, J. Ernwein, D. Errede, S. Errede, M. Escalier, C. Escobar, X. Espinal Curull, B. Esposito, F. Esteves, F. Etienne, A.I. Etienvre, E. Etzion, H. Evans, V.N. Evdokimov, P. Evtoukhovitch, A. Eyring, L. Fabbri, C.W. Fabjan, C. Fabre, P. Faccioli, K. Facius, V. Fadeyev, R.M. Fakhrutdinov, S. Falciano, I. Falleau, A.C. Falou, Y. Fang, M. Fanti, A. Farbin, A. Farilla, J. Farrell, P. Farthouat, D. Fasching, F. Fassi, P. Fassnacht, D. Fassouliotis, F. Fawzi, L. Fayard, F. Fayette, R. Febbraro, O.L. Fedin, I. Fedorko, L. Feld, G. Feldman, L. Feligioni, C. Feng, E.J. Feng, J. Fent, A.B. Fenyuk, J. Ferencei, D. Ferguson, J. Ferland, W. Fernando, S. Ferrag, A. Ferrari, P. Ferrari, R. Ferrari, A. Ferrer, M.L. Ferrer, D. Ferrere, C. Ferretti, F. Ferro, M. Fiascaris, S. Fichet, F. Fiedler, V. Filimonov, A. Filipčič, A. Filippas, F. Filthaut, M. Fincke-Keeler, G. Finocchiaro, L. Fiorini, A. Firan, P. Fischer, M.J. Fisher, S.M. Fisher, V. Flaminio, J. Flammer, M. Flechl, I. Fleck, W. Flegel, P. Fleischmann, S. Fleischmann, C.M. Fleta Corral, F. Fleuret, T. Flick, J. Flix, L.R. Flores Castillo, M.J. Flowerdew, F. Föhlisch, M. Fokitis, T.M. Fonseca Martin, J. Fopma, D.A. Forbush, A. Formica, J.M. Foster, D. Fournier, A. Foussat, A.J. Fowler, H. Fox, P. Francavilla, D. Francis, S. Franz, J.T. Fraser, M. Fraternali, S. Fratianni, J. Freestone, R.S. French, K. Fritsch, D. Froidevaux, J.A. Frost, C. Fukunaga, J. Fulachier, E. Fullana Torregrosa, J. Fuster, C. Gabaldon, S. Gadomski, G. Gagliardi, P. Gagnon, E.J. Gallas, M.V. Gallas, B.J. Gallop, K.K. Gan, F.C. Gannaway, Y.S. Gao, V.A. Gapienko, A. Gaponenko, C. Garciá, M. Garcia-Sciveres, J.E. Garcìa Navarro, V. Garde, R.W. Gardner, N. Garelli, H. Garitaonandia, V.G. Garonne, J. Garvey, C. Gatti, G. Gaudio, O. Gaumer, V. Gautard, P. Gauzzi, I.L. Gavrilenko, C. Gay, J.C. Gayde, E.N. Gazis, E. Gazo, C.N.P. Gee, C. Geich-Gimbel, K. Gellerstedt, C. Gemme, M.H. Genest, S. Gentile, M.A. George, S. George, P. Gerlach, Y. Gernizky, C. Geweniger, H. Ghazlane, V.M. Ghete, P. Ghez, N. Ghodbane, B. Giacobbe, S. Giagu, V. Giakoumopoulou, V. Giangiobbe, F. Gianotti, B. Gibbard, A. Gibson, M.D. Gibson, S.M. Gibson, G.F. Gieraltowski, I. Gil Botella, L.M. Gilbert, M. Gilchriese, O. Gildemeister, V. Gilewsky, A.R. Gillman, D.M. Gingrich, J. Ginzburg, N. Giokaris, M.P. Giordani, C.G. Girard, P.F. Giraud, P. Girtler, D. Giugni, P. Giusti, B.K. Gjelsten, C. Glasman, A. Glazov, K.W. Glitza, G.L. Glonti, K.G. Gnanvo, J. Godlewski, T. Göpfert, C. Gössling, T. Göttfert, S. Goldfarb, D. Goldin, N. Goldschmidt, T. Golling, N.P. Gollub, P.J. Golonka, S.N. Golovnia, A. Gomes, J. Gomes, R. Gonçalo, A. Gongadze, A. Gonidec, S. Gonzalez, S. González de la Hoz, V. González Millán, M.L. Gonzalez Silva, B. Gonzalez-

Pineiro, S. González-Sevilla, M.J. Goodrick, J.J. Goodson, L. Goossens, P.A. Gorbounov, A. Gordeev, H. Gordon, I. Gorelov, G. Gorfine, B. Gorini, E. Gorini, A. Gorišek, E. Gornicki, S.A. Gorokhov, B.T. Gorski, S.V. Goryachev, V.N. Goryachev, M. Gosselink, M.I. Gostkin, M. Gouanère, I. Gough Eschrich, D. Goujdami, M. Goulette, I. Gousakov, J. Gouveia, S. Gowdy, C. Goy, I. Grabowska-Bold, V. Grabski, P. Grafström, C. Grah, K.J. Grahn, F. Grancagnolo, S. Grancagnolo, H. Grassmann, V. Gratchev, H.M. Gray, E. Graziani, B. Green, A. Greenall, D. Greenfield, D. Greenwood, I.M. Gregor, A. Grewal, E. Griesmayer, N. Grigalashvili, C. Grigson, A.A. Grillo, F. Grimaldi, K. Grimm, P.L.Y. Gris, Y. Grishkevich, H. Groenstege, L.S. Groer, J. Grognuz, M. Groh, E. Gross, J. Grosse-Knetter, M.E.M. Grothe, J. Grudzinski, C. Gruse, M. Gruwe, K. Grybel, P. Grybos, E.M. Gschwendtner, V.J. Guarino, C.J. Guicheney, G. Guilhem, T. Guillemin, J. Gunther, B. Guo, A. Gupta, L. Gurriana, V.N. Gushchin, P. Gutierrez, L. Guy, C. Guyot, C. Gwenlan, C.B. Gwilliam, A. Haas, S. Haas, C. Haber, G. Haboubi, R. Hackenburg, E. Hadash, H.K. Hadavand, C. Haeberli, R. Härtel, R. Haggerty, F. Hahn, S. Haider, Z. Hajduk, M. Hakimi, H. Hakobyan, H. Hakobyan, J. Haller, G.D. Hallewell, B. Hallgren, K. Hamacher, A. Hamilton, H. Han, L. Han, K. Hanagaki, M. Hance, P. Hanke, C.J. Hansen, F.H. Hansen, J.R. Hansen, J.B. Hansen, J.D. Hansen, P.H. Hansen, T. Hansl-Kozanecka, G. Hanson, P. Hansson, K. Hara, S. Harder, A. Harel, T. Harenberg, R. Harper, J.C. Hart, R.G.G. Hart, F. Hartjes, N. Hartman, T. Haruyama, A. Harvey, Y. Hasegawa, K. Hashemi, S. Hassani, M. Hatch, R.W. Hatley, T.G. Haubold, D. Hauff, F. Haug, S. Haug, M. Hauschild, R. Hauser, C. Hauviller, M. Havranek, B.M. Hawes, R.J. Hawkings, D. Hawkins, T. Hayler, H.S. Hayward, S.J. Haywood, E. Hazen, M. He, Y.P. He, S.J. Head, V. Hedberg, L. Heelan, F.E.W. Heinemann, M. Heldmann, S. Hellman, C. Helsens, R.C.W. Henderson, P.J. Hendriks, A.M. Henriques Correia, S. Henrot-Versille, F. Henry-Couannier, T. Henß, G. Herten, R. Hertenberger, L. Hervas, M. Hess, N.P. Hessey, A. Hicheur, A. Hidvegi, E. Higón-Rodriguez, D. Hill, J. Hill, J.C. Hill, N. Hill, S.J. Hillier, I. Hinchliffe, D. Hindson, C. Hinkelbein, T.A. Hodges, M.C. Hodgkinson, P. Hodgson, A. Hoecker, M.R. Hoeferkamp, J. Hoffman, A.E. Hoffmann, D. Hoffmann, H.F. Hoffmann, M. Holder, T.I. Hollins, G. Hollyman, A. Holmes, S.O. Holmgren, R. Holt, E. Holtom, T. Holy, R.J. Homer, Y. Homma, P. Homola, W. Honerbach, A. Honma, I. Hooton, T. Horazdovsky, C. Horn, S. Horvat, J.Y. Hostachy, T. Hott, S. Hou, M.A. Houlden, A. Hoummada, J. Hover, D.F. Howell, J. Hrivnac, I. Hruska, T. Hryn'ova, G.S. Huang, Z. Hubacek, F. Hubaut, F. Huegging, B.T. Huffman, E. Hughes, G. Hughes, R.E. Hughes-Jones, W. Hulsbergen, P. Hurst, M. Hurwitz, T. Huse, N. Huseynov, J. Huston, J. Huth, G. Iacobucci, M. Ibbotson, I. Ibragimov, R. Ichimiya, L. Iconomidou-Fayard, J. Idarraga, M. Idzik, P. Iengo, M.C. Iglesias Escudero, O. Igonkina, Y. Ikegami, M. Ikeno, Y. Ilchenko, Y. Ilyushenka, D. Imbault, P. Imbert, M. Imhaeuser, M. Imori, T. Ince, J. Inigo-Golfin, K. Inoue, P. Ioannou, M. Iodice, G. Ionescu, K. Ishii, M. Ishino, Y. Ishizawa, R. Ishmukhametov, C. Issever, H. Ito, A.V. Ivashin, W. Iwanski, H. Iwasaki, J.M. Izen, V. Izzo, J. Jackson, J.N. Jackson, M. Jaekel, S. Jagielski, M. Jahoda, V. Jain, K. Jakobs, J. Jakubek, E. Jansen, P.P.M. Jansweijer, R.C. Jared, G. Jarlskog, S. Jarp, P. Jarron, K. Jelen, I. Jen-La Plante, P. Jenni, A. Jeremie, P. Jez, S. Jézéquel, Y. Jiang, G. Jin, S. Jin, O. Jinnouchi, D. Joffe, L.G. Johansen, M. Johansen, K.E. Johansson, P. Johansson, K.A. Johns, K. Jon-And, M. Jones, R. Jones, R.W.L. Jones, T.W. Jones, T.J. Jones, A. Jones, O. Jonsson, K.K. Joo, D. Joos, M. Joos, C. Joram, S. Jorgensen, J. Joseph, P. Jovanovic, S.S. Junnarkar, V. Juranek, P. Jussel, V.V. Kabachenko, S. Kabana, M. Kaci, A. Kaczmarska, M. Kado, H. Kagan, S. Kagawa, S. Kaiser, E. Kajomovitz, S. Kakurin, L.V. Kalinovskaya, S. Kama, H. Kambara, N. Kanaya, A. Kandasamy, S. Kandasamy, M. Kaneda, V.A. Kantserov, J. Kanzaki, B. Kaplan, A. Kapliy, J. Kaplon, M. Karagounis, M. Karagoz Unel, K. Karr, P. Karst, V. Kartvelishvili, A.N. Karyukhin, L. Kashif, A. Kasmi, R.D. Kass, A. Kastanas, M. Kataoka, Y. Kataoka, E. Katsoufis, S. Katunin, K. Kawagoe, M. Kawai, T. Kawamoto, F. Kayumov, V.A. Kazanin, M.Y. Kazarinov, A. Kazarov, S.I. Kazi, J.R. Keates, R. Keeler, P.T. Keener, R. Kehoe, M. Keil, G.D. Kekelidze, M. Kelly, J. Kennedy, M. Kenyon, O. Kepka, N. Kerschen, B.P. Kerševan, S. Kersten, C. Ketterer, M. Khakzad, F. Khalilzade, H. Khandanyan, A. Khanov, D. Kharchenko, A. Khodinov, A.G. Kholodenko, A. Khomich, V.P. Khomutnikov, G. Khoriauli, N. Khovanskiy, V. Khovanskiy, E. Khramov,

J. Khubua, G. Kieft, J.A. Kierstead, G. Kilvington, H. Kim, H. Kim, S.H. Kim, P. Kind, B.T. King, J. Kirk, G.P. Kirsch, L.E. Kirsch, A.E. Kiryunin, D. Kisielewska, B. Kisielewski, T. Kittelmann, A.M. Kiver, H. Kiyamura, E. Kladiva, J. Klaiber-Lodewigs, K. Kleinknecht, A. Klier, A. Klimentov, C.R. Kline, R. Klingenberg, E.B. Klinkby, T. Klioutchnikova, P.F. Klok, S. Klous, E.E. Kluge, P. Kluit, M. Klute, S. Kluth, N.K. Knecht, E. Kneringer, E. Knezo, J. Knobloch, B.R. Ko, T. Kobayashi, M. Kobel, P. Kodys, A.C. König, S. König, L. Köpke, F. Koetsveld, T. Koffas, E. Koffeman, Z. Kohout, T. Kohriki, T. Kokott, G.M. Kolachev, H. Kolanoski, V. Kolesnikov, I. Koletsou, M. Kollefrath, S. Kolos, S.D. Kolya, A.A. Komar, J.R. Komaragiri, T. Kondo, Y. Kondo, N.V. Kondratyeva, T. Kono, A.I. Kononov, R. Konoplich, S.P. Konovalov, N. Konstantinidis, A. Kootz, S. Koperny, S.V. Kopikov, K. Korcyl, K. Kordas, V. Koreshev, A. Korn, I. Korolkov, V.A. Korotkov, H. Korsmo, O. Kortner, M.E. Kostrikov, V.V. Kostyukhin, M.J. Kotamäki, D. Kotchetkov, S. Kotov, V.M. Kotov, K.Y. Kotov, C. Kourkoumelis, A. Koutsman, S. Kovalenko, R. Kowalewski, H. Kowalski, T.Z. Kowalski, W. Kozanecki, A.S. Kozhin, V. Kral, V. Kramarenko, G. Kramberger, A. Kramer, O. Krasel, M.W. Krasny, A. Krasznahorkay, A. Krepouri, P. Krieger, P. Krivkova, G. Krobath, H. Kroha, J. Krstic, U. Kruchonak, H. Krüger, K. Kruger, Z.V. Krumshteyn, P. Kubik, W. Kubischta, T. Kubota, L.G. Kudin, J. Kudlaty, A. Kugel, T. Kuhl, D. Kuhn, V. Kukhtin, Y. Kulchitsky, N. Kundu, A. Kupco, M. Kupper, H. Kurashige, L.L. Kurchaninov, Y.A. Kurochkin, V. Kus, W. Kuykendall, P. Kuzhir, E.K. Kuznetsova, O. Kvasnicka, R. Kwee, D. La Marra, M. La Rosa, L. La Rotonda, L. Labarga, J.A. Labbe, C. Lacasta, F. Lacava, H. Lacker, D. Lacour, V.R. Lacuesta, E. Ladygin, R. Lafaye, B. Laforge, T. Lagouri, S. Lai, E. Lamanna, M. Lambacher, F. Lambert, W. Lampl, E. Lancon, U. Landgraf, M.P.J. Landon, H. Landsman, R.R. Langstaff, A.J. Lankford, F. Lanni, K. Lantzsch, A. Lanza, V.V. Lapin, S. Laplace, J.F. Laporte, V. Lara, T. Lari, A.V. Larionov, C. Lasseur, W. Lau, P. Laurelli, A. Lavorato, W. Lavrijsen, A.B. Lazarev, A.C. LeBihan, O. LeDortz, C. LeManer, M. LeVine, L. Leahu, M. Leahu, C. Lebel, M. Lechowski, T. LeCompte, F. Ledroit-Guillon, H. Lee, J.S.H. Lee, S.C. Lee, M. Lefebvre, R.P. Lefevre, M. Legendre, A. Leger, B.C. LeGeyt, C. Leggett, M. Lehmacher, G. Lehmann Miotto, M. Lehto, R. Leitner, D. Lelas, D. Lellouch, M. Leltchouk, V. Lendermann, K.J.C. Leney, T. Lenz, G. Lenzen, J. Lepidis, C. Leroy, J.R. Lessard, J. Lesser, C.G. Lester, M. Letheren, A. Leung Fook Cheong, J. Levêque, D. Levin, L.J. Levinson, M.S. Levitski, M. Lewandowska, M. Leyton, J. Li, W. Li, M. Liabline, Z. Liang, Z. Liang, B. Liberti, P. Lichard, W. Liebig, R. Lifshitz, D. Liko, H. Lim, M. Limper, S.C. Lin, A. Lindahl, F. Linde, L. Lindquist, S.W. Lindsay, V. Linhart, A.J. Lintern, A. Liolios, A. Lipniacka, T.M. Liss, A. Lissauer, J. List, A.M. Litke, S. Liu, T. Liu, Y. Liu, M. Livan, A. Lleres, G. Llosá Llácer, S.L. Lloyd, F. Lobkowicz, P. Loch, W.S. Lockman, T. Loddenkoetter, F.K. Loebinger, A. Loginov, C.W. Loh, T. Lohse, K. Lohwasser, M. Lokajicek, J. Loken, S. Lokwitz, M.C. Long, L. Lopes, D. Lopez Mateos, M.J. Losty, X. Lou, K.F. Loureiro, L. Lovas, J. Love, A. Lowe, M. Lozano Fantoba, F. Lu, J. Lu, L. Lu, H.J. Lubatti, S. Lucas, C. Luci, A. Lucotte, A. Ludwig, I. Ludwig, J. Ludwig, F. Luehring, D. Lüke, G. Luijckx, L. Luisa, D. Lumb, L. Luminari, E. Lund, B. Lund-Jensen, B. Lundberg, J. Lundquist, A. Lupi, N. Lupu, G. Lutz, D. Lynn, J. Lynn, J. Lys, V. Lysan, E. Lytken, J.M. López-Amengual, H. Ma, L.L. Ma, M. Maaß en, G. Maccarrone, G.G.R. Mace, D. Macina, R. Mackeprang, A. Macpherson, D. MacQueen, C. Macwaters, R.J. Madaras, W.F. Mader, R. Maenner, T. Maeno, P. Mättig, S. Mättig, C.A. Magrath, Y. Mahalalel, K. Mahboubi, G. Mahout, C. Maidantchik, A. Maio, G.M. Mair, K. Mair, Y. Makida, D. Makowiecki, P. Malecki, V.P. Maleev, F. Malek, D. Malon, S. Maltezos, V. Malychev, S. Malyukov, M. Mambelli, R. Mameghani, J. Mamuzic, A. Manabe, A. Manara, G. Manca, L. Mandelli, I. Mandić, M. Mandl, J. Maneira, M. Maneira, P.S. Mangeard, M. Mangin-Brinet, I.D. Manjavidze, W.A. Mann, S. Manolopoulos, A. Manousakis-Katsikakis, B. Mansoulie, A. Manz, A. Mapelli, L. Mapelli, L. March, J.F. Marchand, M. Marchesotti, M. Marcisovsky, A. Marin, C.N. Marques, F. Marroquim, R. Marshall, Z. Marshall, F.K. Martens, S. Marti i. Garcia, A.J. Martin, B. Martin, B. Martin, F.F. Martin, J.P. Martin, Ph Martin, G. Martinez, C. Martínez Lacambra, V. Martinez Outschoorn, A. Martini, J. Martins, T. Maruyama, F. Marzano, T. Mashimo, R. Mashinistov, J. Masik, A.L. Maslennikov, M. Maß, I. Massa, G. Massaro, N. Massol, M. Mathes, J. Matheson, P. Matricon, H. Matsumoto, H. Matsunaga,

J.M. Maugain, S.J. Maxfield, E.N. May, J.K. Mayer, C. Mayri, R. Mazini, M. Mazzanti, P. Mazzanti, E. Mazzoni, F. Mazzucato, S.P. McKee, R.L. McCarthy, C. McCormick, N.A. McCubbin, J. McDonald, K.W. McFarlane, S. McGarvie, H. McGlone, R.A. McLaren, S.J. McMahon, T.R. McMahon, T.J. McMahon, R.A. McPherson, M. Mechtel, D. Meder-Marouelli, M. Medinnis, R. Meera-Lebbai, C. Meessen, R. Mehdiyev, A. Mehta, K. Meier, H. Meinhard, J. Meinhardt, C. Meirosu, F. Meisel, A. Melamed-Katz, B.R. Mellado Garcia, P. Mendes Jorge, P. Mendez, S. Menke, C. Menot, E. Meoni, D. Merkl, L. Merola, C. Meroni, F.S. Merritt, I. Messmer, J. Metcalfe, S. Meuser, J.P. Meyer, T.C. Meyer, W.T. Meyer, V. Mialkovski, M. Michelotto, L. Micu, R. Middleton, P. Miele, A. Migliaccio, L. Mijović, G. Mikenberg, M. Mikestikova, M. Mikestikova, B. Mikulec, M. Mikuž, D.W. Miller, R.J. Miller, W. Miller, M. Milosavljevic, D.A. Milstead, S. Mima, A.A. Minaenko, M. Minano, I.A. Minashvili, A.I. Mincer, B. Mindur, M. Mineev, L.M. Mir, G. Mirabelli, L. Miralles Verge, S. Misawa, S. Miscetti, A. Misiejuk, A. Mitra, G.Y. Mitrofanov, V.A. Mitsou, P.S. Miyagawa, Y. Miyazaki, J.U. Mjörnmark, S. Mkrtchyan, D. Mladenov, T. Moa, M. Moch, A. Mochizuki, P. Mockett, P. Modesto, S. Moed, K. Mönig, N. Möser, B. Mohn, W. Mohr, S. Mohrdieck-Möck, A.M. Moisseev, R.M. Moles Valls, J. Molina-Perez, A. Moll, G. Moloney, R. Mommsen, L. Moneta, E. Monnier, G. Montarou, S. Montesano, F. Monticelli, R.W. Moore, T.B. Moore, G.F. Moorhead, A. Moraes, J. Morel, A. Moreno, D. Moreno, P. Morettini, D. Morgan, M. Morii, J. Morin, A.K. Morley, G. Mornacchi, M.C. Morone, S.V. Morozov, E.J. Morris, J. Morris, M.C. Morrissey, H.G. Moser, M. Mosidze, A. Moszczynski, S.V. Mouraviev, T. Mouthuy, T.H. Moye, E.J.W. Moyse, J. Mueller, M. Müller, A. Muijs, T.R. Muller, A. Munar, D.J. Munday, K. Murakami, R. Murillo Garcia, W.J. Murray, A.G. Myagkov, M. Myska, K. Nagai, Y. Nagai, K. Nagano, Y. Nagasaka, A.M. Nairz, D. Naito, K. Nakamura, Y. Nakamura, I. Nakano, G. Nanava, A. Napier, M. Nassiakou, I. Nasteva, N.R. Nation, T. Naumann, F. Nauyock, S.K. Nderitu, H.A. Neal, E. Nebot, P. Nechaeva, A. Neganov, A. Negri, S. Negroni, C. Nelson, S. Nemecek, P. Nemethy, A.A. Nepomuceno, M. Nessi, S.Y. Nesterov, L. Neukermans, P. Nevski, F.M. Newcomer, A. Nichols, C. Nicholson, R. Nicholson, R.B. Nickerson, R. Nicolaidou, G. Nicoletti, B. Nicquevert, M. Niculescu, J. Nielsen, T. Niinikoski, M.J. Niinimaki, N. Nikitin, K. Nikolaev, I. Nikolic-Audit, K. Nikolopoulos, H. Nilsen, B.S. Nilsson, P. Nilsson, A. Nisati, R. Nisius, L.J. Nodulman, M. Nomachi, H. Nomoto, J.M. Noppe, M. Nordberg, O. Norniella Francisco, P.R. Norton, J. Novakova, M. Nowak, M. Nozaki, R. Nunes, G. Nunes Hanninger, T. Nunnemann, T. Nyman, P. O'Connor, S.W. O'Neale, D.C. O'Neil, M. O'Neill, V. O'Shea, F.G. Oakham, H. Oberlack, M. Obermaier, P. Oberson, A. Ochi, W. Ockenfels, S. Odaka, I. Odenthal, G.A. Odino, H. Ogren, S.H. Oh, T. Ohshima, H. Ohshita, H. Okawa, M. Olcese, A.G. Olchevski, C. Oliver, J. Oliver, M. Olivo Gomez, A. Olszewski, J. Olszowska, C. Omachi, A. Onea, A. Onofre, C.J. Oram, G. Ordonez, M.J. Oreglia, F. Orellana, Y. Oren, D. Orestano, I.O. Orlov, R.S. Orr, F. Orsini, L.S. Osborne, B. Osculati, C. Osuna, R. Otec, R. Othegraven, B. Ottewell, F. Ould-Saada, A. Ouraou, Q. Ouyang, O.K. Øye, V.E. Ozcan, K. Ozone, N. Ozturk, A. Pacheco Pages, S. Padhi, C. Padilla Aranda, E. Paganis, F. Paige, P.M. Pailler, K. Pajchel, S. Palestini, J. Palla, D. Pallin, M.J. Palmer, Y.B. Pan, N. Panikashvili, V.N. Panin, S. Panitkin, D. Pantea, M. Panuskova, V. Paolone, A. Paoloni, I. Papadopoulos, T. Papadopoulou, I. Park, W. Park, M.A. Parker, S. Parker, C. Parkman, F. Parodi, J.A. Parsons, U. Parzefall, E. Pasqualucci, G. Passardi, A. Passeri, M.S. Passmore, F. Pastore, Fr. Pastore, S. Pataraia, D. Pate, J.R. Pater, S. Patricelli, T. Pauly, E. Pauna, L.S. Peak, S.J.M. Peeters, M. Peez, E. Pei, S.V. Peleganchuk, G. Pellegrini, R. Pengo, J. Pequenao, M. Perantoni, A. Perazzo, A. Pereira, E. Perepelkin, V.J.O. Perera, E. Perez Codina, V. Perez Reale, I. Peric, L. Perini, H. Pernegger, E. Perrin, R. Perrino, P. Perrodo, G. Perrot, P. Perus, V.D. Peshekhonov, E. Petereit, J. Petersen, T.C. Petersen, P.J.F. Petit, C. Petridou, E. Petrolo, F. Petrucci, R. Petti, M. Pezzetti, B. Pfeifer, A. Phan, A.W. Phillips, P.W. Phillips, G. Piacquadio, M. Piccinini, A. Pickford, R. Piegaia, S. Pier, J.E. Pilcher, A.D. Pilkington, M.A. Pimenta Dos Santos, J. Pina, J.L. Pinfold, J. Ping, J. Pinhão, B. Pinto, O. Pirotte, R. Placakyte, A. Placci, M. Plamondon, W.G. Plano, M.A. Pleier, A.V. Pleskach, S. Podkladkin, F. Podlyski, P. Poffenberger, L. Poggioli, M. Pohl, I. Polak, G. Polesello, A. Policicchio, A. Polini, V. Polychronakos, D.M. Pomarede, K. Pommès,

P. Ponsot, L. Pontecorvo, B.G. Pope, R. Popescu, D.S. Popovic, A. Poppleton, J. Popule, X. Portell Bueso, C. Posch, G.E. Pospelov, P. Pospichal, S. Pospisil, M. Postranecky, I.N. Potrap, C.J. Potter, G. Poulard, A. Pousada, J. Poveda, R. Prabhu, P. Pralavorio, S. Prasad, J. Prast, S. Prat, M. Prata, R. Pravahan, T. Preda, K. Pretzl, L. Pribyl, D. Price, L.E. Price, M.J. Price, P.M. Prichard, D. Prieur, M. Primavera, D. Primor, K. Prokofiev, E. Prosso, J. Proudfoot, H. Przysiezniak, C. Puigdengoles, J. Purdham, M. Purohit, P. Puzo, A.N. Pylaev, Y. Pylypchenko, M. Qi, J. Qian, W. Qian, Z. Qian, D. Qing, A. Quadt, D.R. Quarrie, W.B. Quayle, J.J. Rabbers, V. Radeka, J.M. Rafi, F. Ragusa, A.M. Rahimi, D. Rahm, C. Raine, B. Raith, S. Rajagopalan, S. Rajek, H. Rammer, M. Ramstedt, S. Rangod, P.N. Ratoff, T. Raufer, F. Rauscher, E. Rauter, M. Raymond, A.L. Reads, D. Rebuzzi, G.R. Redlinger, K. Reeves, M. Rehak, A. Reichold, E. Reinherz-Aronis, I. Reisinger, D. Reljic, C. Rembser, Z. Ren, S.R.C. Renaudin-Crepe, P. Renkel, B. Rensch, S. Rescia, M. Rescigno, S. Resconi, B. Resende, P. Rewiersma, J. Rey, M. Rey-Campagnolle, E. Rezaie, P. Reznicek, R.A. Richards, J.P. Richer, R.H. Richter, R. Richter, E. Richter-Was, M. Ridel, W. Riegler, S. Rieke, M. Rijpstra, M. Rijssenbeek, A. Rimoldi, R.R. Rios, I. Riu Dachs, M. Rivline, G. Rivoltella, F. Rizatdinova, S.H. Robertson, A. Robichaud-Veronneau, S. Robins, D. Robinson, A. Robson, J.H. Rochford, C. Roda, S. Rodier, S. Roe, O. Røhne, F. Rohrbach, J. Roldán, S. Rolli, J.B. Romance, A. Romaniouk, V.M. Romanov, G. Romeo, L. Roos, E. Ros, S. Rosati, F. Rosenbaum, G.A. Rosenbaum, E.I. Rosenberg, L. Rosselet, L.P. Rossi, L. Rossi, M. Rotaru, J. Rothberg, I. Rottländer, D. Rousseau, A. Rozanov, Y. Rozen, R. Ruber, B. Ruckert, G. Rudolph, F. Rühr, F. Ruggieri, G. Ruggiero, H. Ruiz, A. Ruiz-Martinez, E. Rulikowska-Zarebska, V. Rumiantsev, L. Rumyantsev, K. Runge, O. Runolfsson, N.A. Rusakovich, D.R. Rust, J.P. Rutherfoord, C. Ruwiedel, Y.F. Ryabov, V. Ryadovikov, P. Ryan, G. Rybkine, J. Sá da Costa, A.F. Saavedra, S. Saboumazrag, H. F-W Sadrozinski, R. Sadykov, H. Sakamoto, P. Sala, A. Salamon, M. Saleem, D. Salihagic, J. Salt, O. Saltó Bauza, B.M. Salvachúa Ferrando, D. Salvatore, A. Salzburger, D. Sampsonidis, B.H. Samset, C.A. Sánchez Sánchez, M.A. Sanchis Lozano, E. Sanchis Peris, H. Sandaker, H.G. Sander, M. Sandhoff, S. Sandvoss, D.P.C. Sankey, B. Sanny, S. Sansone, A. Sansoni, C. Santamarina Rios, J. Santander, L. Santi, C. Santoni, R. Santonico, J. Santos, M. Sapinski, J.G. Saraiva, F. Sarri, O. Sasaki, T. Sasaki, N. Sasao, I. Satsounkevitch, D. Sauvage, G. Sauvage, P. Savard, A.Y. Savine, V. Savinov, A. Savoy-Navarro, P. Savva, D.H. Saxon, L.P. Says, C. Sbarra, E. Sbrissa, A. Sbrizzi, D.A. Scannicchio, J. Schaarschmidt, P. Schacht, U. Schäfer, A.C. Schaffer, D. Schaile, M. Schaller, A.G. Schamov, V.A. Schegelsky, D. Scheirich, M. Schernau, M.I. Scherzer, C. Schiavi, H. Schick, J. Schieck, P. Schieferdecker, M. Schioppa, G. Schlager, S. Schlenker, J.L. Schlereth, P. Schmid, M.P. Schmidt, C. Schmitt, K. Schmitt, M. Schmitz, H. Schmücker, T. Schoerner, R.C. Scholte, M. Schott, D. Schouten, M. Schram, A. Schricker, D. Schroff, S. Schuh, H.W. Schuijlenburg, G. Schuler, J. Schultes, H.C. Schultz-Coulon, J. Schumacher, M. Schumacher, Ph Schune, A. Schwartzman, D. Schweiger, Ph Schwemling, C. Schwick, R. Schwienhorst, R. Schwierz, J. Schwindling, W.G. Scott, H. Secker, E. Sedykh, N. Seguin-Moreau, E. Segura, S.C. Seidel, A. Seiden, J.M. Seixas, G. Sekhniaidze, D.M. Seliverstov, B. Selldén, M. Seman, N. Semprini-Cesari, C. Serfon, L. Serin, R. Seuster, H. Severini, M.E. Sevior, K.A. Sexton, A. Sfyrla, T.P. Shah, L. Shan, J.T. Shank, M. Shapiro, P.B. Shatalov, L. Shaver, C. Shaw, T.G. Shears, P. Sherwood, A. Shibata, P. Shield, S. Shilov, M. Shimojima, T. Shin, M. Shiyakova, A. Shmeleva, M. Shoa, M.J. Shochet, M.A. Shupe, P. Sicho, A. Sidoti, A. Siebel, M. Siebel, J. Siegrist, D. Sijacki, J. Silva, S.B. Silverstein, V. Simak, Lj Simic, S. Simion, B. Simmons, M. Simonyan, P. Sinervo, V. Sipica, G. Siragusa, A.N. Sisakyan, S. Sivoklokov, J. Sjölin, P. Skubic, N. Skvorodnev, P. Slattery, T. Slavicek, K. Sliwa, T.J. Sloan, J. Sloper, V. Smakhtin, A. Small, S. Yu Smirnov, Y. Smirnov, L. Smirnova, O. Smirnova, N.A. Smith, B.C. Smith, D.S. Smith, J. Smith, K.M. Smith, B. Smith, M. Smizanska, K. Smolek, A.A. Snesarev, S.W. Snow, J. Snow, J. Snuverink, S. Snyder, M. Soares, S. Soares, R. Sobie, J. Sodomka, M. Söderberg, A. Soffer, C.A. Solans, M. Solar, D. Sole, E. Solfaroli Camillocci, A.A. Solodkov, O.V. Solov'yanov, I. Soloviev, R. Soluk, J. Sondericker, V. Sopko, B. Sopko, M. Sorbi, J. Soret Medel, M. Sosebee, V.V. Sosnovtsev, L. Sospedra Suay, A. Soukharev, J. Soukup, S. Spagnolo, F. Spano, P. Speckmayer, M. Spegel, E. Spencer, R. Spighi, G. Spigo,

F. Spila, E. Spiriti, R. Spiwoks, L. Spogli, M. Spousta, G. Sprachmann, B. Spurlock, R.D. St. Denis, T. Stahl, R.J. Staley, R. Stamen, S.N. Stancu, E. Stanecka, R.W. Stanek, C. Stanescu, S. Stapnes, E.A. Starchenko, P. Staroba, J. Stastny, A. Staude, P. Stavina, M. Stavrianakou, G. Stavropoulos, E. Stefanidis, J.L. Steffens, I. Stekl, H.J. Stelzer, H. Stenzel, G. Stewart, T.D. Stewart, W. Stiller, T. Stockmanns, M. Stodulski, S. Stonjek, A. Stradling, A. Straessner, J. Strandberg, A. Strandlie, M. Strauss, V. Strickland, D. Striegel, P. Strizenec, R. Ströhmer, D.M. Strom, J.A. Strong, R. Stroynowski, B. Stugu, I. Stumer, D. Su, S. Subramania, S.I. Suchkov, Y. Sugaya, T. Sugimoto, M. Suk, V.V. Sulin, S. Sultanov, Z. Sun, B. Sundal, S. Sushkov, G. Susinno, P. Sutcliffe, M.R. Sutton, Yu M. Sviridov, I. Sykora, R.R. Szczygiel, B. Szeless, T. Szymocha, J. Sánchez, D. Ta, S. Taboada Gameiro, M. Tadel, R. Tafirout, A. Taga, H. Takai, R. Takashima, H. Takeda, T. Takeshita, M. Talby, A. Talyshev, M.C. Tamsett, J. Tanaka, K. Tanaka, R. Tanaka, S. Tanaka, S. Tanaka, Y. Tanaka, G.P. Tappern, S. Tapprogge, S. Tarem, F. Tarrade, J. Tarrant, G. Tartarelli, P. Tas, M. Tasevsky, Y. Tayalati, F.E. Taylor, G. Taylor, G.N. Taylor, R.P. Taylor, V. Tcherniatine, F. Tegenfeldt, P. Teixeira-Dias, H. Ten Kate, P.K. Teng, R. Ter-Antonyan, S. Terada, J. Terron, M. Terwort, R.J. Teuscher, C.M. Tevlin, J. Thadome, J. Thion, M. Thioye, A. Thomas, J.P. Thomas, T.L. Thomas, E. Thomas, R.J. Thompson, A.S. Thompson, R.P. Thun, T. Tic, V.O. Tikhomirov, Y.A. Tikhonov, S. Timm, C.J.W.P. Timmermans, P. Tipton, F.J. Tique Aires Viegas, S. Tisserant, M. Titov, J. Tobias, V.M. Tocut, B. Toczek, S. Todorova-Nova, J. Tojo, S. Tokár, K. Tokushuku, L. Tomasek, M. Tomasek, F. Tomasz, M. Tomoto, D. Tompkins, L. Tompkins, K. Toms, A. Tonazzo, G. Tong, A. Tonoyan, C. Topfel, N.D. Topilin, E. Torrence, J.G. Torres Pais, J. Toth, F. Touchard, D.R. Tovey, S.N. Tovey, E.F. Towndrow, T. Trefzger, M. Treichel, J. Treis, L. Tremblet, W. Tribanek, A. Tricoli, I.M. Trigger, G. Trilling, S. Trincaz-Duvoid, M.F. Tripiana, W. Trischuk, Z. Trka, B. Trocmé, C. Troncon, J. C-L Tseng, I. Tsiafis, P.V. Tsiareshka, G. Tsipolitis, E.G. Tskhadadze, I.I. Tsukerman, V. Tsulaia, S. Tsuno, M. Turala, I. Turk Cakir, E. Turlay, P.M. Tuts, M.S. Twomey, M. Tyndel, D. Typaldos, H. Tyrvainen, E. Tzamarioudaki, G. Tzanakos, I. Ueda, M. Uhrmacher, F. Ukegawa, M. Ullán Comes, G. Unal, D.G. Underwood, A. Undrus, G. Unel, Y. Unno, E. Urkovsky, G. Usai, Y. Usov, L. Vacavant, V. Vacek, B. Vachon, S. Vahsen, C. Valderanis, J. Valenta, P. Valente, A. Valero, S. Valkar, J.A. Valls Ferrer, H. Van der Bij, H. van der Graaf, E. van der Kraaij, B. Van Eijk, N. van Eldik, P. van Gemmeren, Z. van Kesteren, I. van Vulpen, R. Van Berg, W. Vandelli, G. Vandoni, A. Vaniachine, F. Vannucci, M. Varanda, F. Varela Rodriguez, R. Vari, E.W. Varnes, D. Varouchas, A. Vartapetian, K.E. Varvell, V.I. Vassilakopoulos, L. Vassilieva, E. Vataga, L. Vaz, F. Vazeille, P. Vedrine, G. Vegni, J.J. Veillet, C. Vellidis, F. Veloso, R. Veness, S. Veneziano, A. Ventura, S. Ventura, V. Vercesi, M. Verducci, W. Verkerke, J.C. Vermeulen, L. Vertogardov, M.C. Vetterli, I. Vichou, T. Vickey, G.H.A. Viehhauser, E. Vigeolas, M. Villa, E.G. Villani, J. Villate, I. Villella, E. Vilucchi, P. Vincent, H. Vincke, M.G. Vincter, V.B. Vinogradov, M. Virchaux, S. Viret, J. Virzi, A. Vitale, I. Vivarelli, R. Vives, F. Vives Vaques, S. Vlachos, H. Vogt, P. Vokac, C.F. Vollmer, M. Volpi, G. Volpini, R. von Boehn-Buchholz, H. von der Schmitt, E. von Toerne, V. Vorobel, A.P. Vorobiev, A.S. Vorozhtsov, S.B. Vorozhtsov, M. Vos, K.C. Voss, R. Voss, J.H. Vossebeld, A.S. Vovenko, N. Vranjes, V. Vrba, M. Vreeswijk, T. Vu Anh, B. Vuaridel, M. Vudragovic, V. Vuillemin, R. Vuillermet, A. Wänanen, H. Wahlen, J. Walbersloh, R. Walker, W. Walkowiak, R. Wall, R.S. Wallny, S. Walsh, C. Wang, J.C. Wang, F. Wappler, A. Warburton, C.P. Ward, G.P. Warner, M. Warren, M. Warsinsky, R. Wastie, P.M. Watkins, A.T. Watson, G. Watts, A.T. Waugh, B.M. Waugh, C. Weaverdyck, M. Webel, G. Weber, J. Weber, M. Weber, P. Weber, A.R. Weidberg, P.M. Weilhammer, J. Weingarten, C. Weiser, H. Wellenstein, H.P. Wellisch, P.S. Wells, A. Wemans, M. Wen, T. Wenaus, S. Wendler, T. Wengler, S. Wenig, N. Wermes, P. Werneke, P. Werner, U. Werthenbach, S.J. Wheeler-Ellis, S.P. Whitaker, A. White, M.J. White, S. White, D. Whittington, F. Wicek, D. Wicke, F.J. Wickens, W. Wiedenmann, M. Wielers, P. Wienemann, M. Wiesmann, M. Wiesmann, T. Wijnen, A. Wildauer, I. Wilhelm, H.G. Wilkens, H.H. Williams, W. Willis, S. Willocq, I. Wilmut, J.A. Wilson, A. Wilson, I. Wingerter-Seez, L. Winton, W. Witzeling, T. Wlodek, E. Woehrling, M.W. Wolter, H. Wolters, B. Wosiek, J. Wotschack, M.J. Woudstra, C. Wright, S.L. Wu, X. Wu, J. Wuestenfeld, R. Wunstorf, S. Xella-Hansen, A. Xiang, S. Xie, Y. Xie, G. Xu, N. Xu, A. Yamamoto, S. Yamamoto,

H. Yamaoka, Y. Yamazaki, Z. Yan, H. Yang, J.C. Yang, S. Yang, U.K. Yang, Y. Yang, Z. Yang, W.M. Yao, Y. Yao, K. Yarradoddi, Y. Yasu, J. Ye, M. Yilmaz, R. Yoosoofmiya, K. Yorita, H. Yoshida, R. Yoshida, C. Young, S.P. Youssef, D. Yu, J. Yu, M. Yu, X. Yu, J. Yuan, A. Yurkewicz, V.G. Zaets, R. Zaidan, A.M. Zaitsev, J. Zajac, Z. Zajacova, A. Yu Zalite, Yo K. Zalite, L. Zanello, P. Zarzhitsky, A. Zaytsev, M. Zdrazil, C. Zeitnitz, M. Zeller, P.F. Zema, C. Zendler, A.V. Zenin, T. Zenis, Z. Zenonos, S. Zenz, D. Zerwas, H. Zhang, J. Zhang, W. Zheng, X. Zhang, L. Zhao, T. Zhao, X. Zhao, Z. Zhao, A. Zhelezko, A. Zhemchugov, S. Zheng, L. Zhichao, B. Zhou, N. Zhou, S. Zhou, Y. Zhou, C.G. Zhu, H.Z. Zhu, X.A. Zhuang, V. Zhuravlov, B. Zilka, N.I. Zimin, S. Zimmermann, M. Ziolkowski, R. Zitoun, L. Zivkovic, V.V. Zmouchko, G. Zobernig, A. Zoccoli, M.M. Zoeller, Y. Zolnierowski, A. Zsenei, M. zur Nedden, and V. Zychacek. The ATLAS experiment at the CERN Large Hadron Collider. J. Instrum. **3**(8), S08003 (2008). https://doi.org/10.1088/1748-0221/3/08/S08003

M. Bähr, S. Gieseke, M.A. Gigg, D. Grellscheid, K. Hamilton, O. Latunde-Dada, S. Plätzer, P. Richardson, M.H. Seymour, A. Sherstnev, B.R. Webber, Herwig++ physics and manual. Eur. Phys. J. C **58**(4), 639–707 (2008). https://doi.org/10.1140/epjc/s10052-008-0798-9

C.A.L. Bailer-Jones, M. Irwin, T. von Hippel, Automated classification of stellar spectra - II. Two-dimensional classification with neural networks and principal components analysis. Mon. Not. R. Astron. Soc. **298**(2), 361–377 (1998). https://doi.org/10.1046/j.1365-8711.1998.01596.x

V. Balakrishnan, D. Champion, E. Barr, M. Kramer, R. Sengar, M. Bailes, Pulsar candidate identification using semi-supervised generative adversarial networks. Mon. Not. R. Astron. Soc. **505**(1), 1180–1194 (2021). https://doi.org/10.1093/mnras/stab1308

V. Balakrishnan, D. Champion, E. Barr, M. Kramer, V. Venkatraman Krishnan, R.P. Eatough, R. Sengar, M. Bailes, Coherent search for binary pulsars across all Five Keplerian parameters in radio observations using the template-bank algorithm. Mon. Not. R. Astron. Soc. **511**(1), 1265–1284 (2022). https://doi.org/10.1093/mnras/stab3746

N.M. Ball, R.J. Brunner, Data mining and machine learning in astronomy. Int. J. Mod. Phys. D **19** (7), 1049–1106 (2010). https://doi.org/10.1142/S0218271810017160

J.K. Banfield, O.I. Wong, K.W. Willett, R.P. Norris, L. Rudnick, S.S. Shabala, B.D. Simmons, C. Snyder, A. Garon, N. Seymour, E. Middelberg, H. Andernach, C.J. Lintott, K. Jacob, A.D. Kapińska, M.Y. Mao, K.L. Masters, M.J. Jarvis, K. Schawinski, E. Paget, R. Simpson, H.R. Klöckner, S. Bamford, T. Burchell, K.E. Chow, G. Cotter, L. Fortson, I. Heywood, T.W. Jones, S. Kaviraj, Á. R. López-Sánchez, W.P. Maksym, K. Polsterer, K. Borden, R.P. Hollow, L. Whyte, Radio Galaxy Zoo: host galaxies and radio morphologies derived from visual inspection. Mon. Not. R. Astron. Soc. **453**(3), 2326–2340 (2015). https://doi.org/10.1093/mnras/stv1688

J.A. Barkai, M.A.W. Verheijen, E.T. Martínez, M.H.F. Wilkinson, A comparative study of source-finding techniques in HI emission line cubes using SoFiA, MTObjects, and supervised deep learning. arXiv e-prints, art. arXiv:2211.12809 (November 2022)

D. Baron, Machine learning in astronomy: a practical overview. arXiv e-prints, art. arXiv:1904.07248 (April 2019)

D. Baron, D. Poznanski, The weirdest SDSS galaxies: results from an outlier detection algorithm. Mon. Not. R. Astron. Soc. **465**(4), 4530–4555 (2017). https://doi.org/10.1093/mnras/stw3021

D.J. Bastien, A.M.M. Scaife, H. Tang, M. Bowles, F. Porter, Structured variational inference for simulating populations of radio galaxies. Mon. Not. R. Astron. Soc. **503**(3), 3351–3370 (2021). https://doi.org/10.1093/mnras/stab588

S.D. Bates, M. Bailes, B.R. Barsdell, N.D.R. Bhat, M. Burgay, S. Burke-Spolaor, D.J. Champion, P. Coster, N. D'Amico, A. Jameson, S. Johnston, M.J. Keith, M. Kramer, L. Levin, A. Lyne, S. Milia, C. Ng, C. Nietner, A. Possenti, B. Stappers, D. Thornton, W. van Straten, The high time resolution universe pulsar survey — VI. An artificial neural network and timing of 75 pulsars. Mon. Not. R. Astron. Soc. **427**(2), 1052–1065 (2012). https://doi.org/10.1111/j.1365-2966.2012.22042.x

B. Becker, M. Vaccari, M. Prescott, T. Grobler, CNN architecture comparison for radio galaxy classification. Mon. Not. R. Astron. Soc. **503**(2), 1828–1846 (2021). https://doi.org/10.1093/mnras/stab325

R. Bellman, R. Kalaba, A mathematical theory of adaptive control processes. Proc. Natl. Acad. Sci. **45**(8), 1288–1290 (1959)

Y. Bengio, A. Courville, P. Vincent, Representation learning: a review and new perspectives. arXiv e-prints, art. arXiv:1206.5538 (June 2012)

S. Bethapudi, S. Desai, Separation of pulsar signals from noise using supervised machine learning algorithms. Astron. Comput. **23**, 15 (2018). https://doi.org/10.1016/j.ascom.2018.02.002

P. Bhambra, B. Joachimi, O. Lahav, Explaining deep learning of galaxy morphology with saliency mapping. Mon. Not. R. Astron. Soc. **511**(4), 5032–5041 (2022). https://doi.org/10.1093/mnras/stac368

S. Bieringer, A. Butter, S. Diefenbacher, E. Eren, F. Gaede, D. Hundhausen, G. Kasieczka, B. Nachman, T. Plehn, M. Trabs, Calomplification - the power of generative calorimeter models. J. Instrum. **17**(9), P09028 (2022). https://doi.org/10.1088/1748-0221/17/09/P09028

D.M. Blei, A. Kucukelbir, J.D. McAuliffe, Variational inference: a review for statisticians. J. Am. Stat. Assoc. **112**(518), 859–877 (2017)

A. Bonaldi, M. Bonato, V. Galluzzi, I. Harrison, M. Massardi, S. Kay, G. De Zotti, M.L. Brown, The tiered radio extragalactic continuum simulation (T-RECS). Mon. Not. R. Astron. Soc. **482** (1), 2–19 (2019). https://doi.org/10.1093/mnras/sty2603

A. Bonaldi, T. An, M. Brüggen, S. Burkutean, B. Coelho, H. Goodarzi, P. Hartley, P.K. Sandhu, C. Wu, L. Yu, M.H. Zhoolideh Haghighi, S. Antón, Z. Bagheri, D. Barbosa, J.P. Barraca, D. Bartashevich, M. Bergano, M. Bonato, J. Brand, A. de Gasperin, F. Giannetti, R. Dodson, P. Jain, S. Jaiswal, B. Lao, B. Liu, E. Liuzzo, Y. Lu, V. Lukic, D. Maia, N. Marchili, M. Massardi, P. Mohan, J.B. Morgado, M. Panwar, P. Prabhakar, V.A.R.M. Ribeiro, K.L.J. Rygl, V. Sabz Ali, E. Saremi, E. Schisano, S. Sheikhnezami, A. Vafaei Sadr, A. Wong, O.I. Wong, Square kilometre array science data challenge 1: analysis and results. Mon. Not. R. Astron. Soc. **500**, 3821 (2021). https://doi.org/https://doi.org/10.1093/mnras/staa3023

K. Borne, Scientific data mining in astronomy. arXiv e-prints, art. arXiv:0911.0505 (November 2009)

W. Borucki, D. Koch, G. Basri, N. Batalha, T. Brown, D. Caldwell, J. Christensen-Dalsgaard, W. Cochran, E. Dunham, T.N. Gautier, J. Geary, R. Gilliland, J. Jenkins, Y. Kondo, D. Latham, J.J. Lissauer, D. Monet, Finding Earth-size planets in the habitable zone: the Kepler Mission, in *Exoplanets: Detection, Formation and Dynamics*, ed. by Y.-S. Sun, S. Ferraz-Mello, J.-L. Zhou, vol. 249 (2008), pp. 17–24. https://doi.org/10.1017/S174392130801630X

A. Boucaud, M. Huertas-Company, C. Heneka, E.E.O. Ishida, N. Sedaghat, R.S. de Souza, B. Moews, H. Dole, M. Castellano, E. Merlin, V. Roscani, A. Tramacere, M. Killedar, A.M.M. Trindade, Collaboration COIN, Photometry of high-redshift blended galaxies using deep learning. Mon. Not. R. Astron. Soc. **491**(2), 2481–2495 (2020). https://doi.org/10.1093/mnras/stz3056

M. Bowles, M. Bromley, M. Allen, A. Scaife, E(2) equivariant self-attention for radio astronomy. arXiv e-prints, art. arXiv:2111.04742 (November 2021a)

M. Bowles, A.M.M. Scaife, F. Porter, H. Tang, D.J. Bastien, Attention-gating for improved radio galaxy classification. Mon. Not. R. Astron. Soc. **501**(3), 4579–4595 (2021b). https://doi.org/10.1093/mnras/staa3946

M. Bowles, H. Tang, E. Vardoulaki, E.L. Alexander, Y. Luo, L. Rudnick, M. Walmsley, F. Porter, A.M.M. Scaife, I.V. Slijepcevic, G. Segal, A new task: deriving semantic class targets for the physical sciences. arXiv e-prints, art. arXiv:2210.14760 (October 2022). https://doi.org/10.48550/arXiv.2210.14760

R. Braun, T. Bourke, J.A. Green, E. Keane, J. Wagg, Advancing astrophysics with the square kilometre array, in *Advancing Astrophysics with the Square Kilometre Array (AASKA14)* (2015), 174 pp. https://doi.org/10.22323/1.215.0174

J. Brehmer, F. Kling, I. Espejo, K. Cranmer, MadMiner: machine learning-based inference for particle physics. arXiv e-prints, art. arXiv:1907.10621 (July 2019). https://doi.org/10.48550/arXiv.1907.10621

L. Breiman, Random forests. Mach. Learn. **45**(1), 5–32 (2001)

H. Bretonnière, A. Boucaud, M. Huertas-Company, Probabilistic segmentation of overlapping galaxies for large cosmological surveys. arXiv e-prints, art. arXiv:2111.15455 (November 2021)

M.M. Breunig, H.-P. Kriegel, R.T. Ng, J. Sander, Lof: identifying density-based local outliers, in *Proceedings of the 2000 ACM SIGMOD international conference on Management of data* (2000), pp. 93–104

J. Bruna, S. Mallat, Invariant scattering convolution networks. arXiv e-prints, art. arXiv:1203.1513 (March 2012)

M.R. Buckley, S.H. Lim, E. Putney, D. Shih, Measuring Galactic dark matter through unsupervised machine learning. Mon. Not. R. Astron. Soc. (2023). https://doi.org/10.1093/mnras/stad843

C.J. Burke, P.D. Aleo, Y.-C. Chen, X. Liu, J.R. Peterson, G.H. Sembroski, J. Yao-Yu Lin, Deblending and classifying astronomical sources with Mask R-CNN deep learning. Mon. Not. R. Astron. Soc. **490**(3), 3952–3965 (2019). https://doi.org/10.1093/mnras/stz2845

A. Butter, S. Diefenbacher, G. Kasieczka, B. Nachman, T. Plehn, GANplifying event samples. SciPost Phys. **10**(6), 139 (2021). https://doi.org/10.21468/SciPostPhys.10.6.139

A. Butter, T. Plehn, S. Schumann, S. Badger, S. Caron, K. Cranmer, F.A. Di Bello, E. Dreyer, S. Forte, S. Ganguly, D. Gonçalves, E. Gross, T. Heimel, G. Heinrich, L. Heinrich, A. Held, S. Höche, J.N. Howard, P. Ilten, J. Isaacson, T. Janßen, S. Jones, M. Kado, M. Kagan, G. Kasieczka, F. Kling, S. Kraml, C. Krause, F. Krauss, K. Kröninger, R.K. Barman, M. Luchmann, V. Magerya, D. Maitre, B. Malaescu, F. Maltoni, T. Martini, O. Mattelaer, B. Nachman, S. Pitz, J. Rojo, M. Schwartz, D. Shih, F. Siegert, R. Stegeman, B. Stienen, J. Thaler, R. Verheyen, D. Whiteson, R. Winterhalder, J. Zupan, Machine learning and LHC event generation. arXiv e-prints, art. arXiv:2203.07460 (March 2022). https://doi.org/10.48550/arXiv.2203.07460

J. Caldeira, W.L.K. Wu, B. Nord, C. Avestruz, S. Trivedi, K.T. Story, DeepCMB: lensing reconstruction of the cosmic microwave background with deep neural networks. Astron. Comput. **28**, 100307 (2019). https://doi.org/10.1016/j.ascom.2019.100307

N. Carion, F. Massa, G. Synnaeve, N. Usunier, A. Kirillov, S. Zagoruyko, End-to-end object detection with transformers. arXiv e-prints, art. arXiv:2005.12872 (May 2020)

G. Carleo, I. Cirac, K. Cranmer, L. Daudet, M. Schuld, N. Tishby, L. Vogt-Maranto, L. Zdeborová, Machine learning and the physical sciences*. Rev. Mod. Phys. **91**(4), 045002 (2019). https://doi.org/10.1103/RevModPhys.91.045002

C. Chatterjee, L. Wen, F. Diakogiannis, K. Vinsen, Extraction of binary black hole gravitational wave signals from detector data using deep learning. Phys. Rev. D **104**(6), 064046 (2021). https://doi.org/10.1103/PhysRevD.104.064046

T. Chen, C. Guestrin, XGBoost: a scalable tree boosting system. arXiv e-prints, art. arXiv:1603.02754 (March 2016)

T. Chen, T. He, Higgs Boson discovery with boosted trees, in *Proceedings of the NIPS 2014 Workshop on High-energy Physics and Machine Learning*, ed. by G. Cowan, C. Germain, I. Guyon, B. Kégl, D. Rousseau, vol. 42. Proceedings of Machine Learning Research (PMLR), Montreal (2015), pp. 69–80. URL https://proceedings.mlr.press/v42/chen14.html

M. Choudhury, A. Datta, A. Chakraborty, Extracting the 21 cm Global signal using artificial neural networks. Mon. Not. R. Astron. Soc. **491**(3), 4031–4044 (2020). https://doi.org/10.1093/mnras/stz3107

CMS Collaboration, S. Chatrchyan, G. Hmayakyan, V. Khachatryan, A.M. Sirunyan, W. Adam, T. Bauer, T. Bergauer, H. Bergauer, M. Dragicevic, J. Erö, M. Friedl, R. Frühwirth, V.M. Ghete, P. Glaser, C. Hartl, N. Hoermann, J. Hrubec, S. Hänsel, M. Jeitler, K. Kastner, M. Krammer, I. Magrans de Abril, M. Markytan, I. Mikulec, B. Neuherz, T. Nöbauer, M. Oberegger, M. Padrta, M. Pernicka, P. Porth, H. Rohringer, S. Schmid, T. Schreiner, R. Stark, H. Steininger, J. Strauss, A. Taurok, D. Uhl, W. Waltenberger, G. Walzel, E. Widl, C.E. Wulz, V. Petrov, V. Prosolovich, V. Chekhovsky, O. Dvornikov, I. Emeliantchik, A. Litomin, V. Makarenko, I. Marfin, V. Mossolov, N. Shumeiko, A. Solin, R. Stefanovitch, J. Suarez Gonzalez, A. Tikhonov, A. Fedorov, M. Korzhik, O. Missevitch, R. Zuyeuski, W. Beaumont, M. Cardaci, E. DeLanghe, E.A. DeWolf, E. Delmeire, S. Ochesanu, M. Tasevsky, P. Van

Mechelen, J. D'Hondt, S. DeWeirdt, O. Devroede, R. Goorens, S. Hannaert, J. Heyninck, J. Maes, M.U. Mozer, S. Tavernier, W. Van Doninck, L. Van Lancker, P. Van Mulders, I. Villella, C. Wastiels, C. Yu, O. Bouhali, O. Charaf, B. Clerbaux, P. DeHarenne, G. DeLentdecker, J.P. Dewulf, S. Elgammal, R. Gindroz, G.H. Hammad, T. Mahmoud, L. Neukermans, M. Pins, R. Pins, S. Rugovac, J. Stefanescu, V. Sundararajan, C. Vander Velde, P. Vanlaer, J. Wickens, M. Tytgat, S. Assouak, J.L. Bonnet, G. Bruno, J. Caudron, B. DeCallatay, J. DeFavereau DeJeneret, S. DeVisscher, P. Demin, D. Favart, C. Felix, B. Florins, E. Forton, A. Giammanco, G. Grégoire, M. Jonckman, D. Kcira, T. Keutgen, V. Lemaitre, D. Michotte, O. Militaru, S. Ovyn, T. Pierzchala, K. Piotrzkowski, V. Roberfroid, X. Rouby, N. Schul, O. Van der Aa, N. Beliy, E. Daubie, P. Herquet, G. Alves, M.E. Pol, M.H.G. Souza, M. Vaz, D. DeJesus Damiao, V. Oguri, A. Santoro, A. Sznajder, E. DeMoraes Gregores, R.L. Iope, S.F. Novaes, T. Tomei, T. Anguelov, G. Antchev, I. Atanasov, J. Damgov, N. Darmenov, L. Dimitrov, V. Genchev, P. Iaydjiev, A. Marinov, S. Piperov, S. Stoykova, G. Sultanov, R. Trayanov, I. Vankov, C. Cheshkov, A. Dimitrov, M. Dyulendarova, I. Glushkov, V. Kozhuharov, L. Litov, M. Makariev, E. Marinova, S. Markov, M. Mateev, I. Nasteva, B. Pavlov, P. Petev, P. Petkov, V. Spassov, Z. Toteva, V. Velev, V. Verguilov, J.G. Bian, G.M. Chen, H.S. Chen, M. Chen, C.H. Jiang, B. Liu, X.Y. Shen, H.S. Sun, J. Tao, J. Wang, M. Yang, Z. Zhang, W.R. Zhao, H.L. Zhuang, Y. Ban, J. Cai, Y.C. Ge, S. Liu, H.T. Liu, L. Liu, S.J. Qian, Q. Wang, Z.H. Xue, Z.C. Yang, Y.L. Ye, J. Ying, P.J. Li, J. Liao, Z.L. Xue, D.S. Yan, H. Yuan, C.A. Carrillo Montoya, J.C. Sanabria, N. Godinovic, I. Puljak, I. Soric, Z. Antunovic, M. Dzelalija, K. Marasovic, V. Brigljevic, K. Kadija, S. Morovic, R. Fereos, C. Nicolaou, A. Papadakis, F. Ptochos, P.A. Razis, D. Tsiakkouri, Z. Zinonos, A. Hektor, M. Kadastik, K. Kannike, E. Lippmaa, M. Müntel, M. Raidal, L. Rebane, P.A. Aarnio, E. Anttila, K. Banzuzi, P. Bulteau, S. Czellar, N. Eiden, C. Eklund, P. Engstrom, A. Heikkinen, A. Honkanen, J. Härkönen, V. Karimäki, H.M. Katajisto, R. Kinnunen, J. Klem, J. Kortesmaa, M. Kotamäki, A. Kuronen, T. Lampén, K. Lassila-Perini, V. Lefébure, S. Lehti, T. Lindén, P.R. Luukka, S. Michal, F. Moura Brigido, T. Mäenpää, T. Nyman, J. Nystén, E. Pietarinen, K. Skog, K. Tammi, E. Tuominen, J. Tuominiemi, D. Ungaro, T.P. Vanhala, L. Wendland, C. Williams, M. Iskanius, A. Korpela, G. Polese, T. Tuuva, G. Bassompierre, A. Bazan, P.Y. David, J. Ditta, G. Drobychev, N. Fouque, J.P. Guillaud, V. Hermel, A. Karneyeu, T. LeFlour, S. Lieunard, M. Maire, P. Mendiburu, P. Nedelec, J.P. Peigneux, M. Schneegans, D. Sillou, J.P. Vialle, M. Anfreville, J.P. Bard, P. Besson, E. Bougamont, M. Boyer, P. Bredy, R. Chipaux, M. Dejardin, D. Denegri, J. Descamps, B. Fabbro, J.L. Faure, S. Ganjour, F.X. Gentit, A. Givernaud, P. Gras, G. Hamel de Monchenault, P. Jarry, C. Jeanney, F. Kircher, M.C. Lemaire, Y. Lemoigne, B. Levesy, E. Locci, J.P. Lottin, I. Mandjavidze, M. Mur, J.P. Pansart, A. Payn, J. Rander, J.M. Reymond, J. Rolquin, F. Rondeaux, A. Rosowsky, J.Y.A. Rousse, Z.H. Sun, J. Tartas, A. Van Lysebetten, P. Venault, P. Verrecchia, M. Anduze, J. Badier, S. Baffioni, M. Bercher, C. Bernet, U. Berthon, J. Bourotte, A. Busata, P. Busson, M. Cerutti, D. Chamont, C. Charlot, C. Collard, A. Debraine, D. Decotigny, L. Dobrzynski, O. Ferreira, Y. Geerebaert, J. Gilly, C. Gregory, L. Guevara Riveros, M. Haguenauer, A. Karar, B. Koblitz, D. Lecouturier, A. Mathieu, G. Milleret, P. Miné, P. Paganini, P. Poilleux, N. Pukhaeva, N. Regnault, T. Romanteau, I. Semeniouk, Y. Sirois, C. Thiebaux, J.C. Vanel, A. Zabi, J.L. Agram, A. Albert, L. Anckenmann, J. Andrea, F. Anstotz, A.M. Bergdolt, J.D. Berst, R. Blaes, D. Bloch, J.M. Brom, J. Cailleret, F. Charles, E. Christophel, G. Claus, J. Coffin, C. Colledani, J. Croix, E. Dangelser, N. Dick, F. Didierjean, F. Drouhin, W. Dulinski, J.P. Ernenwein, R. Fang, J.C. Fontaine, G. Gaudiot, W. Geist, D. Gelé, T. Goeltzenlichter, U. Goerlach, P. Graehling, L. Gross, C. Guo Hu, J.M. Helleboid, T. Henkes, M. Hoffer, C. Hoffmann, J. Hosselet, L. Houchu, Y. Hu, D. Huss, C. Illinger, F. Jeanneau, P. Juillot, T. Kachelhoffer, M.R. Kapp, H. Kettunen, L. Lakehal Ayat, A.C. LeBihan, A. Lounis, C. Maazouzi, V. Mack, P. Majewski, D. Mangeol, J. Michel, S. Moreau, C. Olivetto, A. Pallarès, Y. Patois, P. Pralavorio, C. Racca, Y. Riahi, I. Ripp-Baudot, P. Schmitt, J.P. Schunck, G. Schuster, B. Schwaller, M.H. Sigward, J.L. Sohler, J. Speck, R. Strub, T. Todorov, R. Turchetta, P. Van Hove, D. Vintache, A. Zghiche, M. Ageron, J.E. Augustin, C. Baty, G. Baulieu, M. Bedjidian, J. Blaha, A. Bonnevaux, G. Boudoul, P. Brunet, E. Chabanat, E.C. Chabert, R. Chierici, V. Chorowicz, C. Combaret, D. Contardo, R. Della Negra, P. Depasse,

O. Drapier, M. Dupanloup, T. Dupasquier, H. El Mamouni, N. Estre, J. Fay, S. Gascon, N. Giraud, C. Girerd, G. Guillot, R. Haroutunian, B. Ille, M. Lethuillier, N. Lumb, C. Martin, H. Mathez, G. Maurelli, S. Muanza, P. Pangaud, S. Perries, O. Ravat, E. Schibler, F. Schirra, G. Smadja, S. Tissot, B. Trocme, S. Vanzetto, J.P. Walder, Y. Bagaturia, D. Mjavia, A. Mzhavia, Z. Tsamalaidze, V. Roinishvili, R. Adolphi, G. Anagnostou, R. Brauer, W. Braunschweig, H. Esser, L. Feld, W. Karpinski, A. Khomich, K. Klein, C. Kukulies, K. Lübelsmeyer, J. Olzem, A. Ostaptchouk, D. Pandoulas, G. Pierschel, F. Raupach, S. Schael, A. Schultz von Dratzig, G. Schwering, R. Siedling, M. Thomas, M. Weber, B. Wittmer, M. Wlochal, F. Adamczyk, A. Adolf, G. Altenhöfer, S. Bechstein, S. Bethke, P. Biallass, O. Biebel, M. Bontenackels, K. Bosseler, A. Böhm, M. Erdmann, H. Faissner, B. Fehr, H. Fesefeldt, G. Fetchenhauer, J. Frangenheim, J.H. Frohn, J. Grooten, T. Hebbeker, S. Hermann, E. Hermens, G. Hilgers, K. Hoepfner, C. Hof, E. Jacobi, S. Kappler, M. Kirsch, P. Kreuzer, R. Kupper, H.R. Lampe, D. Lanske, R. Mameghani, A. Meyer, S. Meyer, T. Moers, E. Müller, R. Pahlke, B. Philipps, D. Rein, H. Reithler, W. Reuter, P. Rütten, S. Schulz, H. Schwarthoff, W. Sobek, M. Sowa, T. Stapelberg, H. Szczesny, H. Teykal, D. Teyssier, H. Tomme, W. Tomme, M. Tonutti, O. Tsigenov, J. Tutas, J. Vandenhirtz, H. Wagner, M. Wegner, C. Zeidler, F. Beissel, M. Davids, M. Duda, G. Flügge, M. Giffels, T. Hermanns, D. Heydhausen, S. Kalinin, S. Kasselmann, G. Kaussen, T. Kress, A. Linn, A. Nowack, L. Perchalla, M. Poettgens, O. Pooth, P. Sauerland, A. Stahl, D. Tornier, M.H. Zoeller, U. Behrens, K. Borras, A. Flossdorf, D. Hatton, B. Hegner, M. Kasemann, R. Mankel, A. Meyer, J. Mnich, C. Rosemann, C. Youngman, W.D. Zeuner, F. Bechtel, P. Buhmann, E. Butz, G. Flucke, R.H. Hamdorf, U. Holm, R. Klanner, U. Pein, N. Schirm, P. Schleper, G. Steinbrück, R. Van Staa, R. Wolf, B. Atz, T. Barvich, P. Blüm, F. Boegelspacher, H. Bol, Z.Y. Chen, S. Chowdhury, W. DeBoer, P. Dehm, G. Dirkes, M. Fahrer, U. Felzmann, M. Frey, A. Furgeri, E. Gregoriev, F. Hartmann, F. Hauler, S. Heier, K. Kärcher, B. Ledermann, S. Mueller, Th Müller, D. Neuberger, C. Piasecki, G. Quast, K. Rabbertz, A. Sabellek, A. Scheurer, F.P. Schilling, H.J. Simonis, A. Skiba, P. Steck, A. Theel, W.H. Thümmel, A. Trunov, A. Vest, T. Weiler, C. Weiser, S. Weseler, V. Zhukov, M. Barone, G. Daskalakis, N. Dimitriou, G. Fanourakis, C. Filippidis, T. Geralis, C. Kalfas, K. Karafasoulis, A. Koimas, A. Kyriakis, S. Kyriazopoulou, D. Loukas, A. Markou, C. Markou, N. Mastroyiannopoulos, C. Mavrommatis, J. Mousa, I. Papadakis, E. Petrakou, I. Siotis, K. Theofilatos, S. Tzamarias, A. Vayaki, G. Vermisoglou, A. Zachariadou, L. Gouskos, G. Karapostoli, P. Katsas, A. Panagiotou, C. Papadimitropoulos, X. Aslanoglou, I. Evangelou, P. Kokkas, N. Manthos, I. Papadopoulos, F.A. Triantis, G. Bencze, L. Boldizsar, G. Debreczeni, C. Hajdu, P. Hidas, D. Horvath, P. Kovesarki, A. Laszlo, G. Odor, G. Patay, F. Sikler, G. Veres, G. Vesztergombi, P. Zalan, A. Fenyvesi, J. Imrek, J. Molnar, D. Novak, J. Palinkas, G. Szekely, N. Beni, A. Kapusi, G. Marian, B. Radics, P. Raics, Z. Szabo, Z. Szillasi, Z.L. Trocsanyi, G. Zilizi, H.S. Bawa, S.B. Beri, V. Bhandari, V. Bhatnagar, M. Kaur, J.M. Kohli, A. Kumar, B. Singh, J.B. Singh, S. Arora, S. Bhattacharya, S. Chatterji, S. Chauhan, B.C. Choudhary, P. Gupta, M. Jha, K. Ranjan, R.K. Shivpuri, A.K. Srivastava, R.K. Choudhury, D. Dutta, M. Ghodgaonkar, S. Kailas, S.K. Kataria, A.K. Mohanty, L.M. Pant, P. Shukla, A. Topkar, T. Aziz, Sunanda Banerjee, S. Bose, S. Chendvankar, P.V. Deshpande, M. Guchait, A. Gurtu, M. Maity, G. Majumder, K. Mazumdar, A. Nayak, M.R. Patil, S. Sharma, K. Sudhakar, B.S. Acharya, Sudeshna Banerjee, S. Bheesette, S. Dugad, S.D. Kalmani, V.R. Lakkireddi, N.K. Mondal, N. Panyam, P. Verma, H. Arfaei, M. Hashemi, M. Mohammadi Najafabadi, A. Moshaii, S. Paktinat Mehdiabadi, M. Felcini, M. Grunewald, K. Abadjiev, M. Abbrescia, L. Barbone, P. Cariola, F. Chiumarulo, A. Clemente, A. Colaleo, D. Creanza, N. DeFilippis, M. DePalma, G. DeRobertis, G. Donvito, R. Ferorelli, L. Fiore, M. Franco, D. Giordano, R. Guida, G. Iaselli, N. Lacalamita, F. Loddo, G. Maggi, M. Maggi, N. Manna, B. Marangelli, M.S. Mennea, S. My, S. Natali, S. Nuzzo, G. Papagni, C. Pinto, A. Pompili, G. Pugliese, A. Ranieri, F. Romano, G. Roselli, G. Sala, G. Selvaggi, L. Silvestris, P. Tempesta, R. Trentadue, S. Tupputi, G. Zito, G. Abbiendi, W. Bacchi, C. Battilana, A.C. Benvenuti, M. Boldini, D. Bonacorsi, S. Braibant-Giacomelli, V.D. Cafaro, P. Capiluppi, A. Castro, F.R. Cavallo, C. Ciocca, G. Codispoti, M. Cuffiani, I. D'Antone, G.M. Dallavalle, F. Fabbri, A. Fanfani, S. Finelli, P. Giacomelli, V. Giordano, M. Giunta, C. Grandi, M. Guerzoni,

L. Guiducci, S. Marcellini, G. Masetti, A. Montanari, F.L. Navarria, F. Odorici, A. Paolucci, G. Pellegrini, A. Perrotta, A.M. Rossi, T. Rovelli, G.P. Siroli, G. Torromeo, R. Travaglini, G.P. Veronese, S. Albergo, M. Chiorboli, S. Costa, M. Galanti, G. Gatto Rotondo, N. Giudice, N. Guardone, F. Noto, R. Potenza, M.A. Saizu, G. Salemi, C. Sutera, A. Tricomi, C. Tuve, L. Bellucci, M. Brianzi, G. Broccolo, E. Catacchini, V. Ciulli, C. Civinini, R. D'Alessandro, E. Focardi, S. Frosali, C. Genta, G. Landi, P. Lenzi, A. Macchiolo, F. Maletta, F. Manolescu, C. Marchettini, L. Masetti, S. Mersi, M. Meschini, C. Minelli, S. Paoletti, G. Parrini, E. Scarlini, G. Sguazzoni, L. Benussi, M. Bertani, S. Bianco, M. Caponero, D. Colonna, L. Daniello, F. Fabbri, F. Felli, M. Giardoni, A. La Monaca, B. Ortenzi, M. Pallotta, A. Paolozzi, C. Paris, L. Passamonti, D. Pierluigi, B. Ponzio, C. Pucci, A. Russo, G. Saviano, P. Fabbricatore, S. Farinon, M. Greco, R. Musenich, S. Badoer, L. Berti, M. Biasotto, S. Fantinel, E. Frizziero, U. Gastaldi, M. Gulmini, F. Lelli, G. Maron, S. Squizzato, N. Toniolo, S. Traldi, S. Banfi, R. Bertoni, M. Bonesini, L. Carbone, G.B. Cerati, F. Chignoli, P. D'Angelo, A. DeMin, P. Dini, F.M. Farina, F. Ferri, P. Govoni, S. Magni, M. Malberti, S. Malvezzi, R. Mazza, D. Menasce, V. Miccio, L. Moroni, P. Negri, M. Paganoni, D. Pedrini, A. Pullia, S. Ragazzi, N. Redaelli, M. Rovere, L. Sala, S. Sala, R. Salerno, T. Tabarelli de Fatis, V. Tancini, S. Taroni, A. Boiano, F. Cassese, C. Cassese, A. Cimmino, B. D'Aquino, L. Lista, D. Lomidze, P. Noli, P. Paolucci, G. Passeggio, D. Piccolo, L. Roscilli, C. Sciacca, A. Vanzanella, P. Azzi, N. Bacchetta, L. Barcellan, M. Bellato, M. Benettoni, D. Bisello, E. Borsato, A. Candelori, R. Carlin, L. Castellani, P. Checchia, L. Ciano, A. Colombo, E. Conti, M. Da Rold, F. Dal Corso, M. DeGiorgi, M. DeMattia, T. Dorigo, U. Dosselli, C. Fanin, G. Galet, F. Gasparini, U. Gasparini, A. Giraldo, P. Giubilato, F. Gonella, A. Gresele, A. Griggio, P. Guaita, A. Kaminskiy, S. Karaevskii, V. Khomenkov, D. Kostylev, S. Lacaprara, I. Lazzizzera, I. Lippi, M. Loreti, M. Margoni, R. Martinelli, S. Mattiazzo, M. Mazzucato, A.T. Meneguzzo, L. Modenese, F. Montecassiano, A. Neviani, M. Nigro, A. Paccagnella, D. Pantano, A. Parenti, M. Passaseo, R. Pedrotta, M. Pegoraro, G. Rampazzo, S. Reznikov, P. Ronchese, A. Sancho Daponte, P. Sartori, I. Stavitskiy, M. Tessaro, E. Torassa, A. Triossi, S. Vanini, S. Ventura, L. Ventura, M. Verlato, M. Zago, F. Zatti, P. Zotto, G. Zumerle, P. Baesso, G. Belli, U. Berzano, S. Bricola, A. Grelli, G. Musitelli, R. Nardò, M.M. Necchi, D. Pagano, S.P. Ratti, C. Riccardi, P. Torre, A. Vicini, P. Vitulo, C. Viviani, D. Aisa, S. Aisa, F. Ambroglini, M.M. Angarano, E. Babucci, D. Benedetti, M. Biasini, G.M. Bilei, S. Bizzaglia, M.T. Brunetti, B. Caponeri, B. Checcucci, R. Covarelli, N. Dinu, L. Fanò, L. Farnesini, M. Giorgi, P. Lariccia, G. Mantovani, F. Moscatelli, D. Passeri, A. Piluso, P. Placidi, V. Postolache, R. Santinelli, A. Santocchia, L. Servoli, D. Spiga, P. Azzurri, G. Bagliesi, G. Balestri, A. Basti, R. Bellazzini, L. Benucci, J. Bernardini, L. Berretta, S. Bianucci, T. Boccali, A. Bocci, L. Borrello, F. Bosi, F. Bracci, A. Brez, F. Calzolari, R. Castaldi, U. Cazzola, M. Ceccanti, R. Cecchi, C. Cerri, A.S. Cucoanes, R. Dell'Orso, D. Dobur, S. Dutta, F. Fiori, L. Foà, A. Gaggelli, S. Gennai, A. Giassi, S. Giusti, D. Kartashov, A. Kraan, L. Latronico, F. Ligabue, S. Linari, T. Lomtadze, G.A. Lungu, G. Magazzu, P. Mammini, F. Mariani, G. Martinelli, M. Massa, A. Messineo, A. Moggi, F. Palla, F. Palmonari, G. Petragnani, G. Petrucciani, A. Profeti, F. Raffaelli, D. Rizzi, G. Sanguinetti, S. Sarkar, G. Segneri, D. Sentenac, A.T. Serban, A. Slav, P. Spagnolo, G. Spandre, R. Tenchini, S. Tolaini, G. Tonelli, A. Venturi, P.G. Verdini, M. Vos, L. Zaccarelli, S. Baccaro, L. Barone, A. Bartoloni, B. Borgia, G. Capradossi, F. Cavallari, A. Cecilia, D. D'Angelo, I. Dafinei, D. DelRe, E. Di Marco, M. Diemoz, G. Ferrara, C. Gargiulo, S. Guerra, M. Iannone, E. Longo, M. Montecchi, M. Nuccetelli, G. Organtini, A. Palma, R. Paramatti, F. Pellegrino, S. Rahatlou, C. Rovelli, F. Safai Tehrani, A. Zullo, G. Alampi, N. Amapane, R. Arcidiacono, S. Argiro, M. Arneodo, R. Bellan, F. Benotto, C. Biino, S. Bolognesi, M.A. Borgia, C. Botta, A. Brasolin, N. Cartiglia, R. Castello, G. Cerminara, R. Cirio, M. Cordero, M. Costa, D. Dattola, F. Daudo, G. Dellacasa, N. Demaria, G. Dughera, F. Dumitrache, R. Farano, G. Ferrero, E. Filoni, G. Kostyleva, H.E. Larsen, C. Mariotti, M. Marone, S. Maselli, E. Menichetti, P. Mereu, E. Migliore, G. Mila, V. Monaco, M. Musich, M. Nervo, M.M. Obertino, R. Panero, A. Parussa, N. Pastrone, C. Peroni, G. Petrillo, A. Romero, M. Ruspa, R. Sacchi, M. Scalise, A. Solano, A. Staiano, P.P. Trapani, D. Trocino, V. Vaniev, A. Vilela Pereira, A. Zampieri, S. Belforte, F. Cossutti, G. Della Ricca, B. Gobbo, C. Kavka, A. Penzo,

Y.E. Kim, S.K. Nam, D.H. Kim, G.N. Kim, J.C. Kim, D.J. Kong, S.R. Ro, D.C. Son, S.Y. Park, Y.J. Kim, J.Y. Kim, I.T. Lim, M.Y. Pac, S.J. Lee, S.Y. Jung, J.T. Rhee, S.H. Ahn, B.S. Hong, Y.K. Jeng, M.H. Kang, H.C. Kim, J.H. Kim, T.J. Kim, K.S. Lee, J.K. Lim, D.H. Moon, I.C. Park, S.K. Park, M.S. Ryu, K.S. Sim, K.J. Son, S.J. Hong, Y.I. Choi, H. Castilla Valdez, A. Sanchez Hernandez, S. Carrillo Moreno, A. Morelos Pineda, A. Aerts, P. Van der Stok, H. Weffers, P. Allfrey, R.N.C. Gray, M. Hashimoto, D. Krofcheck, A.J. Bell, N. Bernardino Rodrigues, P.H. Butler, S. Churchwell, R. Knegjens, S. Whitehead, J.C. Williams, Z. Aftab, U. Ahmad, I. Ahmed, W. Ahmed, M.I. Asghar, S. Asghar, G. Dad, M. Hafeez, H.R. Hoorani, I. Hussain, N. Hussain, M. Iftikhar, M.S. Khan, K. Mehmood, A. Osman, H. Shahzad, A.R. Zafar, A. Ali, A. Bashir, A.M. Jan, A. Kamal, F. Khan, M. Saeed, S. Tanwir, M.A. Zafar, J. Blocki, A. Cyz, E. Gladysz-Dziadus, S. Mikocki, M. Rybczynski, J. Turnau, Z. Wlodarczyk, P. Zychowski, K. Bunkowski, M. Cwiok, H. Czyrkowski, R. Dabrowski, W. Dominik, K. Doroba, A. Kalinowski, K. Kierzkowski, M. Konecki, J. Krolikowski, I.M. Kudla, M. Pietrusinski, K. Pozniak, W. Zabolotny, P. Zych, R. Gokieli, L. Goscilo, M. Górski, K. Nawrocki, P. Traczyk, G. Wrochna, P. Zalewski, K.T. Pozniak, R. Romaniuk, W.M. Zabolotny, R. Alemany-Fernandez, C. Almeida, N. Almeida, A.S. Araujo Vila Verde, T. Barata Monteiro, M. Bluj, S. Da Mota Silva, A. David Tinoco Mendes, M. Freitas Ferreira, M. Gallinaro, M. Husejko, A. Jain, M. Kazana, P. Musella, R. Nobrega, J. Rasteiro Da Silva, P.Q. Ribeiro, M. Santos, P. Silva, S. Silva, I. Teixeira, J.P. Teixeira, J. Varela, G. Varner, N. Vaz Cardoso, I. Altsybeev, K. Babich, A. Belkov, I. Belotelov, P. Bunin, S. Chesnevskaya, V. Elsha, Y. Ershov, I. Filozova, M. Finger, Jr. Finger, M., A. Golunov, I. Golutvin, N. Gorbounov, I. Gramenitski, V. Kalagin, A. Kamenev, V. Karjavin, S. Khabarov, V. Khabarov, Y. Kiryushin, V. Konoplyanikov, V. Korenkov, G. Kozlov, A. Kurenkov, A. Lanev, V. Lysiakov, A. Malakhov, I. Melnitchenko, V.V. Mitsyn, K. Moisenz, P. Moisenz, S. Movchan, E. Nikonov, D. Oleynik, V. Palichik, V. Perelygin, A. Petrosyan, E. Rogalev, V. Samsonov, M. Savina, R. Semenov, S. Sergeev, S. Shmatov, S. Shulha, V. Smirnov, D. Smolin, A. Tcheremoukhine, O. Teryaev, E. Tikhonenko, A. Urkinbaev, S. Vasil'ev, A. Vishnevskiy, A. Volodko, N. Zamiatin, A. Zarubin, P. Zarubin, E. Zubarev, N. Bondar, Y. Gavrikov, V. Golovtsov, Y. Ivanov, V. Kim, V. Kozlov, V. Lebedev, G. Makarenkov, F. Moroz, P. Neustroev, G. Obrant, E. Orishchin, A. Petrunin, Y. Shcheglov, A. Shchetkovskiy, V. Sknar, V. Skorobogatov, I. Smirnov, V. Sulimov, V. Tarakanov, L. Uvarov, S. Vavilov, G. Velichko, S. Volkov, A. Vorobyev, D. Chmelev, D. Druzhkin, A. Ivanov, V. Kudinov, O. Logatchev, S. Onishchenko, A. Orlov, V. Sakharov, V. Smetannikov, A. Tikhomirov, S. Zavodthikov, Yu Andreev, A. Anisimov, V. Duk, S. Gninenko, N. Golubev, D. Gorbunov, M. Kirsanov, N. Krasnikov, V. Matveev, A. Pashenkov, A. Pastsyak, V.E. Postoev, A. Sadovski, A. Skassyrskaia, Alexander Solovey, Anatoly Solovey, D. Soloviev, A. Toropin, S. Troitsky, A. Alekhin, A. Baldov, V. Epshteyn, V. Gavrilov, N. Ilina, V. Kaftanov, V. Karpishin, I. Kiselevich, V. Kolosov, M. Kossov, A. Krokhotin, S. Kuleshov, A. Oulianov, A. Pozdnyakov, G. Safronov, S. Semenov, N. Stepanov, V. Stolin, E. Vlasov, V. Zaytsev, E. Boos, M. Dubinin, L. Dudko, A. Ershov, G. Eyyubova, A. Gribushin, V. Ilyin, V. Klyukhin, O. Kodolova, N.A. Kruglov, A. Kryukov, I. Lokhtin, L. Malinina, V. Mikhaylin, S. Petrushanko, L. Sarycheva, V. Savrin, L. Shamardin, A. Sherstnev, A. Snigirev, K. Teplov, I. Vardanyan, A.M. Fomenko, N. Konovalova, V. Kozlov, A.I. Lebedev, N. Lvova, S.V. Rusakov, A. Terkulov, V. Abramov, S. Akimenko, A. Artamonov, A. Ashimova, I. Azhgirey, S. Bitioukov, O. Chikilev, K. Datsko, A. Filine, A. Godizov, P. Goncharov, V. Grishin, A. Inyakin, V. Kachanov, A. Kalinin, A. Khmelnikov, D. Konstantinov, A. Korablev, V. Krychkine, A. Krinitsyn, A. Levine, I. Lobov, V. Lukanin, Y. Mel'nik, V. Molchanov, V. Petrov, V. Petukhov, V. Pikalov, A. Ryazanov, R. Ryutin, V. Shelikhov, V. Skvortsov, S. Slabospitsky, A. Sobol, A. Sytine, V. Talov, L. Tourtchanovitch, S. Troshin, N. Tyurin, A. Uzunian, A. Volkov, S. Zelepoukine, V. Lukyanov, G. Mamaeva, Z. Prilutskaya, I. Rumyantsev, S. Sokha, S. Tataurschikov, I. Vasilyev, P. Adzic, I. Anicin, M. Djordjevic, D. Jovanovic, D. Maletic, J. Puzovic, N. Smiljkovic, E. Aguayo Navarrete, M. Aguilar-Benitez, J. Ahijado Munoz, J.M. Alarcon Vega, J. Alberdi, J. Alcaraz Maestre, M. Aldaya Martin, P. Arce, J.M. Barcala, J. Berdugo, C.L. Blanco Ramos, C. Burgos Lazaro, J. Caballero Bejar, E. Calvo, M. Cerrada, M. Chamizo Llatas, J.J.

Chercoles Catalán, N. Colino, M. Daniel, B. DeLa Cruz, A. Delgado Peris, C. Fernandez Bedoya, A. Ferrando, M.C. Fouz, D. Francia Ferrero, J. Garcia Romero, P. Garcia-Abia, O. Gonzalez Lopez, J.M. Hernandez, M.I. Josa, J. Marin, G. Merino, A. Molinero, J.J. Navarrete, J.C. Oller, J. Puerta Pelayo, J.C. Puras Sanchez, J. Ramirez, L. Romero, C. Villanueva Munoz, C. Willmott, C. Yuste, C. Albajar, J.F. de Trocóniz, I. Jimenez, R. Macias, R.F. Teixeira, J. Cuevas, J. Fernández Menéndez, I. Gonzalez Caballero, J. Lopez-Garcia, H. Naves Sordo, J.M. Vizan Garcia, I.J. Cabrillo, A. Calderon, D. Cano Fernandez, I. Diaz Merino, J. Duarte Campderros, M. Fernandez, J. Fernandez Menendez, C. Figueroa, L.A. Garcia Moral, G. Gomez, F. Gomez Casademunt, J. Gonzalez Sanchez, R. Gonzalez Suarez, C. Jorda, P. Lobelle Pardo, A. Lopez Garcia, A. Lopez Virto, J. Marco, R. Marco, C. Martinez Rivero, P. Martinez Ruiz del Arbol, F. Matorras, P. Orviz Fernandez, A. Patino Revuelta, T. Rodrigo, D. Rodriguez Gonzalez, A. Ruiz Jimeno, L. Scodellaro, M. Sobron Sanudo, I. Vila, R. Vilar Cortabitarte, M. Barbero, D. Goldin, B. Henrich, L. Tauscher, S. Vlachos, M. Wadhwa, D. Abbaneo, S.M. Abbas, I. Ahmed, S. Akhtar, M.I. Akhtar, E. Albert, M. Alidra, S. Ashby, P. Aspell, E. Auffray, P. Baillon, A. Ball, S.L. Bally, N. Bangert, R. Barillère, D. Barney, S. Beauceron, F. Beaudette, G. Benelli, R. Benetta, J.L. Benichou, W. Bialas, A. Bjorkebo, D. Blechschmidt, C. Bloch, P. Bloch, S. Bonacini, J. Bos, M. Bosteels, V. Boyer, A. Branson, H. Breuker, R. Bruneliere, O. Buchmuller, D. Campi, T. Camporesi, A. Caner, E. Cano, E. Carrone, A. Cattai, J.P. Chatelain, M. Chauvey, T. Christiansen, M. Ciganek, S. Cittolin, J. Cogan, A. Conde Garcia, H. Cornet, E. Corrin, M. Corvo, S. Cucciarelli, B. Curé, D. D'Enterria, A. DeRoeck, T. de Visser, C. Delaere, M. Delattre, C. Deldicque, D. Delikaris, D. Deyrail, S. Di Vincenzo, A. Domeniconi, S. Dos Santos, G. Duthion, L.M. Edera, A. Elliott-Peisert, M. Eppard, F. Fanzago, M. Favre, H. Foeth, R. Folch, N. Frank, S. Fratianni, M.A. Freire, A. Frey, A. Fucci, W. Funk, A. Gaddi, F. Gagliardi, M. Gastal, M. Gateau, J.C. Gayde, H. Gerwig, A. Ghezzi, D. Gigi, K. Gill, A.S. Giolo-Nicollerat, J.P. Girod, F. Glege, W. Glessing, R. Gomez-Reino Garrido, R. Goudard, R. Grabit, J.P. Grillet, P. Gutierrez Llamas, E. Gutierrez Mlot, J. Gutleber, R. Hall-wilton, R. Hammarstrom, M. Hansen, J. Harvey, A. Hervé, J. Hill, H.F. Hoffmann, A. Holzner, A. Honma, D. Hufnagel, M. Huhtinen, S.D. Ilie, V. Innocente, W. Jank, P. Janot, P. Jarron, M. Jeanrenaud, P. Jouvel, R. Kerkach, K. Kloukinas, L.J. Kottelat, J.C. Labbé, D. Lacroix, X. Lagrue, C. Lasseur, E. Laure, J.F. Laurens, P. Lazeyras, J.M. LeGoff, M. Lebeau, P. Lecoq, F. Lemeilleur, M. Lenzi, N. Leonardo, C. Leonidopoulos, M. Letheren, M. Liendl, F. Limia-Conde, L. Linssen, C. Ljuslin, B. Lofstedt, R. Loos, J.A. Lopez Perez, C. Lourenco, A. Lyonnet, A. Machard, R. Mackenzie, N. Magini, G. Maire, L. Malgeri, R. Malina, M. Mannelli, A. Marchioro, J. Martin, F. Meijers, P. Meridiani, E. Meschi, T. Meyer, A. Meynet Cordonnier, J.F. Michaud, L. Mirabito, R. Moser, F. Mossiere, J. Muffat-Joly, M. Mulders, J. Mulon, E. Murer, P. Mättig, A. Oh, A. Onnela, M. Oriunno, L. Orsini, J.A. Osborne, C. Paillard, I. Pal, G. Papotti, G. Passardi, A. Patino-Revuelta, V. Patras, B. Perea Solano, E. Perez, G. Perinic, J.F. Pernot, P. Petagna, P. Petiot, P. Petit, A. Petrilli, A. Pfeiffer, C. Piccut, M. Pimiä, R. Pintus, M. Pioppi, A. Placci, L. Pollet, H. Postema, M.J. Price, R. Principe, A. Racz, E. Radermacher, R. Ranieri, G. Raymond, P. Rebecchi, J. Rehn, S. Reynaud, H. Rezvani Naraghi, D. Ricci, M. Ridel, M. Risoldi, P. Rodrigues Simoes Moreira, A. Rohlev, G. Roiron, G. Rolandi, P. Rumerio, O. Runolfsson, V. Ryjov, H. Sakulin, D. Samyn, L.C. Santos Amaral, H. Sauce, E. Sbrissa, P. Scharff-Hansen, P. Schieferdecker, W.D. Schlatter, B. Schmitt, H.G. Schmuecker, M. Schröder, C. Schwick, C. Schäfer, I. Segoni, P. Sempere Roldán, S. Sgobba, A. Sharma, P. Siegrist, C. Sigaud, N. Sinanis, T. Sobrier, P. Sphicas, M. Spiropulu, G. Stefanini, A. Strandlie, F. Szoncsó, B.G. Taylor, O. Teller, A. Thea, E. Tournefier, D. Treille, P. Tropea, J. Troska, E. Tsesmelis, A. Tsirou, J. Valls, I. Van Vulpen, M. Vander Donckt, F. Vasey, M. Vazquez Acosta, L. Veillet, P. Vichoudis, G. Waurick, J.P. Wellisch, P. Wertelaers, M. Wilhelmsson, I.M. Willers, M. Winkler, M. Zanetti, W. Bertl, K. Deiters, P. Dick, W. Erdmann, D. Feichtinger, K. Gabathuler, Z. Hochman, R. Horisberger, Q. Ingram, H.C. Kaestli, D. Kotlinski, S. König, P. Poerschke, D. Renker, T. Rohe, T. Sakhelashvili, A. Starodumov, V. Aleksandrov, F. Behner, I. Beniozef, B. Betev, B. Blau, A.M. Brett, L. Caminada, Z. Chen, N. Chivarov, D. Da Silva Di Calafiori, S. Dambach, G. Davatz, V. Delachenal, R. Della Marina, H. Dimov, G. Dissertori,

M. Dittmar, L. Djambazov, M. Dröge, C. Eggel, J. Ehlers, R. Eichler, M. Elmiger, G. Faber, K. Freudenreich, J.F. Fuchs, G.M. Georgiev, C. Grab, C. Haller, J. Herrmann, M. Hilgers, W. Hintz, Hans Hofer, Heinz Hofer, U. Horisberger, I. Horvath, A. Hristov, C. Humbertclaude, B. Iliev, W. Kastli, A. Kruse, J. Kuipers, U. Langenegger, P. Lecomte, E. Lejeune, G. Leshev, C. Lesmond, B. List, P.D. Luckey, W. Lustermann, J.D. Maillefaud, C. Marchica, A. Maurisset, B. Meier, P. Milenovic, M. Milesi, F. Moortgat, I. Nanov, A. Nardulli, F. Nessi-Tedaldi, B. Panev, L. Pape, F. Pauss, E. Petrov, G. Petrov, M.M. Peynekov, D. Pitzl, T. Punz, P. Riboni, J. Riedlberger, A. Rizzi, F.J. Ronga, P.A. Roykov, U. Röser, D. Schinzel, A. Schöning, A. Sourkov, K. Stanishev, S. Stoenchev, F. Stöckli, H. Suter, P. Trüb, S. Udriot, D.G. Uzunova, I. Veltchev, G. Viertel, H.P. von Gunten, S. Waldmeier-Wicki, R. Weber, M. Weber, J. Weng, M. Wensveen, F. Wittgenstein, K. Zagoursky, E. Alagoz, C. Amsler, V. Chiochia, C. Hoermann, C. Regenfus, P. Robmann, T. Rommerskirchen, A. Schmidt, S. Steiner, D. Tsirigkas, L. Wilke, S. Blyth, Y.H. Chang, E.A. Chen, A. Go, C.C. Hung, C.M. Kuo, S.W. Li, W. Lin, P. Chang, Y. Chao, K.F. Chen, Z. Gao, G.W.S. Hou, Y.B. Hsiung, Y.J. Lei, S.W. Lin, R.S. Lu, J.G. Shiu, Y.M. Tzeng, K. Ueno, Y. Velikzhanin, C.C. Wang, M.Z. Wang, S. Aydin, A. Azman, M.N. Bakirci, S. Basegmez, S. Cerci, I. Dumanoglu, S. Erturk, E. Eskut, A. Kayis Topaksu, H. Kisoglu, P. Kurt, K. Ozdemir, N. Ozdes Koca, H. Ozkurt, S. Ozturk, A. Polatöz, K. Sogut, H. Topakli, M. Vergili, G. Önengüt, H. Gamsizkan, S. Sekmen, M. Serin-Zeyrek, R. Sever, M. Zeyrek, M. Deliomeroglu, E. Gülmez, E. Isiksal, M. Kaya, O. Kaya, S. Ozkorucuklu, N. Sonmez, B. Grinev, V. Lyubynskiy, V. Senchyshyn, L. Levchuk, S. Lukyanenko, D. Soroka, P. Sorokin, S. Zub, A. Anjum, N. Baker, T. Hauer, R. McClatchey, M. Odeh, D. Rogulin, A. Solomonides, J.J. Brooke, R. Croft, D. Cussans, D. Evans, R. Frazier, N. Grant, M. Hansen, R.D. Head, G.P. Heath, H.F. Heath, C. Hill, B. Huckvale, J. Jackson, C. Lynch, C.K. Mackay, S. Metson, S.J. Nash, D.M. Newbold, A.D. Presland, M.G. Probert, E.C. Reid, V.J. Smith, R.J. Tapper, R. Walton, E. Bateman, K.W. Bell, R.M. Brown, B. Camanzi, I.T. Church, D.J.A. Cockerill, J.E. Cole, J.F. Connolly, J.A. Coughlan, P.S. Flower, P. Ford, V.B. Francis, M.J. French, S.B. Galagedera, W. Gannon, A.P.R. Gay, N.I. Geddes, R.J.S. Greenhalgh, R.N.J. Halsall, W.J. Haynes, J.A. Hill, F.R. Jacob, P.W. Jeffreys, L.L. Jones, B.W. Kennedy, A.L. Lintern, A.B. Lodge, A.J. Maddox, Q.R. Morrissey, P. Murray, G.N. Patrick, C.A.X. Pattison, M.R. Pearson, S.P.H. Quinton, G.J. Rogers, J.G. Salisbury, A.A. Shah, C.H. Shepherd-Themistocleous, B.J. Smith, M. Sproston, R. Stephenson, S. Taghavi, I.R. Tomalin, M.J. Torbet, J.H. Williams, W.J. Womersley, S.D. Worm, F. Xing, M. Apollonio, F. Arteche, R. Bainbridge, G. Barber, P. Barrillon, J. Batten, R. Beuselinck, P.M. Brambilla Hall, D. Britton, W. Cameron, D.E. Clark, I.W. Clark, D. Colling, N. Cripps, G. Davies, M. Della Negra, G. Dewhirst, S. Dris, C. Foudas, J. Fulcher, D. Futyan, D.J. Graham, S. Greder, S. Greenwood, G. Hall, J.F. Hassard, J. Hays, G. Iles, V. Kasey, M. Khaleeq, J. Leaver, P. Lewis, B.C. MacEvoy, O. Maroney, E.M. McLeod, D.G. Miller, J. Nash, A. Nikitenko, E. Noah Messomo, M. Noy, A. Papageorgiou, M. Pesaresi, K. Petridis, D.R. Price, X. Qu, D.M. Raymond, A. Rose, S. Rutherford, M.J. Ryan, F. Sciacca, C. Seez, P. Sharp, G. Sidiropoulos, M. Stettler, M. Stoye, J. Striebig, M. Takahashi, H. Tallini, A. Tapper, C. Timlin, L. Toudup, T. Virdee, S. Wakefield, P. Walsham, D. Wardrope, M. Wingham, Y. Zhang, O. Zorba, C. Da Via, I. Goitom, P.R. Hobson, D.C. Imrie, I. Reid, C. Selby, O. Sharif, L. Teodorescu, S.J. Watts, I. Yaselli, E. Hazen, A. Heering, A. Heister, C. Lawlor, D. Lazic, E. Machado, J. Rohlf, L. Sulak, F. Varela Rodriguez, S.X. Wu, A. Avetisyan, T. Bose, L. Christofek, D. Cutts, S. Esen, R. Hooper, G. Landsberg, M. Narain, D. Nguyen, T. Speer, K.V. Tsang, R. Breedon, M. Case, M. Chertok, J. Conway, P.T. Cox, J. Dolen, R. Erbacher, Y. Fisyak, E. Friis, G. Grim, B. Holbrook, W. Ko, A. Kopecky, R. Lander, F.C. Lin, A. Lister, S. Maruyama, D. Pellett, J. Rowe, M. Searle, J. Smith, A. Soha, M. Squires, M. Tripathi, R. Vasquez Sierra, C. Veelken, V. Andreev, K. Arisaka, Y. Bonushkin, S. Chandramouly, D. Cline, R. Cousins, S. Erhan, J. Hauser, M. Ignatenko, C. Jarvis, B. Lisowski, C. Matthey, B. Mohr, J. Mumford, S. Otwinowski, Y. Pischalnikov, G. Rakness, P. Schlein, Y. Shi, B. Tannenbaum, J. Tucker, V. Valuev, R. Wallny, H.G. Wang, X. Yang, Y. Zheng, J. Andreeva, J. Babb, S. Campana, D. Chrisman, R. Clare, J. Ellison, D. Fortin, J.W. Gary, W. Gorn, G. Hanson, G.Y. Jeng, S.C. Kao, J.G. Layter, F. Liu, H. Liu, A. Luthra, G. Pasztor, H. Rick, A. Satpathy, B.C. Shen, R. Stringer, V. Sytnik,

P. Tran, S. Villa, R. Wilken, S. Wimpenny, D. Zer-Zion, J.G. Branson, J.A. Coarasa Perez, E. Dusinberre, R. Kelley, M. Lebourgeois, J. Letts, E. Lipeles, B. Mangano, T. Martin, M. Mojaver, J. Muelmenstaedt, M. Norman, H.P. Paar, A. Petrucci, H. Pi, M. Pieri, A. Rana, M. Sani, V. Sharma, S. Simon, A. White, F. Würthwein, A. Yagil, A. Affolder, A. Allen, C. Campagnari, M. D'Alfonso, A. Dierlamm, J. Garberson, D. Hale, J. Incandela, P. Kalavase, S.A. Koay, D. Kovalskyi, V. Krutelyov, S. Kyre, J. Lamb, S. Lowette, M. Nikolic, V. Pavlunin, F. Rebassoo, J. Ribnik, J. Richman, R. Rossin, Y.S. Shah, D. Stuart, S. Swain, J.R. Vlimant, D. White, M. Witherell, A. Bornheim, J. Bunn, J. Chen, G. Denis, P. Galvez, M. Gataullin, I. Legrand, V. Litvine, Y. Ma, R. Mao, D. Nae, I. Narsky, H.B. Newman, T. Orimoto, C. Rogan, S. Shevchenko, C. Steenberg, X. Su, M. Thomas, V. Timciuc, F. van Lingen, J. Veverka, B.R. Voicu, A. Weinstein, R. Wilkinson, Y. Xia, Y. Yang, L.Y. Zhang, K. Zhu, R.Y. Zhu, T. Ferguson, D.W. Jang, S.Y. Jun, M. Paulini, J. Russ, N. Terentyev, H. Vogel, I. Vorobiev, M. Bunce, J.P. Cumalat, M.E. Dinardo, B.R. Drell, W.T. Ford, K. Givens, B. Heyburn, D. Johnson, U. Nauenberg, K. Stenson, S.R. Wagner, L. Agostino, J. Alexander, F. Blekman, D. Cassel, S. Das, J.E. Duboscq, L.K. Gibbons, B. Heltsley, C.D. Jones, V. Kuznetsov, J.R. Patterson, D. Riley, A. Ryd, S. Stroiney, W. Sun, J. Thom, J. Vaughan, P. Wittich, C.P. Beetz, G. Cirino, V. Podrasky, C. Sanzeni, D. Winn, S. Abdullin, M.A. Afaq, M. Albrow, J. Amundson, G. Apollinari, M. Atac, W. Badgett, J.A. Bakken, B. Baldin, K. Banicz, L.A.T. Bauerdick, A. Baumbaugh, J. Berryhill, P.C. Bhat, M. Binkley, I. Bloch, F. Borcherding, A. Boubekeur, M. Bowden, K. Burkett, J.N. Butler, H.W.K. Cheung, G. Chevenier, F. Chlebana, I. Churin, S. Cihangir, W. Dagenhart, M. Demarteau, D. Dykstra, D.P. Eartly, J.E. Elias, V.D. Elvira, D. Evans, I. Fisk, J. Freeman, I. Gaines, P. Gartung, F.J.M. Geurts, L. Giacchetti, D.A. Glenzinski, E. Gottschalk, T. Grassi, D. Green, C. Grimm, Y. Guo, O. Gutsche, A. Hahn, J. Hanlon, R.M. Harris, T. Hesselroth, S. Holm, B. Holzman, E. James, H. Jensen, M. Johnson, U. Joshi, B. Klima, S. Kossiakov, K. Kousouris, J. Kowalkowski, T. Kramer, S. Kwan, C.M. Lei, M. Leininger, S. Los, L. Lueking, G. Lukhanin, S. Lusin, K. Maeshima, J.M. Marraffino, D. Mason, P. McBride, T. Miao, S. Moccia, N. Mokhov, S. Mrenna, S.J. Murray, C. Newman-Holmes, C. Noeding, V. O'Dell, M. Paterno, D. Petravick, R. Pordes, O. Prokofyev, N. Ratnikova, A. Ronzhin, V. Sekhri, E. Sexton-Kennedy, I. Sfiligoi, T.M. Shaw, E. Skup, R.P. Smith, W.J. Spalding, L. Spiegel, M. Stavrianakou, G. Stiehr, A.L. Stone, I. Suzuki, P. Tan, W. Tanenbaum, L.E. Temple, S. Tkaczyk, L. Uplegger, E.W. Vaandering, R. Vidal, R. Wands, H. Wenzel, J. Whitmore, E. Wicklund, W.M. Wu, Y. Wu, J. Yarba, V. Yarba, F. Yumiceva, J.C. Yun, T. Zimmerman, D. Acosta, P. Avery, V. Barashko, P. Bartalini, D. Bourilkov, R. Cavanaugh, S. Dolinsky, A. Drozdetskiy, R.D. Field, Y. Fu, I.K. Furic, L. Gorn, D. Holmes, B.J. Kim, S. Klimenko, J. Konigsberg, A. Korytov, K. Kotov, P. Levchenko, A. Madorsky, K. Matchev, G. Mitselmakher, Y. Pakhotin, C. Prescott, L. Ramond, P. Ramond, M. Schmitt, B. Scurlock, J. Stasko, H. Stoeck, D. Wang, J. Yelton, V. Gaultney, L. Kramer, L.M. Lebolo, S. Linn, P. Markowitz, G. Martinez, J.L. Rodriguez, T. Adams, A. Askew, O. Atramentov, M. Bertoldi, W.G.D. Dharmaratna, Y. Gershtein, S.V. Gleyzer, S. Hagopian, V. Hagopian, C.J. Jenkins, K.F. Johnson, H. Prosper, D. Simek, J. Thomaston, M. Baarmand, L. Baksay, S. Guragain, M. Hohlmann, H. Mermerkaya, R. Ralich, I. Vodopiyanov, M.R. Adams, I.M. Anghel, L. Apanasevich, O. Barannikova, V.E. Bazterra, R.R. Betts, C. Dragoiu, E.J. Garcia-Solis, C.E. Gerber, D.J. Hofman, R. Hollis, A. Iordanova, S. Khalatian, C. Mironov, E. Shabalina, A. Smoron, N. Varelas, U. Akgun, E.A. Albayrak, A.S. Ayan, R. Briggs, K. Cankocak, W. Clarida, A. Cooper, P. Debbins, F. Duru, M. Fountain, E. McCliment, J.P. Merlo, A. Mestvirishvili, M.J. Miller, A. Moeller, C.R. Newsom, E. Norbeck, J. Olson, Y. Onel, L. Perera, I. Schmidt, S. Wang, T. Yetkin, E.W. Anderson, H. Chakir, J.M. Hauptman, J. Lamsa, B.A. Barnett, B. Blumenfeld, C.Y. Chien, G. Giurgiu, A. Gritsan, D.W. Kim, C.K. Lae, P. Maksimovic, M. Swartz, N. Tran, P. Baringer, A. Bean, J. Chen, D. Coppage, O. Grachov, M. Murray, V. Radicci, J.S. Wood, V. Zhukova, D. Bandurin, T. Bolton, K. Kaadze, W.E. Kahl, Y. Maravin, D. Onoprienko, R. Sidwell, Z. Wan, B. Dahmes, J. Gronberg, J. Hollar, D. Lange, D. Wright, C.R. Wuest, D. Baden, R. Bard, S.C. Eno, D. Ferencek, N.J. Hadley, R.G. Kellogg, M. Kirn, S. Kunori, E. Lockner, F. Ratnikov, F. Santanastasio, A. Skuja, T. Toole, L. Wang, M. Wetstein, B. Alver, M. Ballintijn, G. Bauer, W. Busza, G. Gomez

Ceballos, K.A. Hahn, P. Harris, M. Klute, I. Kravchenko, W. Li, C. Loizides, T. Ma, S. Nahn, C. Paus, S. Pavlon, J. Piedra Gomez, C. Roland, G. Roland, M. Rudolph, G. Stephans, K. Sumorok, S. Vaurynovich, E.A. Wenger, B. Wyslouch, D. Bailleux, S. Cooper, P. Cushman, A. DeBenedetti, A. Dolgopolov, P.R. Dudero, R. Egeland, G. Franzoni, W.J. Gilbert, D. Gong, J. Grahl, J. Haupt, K. Klapoetke, I. Kronkvist, Y. Kubota, J. Mans, R. Rusack, S. Sengupta, B. Sherwood, A. Singovsky, P. Vikas, J. Zhang, M. Booke, L.M. Cremaldi, R. Godang, R. Kroeger, M. Reep, J. Reidy, D.A. Sanders, P. Sonnek, D. Summers, S. Watkins, K. Bloom, B. Bockelman, D.R. Claes, A. Dominguez, M. Eads, M. Furukawa, J. Keller, T. Kelly, C. Lundstedt, S. Malik, G.R. Snow, D. Swanson, K.M. Ecklund, I. Iashvili, A. Kharchilava, A. Kumar, M. Strang, G. Alverson, E. Barberis, O. Boeriu, G. Eulisse, T. McCauley, Y. Musienko, S. Muzaffar, I. Osborne, S. Reucroft, J. Swain, L. Taylor, L. Tuura, B. Gobbi, M. Kubantsev, A. Kubik, R.A. Ofierzynski, M. Schmitt, E. Spencer, S. Stoynev, M. Szleper, M. Velasco, S. Won, K. Andert, B. Baumbaugh, B.A. Beiersdorf, L. Castle, J. Chorny, A. Goussiou, M. Hildreth, C. Jessop, D.J. Karmgard, T. Kolberg, J. Marchant, N. Marinelli, M. McKenna, R. Ruchti, M. Vigneault, M. Wayne, D. Wiand, B. Bylsma, L.S. Durkin, J. Gilmore, J. Gu, P. Killewald, T.Y. Ling, C.J. Rush, V. Sehgal, G. Williams, N. Adam, S. Chidzik, P. Denes, P. Elmer, A. Garmash, D. Gerbaudo, V. Halyo, J. Jones, D. Marlow, J. Olsen, P. Piroué, D. Stickland, C. Tully, J.S. Werner, T. Wildish, S. Wynhoff, Z. Xie, X.T. Huang, A. Lopez, H. Mendez, J.E. Ramirez Vargas, A. Zatserklyaniy, A. Apresyan, K. Arndt, V.E. Barnes, G. Bolla, D. Bortoletto, A. Bujak, A. Everett, M. Fahling, A.F. Garfinkel, L. Gutay, N. Ippolito, Y. Kozhevnikov, A.T. Laasanen, C. Liu, V. Maroussov, S. Medved, P. Merkel, D.H. Miller, J. Miyamoto, N. Neumeister, A. Pompos, A. Roy, A. Sedov, I. Shipsey, V. Cuplov, N. Parashar, P. Bargassa, S.J. Lee, J.H. Liu, D. Maronde, M. Matveev, T. Nussbaum, B.P. Padley, J. Roberts, A. Tumanov, A. Bodek, H. Budd, J. Cammin, Y.S. Chung, P. DeBarbaro, R. Demina, G. Ginther, Y. Gotra, S. Korjenevski, D.C. Miner, W. Sakumoto, P. Slattery, M. Zielinski, A. Bhatti, L. Demortier, K. Goulianos, K. Hatakeyama, C. Mesropian, E. Bartz, S.H. Chuang, J. Doroshenko, E. Halkiadakis, P.F. Jacques, D. Khits, A. Lath, A. Macpherson, R. Plano, K. Rose, S. Schnetzer, S. Somalwar, R. Stone, T.L. Watts, G. Cerizza, M. Hollingsworth, J. Lazoflores, G. Ragghianti, S. Spanier, A. York, A. Aurisano, A. Golyash, T. Kamon, C.N. Nguyen, J. Pivarski, A. Safonov, D. Toback, M. Weinberger, N. Akchurin, L. Berntzon, K.W. Carrell, K. Gumus, C. Jeong, H. Kim, S.W. Lee, B.G. McGonagill, Y. Roh, A. Sill, M. Spezziga, R. Thomas, I. Volobouev, E. Washington, R. Wigmans, E. Yazgan, T. Bapty, D. Engh, C. Florez, W. Johns, T. Keskinpala, E. Luiggi Lopez, S. Neema, S. Nordstrom, S. Pathak, P. Sheldon, D. Andelin, M.W. Arenton, M. Balazs, M. Buehler, S. Conetti, B. Cox, R. Hirosky, M. Humphrey, R. Imlay, A. Ledovskoy, II Phillips, D., H. Powell, M. Ronquest, R. Yohay, M. Anderson, Y.W. Baek, J.N. Bellinger, D. Bradley, P. Cannarsa, D. Carlsmith, I. Crotty, S. Dasu, F. Feyzi, T. Gorski, L. Gray, K.S. Grogg, M. Grothe, M. Jaworski, P. Klabbers, J. Klukas, A. Lanaro, C. Lazaridis, J. Leonard, R. Loveless, M. Magrans de Abril, A. Mohapatra, G. Ott, W.H. Smith, M. Weinberg, D. Wenman, G.S. Atoian, S. Dhawan, V. Issakov, H. Neal, A. Poblaguev, M.E. Zeller, G. Abdullaeva, A. Avezov, M.I. Fazylov, E.M. Gasanov, A. Khugaev, Y.N. Koblik, M. Nishonov, K. Olimov, A. Umaraliev, B.S. Yuldashev, The CMS experiment at the CERN LHC. J. Instrum. **3**(8), S08004 (2008). https://doi.org/10.1088/1748-0221/3/08/S08004

J. Collins, K. Howe, B. Nachman, Anomaly detection for resonant new physics with machine learning. Phys. Rev. Lett. **121**(24), 241803 (2018). https://doi.org/10.1103/PhysRevLett.121.241803

J.H. Collins, K. Howe, B. Nachman, Extending the search for new resonances with machine learning. Phys. Rev. D **99**(1), 014038 (2019). https://doi.org/10.1103/PhysRevD.99.014038

J.M. Cordes, P.C.C. Freire, D.R. Lorimer, F. Camilo, D.J. Champion, D.J. Nice, R. Ramachandran, J.W.T. Hessels, W. Vlemmings, J. van Leeuwen, S.M. Ransom, N.D.R. Bhat, Z. Arzoumanian, M.A. McLaughlin, V.M. Kaspi, L. Kasian, J.S. Deneva, B. Reid, S. Chatterjee, J.L. Han, D.C. Backer, I.H. Stairs, A.A. Deshpande, C.A. Faucher-Giguère, Arecibo pulsar survey using ALFA. I. Survey strategy and first discoveries. Astrophys. J. **637**(1), 446–455 (2006). https://doi.org/10.1086/498335

A. Dabbech, M. Terris, A. Jackson, M. Ramatsoku, O.M. Smirnov, Y. Wiaux, First AI for deep super-resolution wide-field imaging in radio astronomy: unveiling structure in ESO 137-006. Astrophys. J. Lett. **939**(1), L4 (2022). https://doi.org/10.3847/2041-8213/ac98af

R.T. D'Agnolo, A. Wulzer, Learning new physics from a machine. Phys. Rev. D **99**(1), 015014 (2019). https://doi.org/10.1103/PhysRevD.99.015014

E. Dai, J. Chen, Graph-augmented normalizing flows for anomaly detection of multiple time series. arXiv e-prints, art. arXiv:2202.07857 (February 2022)

G.R. Davis, A. Chrysostomou, C. Taljaard, Ska1 operational concept document. SKA Document Number SKA-TEL-SKO-0000307. Revision **2**, 2016-12 (2016)

D.R. DeBoer, A.R. Parsons, J.E. Aguirre, P. Alexander, Z.S. Ali, A.P. Beardsley, G. Bernardi, J.D. Bowman, R.F. Bradley, C.L. Carilli, C. Cheng, E. de Lera Acedo, J.S. Dillon, A. Ewall-Wice, G. Fadana, N. Fagnoni, R. Fritz, S.R. Furlanetto, B. Glendenning, B.Greig, J. Grobbelaar, B.J. Hazelton, J.N. Hewitt, J. Hickish, D.C. Jacobs, A. Julius, M. Kariseb, S.A. Kohn, T. Lekalake, A. Liu, A. Loots, D. MacMahon, L. Malan, C. Malgas, M. Maree, Z. Martinot, N. Mathison, E. Matsetela, A. Mesinger, M.F. Morales, A.R. Neben, N. Patra, S. Pieterse, J.C. Pober, N. Razavi-Ghods, J. Ringuette, J. Robnett, K. Rosie, R. Sell, C. Smith, A. Syce, M. Tegmark, N. Thyagarajan, P.K.G. Williams, H. Zheng, Hydrogen epoch of reionization array (HERA). Publ. Astron. Soc. Pac. **129**(974), 045001 (2017). https://doi.org/10.1088/1538-3873/129/974/045001

T.J. Deeming, Stellar spectral classification: I. Application of component analysis. Mon. Not. R. Astron. Soc. **127**(6), 493–516 (1964). ISSN 0035-8711. https://doi.org/10.1093/mnras/127.6.493

A.M. Delgado, D. Wadekar, B. Hadzhiyska, S. Bose, L. Hernquist, S. Ho, Modelling the galaxy-halo connection with machine learning. Mon. Not. R. Astron. Soc. **515**(2), 2733–2746 (2022). https://doi.org/10.1093/mnras/stac1951

M. Delli Veneri, Ł. Tychoniec, F. Guglielmetti, G. Longo, E. Villard, 3D detection and characterization of ALMA sources through deep learning. Mon. Not. R. Astron. Soc. **518** (3), 3407–3427 (2023). https://doi.org/10.1093/mnras/stac3314

J.S. Denker, Y. LeCun, Transforming neural-net output levels to probability distributions, *Advances in Neural Information Processing Systems* (1991)

J. Denker, D. Schwartz, B. Wittner, S. Solla, R. Howard, L. Jackel, J. Hopfield, Large automatic learning, rule extraction, and generalization. Complex Syst. **1**(5), 877–922 (1987)

P.E. Dewdney, P.J. Hall, R.T. Schilizzi, T.J.L.W. Lazio, The square kilometre array. IEEE Proc. **97**(8), 1482–1496 (2009). https://doi.org/10.1109/JPROC.2009.2021005

S.G. Djorgovski, A.A. Mahabal, M.J. Graham, K. Polsterer, A. Krone-Martins, Applications of AI in astronomy. arXiv e-prints, art. arXiv:2212.01493 (December 2022)

L. Doorenbos, O. Torbaniuk, S. Cavuoti, M. Paolillo, G. Longo, M. Brescia, R. Sznitman, P. Márquez-Neila, ULISSE: A tool for one-shot sky exploration and its application for detection of active galactic nuclei. Astron. Astrophys. **666**, A171 (2022). https://doi.org/10.1051/0004-6361/202243900

F. Doshi-Velez, B. Kim, Towards a rigorous science of interpretable machine learning. arXiv e-prints, art. arXiv:1702.08608 (February 2017)

J. Duarte, S. Han, P. Harris, S. Jindariani, E. Kreinar, B. Kreis, J. Ngadiuba, M. Pierini, R. Rivera, N. Tran, Z. Wu, Fast inference of deep neural networks in FPGAs for particle physics. J. Instrum. **13**(7), P07027 (2018). https://doi.org/10.1088/1748-0221/13/07/P07027

R.P. Eatough, N. Molkenthin, M. Kramer, A. Noutsos, M.J. Keith, B.W. Stappers, A.G. Lyne, Selection of radio pulsar candidates using artificial neural networks. Mon. Not. R. Astron. Soc. **407**(4), 2443–2450 (2010). https://doi.org/10.1111/j.1365-2966.2010.17082.x

M. Erdmann, J. Glombitza, G. Kasieczka, U. Klemradt, *Deep Learning for Physics Research* (World Scientific, Singapore, 2021)

M. Ester, H.-P. Kriegel, J. Sander, X. Xu, A density-based algorithm for discovering clusters in large spatial databases with noise, in *KDD* (1996)

P.A. Estévez, J.C. Prıncipe, P. Zegers, *Advances in Self-Organizing Maps* (Springer, Berlin, 2013)

Euclid Collaboration, A. Humphrey et al., Euclid preparation: XXII. Selection of Quiescent Galaxies from Mock Photometry using Machine Learning. arXiv e-prints, art. arXiv:2209.13074 (September 2022)

L. Evans, P. Bryant, LHC machine. J. Instrum. **3**(8), S08001 (2008). https://doi.org/10.1088/1748-0221/3/08/S08001

M. Farina, Y. Nakai, D. Shih, Searching for new physics with deep autoencoders. Phys. Rev. D **101**(7), 075021 (2020). https://doi.org/10.1103/PhysRevD.101.075021

E.D. Feigelson, G. Jogesh Babu, *Statistical Methods for Astronomy* (Springer Netherlands, Dordrecht, 2013), pp. 445–480. ISBN 978-94-007-5618-2. https://doi.org/10.1007/978-94-007-5618-2_10

E.D. Feigelson, R.S. De Souza, E.E.O. Ishida, G.J. Babu, Twenty-first-century statistical and computational challenges in astrophysics. Annu. Rev. Stat. Appl. **8**, 493–517 (2021)

C.J. Fluke, C. Jacobs, Surveying the reach and maturity of machine learning and artificial intelligence in astronomy. WIREs Data Min. Knowl. Discov. **10**(2), e1349 (2020). https://doi.org/10.1002/widm.1349

K. Fukushima, Neocognitron: a self-organizing neural network model for a mechanism of pattern recognition unaffected by shift in position. Biol. Cybern. **36**, 193–202 (2004)

Y. Gal, Z. Ghahramani, Dropout as a bayesian approximation: Representing model uncertainty in deep learning, in *International Conference on Machine Learning*. PMLR (2016), pp. 1050–1059

T.J. Galvin, M. Huynh, R.P. Norris, X.R. Wang, E. Hopkins, O.I. Wong, S. Shabala, L. Rudnick, M.J. Alger, K.L. Polsterer, Radio galaxy zoo: knowledge transfer using rotationally invariant self-organizing maps. Publ. Astron. Soc. Pac. **131**(1004), 108009 (2019). https://doi.org/10.1088/1538-3873/ab150b

T.J. Galvin, M.T. Huynh, R.P. Norris, X.R. Wang, E. Hopkins, K. Polsterer, N.O. Ralph, A.N. O'Brien, G.H. Heald, Cataloguing the radio-sky with unsupervised machine learning: a new approach for the SKA era. Mon. Not. R. Astron. Soc. **497**(3), 2730–2758 (2020). https://doi.org/10.1093/mnras/staa1890

H. Gan, L.V.E. Koopmans, F.G. Mertens, M. Mevius, A.R. Offringa, B. Ciardi, B.K. Gehlot, R. Ghara, A. Ghosh, S.K. Giri, I.T. Iliev, G. Mellema, V.N. Pandey, S. Zaroubi, Statistical analysis of the causes of excess variance in the 21 cm signal power spectra obtained with the low-frequency array. Astron. Astrophys. **663**, A9 (2022). https://doi.org/10.1051/0004-6361/202142945

J. Gawlikowski, C.R.N. Tassi, M. Ali, J. Lee, M. Humt, J. Feng, A. Kruspe, R. Triebel, P. Jung, R. Roscher et al., A survey of uncertainty in deep neural networks. arXiv preprint arXiv:2107.03342 (2021)

J.E. Geach, Unsupervised self-organized mapping: a versatile empirical tool for object selection, classification and redshift estimation in large surveys. Mon. Not. R. Astron. Soc. **419**(3), 2633–2645 (2012). https://doi.org/10.1111/j.1365-2966.2011.19913.x

M. Gell-Mann, *The Quark and the Jaguar: Adventures in the Simple and the Complex* (Henry Holt and Company, New York, 1994)

M. Gell-Mann, S. Lloyd, Information measures, effective complexity, and total information. Complexity **2**(1), 44–52 (1996)

C. Gheller, F. Vazza, Convolutional deep denoising autoencoders for radio astronomical images. Mon. Not. R. Astron. Soc. **509**(1), 990–1009 (2022). https://doi.org/10.1093/mnras/stab3044

C. Gheller, F. Vazza, A. Bonafede, Deep learning based detection of cosmological diffuse radio sources. Mon. Not. R. Astron. Soc. **480**(3), 3749–3761 (2018). https://doi.org/10.1093/mnras/sty2102

N. Gillet, A. Mesinger, B. Greig, A. Liu, G. Ucci, Deep learning from 21-cm tomography of the cosmic dawn and reionization. Mon. Not. R. Astron. Soc. **484**(1), 282–293 (2019). https://doi.org/10.1093/mnras/stz010

R. Girshick, Fast R-CNN. arXiv e-prints, art. arXiv:1504.08083 (April 2015)

R. Girshick, J. Donahue, T. Darrell, J. Malik, Rich feature hierarchies for accurate object detection and semantic segmentation. arXiv e-prints, art. arXiv:1311.2524 (November 2013)

T. Gleisberg, S. Höche, F. Krauss, M. Schönherr, S. Schumann, F. Siegert, J. Winter, Event generation with SHERPA 1.1. J. High Energy Phys. **2009**(2), 007 (2009). https://doi.org/10.1088/1126-6708/2009/02/007

R.E. González, R.P. Muñoz, C.A. Hernández, Galaxy detection and identification using deep learning and data augmentation. Astron. Comput. **25**, 103–109 (2018). https://doi.org/10.1016/j.ascom.2018.09.004

I.J. Goodfellow, J. Pouget-Abadie, M. Mirza, B. Xu, D. Warde-Farley, S. Ozair, A. Courville, Y. Bengio, Generative adversarial networks. arXiv e-prints, art. arXiv:1406.2661 (June 2014)

B. Greig, Y.-S. Ting, A.A. Kaurov, Exploring the cosmic 21-cm signal from the epoch of reionization using the wavelet scattering transform. Mon. Not. R. Astron. Soc. **513**(2), 1719–1741 (2022). https://doi.org/10.1093/mnras/stac977

C. Guo, G. Pleiss, Y. Sun, K.Q. Weinberger, On calibration of modern neural networks, in *International Conference on Machine Learning*. PMLR (2017), pp. 1321–1330

P. Guo, F. Duan, P. Wang, Y. Yao, Q. Yin, X. Xin, D. Li, L. Qian, S. Wang, Z. Pan, L. Zhang, Pulsar candidate classification using generative adversary networks. Mon. Not. R. Astron. Soc. **490** (4), 5424–5439 (2019). https://doi.org/10.1093/mnras/stz2975

N. Gupta, M. Huynh, R.P. Norris, R. Wang, A.M. Hopkins, H. Andernach, B.S. Koribalski, T.J. Galvin, Discovery of peculiar radio morphologies with ASKAP using unsupervised machine learning. arXiv e-prints, art. arXiv:2208.13997 (August 2022)

A. Hallin, J. Isaacson, G. Kasieczka, C. Krause, B. Nachman, T. Quadfasel, M. Schlaffer, D. Shih, M. Sommerhalder, Classifying anomalies through outer density estimation. Phys. Rev. D **106** (5), 055006 (2022a). https://doi.org/10.1103/PhysRevD.106.055006

A. Hallin, G. Kasieczka, T. Quadfasel, D. Shih, M. Sommerhalder, Resonant anomaly detection without background sculpting. arXiv e-prints, art. arXiv:2210.14924 (October 2022b). https://doi.org/10.48550/arXiv.2210.14924

P. Hartley, A. Bonaldi, R. Braun, J.N.H.S Aditya, S. Aicardi, L. Alegre, A. Chakraborty, X. Chen, S. Choudhuri, A.O. Clarke, J. Coles, J.S. Collinson, D. Cornu, L. Darriba, M.D. Veneri, J. Forbrich, B. Fraga, A. Galan, J. Garrido, F. Gubanov, H. Håkansson, M.J. Hardcastle, C. Heneka, D. Herranz, K.M. Hess, M. Jagannath, S. Jaiswal, R.J. Jurek, D. Korber, S. Kitaeff, D. Kleiner, B. Lao, X. Lu, A. Mazumder, J. Moldón, R. Mondal, S. Ni, M. Önnheim, M. Parra, N. Patra, A. Peel, P. Salomé, S. Sánchez-Expósito, M. Sargent, B. Semelin, P. Serra, A.K. Shaw, A.X. and Shen, A. Sjöberg, L. Smith, A. Soroka, V. Stolyarov, E. Tolley, M.C Toribio, J.M. van der Hulst, A.V. Sadr, L. Verdes-Montenegro, T. Westmeier, K. Yu, L. Yu, L. Zhang, X. Zhang, Y. Zhang, A. Alberdi, M. Ashdown, C.R. Bom, M. Brüggen, J. Cannon, R. Chen, F. Combes, J. Conway, F. Courbin, J. Ding, G. Fourestey, J. Freundlich, L. Gao, C. Gheller, Q. Guo, E. Gustavsson, M. Jirstrand, M.G. Jones, G. Józsa, P. Kamphuis, J.-P. Kneib, M. Lindqvist, B. Liu, Y. Liu, Y. Mao, A. Marchal, I. Márquez, A. Meshcheryakov, M. Olberg, N. Oozeer, M. Pandey-Pommier, W. Pei, B. Peng, J. Sabater, A. Sorgho, J.L. Starck, C. Tasse, A. Wang, Y. Wang, H. Xi, X. Yang, H. Zhang, J. Zhang, M. Zhao, S. Zuo, SKA Science Data Challenge 2: analysis and results. Mon. Not. R. Astron. Soc. **523**(2), 1967–1993 (2023)

D.M. Hawkins, *Identification of Outliers*, vol. 11 (Springer, Berlin, 1980)

M.A. Hayat, G. Stein, P. Harrington, Z. Lukić, M. Mustafa, Self-supervised representation learning for astronomical images. Astrophys. J. Lett. **911**(2), L33 (2021). https://doi.org/10.3847/2041-8213/abf2c7

K. He, G. Gkioxari, P. Dollár, R. Girshick, Mask R-CNN. arXiv e-prints, art. arXiv:1703.06870 (March 2017)

T. Heimel, G. Kasieczka, T. Plehn, J. Thompson, QCD or what? SciPost Phys. **6**(3), 030 (2019). https://doi.org/10.21468/SciPostPhys.6.3.030

S. Heinis, S. Kumar, S. Gezari, W.S. Burgett, K.C. Chambers, P.W. Draper, H. Flewelling, N. Kaiser, E.A. Magnier, N. Metcalfe, C. Waters, Of genes and machines: application of a combination of machine learning tools to astronomy datasets. Astrophys. J. **821**(2), 86 (2016). https://doi.org/10.3847/0004-637X/821/2/86

C. Heneka, A. Cooray, Optimal survey parameters: Ly α and H α intensity mapping for synergy with the 21-cm signal during reionization. Mon. Not. R. Astron. Soc. **506**(2), 1573–1584 (2021). https://doi.org/10.1093/mnras/stab1842

C. Heneka, A. Mesinger, The spin-temperature dependence of the 21-cm-LAE cross-correlation. Mon. Not. R. Astron. Soc. **496**(1), 581–589 (2020). https://doi.org/10.1093/mnras/staa1517

C. Heneka, A. Cooray, C. Feng, Probing the intergalactic medium with Lyα and 21 cm fluctuations. Astrophys. J. **848**(1), 52 (2017). https://doi.org/10.3847/1538-4357/aa8eed

G.E. Hinton, R.R. Salakhutdinov, Reducing the dimensionality of data with neural networks. Science **313**(5786), 504–507 (2006). https://doi.org/10.1126/science.1127647

G.E Hinton, D. Van Camp, Keeping the neural networks simple by minimizing the description length of the weights, in *Proceedings of the Sixth Annual Conference on Computational Learning Theory* (1993), pp. 5–13

M. Ho, M. Ntampaka, M.M. Rau, M. Chen, A. Lansberry, F. Ruehle, H. Trac, The dynamical mass of the Coma cluster from deep learning. Nat. Astron. (2022). https://doi.org/10.1038/s41550-022-01711-1

M. Hobson, P. Graff, F. Feroz, A. Lasenby, Machine-learning in astronomy. Proc. Int. Astron. Union **10**(S306), 279–287 (2014). https://doi.org/10.1017/S1743921314013672

H. Hotelling, Analysis of a complex of statistical variables into principal components. J. Educ. Psychol. **24**(6), 417 (1933)

B. Hoyle, M.M. Rau, R. Zitlau, S. Seitz, J. Weller, Feature importance for machine learning redshifts applied to SDSS galaxies. Mon. Not. R. Astron. Soc. **449**(2), 1275–1283 (2015). https://doi.org/10.1093/mnras/stv373

R.L. Hu, J. Karnowski, R. Fadely, J.-P. Pommier, Image segmentation to distinguish between overlapping human chromosomes. arXiv e-prints, art. arXiv:1712.07639 (December 2017)

G. Huang, Y. Li, G. Pleiss, Z. Liu, J.E. Hopcroft, K.Q. Weinberger, Snapshot ensembles: train 1, get m for free. arXiv preprint arXiv:1704.00109 (2017)

E. Hüllermeier, W. Waegeman. Aleatoric and epistemic uncertainty in machine learning: An introduction to concepts and methods. Mach. Learn. **110**(3), 457–506 (2021)

A. Hutter, C.M. Trott, P. Dayal, Survey parameters for detecting 21-cm-Ly α emitter cross-correlations with the Square Kilometre Array. Mon. Not. R. Astron. Soc. **479**(1), L129–L133 (2018). https://doi.org/10.1093/mnrasl/sly115

A. Hutter, C.A. Watkinson, J. Seiler, P. Dayal, M. Sinha, D.J. Croton, The 21 cm bispectrum during reionization: a tracer of the ionization topology. Mon. Not. R. Astron. Soc. **492**(1), 653–667 (2020). https://doi.org/10.1093/mnras/stz3139

A.P. Igoshev, S.B. Popov, Gaussian mixture models and the population synthesis of radio pulsars. Mon. Not. R. Astron. Soc. **434**(3), 2229–2237 (2013). https://doi.org/10.1093/mnras/stt1167

S. Ioffe, C. Szegedy, Batch normalization: accelerating deep network training by reducing internal covariate shift, in *International Conference on Machine Learning*, PMLR (2015), pp. 448–456

Z. Ivezic, J.A. Tyson, T. Axelrod, D. Burke, C.F. Claver, K.H. Cook, S.M. Kahn, R.H. Lupton, D.G. Monet, P.A. Pinto, M.A. Strauss, C.W. Stubbs, L. Jones, A. Saha, R. Scranton, C. Smith, LSST Collaboration, LSST: from science drivers to reference design and anticipated data products, In *American Astronomical Society Meeting Abstracts #213*, vol. 213. American Astronomical Society Meeting Abstracts (January 2009), p. 460.03

G. James, D. Witten, T. Hastie, R. Tibshirani, *An Introduction to Statistical Learning*, vol. 112 (Springer, Berlin, 2013)

N. Jeffrey, J. Alsing, F. Lanusse, Likelihood-free inference with neural compression of DES SV weak lensing map statistics. Mon. Not. R. Astron. Soc. **501**(1), 954–969 (2021). https://doi.org/10.1093/mnras/staa3594

J.M. Jenkins, D.A. Caldwell, H. Chandrasekaran, J.D. Twicken, S.T. Bryson, E.V. Quintana, B.D. Clarke, J. Li, C. Allen, P. Tenenbaum, H. Wu, T.C. Klaus, C.K. Middour, M.T. Cote, S. McCauliff, F.R. Girouard, J.P. Gunter, B. Wohler, J. Sommers, J.R. Hall, A.K.M.K. Uddin, M.S. Wu, P.A. Bhavsar, J. Van Cleve, D.L. Pletcher, J.A. Dotson, M.R. Haas, R.L. Gilliland, D.G. Koch, W.J. Borucki, Overview of the Kepler science processing pipeline. Astrophys. J. Lett. **713**(2), L87–L91 (2010). https://doi.org/10.1088/2041-8205/713/2/L87

W.D. Jennings, C.A. Watkinson, F.B. Abdalla, J.D. McEwen, Evaluating machine learning techniques for predicting power spectra from reionization simulations. Mon. Not. R. Astron. Soc. **483**(3), 2907–2922 (2019). https://doi.org/10.1093/mnras/sty3168

S. Johnston, A.G. Lyne, R.N. Manchester, D.A. Kniffen, N. D'Amico, J. Lim, M. Ashworth, A high-frequency survey of the southern galactic plane for pulsars. Mon. Not. R. Astron. Soc. **255**, 401–411 (1992). https://doi.org/10.1093/mnras/255.3.401

J. Jonas, MeerKAT Team, The MeerKAT radio telescope, in *MeerKAT Science: On the Pathway to the SKA*, p. 1 (January 2016). https://doi.org/10.22323/1.277.0001

L.V. Jospin, H. Laga, F. Boussaid, W. Buntine, M. Bennamoun, Hands-on bayesian neural networks–a tutorial for deep learning users. IEEE Comput. Intell. Mag. **17**(2), 29–48 (2022)

G. Karagiorgi, G. Kasieczka, S. Kravitz, B. Nachman, D. Shih, Machine learning in the search for new fundamental physics. Nat. Rev. Phys. **4**(6), 399–412 (2022). https://doi.org/10.1038/s42254-022-00455-1

G. Kasieczka, T. Plehn, A. Butter, K. Cranmer, D. Debnath, B.M. Dillon, M. Fairbairn, D.A. Faroughy, W. Fedorko, C. Gay, L. Gouskos, J.F. Kamenik, P. Komiske, S. Leiss, A. Lister, S. Macaluso, E. Metodiev, L. Moore, B. Nachman, K. Nordström, J. Pearkes, H. Qu, Y. Rath, M. Rieger, D. Shih, J. Thompson, S. Varma, The machine learning landscape of top taggers. SciPost Phys. **7**(1), 014 (2019). https://doi.org/10.21468/SciPostPhys.7.1.014

G. Kasieczka, B. Nachman, D. Shih, O. Amram, A. Andreassen, K. Benkendorfer, B. Bortolato, G. Brooijmans, F. Canelli, J.H. Collins, B. Dai, F.F. De Freitas, B.M. Dillon, I.-M. Dinu, Z. Dong, J. Donini, J. Duarte, D.A. Faroughy, J. Gonski, P. Harris, A. Kahn, J.F. Kamenik, C.K. Khosa, P. Komiske, L. Le Pottier, P. Martín-Ramiro, A. Matevc, E. Metodiev, V. Mikuni, C.W. Murphy, I. Ochoa, S.E. Park, M. Pierini, D. Rankin, V. Sanz, N. Sarda, U. Seljak, A. Smolkovic, G. Stein, C.M. Suarez, M. Szewc, J. Thaler, S. Tsan, S.-M. Udrescu, L. Vaslin, J.-R. Vlimant, D. Williams, M. Yunus, The LHC Olympics 2020 a community challenge for anomaly detection in high energy physics. Rep. Prog. Phys. **84**(12), 124201 (2021). https://doi.org/10.1088/1361-6633/ac36b9

M.J. Keith, A. Jameson, W. van Straten, M. Bailes, S. Johnston, M. Kramer, A. Possenti, S.D. Bates, N.D.R. Bhat, M. Burgay, S. Burke-Spolaor, N. D'Amico, L. Levin, Peter L. McMahon, S. Milia, B.W. Stappers, The high time resolution universe pulsar survey - I. System configuration and initial discoveries. Mon. Not. R. Astron. Soc. **409**(2), 619–627 (2010). https://doi.org/10.1111/j.1365-2966.2010.17325.x

B.C. Kelly, T.A. McKay, Morphological classification of galaxies by shapelet decomposition in the sloan digital sky survey. Astrophys. J. **127**(2), 625 (2004). https://doi.org/10.1086/380934

A. Kendall, Y. Gal, What uncertainties do we need in bayesian deep learning for computer vision?, in *Advances in Neural Information Processing Systems*, vol. 30 (2017)

J. Kerrigan, P. La Plante, S. Kohn, J.C. Pober, J. Aguirre, Z. Abdurashidova, P. Alexander, Z.S. Ali, Y. Balfour, A.P. Beardsley, G. Bernardi, J.D. Bowman, R.F. Bradley, J. Burba, C.L. Carilli, C. Cheng, D.R. DeBoer, M. Dexter, E. de Lera Acedo, J.S. Dillon, J. Estrada, A. Ewall-Wice, N. Fagnoni, R. Fritz, S.R. Furlanetto, B. Glendenning, B. Greig, J. Grobbelaar, D. Gorthi, Z. Halday, B.J. Hazelton, J. Hickish, D.C. Jacobs, A. Julius, N.S. Kern, P. Kittiwisit, M. Kolopanis, A. Lanman, T. Lekalake, A. Liu, D. MacMahon, L. Malan, C. Malgas, M. Maree, Z.E. Martinot, E. Matsetela, A. Mesinger, M. Molewa, M.F. Morales, T. Mosiane, A.R. Neben, A.R. Parsons, N. Patra, S. Pieterse, N. Razavi-Ghods, J. Ringuette, J. Robnett, K. Rosie, P. Sims, C. Smith, A. Syce, N. Thyagarajan, P.K.G. Williams, H. Zheng, Optimizing sparse RFI prediction using deep learning. Mon. Not. R. Astron. Soc. **488**(2), 2605–2615 (2019). https://doi.org/10.1093/mnras/stz1865

D.P. Kingma, P. Dhariwal, Glow: generative flow with invertible 1x1 convolutions. arXiv e-prints, art. arXiv:1807.03039 (July 2018)

B. Knispel, Pulsar discoveries by volunteer distributed computing and the strongest continuous gravitational wave signal. PhD thesis, Albert-Einstein-Institut, Max-Planck-Institut für Gravitationsphysik, Hannover; Institut für Gravitationsphysik, Leibniz Universität Hannover, Hannover (July 2011)

B. Knispel, R.P. Eatough, H. Kim, E.F. Keane, B. Allen, D. Anderson, C. Aulbert, O. Bock, F. Crawford, H.B. Eggenstein, H. Fehrmann, D. Hammer, M. Kramer, A.G. Lyne, B. Machenschalk, R.B. Miller, M.A. Papa, D. Rastawicki, J. Sarkissian, X. Siemens, B.W. Stappers, Einstein@Home discovery of 24 pulsars in the Parkes multi-beam pulsar survey. Astrophys. J. **774**(2), 93 (2013). https://doi.org/10.1088/0004-637X/774/2/93

T. Kohonen, Self-organized formation of topologically correct feature maps. Biol. Cybern. **43**(1), 59–69 (1982)

T. Kohonen, The self-organizing map. Proc. IEEE **78**(9), 1464–1480 (1990)

M.A. Kuhn, E.D. Feigelson, Mixture models in astronomy. arXiv e-prints, art. arXiv:1711.11101 (November 2017)

J. Kummer, L. Rustige, F. Griese, K. Borras, M. Brüggen, P.L.S. Connor, F. Gaede, G. Kasieczka, P. Schleper, Radio galaxy classification with wGAN-supported augmentation. arXiv e-prints, art. arXiv:2206.15131 (June 2022)

L. Künkel, R.M. Thomas, J.P.W. Verbiest, Detecting pulsars with neural networks: a proof of concept. Mon. Not. R. Astron. Soc. **506**(1), 1111–1127 (2021). https://doi.org/10.1093/mnras/stab1628

O. Lahav, A. Nairn, L. Sodré Jr., M.C. Storrie-Lombardi, Neural computation as a tool for galaxy classification: methods and examples. Mon. Not. R. Astron. Soc. **283**(1), 207–221 (1996)

S. Laine, T. Karras, J. Lehtinen, T. Aila, High-quality self-supervised deep image denoising, in *Advances in Neural Information Processing Systems*, vol. 32 (2019)

B. Lakshminarayanan, A. Pritzel, C. Blundell, Simple and scalable predictive uncertainty estimation using deep ensembles, in *Advances in Neural Information Processing Systems*, vol. 30 (2017)

B. Lao, T. An, A. Wang, Z. Xu, S. Guo, W. Lv, X. Wu, Y. Zhang, Artificial intelligence for celestial object census: the latest technology meets the oldest science. Sci. Bull. **66**(21), 2145–2147 (2021). https://doi.org/10.1016/j.scib.2021.07.015

Y. LeCun, L. Bottou, Y. Bengio, P. Haffner, Gradient-based learning applied to document recognition. Proc. IEEE **86**, 2278–2324 (1998)

K.J. Lee, L. Guillemot, Y.L. Yue, M. Kramer, D.J. Champion, Application of the Gaussian mixture model in pulsar astronomy - pulsar classification and candidates ranking for the Fermi 2FGL catalogue. Mon. Not. R. Astron. Soc. **424**(4), 2832–2840 (2012). https://doi.org/10.1111/j.1365-2966.2012.21413.x

J. Lee, M. Humt, J. Feng, R. Triebel, Estimating model uncertainty of neural networks in sparse information form, in *International Conference on Machine Learning*, PMLR (2020), pp. 5702–5713

M. Leoni, E.E.O. Ishida, J. Peloton, A. Möller, Fink: early supernovae Ia classification using active learning. Astron. Astrophys. **663**, A13 (2022). https://doi.org/10.1051/0004-6361/202142715

D. Li, P. Wang, L. Qian, M. Krco, P. Jiang, Y. Yue, C. Jin, Y. Zhu, Z. Pan, R. Nan, A. Dunning, FAST in space: considerations for a multibeam, multipurpose survey using China's 500-m aperture spherical radio telescope (FAST). IEEE Microwave Mag. **19**(3), 112–119 (2018). https://doi.org/10.1109/MMM.2018.2802178

J. Li, P. Tenenbaum, J.D. Twicken, C.J. Burke, J.M. Jenkins, E.V. Quintana, J.F. Rowe, S.E. Seader, Kepler data validation II-Transit model fitting and multiple-planet search. Publ. Astron. Soc. Pac. **131**(996), 024506 (2019). https://doi.org/10.1088/1538-3873/aaf44d

W. Liang, G.A. Tadesse, D. Ho, F.-F. Li, M. Zaharia, C. Zhang, J. Zou, Advances, challenges and opportunities in creating data for trustworthy AI. Nat. Mach. Intell. **4**(8), 669–677 (2022). ISSN 2522-5839. https://doi.org/10.1038/s42256-022-00516-1

Lightkurve Collaboration, J.V.d.M. Cardoso, C. Hedges, M. Gully-Santiago, N. Saunders, A.M. Cody, T. Barclay, O. Hall, S. Sagear, E. Turtelboom, J. Zhang, A. Tzanidakis, K. Mighell, J. Coughlin, K. Bell, Z. Berta-Thompson, P. Williams, J. Dotson, G. Barentsen, Lightkurve: Kepler and TESS time series analysis in Python. Astrophysics Source Code Library (December 2018)

S.-C. Lin, Y. Su, G. Liang, Y. Zhang, N. Jacobs, Y. Zhang, Estimating cluster masses from SDSS multiband images with transfer learning. Mon. Not. R. Astron. Soc. **512**(3), 3885–3894 (2022). https://doi.org/10.1093/mnras/stac725

C.J. Lintott, K. Schawinski, A. Slosar, K. Land, S. Bamford, D. Thomas, M. Jordan Raddick, R.C. Nichol, A. Szalay, D. Andreescu, P. Murray, J. Vandenberg, Galaxy Zoo: morphologies derived from visual inspection of galaxies from the Sloan Digital Sky Survey. Mon. Not. R. Astron. Soc. **389**(3), 1179–1189 (2008). https://doi.org/10.1111/j.1365-2966.2008.13689.x

A. Liu, J. Richard Shaw, Data analysis for precision 21 cm cosmology. Publ. Astron. Soc. Pac. **132**(1012), 062001 (2020). https://doi.org/10.1088/1538-3873/ab5bfd

F.T. Liu, K.M. Ting, Z.-H. Zhou, Isolation forest, in *2008 Eighth IEEE International Conference on Data Mining* (IEEE, 2008), pp. 413–422

M. Lochner, B.A. Bassett, ASTRONOMALY: personalised active anomaly detection in astronomical data. Astron. Comput. **36**, 100481 (2021). https://doi.org/10.1016/j.ascom.2021.100481

C.J. Lonsdale, R.J. Cappallo, M.F. Morales, F.H. Briggs, L. Benkevitch, J.D. Bowman, J.D. Bunton, S. Burns, B.E. Corey, L. Desouza, S.S. Doeleman, M. Derome, A. Deshpande, M.R. Gopala, L.J. Greenhill, D.E. Herne, J.N. Hewitt, P.A. Kamini, J.C. Kasper, B.B. Kincaid, J. Kocz, E. Kowald, E. Kratzenberg, D. Kumar, M.J. Lynch, S. Madhavi, M. Matejek, D.A. Mitchell, E. Morgan, D. Oberoi, S. Ord, J. Pathikulangara, T. Prabu, A. Rogers, A. Roshi, J.E. Salah, R.J. Sault, N.U. Shankar, K.S. Srivani, J. Stevens, S. Tingay, A. Vaccarella, M. Waterson, R.B. Wayth, R.L. Webster, A.R. Whitney, A. Williams, C. Williams, The Murchison widefield array: design overview. IEEE Proc. **97**(8), 1497–1506 (2009). https://doi.org/10.1109/JPROC.2009.2017564

L. Lucie-Smith, H.V. Peiris, A. Pontzen, B. Nord, J. Thiyagalingam, D. Piras, Discovering the building blocks of dark matter halo density profiles with neural networks. Phys. Rev. D **105** (10), 103533 (2022). https://doi.org/10.1103/PhysRevD.105.103533

V. Lukic, M. Brüggen, J.K. Banfield, O.I. Wong, L. Rudnick, R.P. Norris, B. Simmons, Radio Galaxy Zoo: compact and extended radio source classification with deep learning. Mon. Not. R. Astron. Soc. **476**(1), 246–260 (2018). https://doi.org/10.1093/mnras/sty163

V. Lukic, F. de Gasperin, M. Brüggen, ConvoSource: radio-astronomical source-finding with convolutional neural networks. Galaxies **8**(1), 3 (2019). https://doi.org/10.3390/galaxies8010003

S. Lundberg, S.-I. Lee, A unified approach to interpreting model predictions. arXiv e-prints, art. arXiv:1705.07874 (May 2017)

R.J. Lyon, B.W. Stappers, S. Cooper, J.M. Brooke, J.D. Knowles, Fifty years of pulsar candidate selection: from simple filters to a new principled real-time classification approach. Mon. Not. R. Astron. Soc. **459**, 1104 (2016). https://doi.org/10.1093/mnras/stw656

D.J.C. MacKay, A practical Bayesian framework for backpropagation networks. Neural Comput. **4**(3), 448–472 (1992)

M.A.B. Mahmoud, P. Guo, Learning from imbalanced pulsar data by combine DCGAN and PILAE algorithm. New Astron. **85**, 101561 (2021). https://doi.org/10.1016/j.newast.2020.101561

T.L. Makinen, L. Lancaster, F. Villaescusa-Navarro, P. Melchior, S. Ho, L. Perreault-Levasseur, D.N. Spergel, deep21: a deep learning method for 21 cm foreground removal. J. Cosmol. Astropart. Phys. **2021**(4), 081 (2021). https://doi.org/10.1088/1475-7516/2021/04/081

A. Malinin, M. Gales, Predictive uncertainty estimation via prior networks, in *Advances in Neural Information Processing Systems*, vol. 31 (2018)

R.N. Manchester, A.G. Lyne, J.H. Taylor, J.M. Durdin, M.I. Large, A.G. Little, The second Molonglo pulsar survey - discovery of 155 pulsars. Mon. Not. R. Astron. Soc. **185**, 409–421 (1978). https://doi.org/10.1093/mnras/185.2.409

R.N. Manchester, A.G. Lyne, F. Camilo, J.F. Bell, V.M. Kaspi, N. D'Amico, N.P.F. McKay, F. Crawford, I.H. Stairs, A. Possenti, M. Kramer, D.C. Sheppard, The Parkes multi-beam pulsar survey - I. Observing and data analysis systems, discovery and timing of 100 pulsars. Mon. Not. R. Astron. Soc. **328**(1), 17–35 (2001). https://doi.org/10.1046/j.1365-8711.2001.04751.x

Z.E. Mariet, R. Jenatton, F. Wenzel, D. Tran, Distilling ensembles improves uncertainty estimates, in *Third Symposium on Advances in Approximate Bayesian Inference*, 2020

T. Marill, D. Green, On the effectiveness of receptors in recognition systems. IEEE Trans. Inf. Theory **9**(1), 11–17 (1963)

J.R. Martínez-Galarza, F.B. Bianco, D. Crake, K. Tirumala, A.A. Mahabal, M.J. Graham, D. Giles, A method for finding anomalous astronomical light curves and their analogues. Mon. Not. R. Astron. Soc. **508**(4), 5734–5756 (2021). https://doi.org/10.1093/mnras/stab2588

M. Mesarcik, A.-J. Boonstra, E. Ranguelova, R.V. van Nieuwpoort, Learning to detect RFI in radio astronomy without seeing it. Mon. Not. R. Astron. Soc. (2022). https://doi.org/10.1093/mnras/stac2503

A. Mesinger, *Understanding the Epoch of Cosmic Reionization*, vol. 423 (Springer, Berlin, 2016)

A. Mesinger, S. Furlanetto, R. Cen, 21CMFAST: a fast, seminumerical simulation of the high-redshift 21-cm signal. Mon. Not. R. Astron. Soc. **411**(2), 955–972 (2011). https://doi.org/10.1111/j.1365-2966.2010.17731.x

D. Mohan, A.M.M. Scaife, F. Porter, M. Walmsley, M. Bowles, Quantifying uncertainty in deep learning approaches to radio galaxy classification. Mon. Not. R. Astron. Soc. **511**(3), 3722–3740 (2022). https://doi.org/10.1093/mnras/stac223

G. Montavon, A. Binder, S. Lapuschkin, W. Samek, K.-R. Müller, *Layer-Wise Relevance Propagation: An Overview* (Springer International Publishing, Cham, 2019), pp. 193–209. ISBN 978-3-030-28954-6. https://doi.org/10.1007/978-3-030-28954-6_10

V. Morello, E.D. Barr, M. Bailes, C.M. Flynn, E.F. Keane, W. van Straten, SPINN: a straightforward machine learning solution to the pulsar candidate selection problem. Mon. Not. R. Astron. Soc. **443**(2), 1651–1662 (2014). https://doi.org/10.1093/mnras/stu1188

R.I.J. Mostert, K.J. Duncan, H.J.A. Röttgering, K.L. Polsterer, P.N. Best, M. Brienza, M. Brüggen, M.J. Hardcastle, N. Jurlin, B. Mingo, R. Morganti, T. Shimwell, D. Smith, W.L. Williams, Unveiling the rarest morphologies of the LOFAR Two-metre Sky Survey radio source population with self-organised maps. Astron. Astrophys. **645**, A89 (2021). https://doi.org/10.1051/0004-6361/202038500

S. Murray, B. Greig, A. Mesinger, J. Muñoz, Y. Qin, J. Park, C. Watkinson, 21cmFAST v3: a Python-integrated C code for generating 3D realizations of the cosmic 21cm signal. J. Open Source Softw. **5**(54), 2582 (2020). https://doi.org/10.21105/joss.02582

B. Nachman, D. Shih, Anomaly detection with density estimation. Phys. Rev. D **101**, 075042 (2020). https://doi.org/10.1103/PhysRevD.101.075042

F. Nammour, U. Akhaury, J.N. Girard, F. Lanusse, F. Sureau, C. Ben Ali, J.L. Starck, ShapeNet: shape constraint for galaxy image deconvolution. Astron. Astrophys. **663**, A69 (2022). https://doi.org/10.1051/0004-6361/202142626

S. Neutsch, C. Heneka, M. Brüggen, Inferring astrophysics and dark matter properties from 21 cm tomography using deep learning. Mon. Not. R. Astron. Soc. **511**(3), 3446–3462 (2022). https://doi.org/10.1093/mnras/stac218

S. Ni, Y. Li, L.-Y. Gao, X. Zhang, Eliminating primary beam effect in foreground subtraction of neutral hydrogen intensity mapping survey with deep learning. Astrophys. J. **934**(1), 83 (2022). https://doi.org/10.3847/1538-4357/ac7a34

J. Niebling, B. Baasch, A. Kruspe, Analysis of railway track irregularities with convolutional autoencoders and clustering algorithms, in *European Dependable Computing Conference* (Springer, Berlin, 2020), pp. 78–89

R.P. Norris, Discovering the unexpected in astronomical survey data. Publ. Astron. Soc. Aust. **34**, e007 (2017). https://doi.org/10.1017/pasa.2016.63

R.P. Norris, A.M. Hopkins, J. Afonso, S. Brown, J.J. Condon, L. Dunne, I. Feain, R. Hollow, M. Jarvis, M. Johnston-Hollitt, E. Lenc, E. Middelberg, P. Padovani, I. Prandoni, L. Rudnick, N. Seymour, G. Umana, H. Andernach, D.M. Alexander, P.N. Appleton, D. Bacon, J. Banfield, W. Becker, M.J.I. Brown, P. Ciliegi, C. Jackson, S. Eales, A.C. Edge, B.M. Gaensler, G. Giovannini, C.A. Hales, P. Hancock, M.T. Huynh, E. Ibar, R.J. Ivison, R. Kennicutt, Amy E. Kimball, A.M. Koekemoer, B.S. Koribalski, Á. R. López-Sánchez, M.Y. Mao, T. Murphy, H. Messias, K.A. Pimbblet, A. Raccanelli, K.E. Randall, T.H. Reiprich, I.G. Roseboom, H. Röttgering, D.J. Saikia, R.G. Sharp, O.B. Slee, Ian Smail, M.A. Thompson, J.S. Urquhart,

J.V. Wall, G.B. Zhao, EMU: evolutionary map of the Universe. Publ. Astron. Soc. Aust. **28**(3), 215–248 (2011). https://doi.org/10.1071/AS11021
R.P. Norris, H.T. Intema, A.D. Kapińska, B.S. Koribalski, E. Lenc, L. Rudnick, R.Z.E. Alsaberi, C. Anderson, G.E. Anderson, E. Crawford, R. Crocker, J. English, M.D. Filipović, T.J. Galvin, A.M. Hopkins, N. Hurley-Walker, S. Inoue, K. Luken, P.J. Macgregor, P. Manojlović, J. Marvil, A.N. O'Brien, L. Park, W. Raja, D. Shobhana, T. Venturi, J.D. Collier, C. Hale, A. Hotan, V. Moss, M. Whiting, Unexpected circular radio objects at high Galactic latitude. Publ. Astron. Soc. Aust. **38**, e003 (2021). https://doi.org/10.1017/pasa.2020.52
K. Ntwaetsile, J.E. Geach, Rapid sorting of radio galaxy morphology using Haralick features. Mon. Not. R. Astron. Soc. **502**(3), 3417–3425 (2021). https://doi.org/10.1093/mnras/stab271
P.V. Padmanabh, E.D. Barr, S.S. Sridhar, M.R. Rugel, A. Damas-Segovia, A.M. Jacob, V. Balakrishnan, M. Berezina, M.C.i Bernadich, A. Brunthaler, D.J. Champion, P.C.C. Freire, S. Khan, H.R. Klöckner, M. Kramer, Y.K. Ma, S.A. Mao, Y.P. Men, K.M. Menten, S. Sengupta, V. Venkatraman Krishnan, O. Wucknitz, F. Wyrowski, M.C. Bezuidenhout, S. Buchner, M. Burgay, W. Chen, C.J. Clark, L. Künkel, L. Nieder, B. Stappers, L.S. Legodi, M.M. Nyamai, The MPIfR-MeerKAT Galactic Plane survey I – system setup and early results. arXiv e-prints, art. arXiv:2303.09231 (March 2023). https://doi.org/10.48550/arXiv.2303.09231
B. Panes, C. Eckner, L. Hendriks, S. Caron, K. Dijkstra, G. Jóhannesson, R. Ruiz de Austri, G. Zaharijas, Identification of point sources in gamma rays using U-shaped convolutional neural networks and a data challenge. Astron. Astrophys. **656**, A62 (2021). https://doi.org/10.1051/0004-6361/202141193
G. Papamakarios, E. Nalisnick, D. Jimenez Rezende, S. Mohamed, B. Lakshminarayanan, Normalizing flows for probabilistic modeling and inference. arXiv e-prints, art. arXiv:1912.02762 (December 2019). https://doi.org/10.48550/arXiv.1912.02762
J. Pata, J. Duarte, F. Mokhtar, E. Wulff, J. Yoo, J.-R. Vlimant, M. Pierini, M. Girone, Machine learning for particle flow reconstruction at CMS. J. Phys. Conf. Ser. **2438**, 012100 (2023). https://doi.org/10.1088/1742-6596/2438/1/012100
K. Pearson, Liii. On lines and planes of closest fit to systems of points in space. Lond. Edinb. Dublin Philos. Mag. J. Sci. **2**(11), 559–572 (1901)
F. Pedregosa, G. Varoquaux, A. Gramfort, V. Michel, B. Thirion, O. Grisel, M. Blondel, A. Müller, J. Nothman, G. Louppe, P. Prettenhofer, R. Weiss, V. Dubourg, J. Vanderplas, A. Passos, D. Cournapeau, M. Brucher, M. Perrot, É. Duchesnay, Scikit-learn: machine learning in Python. arXiv e-prints, art. arXiv:1201.0490 (January 2012)
M.A. Peth, J.M. Lotz, P.E. Freeman, C. McPartland, S. Alireza Mortazavi, G.F. Snyder, G. Barro, N.A. Grogin, Y. Guo, S. Hemmati et al., Beyond spheroids and discs: classifications of Candels galaxy structure at 1.4< z< 2 via principal component analysis. Mon. Not. R. Astron. Soc. **458** (1), 963–987 (2016)
J.C. Pober, A.R. Parsons, D.R. DeBoer, P. McDonald, M. McQuinn, J.E. Aguirre, Z. Ali, R.F. Bradley, T.-C. Chang, M.F. Morales, The Baryon acoustic oscillation broadband and broadbeam array: design overview and sensitivity forecasts. Astronom. J. **145**(3), 65 (2013). https://doi.org/10.1088/0004-6256/145/3/65
A.A. Pol, G. Cerminara, C. Germain, M. Pierini, A. Seth, Detector monitoring with artificial neural networks at the CMS experiment at the CERN Large Hadron Collider. arXiv e-prints, art. arXiv:1808.00911 (July 2018). https://doi.org/10.48550/arXiv.1808.00911
K.L. Polsterer, F. Gieseke, C. Igel, Automatic galaxy classification via machine learning techniques: Parallelized Rotation/Flipping INvariant Kohonen Maps (PINK), in *Astronomical Data Analysis Software an Systems XXIV (ADASS XXIV)*, ed. by A.R. Taylor, E. Rosolowsky, vol. 495. Astronomical Society of the Pacific Conference Series (September 2015), p. 81
D. Prelogović, A. Mesinger, S. Murray, G. Fiameni, N. Gillet, Machine learning astrophysics from 21 cm lightcones: impact of network architectures and signal contamination. Mon. Not. R. Astron. Soc. **509**(3), 3852–3867 (2022). https://doi.org/10.1093/mnras/stab3215
D.D. Proctor, A selection of giant radio sources from NVSS. Astrophys. J. Suppl. **224**(2), 18 (2016). https://doi.org/10.3847/0067-0049/224/2/18

H. Qu, L. Gouskos, Jet tagging via particle clouds. Phys. Rev. D **101**(5), 056019 (2020). https://doi.org/10.1103/PhysRevD.101.056019

H. Qu, C. Li, S. Qian, Particle transformer for jet tagging. arXiv e-prints, art. arXiv:2202.03772 (February 2022). https://doi.org/10.48550/arXiv.2202.03772

M. Rafieferantsoa, S. Andrianomena, R. Davé, Predicting the neutral hydrogen content of galaxies from optical data using machine learning. Mon. Not. R. Astron. Soc. **479**(4), 4509–4525 (2018). https://doi.org/10.1093/mnras/sty1777

J.A. Raine, S. Klein, D. Sengupta, T. Golling, CURTAINs for your sliding window: constructing unobserved regions by transforming adjacent intervals. arXiv e-prints, art. arXiv:2203.09470 (March 2022). https://doi.org/10.48550/arXiv.2203.09470

N.O. Ralph, R.P. Norris, G. Fang, L.A.F. Park, T.J. Galvin, M.J. Alger, H. Andernach, C. Lintott, L. Rudnick, S. Shabala, O.I. Wong, Radio Galaxy Zoo: unsupervised clustering of convolutionally auto-encoded radio-astronomical images. Publ. Astron. Soc. Pac. **131**(1004), 108011 (2019). https://doi.org/10.1088/1538-3873/ab213d

P. Re Fiorentin, C.A.L. Bailer-Jones , Y.S. Lee, T.C. Beers, T. Sivarani, R. Wilhelm, C. Allende Prieto, J.E. Norris, Estimation of stellar atmospheric parameters from sdss/segue spectra. Astron. Astrophys. **467**(3), 1373–1387 (2007). https://doi.org/10.1051/0004-6361:20077334

J. Redmon, A. Farhadi, YOLO9000: better, faster, stronger. arXiv e-prints, art. arXiv:1612.08242 (December 2016)

J. Redmon, S. Divvala, R. Girshick, A. Farhadi, You only look once: unified, real-time object detection. arXiv e-prints, art. arXiv:1506.02640 (June 2015)

S. Ren, K. He, R. Girshick, J. Sun, Faster R-CNN: towards real-time object detection with region proposal networks. arXiv e-prints, art. arXiv:1506.01497 (June 2015)

S. Rezaei, J.P. McKean, M. Biehl, A. Javadpour, DECORAS: detection and characterization of radio-astronomical sources using deep learning. Mon. Not. R. Astron. Soc. **510**(4), 5891–5907 (2022). https://doi.org/10.1093/mnras/stab3519

G.R. Ricker, J.N. Winn, R. Vanderspek, D.W. Latham, G.Á. Bakos, J.L. Bean, Z.K. Berta-Thompson, T.M. Brown, L. Buchhave, N.R. Butler, R.P. Butler, W.J. Chaplin, D. Charbonneau, J. Christensen-Dalsgaard, M. Clampin, D. Deming, J. Doty, N. De Lee, C. Dressing, E.W. Dunham, M. Endl, F. Fressin, J. Ge, T. Henning, M.J. Holman, A.W. Howard, S. Ida, J.M. Jenkins, G. Jernigan, J.A. Johnson, L. Kaltenegger, N. Kawai, H. Kjeldsen, G. Laughlin, A.M. Levine, D. Lin, J.J. Lissauer, P. MacQueen, G. Marcy, P.R. McCullough, T.D. Morton, N. Narita, M. Paegert, E. Palle, F. Pepe, J. Pepper, A. Quirrenbach, S.A. Rinehart, D. Sasselov, B. Sato, S. Seager, A. Sozzetti, K.G. Stassun, P. Sullivan, A. Szentgyorgyi, G. Torres, S. Udry, J. Villasenor, Transiting exoplanet survey satellite (TESS). J. Astron. Telesc. Instrum. Syst. **1**, 014003 (2015). https://doi.org/10.1117/1.JATIS.1.1.014003

H. Ritter, A. Botev, D. Barber, A scalable laplace approximation for neural networks, in *6th International Conference on Learning Representations, ICLR 2018-Conference Track Proceedings*, vol. 6. International Conference on Representation Learning (2018).

B.J. Robinson, Science with the Australia telescope. Publ. Astron. Soc. Aust. **7**, 220 (1987). https://doi.org/10.1017/S1323358000022256

O. Ronneberger, P. Fischer, T. Brox, U-Net: convolutional networks for biomedical image segmentation. arXiv e-prints, art. arXiv:1505.04597 (May 2015)

T. Rothschild, D. Nagai, H. Aung, S.B. Green, M. Ntampaka, J. ZuHone, Emulating Sunyaev-Zeldovich images of galaxy clusters using autoencoders. Mon. Not. R. Astron. Soc. **513**(1), 333–344 (2022). https://doi.org/10.1093/mnras/stac438

L. Rudnick, Radio galaxy classification: #Tags, Not Boxes. Galaxies **9**(4), 85 (2021). https://doi.org/10.3390/galaxies9040085

D. Ruhe, M. Kuiack, A. Rowlinson, R. Wijers, P. Forré, Detecting dispersed radio transients in real time using convolutional neural networks. Astron. Comput. **38**, 100512 (2022). https://doi.org/10.1016/j.ascom.2021.100512

J. Sabater, S. Sánchez-Expósito, P. Best, J. Garrido, L. Verdes-Montenegro, D. Lezzi, Calibration of LOFAR data on the cloud. Astron. Comput. **19**, 75–89 (2017). https://doi.org/10.1016/j.ascom.2017.04.001

M. Sadeghi, M. Javaherian, H. Miraghaei, Morphological-based classifications of radio galaxies using supervised machine-learning methods associated with image moments. Astronom. J. **161** (2), 94 (2021). https://doi.org/10.3847/1538-3881/abd314

S. Sanidas, S. Cooper, C.G. Bassa, J.W.T. Hessels, V.I. Kondratiev, D. Michilli, B.W. Stappers, C.M. Tan, J. van Leeuwen, L. Cerrigone, R.A. Fallows, M. Iacobelli, E. Orrú, R.F. Pizzo, A. Shulevski, M.C. Toribio, S. ter Veen, P. Zucca, L. Bondonneau, J.M. Grießmeier, A. Karastergiou, M. Kramer, C. Sobey, The LOFAR Tied-Array All-Sky Survey (LOTAAS): survey overview and initial pulsar discoveries. Astron. Astrophys. **626**, A104 (2019). https://doi.org/10.1051/0004-6361/201935609

A.M.M. Scaife, F. Porter, Fanaroff-Riley classification of radio galaxies using group-equivariant convolutional neural networks. Mon. Not. R. Astron. Soc. **503**(2), 2369–2379 (2021). https://doi.org/10.1093/mnras/stab530

K. Schmidt, F. Geyer, S. Fröse, P.S. Blomenkamp, M. Brüggen, F. de Gasperin, D. Elsässer, W. Rhode, Deep learning-based imaging in radio interferometry. Astron. Astrophys. **664**, A134 (2022). https://doi.org/10.1051/0004-6361/202142113

C.J. Schmit, J.R. Pritchard, Emulation of reionization simulations for Bayesian inference of astrophysics parameters using neural networks. Mon. Not. R. Astron. Soc. **475**(1), 1213–1223 (2018). https://doi.org/10.1093/mnras/stx3292

S. Schneider, E. Rusak, L. Eck, O. Bringmann, W. Brendel, M. Bethge, Improving robustness against common corruptions by covariate shift adaptation, in *Advances in Neural Information Processing Systems*, vol. 33 (2020), pp. 11539–11551

B. Schölkopf, R.C. Williamson, A. Smola, J. Shawe-Taylor, J. Platt, Support vector method for novelty detection, in *Advances in Neural Information Processing Systems*, vol. 12 (1999)

G. Segal, D. Parkinson, R.P. Norris, J. Swan, Identifying complex sources in large astronomical data using a coarse-grained complexity measure. Publ. Astron. Soc. Pac. **131**(1004), 108007 (2019). https://doi.org/10.1088/1538-3873/ab0068

G. Segal, D. Parkinson, R. Norris, A.M. Hopkins, H. Andernach, E.L. Alexander, E. Carretti, B.S. Koribalski, L.S. Legodi, S. Leslie, Y. Luo, J.C.S. Pierce, H. Tang, E. Vardoulaki, T. Vernstrom, Identifying anomalous radio sources in the Evolutionary Map of the Universe Pilot Survey using a complexity-based approach. Mon. Not. R. Astron. Soc. **521**(1), 1429–1447 (2023). https://doi.org/10.1093/mnras/stad537

A. Shalaginov, K. Franke, A new method for an optimal som size determination in neuro-fuzzy for the digital forensics applications (2015), pp. 549–563. ISBN 978-3-319-19221-5. https://doi.org/10.1007/978-3-319-19222-2_46

H. Shen, D. George, E.A. Huerta, Z. Zhao, Denoising gravitational waves using deep learning with recurrent denoising autoencoders. arXiv e-prints, art. arXiv:1711.09919 (November 2017)

D. Shih, M.R. Buckley, L. Necib, J. Tamanas, VIA MACHINAE: Searching for stellar streams using unsupervised machine learning. Mon. Not. R. Astron. Soc. **509**(4), 5992–6007 (2022). https://doi.org/10.1093/mnras/stab3372

H. Shimabukuro, S. Yoshiura, K. Takahashi, S. Yokoyama, K. Ichiki, 21 cm line bispectrum as a method to probe cosmic dawn and epoch of reionization. Mon. Not. R. Astron. Soc. **458**(3), 3003–3011 (2016). https://doi.org/10.1093/mnras/stw482

T.W. Shimwell, H.J.A. Röttgering, P.N. Best, W.L. Williams, T.J. Dijkema, F. de Gasperin, M.J. Hardcastle, G.H. Heald, D.N. Hoang, A. Horneffer, H. Intema, E.K. Mahony, S. Mandal, A.P. Mechev, L. Morabito, J.B.R. Oonk, D. Rafferty, E. Retana-Montenegro, J. Sabater, C. Tasse, R.J. van Weeren, M. Brüggen, G. Brunetti, K.T. Chyży, J.E. Conway, M. Haverkorn, N. Jackson, M.J. Jarvis, J.P. McKean, G.K. Miley, R. Morganti, G.J. White, M.W. Wise, I.M. van Bemmel, R. Beck, M. Brienza, A. Bonafede, G. Calistro Rivera, R. Cassano, A.O. Clarke, D. Cseh, A. Deller, A. Drabent, W. van Driel, D. Engels, H. Falcke, C. Ferrari, S. Fröhlich, M.A. Garrett, J.J. Harwood, V. Heesen, M. Hoeft, C. Horellou, F.P. Israel, A.D. Kapińska, M. Kunert-Bajraszewska, D.J. McKay, N.R. Mohan, E. Orrú, R.F. Pizzo, I. Prandoni, D.J. Schwarz, A. Shulevski, M. Sipior, D.J.B. Smith, S.S. Sridhar, M. Steinmetz, A. Stroe, E. Varenius, P.P. van der Werf, J.A. Zensus, J.T.L. Zwart, The LOFAR Two-metre Sky Survey. I. Survey

description and preliminary data release. Astron. Astrophys. **598**, A104 (2017). https://doi.org/10.1051/0004-6361/201629313
T.W. Shimwell, M.J. Hardcastle, C. Tasse, P.N. Best, H.J.A. Röttgering, W.L. Williams, A. Botteon, A. Drabent, A. Mechev, A. Shulevski, R.J. van Weeren, L. Bester, M. Brüggen, G. Brunetti, J.R. Callingham, K.T. Chyży, J.E. Conway, T.J. Dijkema, K. Duncan, F. de Gasperin, C.L. Hale, M. Haverkorn, B. Hugo, N. Jackson, M. Mevius, G.K. Miley, L.K. Morabito, R. Morganti, A. Offringa, J.B.R. Oonk, D. Rafferty, J. Sabater, D.J.B. Smith, D.J. Schwarz, O. Smirnov, S.P. O'Sullivan, H. Vedantham, G.J. White, J.G. Albert, L. Alegre, B. Asabere, D.J. Bacon, A. Bonafede, E. Bonnassieux, M. Brienza, M. Bilicki, M. Bonato, G. Calistro Rivera, R. Cassano, R. Cochrane, J.H. Croston, V. Cuciti, D. Dallacasa, A. Danezi, R.J. Dettmar, G. Di Gennaro, H.W. Edler, T.A. Enßlin, K.L. Emig, T.M.O. Franzen, C. García-Vergara, Y.G. Grange, G. Gürkan, M. Hajduk, G. Heald, V. Heesen, D.N. Hoang, M. Hoeft, C. Horellou, M. Iacobelli, M. Jamrozy, V. Jelić, R. Kondapally, P. Kukreti, M. Kunert-Bajraszewska, M. Magliocchetti, V. Mahatma, K. Małek, S. Mandal, F. Massaro, Z. Meyer-Zhao, B. Mingo, R.I.J. Mostert, D.G. Nair, S.J. Nakoneczny, B. Nikiel-Wroczyński, E. Orrú, U. Pajdosz-Śmierciak, T. Pasini, I. Prandoni, H.E. van Piggelen, K. Rajpurohit, E. Retana-Montenegro, C.J. Riseley, A. Rowlinson, A. Saxena, C. Schrijvers, F. Sweijen, T.M. Siewert, R. Timmerman, M. Vaccari, J. Vink, J.L. West, A. Wołowska, X. Zhang, J. Zheng, The LOFAR two-metre sky survey. V. Second data release. Astron. Astrophys. **659**, A1 (2022). https://doi.org/10.1051/0004-6361/202142484
J. Shlomi, P. Battaglia, J.-R. Vlimant, Graph neural networks in particle physics. arXiv e-prints, art. arXiv:2007.13681 (July 2020). https://doi.org/10.48550/arXiv.2007.13681
K. Simonyan, A. Vedaldi, A. Zisserman, deep inside convolutional networks: visualising image classification models and saliency maps. arXiv e-prints, art. arXiv:1312.6034 (December 2013)
A.M. Sirunyan, A. Tumasyan, W. Adam, F. Ambrogi, T. Bergauer, M. Dragicevic, J. Erö, A. Escalante Del Valle, M. Flechl, R. Frühwirth, M. Jeitler, N. Krammer, I. Krätschmer, D. Liko, T. Madlener, I. Mikulec, N. Rad, J. Schieck, R. Schöfbeck, M. Spanring, W. Waltenberger, C.E. Wulz, M. Zarucki, V. Drugakov, V. Mossolov, J. Suarez Gonzalez, M.R. Darwish, E.A. De Wolf, D. Di Croce, X. Janssen, A. Lelek, M. Pieters, H. Rejeb Sfar, H. Van Haevermaet, P. Van Mechelen, S. Van Putte, N. Van Remortel, F. Blekman, E.S. Bols, S.S. Chhibra, J. D'Hondt, J. De Clercq, D. Lontkovskyi, S. Lowette, I. Marchesini, S. Moortgat, Q. Python, K. Skovpen, S. Tavernier, W. Van Doninck, P. Van Mulders, D. Beghin, B. Bilin, B. Clerbaux, G. De Lentdecker, H. Delannoy, B. Dorney, L. Favart, A. Grebenyuk, A.K. Kalsi, L. Moureaux, A. Popov, N. Postiau, E. Starling, L. Thomas, C. Vander Velde, P. Vanlaer, D. Vannerom, T. Cornelis, D. Dobur, I. Khvastunov, M. Niedziela, C. Roskas, M. Tytgat, W. Verbeke, B. Vermassen, M. Vit, O. Bondu, G. Bruno, C. Caputo, P. David, C. Delaere, M. Delcourt, A. Giammanco, V. Lemaitre, J. Prisciandaro, A. Saggio, M. Vidal Marono, P. Vischia, J. Zobec, F.L. Alves, G.A. Alves, G. Correia Silva, C. Hensel, A. Moraes, P. Rebello Teles, E. Belchior Batista Das Chagas, W. Carvalho, J. Chinellato, E. Coelho, E.M. Da Costa, G.G. Da Silveira, D. De Jesus Damiao, C. De Oliveira Martins, S. Fonseca De Souza, L.M. Huertas Guativa, H. Malbouisson, J. Martins, D. Matos Figueiredo, M. Medina Jaime, M. Melo De Almeida, C. Mora Herrera, L. Mundim, H. Nogima, W.L. Prado Da Silva, L.J. Sanchez Rosas, A. Santoro, A. Sznajder, M. Thiel, E.J. Tonelli Manganote, F. Torres Da Silva De Araujo, A. Vilela Pereira, C.A. Bernardes, L. Calligaris, T.R. Fernandez Perez Tomei, E.M. Gregores, D.S. Lemos, P.G. Mercadante, S.F. Novaes, SandraS. Padula, A. Aleksandrov, G. Antchev, R. Hadjiiska, P. Iaydjiev, M. Misheva, M. Rodozov, M. Shopova, G. Sultanov, M. Bonchev, A. Dimitrov, T. Ivanov, L. Litov, B. Pavlov, P. Petkov, A. Petrov, W. Fang, X. Gao, L. Yuan, M. Ahmad, Z. Hu, Y. Wang, G.M. Chen, H.S. Chen, M. Chen, C.H. Jiang, D. Leggat, H. Liao, Z. Liu, A. Spiezia, J. Tao, E. Yazgan, H. Zhang, S. Zhang, J. Zhao, A. Agapitos, Y. Ban, G. Chen, A. Levin, J. Li, L. Li, Q. Li, Y. Mao, S.J. Qian, D. Wang, Q. Wang, M. Xiao, C. Avila, A. Cabrera, C. Florez, C.F. González Hernández, M.A. Segura Delgado, J. Mejia Guisao, J.D. Ruiz Alvarez, C.A. Salazar González, N. Vanegas Arbelaez, D. Giljanović, N. Godinovic, D. Lelas, I. Puljak, T. Sculac, Z. Antunovic, M. Kovac, V. Brigljevic, D. Ferencek, K. Kadija, B. Mesic, M. Roguljic, A. Starodumov, T. Susa, M.W.

Ather, A. Attikis, E. Erodotou, A. Ioannou, M. Kolosova, S. Konstantinou, G. Mavromanolakis, J. Mousa, C. Nicolaou, F. Ptochos, P.A. Razis, H. Rykaczewski, D. Tsiakkouri, M. Finger, Jr. Finger, M., A. Kveton, J. Tomsa, E. Ayala, E. Carrera Jarrin, H. Abdalla, S. Khalil, S. Bhowmik, A. Carvalho Antunes De Oliveira, R.K. Dewanjee, K. Ehataht, M. Kadastik, M. Raidal, C. Veelken, P. Eerola, L. Forthomme, H. Kirschenmann, K. Osterberg, M. Voutilainen, F. Garcia, J. Havukainen, J.K. Heikkilä, V. Karimäki, M.S. Kim, R. Kinnunen, T. Lampén, K. Lassila-Perini, S. Laurila, S. Lehti, T. Lindén, P. Luukka, T. Mäenpää, H. Siikonen, E. Tuominen, J. Tuominiemi, T. Tuuva, M. Besancon, F. Couderc, M. Dejardin, D. Denegri, B. Fabbro, J.L. Faure, F. Ferri, S. Ganjour, A. Givernaud, P. Gras, G. Hamel de Monchenault, P. Jarry, C. Leloup, B. Lenzi, E. Locci, J. Malcles, J. Rander, A. Rosowsky, M. Ö. Sahin, A. Savoy-Navarro, M. Titov, G.B. Yu, S. Ahuja, C. Amendola, F. Beaudette, P. Busson, C. Charlot, B. Diab, G. Falmagne, R. Granier de Cassagnac, I. Kucher, A. Lobanov, C. Martin Perez, M. Nguyen, C. Ochando, P. Paganini, J. Rembser, R. Salerno, J.B. Sauvan, Y. Sirois, A. Zabi, A. Zghiche, J.L. Agram, J. Andrea, D. Bloch, G. Bourgatte, J.M. Brom, E.C. Chabert, C. Collard, E. Conte, J.C. Fontaine, D. Gelé, U. Goerlach, M. Jansová, A.C. Le Bihan, N. Tonon, P. Van Hove, S. Gadrat, S. Beauceron, C. Bernet, G. Boudoul, C. Camen, A. Carle, N. Chanon, R. Chierici, D. Contardo, P. Depasse, H. El Mamouni, J. Fay, S. Gascon, M. Gouzevitch, B. Ille, Sa. Jain, F. Lagarde, I.B. Laktineh, H. Lattaud, A. Lesauvage, M. Lethuillier, L. Mirabito, S. Perries, V. Sordini, L. Torterotot, G. Touquet, M. Vander Donckt, S. Viret, T. Toriashvili, I. Bagaturia, C. Autermann, L. Feld, K. Klein, M. Lipinski, D. Meuser, A. Pauls, M. Preuten, M.P. Rauch, J. Schulz, M. Teroerde, B. Wittmer, M. Erdmann, B. Fischer, S. Ghosh, T. Hebbeker, K. Hoepfner, H. Keller, L. Mastrolorenzo, M. Merschmeyer, A. Meyer, P. Millet, G. Mocellin, S. Mondal, S. Mukherjee, D. Noll, A. Novak, T. Pook, A. Pozdnyakov, T. Quast, M. Radziej, Y. Rath, H. Reithler, J. Roemer, A. Schmidt, S.C. Schuler, A. Sharma, S. Wiedenbeck, S. Zaleski, G. Flügge, W. Haj Ahmad, O. Hlushchenko, T. Kress, T. Müller, A. Nowack, C. Pistone, O. Pooth, D. Roy, H. Sert, A. Stahl, M. Aldaya Martin, P. Asmuss, I. Babounikau, H. Bakhshiansohi, K. Beernaert, O. Behnke, A. Bermúdez Martínez, D. Bertsche, A.A. Bin Anuar, K. Borras, V. Botta, A. Campbell, A. Cardini, P. Connor, S. Consuegra Rodríguez, C. Contreras-Campana, V. Danilov, A. De Wit, M.M. Defranchis, C. Diez Pardos, D. Domínguez Damiani, G. Eckerlin, D. Eckstein, T. Eichhorn, A. Elwood, E. Eren, E. Gallo, A. Geiser, A. Grohsjean, M. Guthoff, M. Haranko, A. Harb, A. Jafari, N.Z. Jomhari, H. Jung, A. Kasem, M. Kasemann, H. Kaveh, J. Keaveney, C. Kleinwort, J. Knolle, D. Krücker, W. Lange, T. Lenz, J. Lidrych, K. Lipka, W. Lohmann, R. Mankel, I.A. Melzer-Pellmann, A.B. Meyer, M. Meyer, M. Missiroli, J. Mnich, A. Mussgiller, V. Myronenko, D. Pérez Adán, S.K. Pflitsch, D. Pitzl, A. Raspereza, A. Saibel, M. Savitskyi, V. Scheurer, P. Schütze, C. Schwanenberger, R. Shevchenko, A. Singh, H. Tholen, O. Turkot, A. Vagnerini, M. Van De Klundert, R. Walsh, Y. Wen, K. Wichmann, C. Wissing, O. Zenaiev, R. Zlebcik, R. Aggleton, S. Bein, L. Benato, A. Benecke, V. Blobel, T. Dreyer, A. Ebrahimi, F. Feindt, A. Fröhlich, C. Garbers, E. Garutti, D. Gonzalez, P. Gunnellini, J. Haller, A. Hinzmann, A. Karavdina, G. Kasieczka, R. Klanner, R. Kogler, N. Kovalchuk, S. Kurz, V. Kutzner, J. Lange, T. Lange, A. Malara, J. Multhaup, C.E.N. Niemeyer, A. Perieanu, A. Reimers, O. Rieger, C. Scharf, P. Schleper, S. Schumann, J. Schwandt, J. Sonneveld, H. Stadie, G. Steinbrück, F.M. Stober, B. Vormwald, I. Zoi, M. Akbiyik, C. Barth, M. Baselga, S. Baur, T. Berger, E. Butz, R. Caspart, T. Chwalek, W. De Boer, A. Dierlamm, K. El Morabit, N. Faltermann, M. Giffels, P. Goldenzweig, A. Gottmann, M.A. Harrendorf, F. Hartmann, U. Husemann, S. Kudella, S. Mitra, M.U. Mozer, D. Müller, Th. Müller, M. Musich, A. Nürnberg, G. Quast, K. Rabbertz, M. Schröder, I. Shvetsov, H.J. Simonis, R. Ulrich, M. Wassmer, M. Weber, C. Wöhrmann, R. Wolf, S. Wozniewski, G. Anagnostou, P. Asenov, G. Daskalakis, T. Geralis, A. Kyriakis, D. Loukas, G. Paspalaki, M. Diamantopoulou, G. Karathanasis, P. Kontaxakis, A. Manousakis-katsikakis, A. Panagiotou, I. Papavergou, N. Saoulidou, A. Stakia, K. Theofilatos, K. Vellidis, E. Vourliotis, G. Bakas, K. Kousouris, I. Papakrivopoulos, G. Tsipolitis, I. Evangelou, C. Foudas, P. Gianneios, P. Katsoulis, P. Kokkas, S. Mallios, K. Manitara, N. Manthos, I. Papadopoulos, J. Strologas, F.A. Triantis, D. Tsitsonis, M. Bartók, R. Chudasama, M. Csanad, P. Major, K. Mandal, A. Mehta, M.I. Nagy, G. Pasztor, O. Surányi, G.I. Veres, G. Bencze,

C. Hajdu, D. Horvath, F. Sikler, T. Á. Vámi, V. Veszpremi, G. Vesztergombi, N. Beni, S. Czellar, J. Karancsi, J. Molnar, Z. Szillasi, P. Raics, D. Teyssier, Z.L. Trocsanyi, B. Ujvari, T. Csorgo, W.J. Metzger, F. Nemes, T. Novak, S. Choudhury, J.R. Komaragiri, P.C. Tiwari, S. Bahinipati, C. Kar, G. Kole, P. Mal, V.K. Muraleedharan Nair Bindhu, A. Nayak, D.K. Sahoo, S.K. Swain, S. Bansal, S.B. Beri, V. Bhatnagar, S. Chauhan, N. Dhingra, R. Gupta, A. Kaur, M. Kaur, S. Kaur, P. Kumari, M. Lohan, M. Meena, K. Sandeep, S. Sharma, J.B. Singh, A.K. Virdi, A. Bhardwaj, B.C. Choudhary, R.B. Garg, M. Gola, S. Keshri, Ashok Kumar, M. Naimuddin, P. Priyanka, K. Ranjan, Aashaq Shah, R. Sharma, R. Bhardwaj, M. Bharti, R. Bhattacharya, S. Bhattacharya, U. Bhawandeep, D. Bhowmik, S. Dutta, S. Ghosh, B. Gomber, M. Maity, K. Mondal, S. Nandan, A. Purohit, P.K. Rout, G. Saha, S. Sarkar, T. Sarkar, M. Sharan, B. Singh, S. Thakur, P.K. Behera, P. Kalbhor, A. Muhammad, P.R. Pujahari, A. Sharma, A.K. Sikdar, D. Dutta, V. Jha, V. Kumar, D.K. Mishra, P.K. Netrakanti, L.M. Pant, P. Shukla, T. Aziz, M.A. Bhat, S. Dugad, G.B. Mohanty, N. Sur, S. Banerjee, S. Bhattacharya, S. Chatterjee, P. Das, M. Guchait, S. Karmakar, S. Kumar, G. Majumder, K. Mazumdar, N. Sahoo, S. Sawant, S. Dube, B. Kansal, A. Kapoor, K. Kothekar, S. Pandey, A. Rane, A. Rastogi, S. Sharma, S. Chenarani, E. Eskandari Tadavani, S.M. Etesami, M. Khakzad, M. Mohammadi Najafabadi, M. Naseri, F. Rezaei Hosseinabadi, M. Felcini, M. Grunewald, M. Abbrescia, R. Aly, C. Calabria, A. Colaleo, D. Creanza, L. Cristella, N. De Filippis, M. De Palma, A. Di Florio, W. Elmetenawee, L. Fiore, A. Gelmi, G. Iaselli, M. Ince, S. Lezki, G. Maggi, M. Maggi, J.A. Merlin, G. Miniello, S. My, S. Nuzzo, A. Pompili, G. Pugliese, R. Radogna, A. Ranieri, G. Selvaggi, L. Silvestris, F.M. Simone, R. Venditti, P. Verwilligen, G. Abbiendi, C. Battilana, D. Bonacorsi, L. Borgonovi, S. Braibant-Giacomelli, R. Campanini, P. Capiluppi, A. Castro, F.R. Cavallo, C. Ciocca, G. Codispoti, M. Cuffiani, G.M. Dallavalle, F. Fabbri, A. Fanfani, E. Fontanesi, P. Giacomelli, C. Grandi, L. Guiducci, F. Iemmi, S. Lo Meo, S. Marcellini, G. Masetti, F.L. Navarria, A. Perrotta, F. Primavera, A.M. Rossi, T. Rovelli, G.P. Siroli, N. Tosi, S. Albergo, S. Costa, A. Di Mattia, R. Potenza, A. Tricomi, C. Tuve, G. Barbagli, A. Cassese, R. Ceccarelli, V. Ciulli, C. Civinini, R. D'Alessandro, F. Fiori, E. Focardi, G. Latino, P. Lenzi, M. Meschini, S. Paoletti, G. Sguazzoni, L. Viliani, L. Benussi, S. Bianco, D. Piccolo, M. Bozzo, F. Ferro, R. Mulargia, E. Robutti, S. Tosi, A. Benaglia, A. Beschi, F. Brivio, V. Ciriolo, M.E. Dinardo, P. Dini, S. Gennai, A. Ghezzi, P. Govoni, L. Guzzi, M. Malberti, S. Malvezzi, D. Menasce, F. Monti, L. Moroni, M. Paganoni, D. Pedrini, S. Ragazzi, T. Tabarelli de Fatis, D. Valsecchi, D. Zuolo, S. Buontempo, N. Cavallo, A. De Iorio, A. Di Crescenzo, F. Fabozzi, F. Fienga, G. Galati, A.O.M. Iorio, L. Lista, S. Meola, P. Paolucci, B. Rossi, C. Sciacca, E. Voevodina, P. Azzi, N. Bacchetta, D. Bisello, A. Boletti, A. Bragagnolo, R. Carlin, P. Checchia, P. De Castro Manzano, T. Dorigo, U. Dosselli, F. Gasparini, U. Gasparini, A. Gozzelino, S.Y. Hoh, P. Lujan, M. Margoni, A.T. Meneguzzo, J. Pazzini, M. Presilla, P. Ronchese, R. Rossin, F. Simonetto, A. Tiko, M. Tosi, M. Zanetti, P. Zotto, G. Zumerle, A. Braghieri, D. Fiorina, P. Montagna, S.P. Ratti, V. Re, M. Ressegotti, C. Riccardi, P. Salvini, I. Vai, P. Vitulo, M. Biasini, G.M. Bilei, D. Ciangottini, L. Fanò, P. Lariccia, R. Leonardi, E. Manoni, G. Mantovani, V. Mariani, M. Menichelli, A. Rossi, A. Santocchia, D. Spiga, K. Androsov, P. Azzurri, G. Bagliesi, V. Bertacchi, L. Bianchini, T. Boccali, R. Castaldi, M.A. Ciocci, R. Dell'Orso, S. Donato, G. Fedi, L. Giannini, A. Giassi, M.T. Grippo, F. Ligabue, E. Manca, G. Mandorli, A. Messineo, F. Palla, A. Rizzi, G. Rolandi, S. Roy Chowdhury, A. Scribano, P. Spagnolo, R. Tenchini, G. Tonelli, N. Turini, A. Venturi, P.G. Verdini, F. Cavallari, M. Cipriani, D. Del Re, E. Di Marco, M. Diemoz, E. Longo, P. Meridiani, G. Organtini, F. Pandolfi, R. Paramatti, C. Quaranta, S. Rahatlou, C. Rovelli, F. Santanastasio, L. Soffi, N. Amapane, R. Arcidiacono, S. Argiro, M. Arneodo, N. Bartosik, R. Bellan, A. Bellora, C. Biino, A. Cappati, N. Cartiglia, S. Cometti, M. Costa, R. Covarelli, N. Demaria, B. Kiani, F. Legger, C. Mariotti, S. Maselli, E. Migliore, V. Monaco, E. Monteil, M. Monteno, M.M. Obertino, G. Ortona, L. Pacher, N. Pastrone, M. Pelliccioni, G.L. Pinna Angioni, A. Romero, M. Ruspa, R. Salvatico, V. Sola, A. Solano, D. Soldi, A. Staiano, D. Trocino, S. Belforte, V. Candelise, M. Casarsa, F. Cossutti, A. Da Rold, G. Della Ricca, F. Vazzoler, A. Zanetti, B. Kim, D.H. Kim, G.N. Kim, J. Lee, S.W. Lee, C.S. Moon, Y.D. Oh, S.I. Pak, S. Sekmen, D.C. Son, Y.C. Yang, H. Kim, D.H. Moon, G. Oh, B. Francois, T.J. Kim, J. Park, S. Cho, S. Choi,

Y. Go, S. Ha, B. Hong, K. Lee, K.S. Lee, J. Lim, J. Park, S.K. Park, Y. Roh, J. Yoo, J. Goh, H.S. Kim, J. Almond, J.H. Bhyun, J. Choi, S. Jeon, J. Kim, J.S. Kim, H. Lee, K. Lee, S. Lee, K. Nam, M. Oh, S.B. Oh, B.C. Radburn-Smith, U.K. Yang, H.D. Yoo, I. Yoon, D. Jeon, J.H. Kim, J.S.H. Lee, I.C. Park, Y. Choi, C. Hwang, Y. Jeong, J. Lee, Y. Lee, I. Yu, V. Veckalns, V. Dudenas, A. Juodagalvis, A. Rinkevicius, G. Tamulaitis, J. Vaitkus, Z.A. Ibrahim, F. Mohamad Idris, W.A.T. Wan Abdullah, M.N. Yusli, Z. Zolkapli, J.F. Benitez, A. Castaneda Hernandez, J.A. Murillo Quijada, L. Valencia Palomo, H. Castilla-Valdez, E. De La Cruz-Burelo, I. Heredia-De La Cruz, R. Lopez-Fernandez, A. Sanchez-Hernandez, S. Carrillo Moreno, C. Oropeza Barrera, M. Ramirez-Garcia, F. Vazquez Valencia, J. Eysermans, I. Pedraza, H.A. Salazar Ibarguen, C. Uribe Estrada, A. Morelos Pineda, J. Mijuskovic, N. Raicevic, D. Krofcheck, S. Bheesette, P.H. Butler, A. Ahmad, M. Ahmad, Q. Hassan, H.R. Hoorani, W.A. Khan, M.A. Shah, M. Shoaib, M. Waqas, V. Avati, L. Grzanka, M. Malawski, H. Bialkowska, M. Bluj, B. Boimska, M. Górski, M. Kazana, M. Szleper, P. Zalewski, K. Bunkowski, A. Byszuk, K. Doroba, A. Kalinowski, M. Konecki, J. Krolikowski, M. Olszewski, M. Walczak, M. Araujo, P. Bargassa, D. Bastos, A. Di Francesco, P. Faccioli, B. Galinhas, M. Gallinaro, J. Hollar, N. Leonardo, T. Niknejad, J. Seixas, K. Shchelina, G. Strong, O. Toldaiev, J. Varela, S. Afanasiev, P. Bunin, M. Gavrilenko, I. Golutvin, I. Gorbunov, A. Kamenev, V. Karjavine, A. Lanev, A. Malakhov, V. Matveev, P. Moisenz, V. Palichik, V. Perelygin, M. Savina, S. Shmatov, S. Shulha, N. Skatchkov, V. Smirnov, N. Voytishin, A. Zarubin, L. Chtchipounov, V. Golovtcov, Y. Ivanov, V. Kim, E. Kuznetsova, P. Levchenko, V. Murzin, V. Oreshkin, I. Smirnov, D. Sosnov, V. Sulimov, L. Uvarov, A. Vorobyev, Yu. Andreev, A. Dermenev, S. Gninenko, N. Golubev, A. Karneyeu, M. Kirsanov, N. Krasnikov, A. Pashenkov, D. Tlisov, A. Toropin, V. Epshteyn, V. Gavrilov, N. Lychkovskaya, A. Nikitenko, V. Popov, I. Pozdnyakov, G. Safronov, A. Spiridonov, A. Stepennov, M. Toms, E. Vlasov, A. Zhokin, T. Aushev, O. Bychkova, M. Chadeeva, P. Parygin, E. Popova, V. Rusinov, V. Andreev, M. Azarkin, I. Dremin, M. Kirakosyan, A. Terkulov, A. Belyaev, E. Boos, M. Dubinin, L. Dudko, A. Ershov, A. Gribushin, V. Klyukhin, O. Kodolova, I. Lokhtin, S. Obraztsov, S. Petrushanko, V. Savrin, A. Snigirev, A. Barnyakov, V. Blinov, T. Dimova, L. Kardapoltsev, Y. Skovpen, I. Azhgirey, I. Bayshev, S. Bitioukov, V. Kachanov, D. Konstantinov, P. Mandrik, V. Petrov, R. Ryutin, S. Slabospitskii, A. Sobol, S. Troshin, N. Tyurin, A. Uzunian, A. Volkov, A. Babaev, A. Iuzhakov, V. Okhotnikov, V. Borchsh, V. Ivanchenko, E. Tcherniaev, P. Adzic, P. Cirkovic, M. Dordevic, P. Milenovic, J. Milosevic, M. Stojanovic, M. Aguilar-Benitez, J. Alcaraz Maestre, A. Álvarez Fernández, I. Bachiller, M. Barrio Luna, CristinaF. Bedoya, J.A. Brochero Cifuentes, C.A. Carrillo Montoya, M. Cepeda, M. Cerrada, N. Colino, B. De La Cruz, A. Delgado Peris, J.P. Fernández Ramos, J. Flix, M.C. Fouz, O. Gonzalez Lopez, S. Goy Lopez, J.M. Hernandez, M.I. Josa, D. Moran, Á. Navarro Tobar, A. Pérez-Calero Yzquierdo, J. Puerta Pelayo, I. Redondo, L. Romero, S. Sánchez Navas, M.S. Soares, A. Triossi, C. Willmott, C. Albajar, J.F. de Trocóniz, R. Reyes-Almanza, B. Alvarez Gonzalez, J. Cuevas, C. Erice, J. Fernandez Menendez, S. Folgueras, I. Gonzalez Caballero, J.R. González Fernández, E. Palencia Cortezon, V. Rodríguez Bouza, S. Sanchez Cruz, I.J. Cabrillo, A. Calderon, B. Chazin Quero, J. Duarte Campderros, M. Fernandez, P.J. Fernández Manteca, A. García Alonso, G. Gomez, C. Martinez Rivero, P. Martinez Ruiz del Arbol, F. Matorras, J. Piedra Gomez, C. Prieels, T. Rodrigo, A. Ruiz-Jimeno, L. Russo, L. Scodellaro, I. Vila, J.M. Vizan Garcia, D.U.J. Sonnadara, W.G.D. Dharmaratna, N. Wickramage, D. Abbaneo, B. Akgun, E. Auffray, G. Auzinger, J. Baechler, P. Baillon, A.H. Ball, D. Barney, J. Bendavid, M. Bianco, A. Bocci, P. Bortignon, E. Bossini, E. Brondolin, T. Camporesi, A. Caratelli, G. Cerminara, E. Chapon, G. Cucciati, D. d'Enterria, A. Dabrowski, N. Daci, V. Daponte, A. David, O. Davignon, A. De Roeck, M. Deile, M. Dobson, M. Dünser, N. Dupont, A. Elliott-Peisert, N. Emriskova, F. Fallavollita, D. Fasanella, S. Fiorendi, G. Franzoni, J. Fulcher, W. Funk, S. Giani, D. Gigi, K. Gill, F. Glege, L. Gouskos, M. Gruchala, M. Guilbaud, D. Gulhan, J. Hegeman, C. Heidegger, Y. Iiyama, V. Innocente, T. James, P. Janot, O. Karacheban, J. Kaspar, J. Kieseler, M. Krammer, N. Kratochwil, C. Lange, P. Lecoq, C. Lourenço, L. Malgeri, M. Mannelli, A. Massironi, F. Meijers, S. Mersi, E. Meschi, F. Moortgat, M. Mulders, J. Ngadiuba, J. Niedziela, S. Nourbakhsh, S. Orfanelli, L. Orsini, F. Pantaleo,

L. Pape, E. Perez, M. Peruzzi, A. Petrilli, G. Petrucciani, A. Pfeiffer, M. Pierini, F.M. Pitters, D. Rabady, A. Racz, M. Rieger, M. Rovere, H. Sakulin, J. Salfeld-Nebgen, C. Schäfer, C. Schwick, M. Selvaggi, A. Sharma, P. Silva, W. Snoeys, P. Sphicas, J. Steggemann, S. Summers, V.R. Tavolaro, D. Treille, A. Tsirou, G.P. Van Onsem, A. Vartak, M. Verzetti, W.D. Zeuner, L. Caminada, K. Deiters, W. Erdmann, R. Horisberger, Q. Ingram, H.C. Kaestli, D. Kotlinski, U. Langenegger, T. Rohe, S.A. Wiederkehr, M. Backhaus, P. Berger, N. Chernyavskaya, G. Dissertori, M. Dittmar, M. Donegà, C. Dorfer, T.A. Gómez Espinosa, C. Grab, D. Hits, W. Lustermann, R.A. Manzoni, M.T. Meinhard, F. Micheli, P. Musella, F. Nessi-Tedaldi, F. Pauss, G. Perrin, L. Perrozzi, S. Pigazzini, M.G. Ratti, M. Reichmann, C. Reissel, T. Reitenspiess, B. Ristic, D. Ruini, D.A. Sanz Becerra, M. Schönenberger, L. Shchutska, M.L. Vesterbacka Olsson, R. Wallny, D.H. Zhu, T.K. Aarrestad, C. Amsler, C. Botta, D. Brzhechko, M.F. Canelli, A. De Cosa, R. Del Burgo, B. Kilminster, S. Leontsinis, V.M. Mikuni, I. Neutelings, G. Rauco, P. Robmann, K. Schweiger, C. Seitz, Y. Takahashi, S. Wertz, A. Zucchetta, T.H. Doan, C.M. Kuo, W. Lin, A. Roy, S.S. Yu, P. Chang, Y. Chao, K.F. Chen, P.H. Chen, W.S. Hou, Y.y. Li, R.S. Lu, E. Paganis, A. Psallidas, A. Steen, B. Asavapibhop, C. Asawatangtrakuldee, N. Srimanobhas, N. Suwonjandee, A. Bat, F. Boran, A. Celik, S. Damarseckin, Z.S. Demiroglu, F. Dolek, C. Dozen, I. Dumanoglu, G. Gokbulut, EmineGurpinar Guler, Y. Guler, I. Hos, C. Isik, E.E. Kangal, O. Kara, A. Kayis Topaksu, U. Kiminsu, G. Onengut, K. Ozdemir, S. Ozturk, A.E. Simsek, U.G. Tok, S. Turkcapar, I.S. Zorbakir, C. Zorbilmez, B. Isildak, G. Karapinar, M. Yalvac, I.O. Atakisi, E. Gülmez, M. Kaya, O. Kaya, Ö. Özçelik, S. Tekten, E.A. Yetkin, A. Cakir, K. Cankocak, Y. Komurcu, S. Sen, S. Cerci, B. Kaynak, S. Ozkorucuklu, D. Sunar Cerci, B. Grynyov, L. Levchuk, E. Bhal, S. Bologna, J.J. Brooke, D. Burns, E. Clement, D. Cussans, H. Flacher, J. Goldstein, G.P. Heath, H.F. Heath, L. Kreczko, B. Krikler, S. Paramesvaran, B. Penning, T. Sakuma, S. Seif El Nasr-Storey, V.J. Smith, J. Taylor, A. Titterton, K.W. Bell, A. Belyaev, C. Brew, R.M. Brown, D.J.A. Cockerill, J.A. Coughlan, K. Harder, S. Harper, J. Linacre, K. Manolopoulos, D.M. Newbold, E. Olaiya, D. Petyt, T. Reis, T. Schuh, C.H. Shepherd-Themistocleous, A. Thea, I.R. Tomalin, T. Williams, R. Bainbridge, P. Bloch, J. Borg, S. Breeze, O. Buchmuller, A. Bundock, GurpreetSingh CHAHAL, D. Colling, P. Dauncey, G. Davies, M. Della Negra, R. Di Maria, P. Everaerts, G. Hall, G. Iles, M. Komm, L. Lyons, A.M. Magnan, S. Malik, A. Martelli, V. Milosevic, A. Morton, J. Nash, V. Palladino, M. Pesaresi, D.M. Raymond, A. Richards, A. Rose, E. Scott, C. Seez, A. Shtipliyski, M. Stoye, T. Strebler, A. Tapper, K. Uchida, T. Virdee, N. Wardle, D. Winterbottom, A.G. Zecchinelli, S.C. Zenz, J.E. Cole, P.R. Hobson, A. Khan, P. Kyberd, C.K. Mackay, I.D. Reid, L. Teodorescu, S. Zahid, A. Brinkerhoff, K. Call, B. Caraway, J. Dittmann, K. Hatakeyama, C. Madrid, B. McMaster, N. Pastika, C. Smith, R. Bartek, A. Dominguez, R. Uniyal, A.M. Vargas Hernandez, A. Buccilli, S.I. Cooper, C. Henderson, P. Rumerio, C. West, A. Albert, D. Arcaro, Z. Demiragli, D. Gastler, C. Richardson, J. Rohlf, D. Sperka, D. Spitzbart, I. Suarez, L. Sulak, D. Zou, G. Benelli, B. Burkle, X. Coubez, D. Cutts, Y.t. Duh, M. Hadley, U. Heintz, J.M. Hogan, K.H.M. Kwok, E. Laird, G. Landsberg, K.T. Lau, J. Lee, M. Narain, S. Sagir, R. Syarif, E. Usai, W.Y. Wong, D. Yu, W. Zhang, R. Band, C. Brainerd, R. Breedon, M. Calderon De La Barca Sanchez, M. Chertok, J. Conway, R. Conway, P.T. Cox, R. Erbacher, C. Flores, G. Funk, F. Jensen, W. Ko, O. Kukral, R. Lander, M. Mulhearn, D. Pellett, J. Pilot, M. Shi, D. Taylor, K. Tos, M. Tripathi, Z. Wang, F. Zhang, M. Bachtis, C. Bravo, R. Cousins, A. Dasgupta, A. Florent, J. Hauser, M. Ignatenko, N. Mccoll, W.A. Nash, S. Regnard, D. Saltzberg, C. Schnaible, B. Stone, V. Valuev, K. Burt, Y. Chen, R. Clare, J.W. Gary, S.M.A. Ghiasi Shirazi, G. Hanson, G. Karapostoli, O.R. Long, M. Olmedo Negrete, M.I. Paneva, W. Si, L. Wang, S. Wimpenny, B.R. Yates, Y. Zhang, J.G. Branson, P. Chang, S. Cittolin, S. Cooperstein, N. Deelen, M. Derdzinski, R. Gerosa, D. Gilbert, B. Hashemi, D. Klein, V. Krutelyov, J. Letts, M. Masciovecchio, S. May, S. Padhi, M. Pieri, V. Sharma, M. Tadel, F. Würthwein, A. Yagil, G. Zevi Della Porta, N. Amin, R. Bhandari, C. Campagnari, M. Citron, V. Dutta, M. Franco Sevilla, J. Incandela, B. Marsh, H. Mei, A. Ovcharova, H. Qu, J. Richman, U. Sarica, D. Stuart, S. Wang, D. Anderson, A. Bornheim, O. Cerri, I. Dutta, J.M. Lawhorn, N. Lu, J. Mao, H.B. Newman, T.Q. Nguyen, J. Pata, M. Spiropulu, J.R. Vlimant, S. Xie, Z. Zhang, R.Y. Zhu,

M.B. Andrews, T. Ferguson, T. Mudholkar, M. Paulini, M. Sun, I. Vorobiev, M. Weinberg, J.P. Cumalat, W.T. Ford, E. MacDonald, T. Mulholland, R. Patel, A. Perloff, K. Stenson, K.A. Ulmer, S.R. Wagner, J. Alexander, Y. Cheng, J. Chu, A. Datta, A. Frankenthal, K. Mcdermott, J.R. Patterson, D. Quach, A. Ryd, S.M. Tan, Z. Tao, J. Thom, P. Wittich, M. Zientek, S. Abdullin, M. Albrow, M. Alyari, G. Apollinari, A. Apresyan, A. Apyan, S. Banerjee, L.A.T. Bauerdick, A. Beretvas, D. Berry, J. Berryhill, P.C. Bhat, K. Burkett, J.N. Butler, A. Canepa, G.B. Cerati, H.W.K. Cheung, F. Chlebana, M. Cremonesi, J. Duarte, V.D. Elvira, J. Freeman, Z. Gecse, E. Gottschalk, L. Gray, D. Green, S. Grünendahl, O. Gutsche, J. Hanlon, R.M. Harris, S. Hasegawa, R. Heller, J. Hirschauer, B. Jayatilaka, S. Jindariani, M. Johnson, U. Joshi, T. Klijnsma, B. Klima, M.J. Kortelainen, B. Kreis, S. Lammel, J. Lewis, D. Lincoln, R. Lipton, M. Liu, T. Liu, J. Lykken, K. Maeshima, J.M. Marraffino, D. Mason, P. McBride, P. Merkel, S. Mrenna, S. Nahn, V. O'Dell, V. Papadimitriou, K. Pedro, C. Pena, G. Rakness, F. Ravera, A. Reinsvold Hall, L. Ristori, B. Schneider, E. Sexton-Kennedy, N. Smith, A. Soha, W.J. Spalding, L. Spiegel, S. Stoynev, J. Strait, N. Strobbe, L. Taylor, S. Tkaczyk, N.V. Tran, L. Uplegger, E.W. Vaandering, C. Vernieri, R. Vidal, M. Wang, H.A. Weber, D. Acosta, P. Avery, D. Bourilkov, L. Cadamuro, V. Cherepanov, F. Errico, R.D. Field, S.V. Gleyzer, D. Guerrero, B.M. Joshi, M. Kim, J. Konigsberg, A. Korytov, K.H. Lo, K. Matchev, N. Menendez, G. Mitselmakher, D. Rosenzweig, K. Shi, J. Wang, S. Wang, X. Zuo, Y.R. Joshi, T. Adams, A. Askew, S. Hagopian, V. Hagopian, K.F. Johnson, R. Khurana, T. Kolberg, G. Martinez, T. Perry, H. Prosper, C. Schiber, R. Yohay, J. Zhang, M.M. Baarmand, M. Hohlmann, D. Noonan, M. Rahmani, M. Saunders, F. Yumiceva, M.R. Adams, L. Apanasevich, R.R. Betts, R. Cavanaugh, X. Chen, S. Dittmer, O. Evdokimov, C.E. Gerber, D.A. Hangal, D.J. Hofman, C. Mills, T. Roy, M.B. Tonjes, N. Varelas, J. Viinikainen, H. Wang, X. Wang, Z. Wu, M. Alhusseini, B. Bilki, K. Dilsiz, S. Durgut, R.P. Gandrajula, M. Haytmyradov, V. Khristenko, O.K. Köseyan, J.P. Merlo, A. Mestvirishvili, A. Moeller, J. Nachtman, H. Ogul, Y. Onel, F. Ozok, A. Penzo, C. Snyder, E. Tiras, J. Wetzel, B. Blumenfeld, A. Cocoros, N. Eminizer, A.V. Gritsan, W.T. Hung, S. Kyriacou, P. Maksimovic, C. Mantilla, J. Roskes, M. Swartz, C. Baldenegro Barrera, P. Baringer, A. Bean, S. Boren, J. Bowen, A. Bylinkin, T. Isidori, S. Khalil, J. King, G. Krintiras, A. Kropivnitskaya, C. Lindsey, D. Majumder, W. Mcbrayer, N. Minafra, M. Murray, C. Rogan, C. Royon, S. Sanders, E. Schmitz, J.D. Tapia Takaki, Q. Wang, J. Williams, G. Wilson, S. Duric, A. Ivanov, K. Kaadze, D. Kim, Y. Maravin, D.R. Mendis, T. Mitchell, A. Modak, A. Mohammadi, F. Rebassoo, D. Wright, A. Baden, O. Baron, A. Belloni, S.C. Eno, Y. Feng, N.J. Hadley, S. Jabeen, G.Y. Jeng, R.G. Kellogg, A.C. Mignerey, S. Nabili, F. Ricci-Tam, M. Seidel, Y.H. Shin, A. Skuja, S.C. Tonwar, K. Wong, D. Abercrombie, B. Allen, R. Bi, S. Brandt, W. Busza, I.A. Cali, M. D'Alfonso, G. Gomez Ceballos, M. Goncharov, P. Harris, D. Hsu, M. Hu, M. Klute, D. Kovalskyi, Y.J. Lee, P.D. Luckey, B. Maier, A.C. Marini, C. Mcginn, C. Mironov, S. Narayanan, X. Niu, C. Paus, D. Rankin, C. Roland, G. Roland, Z. Shi, G.S.F. Stephans, K. Sumorok, K. Tatar, D. Velicanu, J. Wang, T.W. Wang, B. Wyslouch, R.M. Chatterjee, A. Evans, S. Guts, P. Hansen, J. Hiltbrand, Sh. Jain, Y. Kubota, Z. Lesko, J. Mans, M. Revering, R. Rusack, R. Saradhy, N. Schroeder, M.A. Wadud, J.G. Acosta, S. Oliveros, K. Bloom, S. Chauhan, D.R. Claes, C. Fangmeier, L. Finco, F. Golf, R. Kamalieddin, I. Kravchenko, J.E. Siado, G.R. Snow, B. Stieger, W. Tabb, G. Agarwal, C. Harrington, I. Iashvili, A. Kharchilava, C. McLean, D. Nguyen, A. Parker, J. Pekkanen, S. Rappoccio, B. Roozbahani, G. Alverson, E. Barberis, C. Freer, Y. Haddad, A. Hortiangtham, G. Madigan, B. Marzocchi, D.M. Morse, T. Orimoto, L. Skinnari, A. Tishelman-Charny, T. Wamorkar, B. Wang, A. Wisecarver, D. Wood, S. Bhattacharya, J. Bueghly, A. Gilbert, T. Gunter, K.A. Hahn, N. Odell, M.H. Schmitt, K. Sung, M. Trovato, M. Velasco, R. Bucci, N. Dev, R. Goldouzian, M. Hildreth, K. Hurtado Anampa, C. Jessop, D.J. Karmgard, K. Lannon, W. Li, N. Loukas, N. Marinelli, I. Mcalister, F. Meng, Y. Musienko, R. Ruchti, P. Siddireddy, G. Smith, S. Taroni, M. Wayne, A. Wightman, M. Wolf, A. Woodard, J. Alimena, B. Bylsma, L.S. Durkin, B. Francis, C. Hill, W. Ji, A. Lefeld, T.Y. Ling, B.L. Winer, G. Dezoort, P. Elmer, J. Hardenbrook, N. Haubrich, S. Higginbotham, A. Kalogeropoulos, S. Kwan, D. Lange, M.T. Lucchini, J. Luo, D. Marlow, K. Mei, I. Ojalvo, J. Olsen, C. Palmer, P. Piroué, D. Stickland, C. Tully, S. Malik, S. Norberg, A. Barker, V.E. Barnes, R. Chawla,

S. Das, L. Gutay, M. Jones, A.W. Jung, A. Khatiwada, B. Mahakud, D.H. Miller, G. Negro, N. Neumeister, C.C. Peng, S. Piperov, H. Qiu, J.F. Schulte, N. Trevisani, F. Wang, R. Xiao, W. Xie, T. Cheng, J. Dolen, N. Parashar, A. Baty, U. Behrens, S. Dildick, K.M. Ecklund, S. Freed, F.J.M. Geurts, M. Kilpatrick, Arun Kumar, W. Li, B.P. Padley, R. Redjimi, J. Roberts, J. Rorie, W. Shi, A.G. Stahl Leiton, Z. Tu, A. Zhang, A. Bodek, P. de Barbaro, R. Demina, J.L. Dulemba, C. Fallon, T. Ferbel, M. Galanti, A. Garcia-Bellido, O. Hindrichs, A. Khukhunaishvili, E. Ranken, R. Taus, B. Chiarito, J.P. Chou, A. Gandrakota, Y. Gershtein, E. Halkiadakis, A. Hart, M. Heindl, E. Hughes, S. Kaplan, I. Laflotte, A. Lath, R. Montalvo, K. Nash, M. Osherson, H. Saka, S. Salur, S. Schnetzer, S. Somalwar, R. Stone, S. Thomas, H. Acharya, A.G. Delannoy, S. Spanier, O. Bouhali, M. Dalchenko, M. De Mattia, A. Delgado, R. Eusebi, J. Gilmore, T. Huang, T. Kamon, H. Kim, S. Luo, S. Malhotra, D. Marley, R. Mueller, D. Overton, L. Perniè, D. Rathjens, A. Safonov, N. Akchurin, J. Damgov, F. De Guio, V. Hegde, S. Kunori, K. Lamichhane, S.W. Lee, T. Mengke, S. Muthumuni, T. Peltola, S. Undleeb, I. Volobouev, Z. Wang, A. Whitbeck, S. Greene, A. Gurrola, R. Janjam, W. Johns, C. Maguire, A. Melo, H. Ni, K. Padeken, F. Romeo, P. Sheldon, S. Tuo, J. Velkovska, M. Verweij, M.W. Arenton, P. Barria, B. Cox, G. Cummings, J. Hakala, R. Hirosky, M. Joyce, A. Ledovskoy, C. Neu, B. Tannenwald, Y. Wang, E. Wolfe, F. Xia, R. Harr, P.E. Karchin, N. Poudyal, J. Sturdy, P. Thapa, T. Bose, J. Buchanan, C. Caillol, D. Carlsmith, S. Dasu, I. De Bruyn, L. Dodd, C. Galloni, H. He, M. Herndon, A. Hervé, U. Hussain, A. Lanaro, A. Loeliger, K. Long, R. Loveless, J. Madhusudanan Sreekala, A. Mallampalli, D. Pinna, T. Ruggles, A. Savin, V. Sharma, W.H. Smith, D. Teague, S. Trembath-reichert, Identification of heavy, energetic, hadronically decaying particles using machine-learning techniques. J. Instrum. **15** (6), P06005 (2020). https://doi.org/10.1088/1748-0221/15/06/P06005

T. Sjöstrand, S. Ask, J.R. Christiansen, R. Corke, N. Desai, P. Ilten, S. Mrenna, S. Prestel, C.O. Rasmussen, P.Z. Skands, An introduction to PYTHIA 8.2. Comput. Phys. Commun. **191**, 159–177 (2015). https://doi.org/10.1016/j.cpc.2015.01.024

I.V. Slijepcevic, A.M.M. Scaife, M. Walmsley, M. Bowles, O. Ivy Wong, S.S. Shabala, H. Tang, Radio Galaxy Zoo: using semi-supervised learning to leverage large unlabelled datasets for radio galaxy classification under dataset shift. Mon. Not. R. Astron. Soc. **514**(2), 2599–2613 (2022a). https://doi.org/10.1093/mnras/stac1135

I.V. Slijepcevic, A.M.M. Scaife, M. Walmsley, M. Bowles, Learning useful representations for radio astronomy "in the wild" with contrastive learning. arXiv e-prints, art. arXiv:2207.08666 (July 2022b)

M. Spinelli, I.P. Carucci, S. Cunnington, S.E. Harper, M.O. Irfan, J. Fonseca, A. Pourtsidou, L. Wolz, SKAO H I intensity mapping: blind foreground subtraction challenge. Mon. Not. R. Astron. Soc. **509**(2), 2048–2074 (2022). https://doi.org/10.1093/mnras/stab3064

J.-L. Starck, J. Fadili, F. Murtagh, The undecimated wavelet decomposition and its reconstruction. IEEE Trans. Image Process. **16**(2), 297–309 (2007). https://doi.org/10.1109/TIP.2006.887733

M. Sugiyama, M. Krauledat, K.-R. Müller, Covariate shift adaptation by importance weighted cross validation. J. Mach. Learn. Res. **8**(5) (2007)

C. Tan, F. Sun, T. Kong, W. Zhang, C. Yang, C. Liu, A survey on deep transfer learning, in *International Conference on Artificial Neural Networks* (Springer, Berlin, 2018), pp. 270–279

H. Tang, A.M.M. Scaife, O.I. Wong, S.S. Shabala, Radio Galaxy Zoo: giant radio galaxy classification using multidomain deep learning. Mon. Not. R. Astron. Soc. **510**(3), 4504–4524 (2022). https://doi.org/10.1093/mnras/stab3553

M. Terris, A. Dabbech, C. Tang, Y. Wiaux, Image reconstruction algorithms in radio interferometry: From handcrafted to learned regularization denoisers. Mon. Not. R. Astron. Soc. **518**(1), 604–622 (2023). https://doi.org/10.1093/mnras/stac2672

N. Tishby, E. Levin, S. Solla, Consistent inference of probabilities in layered networks: predictions and generalizations, in *International 1989 Joint Conference on Neural Networks*, vol. 2 (1989), pp. 403–409. https://doi.org/10.1109/IJCNN.1989.118274

G. Traven, G. Matijevič, T. Zwitter, M. Žerjal, J. Kos, M. Asplund, J. Bland-Hawthorn, A.R Casey, G. De Silva, K. Freeman et al., The galah survey: classification and diagnostics with t-sne reduction of spectral information. Astrophys. J. Suppl. Ser. **228**(2), 24 (2017)

C.M. Trott, J.C. Pober, The status of 21cm interferometric experiments. arXiv e-prints, art. arXiv:1909.12491 (September 2019)

D. Tuccillo, M. Huertas-Company, E. Decencière, S. Velasco-Forero, H. Domínguez Sánchez, P. Dimauro, Deep learning for galaxy surface brightness profile fitting. Mon. Not. R. Astron. Soc. **475**(1), 894–909 (2018). https://doi.org/10.1093/mnras/stx3186

S. Tuli, G. Casale, N.R. Jennings, TranAD: deep transformer networks for anomaly detection in multivariate time series data. arXiv e-prints, art. arXiv:2201.07284 (January 2022)

J.D. Twicken, H. Chandrasekaran, J.M. Jenkins, J.P. Gunter, F. Girouard, T.C. Klaus, Presearch data conditioning in the Kepler Science Operations Center pipeline, in Software and Cyberinfrastructure for Astronomy, ed. by N.M. Radziwill, A. Bridger, vol. 7740. Society of Photo-Optical Instrumentation Engineers (SPIE) Conference Series (July 2010), pp. 77401U. https://doi.org/10.1117/12.856798

S.-M. Udrescu M. Tegmark, AI Feynman: a physics-inspired method for symbolic regression. Sci. Adv. **6**(16), eaay2631 (2020). https://doi.org/10.1126/sciadv.aay2631

A. Vafaei Sadr, B.A. Bassett, N. Oozeer, Y. Fantaye, C. Finlay, Deep learning improves identification of Radio Frequency Interference. Mon. Not. R. Astron. Soc. **499**(1), 379–390 (2020). https://doi.org/10.1093/mnras/staa2724

L. Van der Maaten, G. Hinton, Visualizing data using t-sne. J. Mach. Learn. Res. **9**(11) (2008)

V. Vapnik, *The Nature of Statistical Learning Theory* (Springer Science & Business Media, Berlin, 1999)

A. Vaswani, N. Shazeer, N. Parmar, J. Uszkoreit, L. Jones, A.N. Gomez, L. Kaiser, I. Polosukhin, Attention is all you need. arXiv e-prints, art. arXiv:1706.03762 (June 2017)

C. Vonesch, T. Blu, M. Unser, Generalized daubechies wavelet families. IEEE Trans. Signal Process. **55**(9):4415–4429 (2007)

M. Walmsley, A.M.M. Scaife, C. Lintott, M. Lochner, V. Etsebeth, T. Géron, H. Dickinson, L. Fortson, S. Kruk, K.L. Masters, K.B. Mantha, B.D. Simmons, Practical galaxy morphology tools from deep supervised representation learning. Mon. Not. R. Astron. Soc. **513**(2), 1581–1599 (2022). https://doi.org/10.1093/mnras/stac525

M. Wang, W. Deng, Deep visual domain adaptation: a survey. Neurocomputing **312**, 135–153 (2018)

H. Wang, W. Zhu, P. Guo, D. Li, S. Feng, Q. Yin, C. Miao, Z. Tao, Z. Pan, P. Wang, X. Zheng, X. Deng, Z. Liu, X. Xie, X. Yu, S.You, H. Zhang, Pulsar candidate selection using ensemble networks for FAST drift-scan survey. Sci. China Phys. Mech. Astron. **62**(5):959507 (2019). https://doi.org/10.1007/s11433-018-9388-3

F. Wang, T.R. Henninen, D. Keller, R. Erni, Self-supervised denoising for massive noisy images. arXiv e-prints, art. arXiv:2110.11911 (October 2021)

S. Webb, M. Lochner, D. Muthukrishna, J. Cooke, C. Flynn, A. Mahabal, S. Goode, I. Andreoni, T. Pritchard, T.M.C. Abbott, Unsupervised machine learning for transient discovery in deeper, wider, faster light curves. Mon. Not. R. Astron. Soc. **498**(3), 3077–3094 (2020). https://doi.org/10.1093/mnras/staa2395

A.H. Wright, H. Hildebrandt J.L. van den Busch, C. Heymans, Photometric redshift calibration with self-organising maps. Astron. Astrophys. **637**, A100 (2020). https://doi.org/10.1051/0004-6361/201936782

C. Wu, O.I. Wong, L. Rudnick, S.S. Shabala, M.J. Alger, J.K. Banfield, C.S. Ong, S.V. White, A.F. Garon, R.P. Norris, H. Andernach, J. Tate, V. Lukic, H. Tang, K. Schawinski, F.I. Diakogiannis, Radio Galaxy Zoo: CLARAN - a deep learning classifier for radio morphologies. Mon. Not. R. Astron. Soc. **482**(1), 1211–1230 (2019). https://doi.org/10.1093/mnras/sty2646

S. Yatawatta, I.M. Avruch, Deep reinforcement learning for smart calibration of radio telescopes. Mon. Not. R. Astron. Soc. **505**(2), 2141–2150 (2021). https://doi.org/10.1093/mnras/stab1401

Z. Yi, J. Li, W. Du, M. Liu, Z. Liang, Y. Xing, J. Pan, Y. Bu, X. Kong, H. Wu, Automatic detection of low surface brightness galaxies from Sloan Digital Sky Survey images. Mon. Not. R. Astron. Soc. **513**(3), 3972–3981 (2022). https://doi.org/10.1093/mnras/stac775

D.G. York, J. Adelman Jr., J.E. Anderson, S.F. Anderson, J. Annis, N.A. Bahcall, J.A. Bakken, R. Barkhouser, S. Bastian, E. Berman, W.N. Boroski, S. Bracker, C. Briegel, J.W. Briggs,

J. Brinkmann, R. Brunner, S. Burles, L. Carey, M.A. Carr, F.J. Castander, B. Chen, P.L. Colestock, A.J. Connolly, J.H. Crocker, I. Csabai, P.C. Czarapata, J.E. Davis, M. Doi, T. Dombeck, D. Eisenstein, N. Ellman, B.R. Elms, M.L. Evans, X. Fan, G.R. Federwitz, L. Fiscelli, S. Friedman, J.A. Frieman, M. Fukugita, B. Gillespie, J.E. Gunn, V.K. Gurbani, E. de Haas, M. Haldeman, F.H. Harris, J. Hayes, T.M. Heckman, G.S. Hennessy, R.B. Hindsley, S. Holm, D.J. Holmgren, C.-h. Huang, C. Hull, D. Husby, S.-I. Ichikawa, T. Ichikawa, Ž. Ivezić, S. Kent, R.S.J. Kim, E. Kinney, M. Klaene, A.N. Kleinman, S. Kleinman, G.R. Knapp, J. Korienek, R.G. Kron, P.Z. Kunszt, D.Q. Lamb, B. Lee, R.F. Leger, S. Limmongkol, C. Lindenmeyer, D.C. Long, C. Loomis, J. Loveday, R. Lucinio, R.H. Lupton, B. MacKinnon, E.J. Mannery, P.M. Mantsch, B. Margon, P. McGehee, T.A. McKay, A. Meiksin, A. Merelli, D.G. Monet, J.A. Munn, V.K. Narayanan, T. Nash, E. Neilsen, R. Neswold, H.J. Newberg, R.C. Nichol, T. Nicinski, M. Nonino, N. Okada, S. Okamura, J.P. Ostriker, R. Owen, A.G. Pauls, J. Peoples, R.L. Peterson, D. Petravick, J.R. Pier, A. Pope, R. Pordes, A. Prosapio, R. Rechenmacher, T.R. Quinn, G.T. Richards, M.W. Richmond, C.H. Rivetta, C.M. Rockosi, K. Ruthmansdorfer, D. Sandford, D.J. Schlegel, D.P. Schneider, M. Sekiguchi, G. Sergey, K. Shimasaku, W.A. Siegmund, S. Smee, J.A. Smith, S. Snedden, R. Stone, C. Stoughton, M.A. Strauss, C. Stubbs, M. SubbaRao, A.S. Szalay, I. Szapudi, G.P. Szokoly, A.R. Thakar, C. Tremonti, D.L. Tucker, A. Uomoto, D. Vanden Berk, M.S. Vogeley, P. Waddell, S.-i. Wang, M. Watanabe, D.H. Weinberg, B. Yanny, N. Yasuda, SDSS Collaboration, The Sloan digital sky survey: technical summary. Astron. J. **120**(3), 1579–1587 (2000). https://doi.org/10.1086/301513

H. Yu, X. Hou, Hierarchical clustering in astronomy. Astron. Comput. **41**, 100662 (2022). https://doi.org/10.1016/j.ascom.2022.100662

S. Yu, B. Park, J. Jeong, Deep iterative down-up cnn for image denoising, in *Proceedings of the IEEE/CVF Conference on Computer Vision and Pattern Recognition Workshops* (2019)

X. Zhang, Y. Wang, W. Zhang, Y. Sun, S. He, G. Contardo, F. Villaescusa-Navarro, S. Ho, From dark matter to galaxies with convolutional networks. arXiv e-prints, art. arXiv:1902.05965 (February 2019)

C. Zhang, C. Wang, G. Hobbs, C.J. Russell, D. Li, S.B. Zhang, S. Dai, J.W. Wu, Z.C. Pan, W.W. Zhu, L. Toomey, Z.Y. Ren, Applying saliency-map analysis in searches for pulsars and fast radio bursts. Astron. Astrophys. **642**, A26 (2020). https://doi.org/10.1051/0004-6361/201937234

Z. Zhang, B. Jiang, Y. Zhang, Automatic detection and classification of radio galaxy images by deep learning. Publ. Astron. Soc. Pac. **134**(1036), 064503 (2022). https://doi.org/10.1088/1538-3873/ac67b1

Y. Zhao, Z. Nasrullah, Z. Li, PyOD: a Python toolbox for scalable outlier detection. arXiv e-prints, art. arXiv:1901.01588 (January 2019)

H. Zhao, J. Jin, Y. Liu, Y. Shen, Y. Jiang, AdaBoost-MICNN: a new network framework for pulsar candidate selection. Mon. Not. R. Astron. Soc. **513**(2), 2869–2883 (2022a). https://doi.org/10.1093/mnras/stac619

X. Zhao, Y. Mao, B.D. Wandelt, Implicit likelihood inference of reionization parameters from the 21 cm power spectrum. Astrophys. J. **933**(2), 236 (2022b). https://doi.org/10.3847/1538-4357/ac778e

B. Zhou, A. Khosla, A. Lapedriza, A. Oliva, A. Torralba, Learning deep features for discriminative localization. arXiv e-prints, art. arXiv:1512.04150 (December 2015)

W.W. Zhu, A. Berndsen, E.C. Madsen, M. Tan, I.H. Stairs, A. Brazier, P. Lazarus, R. Lynch, P. Scholz, K. Stovall, S.M. Ransom, S. Banaszak, C.M. Biwer, S. Cohen, L.P. Dartez, J. Flanigan, G. Lunsford, J.G. Martinez, A. Mata, M. Rohr, A. Walker, B. Allen, N.D.R. Bhat, S. Bogdanov, F. Camilo, S. Chatterjee, J.M. Cordes, F. Crawford, J.S. Deneva, G. Desvignes, R.D. Ferdman, P.C.C. Freire, J.W.T. Hessels, F.A. Jenet, D.L. Kaplan, V.M. Kaspi, B. Knispel, K.J. Lee, J. van Leeuwen, y Lyne, M.A. McLaughlin, X. Siemens, L.G. Spitler, A. Venkataraman, Searching for pulsars using image pattern recognition. Astrophys. J. **781**(2), 117 (2014). https://doi.org/10.1088/0004-637X/781/2/117

Chapter 11
Tools for Radio/Multi-Wavelength Big Data Interrogation

Lucia Marchetti, Thijs van der Hulst, Thomas H. Jarrett, Angus Comrie, Alexander Sivitilli, and Kechil Kirkham

The authors and editors of this book wish to express their deepest gratitude to Tom Jarrett, who sadly passed away before its publication, for their invaluable contributions to this work.

In this chapter, we aim to provide an overview of some of the most commonly used software and tools exploited by astronomers to visualise and interrogate Radio data. Sometimes these tools are shared across the field of astronomy and shared through multiple types of analysis. In this case, we will try to provide examples drawn from multi-wavelength astronomy.

L. Marchetti (✉)
Department of Astronomy, University of Cape Town, Cape Town, South Africa

Institute of Radioastronomy, INAF, Bologna, Italy

The Inter-University Institute for Data Intensive Astronomy (IDIA), Department of Astronomy, University of Cape Town, Cape Town, South Africa
e-mail: lucia.marchetti@uct.ac.za

T. van der Hulst
Kapteyn Astronomical Institute, University of Groningen, Groningen, The Netherlands
e-mail: vdhulst@astro.rug.nl

T. H. Jarrett
Department of Astronomy, University of Cape Town, Cape Town, South Africa

A. Comrie · K. Kirkham
The Inter-University Institute for Data Intensive Astronomy (IDIA), Department of Astronomy, University of Cape Town, Cape Town, South Africa
e-mail: angus@idia.ac.za; kechil@idia.ac.za

A. Sivitilli
Department of Astronomy, University of Cape Town, Cape Town, South Africa

The Inter-University Institute for Data Intensive Astronomy (IDIA), Department of Astronomy, University of Cape Town, Cape Town, South Africa
e-mail: alexander.sivitilli@uct.ac.za

E. Vardoulaki et al. (eds.), *Data-Intensive Radio Astronomy*, Astrophysics and Space Science Library 472, https://doi.org/10.1007/978-3-031-58468-8_11

In the last part of this chapter, we will then focus our attention on describing novel software/tools that have been developed in recent years and that will possibly represent the new paradigm for (Radio) data visualisation in astronomy. These tools are designed with the precise aim to solve new problems posed by e.g. the SKA precursors and, eventually, the SKA telescope, such as the need to efficiently visualise multi-parameters data of very big size (e.g. terabyte scale) and possibly stored in a shared cloud where multiple users are working at the same time.

11.1 Conventional Visualisation Tools in Radio and Multi-Wavelength Astronomy

Since the early days of spectral line synthesis, a number of data display systems have been developed to view 3D data. The main strength was the ability to display the frequency (i.e. velocity) behaviour as a movie of consecutive channels, later followed by constructing and displaying 2D cuts through the data cube along a pre-specified line in the image plane (the so-called Position-Velocity (PV) diagrams). Originally these display systems or viewers are embedded in a data analysis software package such as GIPSY (van der Hulst et al. 1992), AIPS (Greisen 1990) and CASA (McMullin et al. 2007). A few stand-alone display systems have been developed, the most notable of these is the suite of programs bundled in the Karma package (Gooch 1996). Karma routines accept various input formats (most notably FITS) and can display images (*kvis*), data cubes (in three orientations xyz, xzy and zyx), produce PV-slices (*kpvslice*) along any preferred direction, plot contours of images/channels on itself or on another image of the same region of the sky. The Karma routines are still used to date by the majority of radio astronomers analysing spectral line data cubes as it still is quite complete and unique in its capabilities.

Other developments like Karma are DS9 (Joye and Mandel 2003), which evolved from SAOImage and, more recently, CARTA (Comrie et al. 2021). The new display system in GIPSY, VISIONS, has many of the features of Karma and can be run semi-standalone (it requires a minimal installation of GIPSY for data access). Most recently attention has also been given to volume rendered 3D data which has the added advantage that it becomes possible to view the 3D shapes of the signal in the data and is very powerful for finding faint features in e.g. the outer parts of galaxies. Two examples are SlicerAstro (Punzo et al. 2015, 2016, 2017) and Frelled (Taylor 2015a,b). We will describe a few of these in more detail below to give a flavour of their capabilities.

Karma This is perhaps one of the oldest stand alone visualisation packages for radio astronomical data, able to read FITS, Graphics Interchange Format GIF and PNM/PPM/PGM files and the file formats of various software packages (MIRIAD, GIPSY, DRAO, AIPS, AIPS++, IRAF). Karma consists of a suite of nine applications which can be run separately. The two most widely used modules are *kvis* and *kpvslice* which display a data cube in movie mode and allow a position-velocity diagram display along a user defined axis in the image plane. *kvis* is able to plot contours on an image. These could be for example the emission in individual

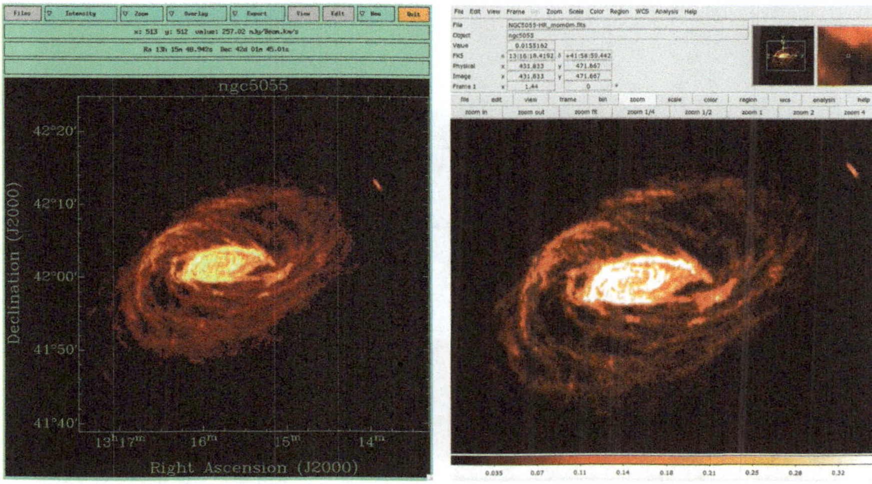

Fig. 11.1 *NGC 5055 viewed in* ***kvis*** *and* ***DS9***. Shown is the HI distribution in NGC 5055 in respectively *kvis* (left panel) and *DS9* (right panel). The colour scheme (heat) and transfer functions have been chosen such that the same features show up in either display. The figures shown are screenshots from *kvis* and *DS9* respectively

channels in the data cube on top of an optical image. The two images, or image and data cube do not have to be on the same projection grid; as long as their world coordinates are described in the header of the data files *kvis* will perform a proper projection before plotting the contours. A display of the HI image of NGC 5055 (data taken from the Halogas survey[1]) in *kvis* is shown in the left panel of Fig. 11.1.

DS9 The sequel to SAOImage which has many of the same properties as *kvis* and can display both individual image and data cubes. In the case of data cubes it also has a movie mode. It is less versatile than *kvis* and can handle only FITS files and FITS tables directly (though it can handle the fairly primitive NRRD files). It is mostly used for examining and analysing optical image data (including 2D spectra), less so for radio astronomical data. A display of the HI image of NGC 5055 (data taken from the Halogas survey) in *DS9* is shown in the right panel of Fig. 11.1.

SlicerAstro This a 3D rendering application, inspired by *kvis*, embedded as an extension package in the general 3D imaging package Slicer[2] developed mostly for medical imaging applications. SlicerAstro reads FITS files and has the following basic features: (1) coupled 2-D/3-D visualisation; (2) interactive and automated filtering, including an optimised, intensity gradient driven smoothing algorithm; (3) interactive and flexible 3-D masking; (4) and interactive 3-D modelling of galaxies

[1] https://www.astron.nl/halogas/.

[2] http://www.slicer.org.

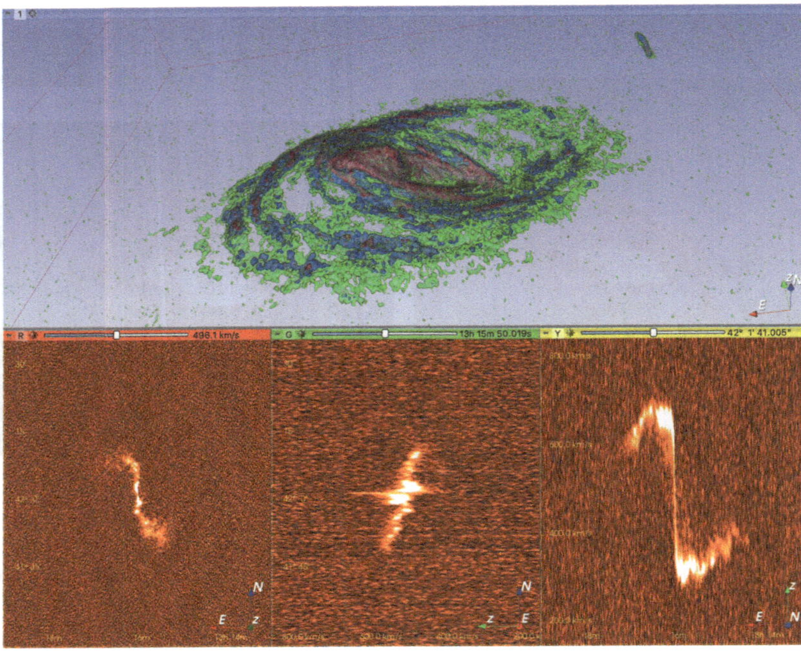

Fig. 11.2 *NGC5055 viewed in* ***SlicerAstro***. Shown is a screenshot of a standard 3D plus 3 2D panel display mode. The top panel shows the 3D rendered view of the HI emission in NGC 5055. The bottom three panels show from left to right respectively a RA-Dec plane, a Velocity-Dec plane and a RA-Velocity plane in the data cube centred at the cube central pixel (the galaxy centre in this case). Sliders at the top of these panels allow navigation through Velocity, RA and Dec respectively

using the ^{3D}Barolo code (Di Teodoro and Fraternali 2015a,b). SlicerAstro can also produce overlays and markups of/on images taking full account of possible differences in projection system of the input datasets. SlicerAstro is available from https://www.astro.rug.nl/~slicerastro. A 3D display of the HI image of NGC 5055 (data taken from the Halogas survey) in *SlicerAstro* is shown in the top panel of Fig. 11.2. The three panels at the bottom show an RA-Dec, velocity-Dec, and RA-velocity plane in the data cube respectively. Sliders at the top allow navigation through the data cube.

Frelled Another 3D rendering package that basically is a set of Python scripts for Blender, an open-source 3D graphics package. The scripts allow importing of 3-D FITS files into Blender, where they can be viewed from any angle in real time. Frelled was developed originally to able to view and easily, rapidly, and interactively mask regions of 3D datasets. The masking capability in Frelled is therefore well developed. Frelled is also able to overlay contour plots of other images. It also has the unique capability of being able to display two FITS files at the same time. A

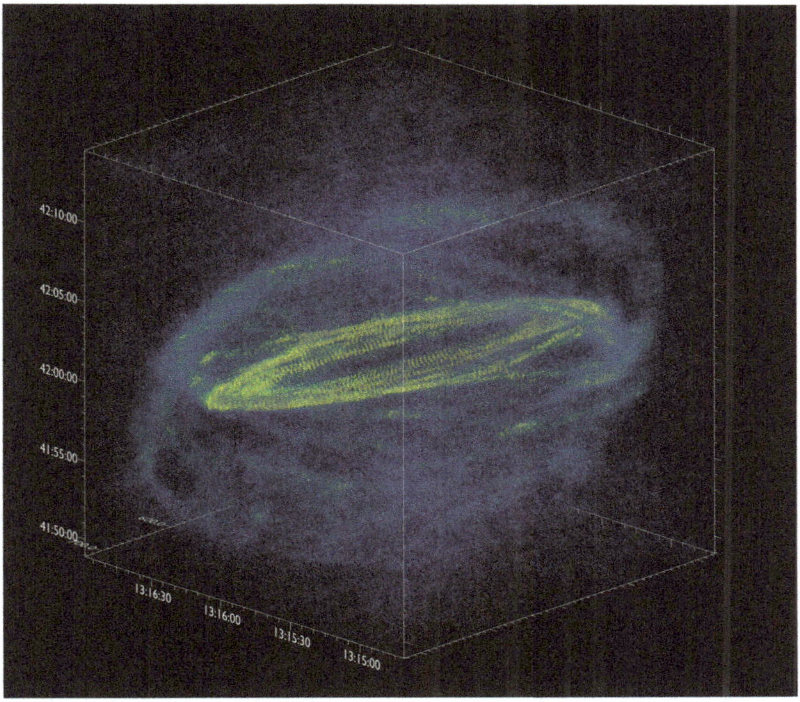

Fig. 11.3 *NGC 5055 viewed in* ***Frelled***. this figure, kindly provided by Rhys Taylor (http://www.rhysy.net), shows a 3D rendering of NGC 5055 by *Frelled* in *Blender*

3D display of the HI image of NGC 5055 (data taken from the Halogas survey) in *Frelled* is shown in the top panel of Fig. 11.3.

11.2 Novel Approaches to Data Visualisation: Tackling Big and Complex Data

11.2.1 Visualisation Tools Running in the Cloud: CARTA

Moving towards the era of the SKA, one of the problems facing the astronomical community is how to analyse terabytes of Radio data efficiently. Already in the era of the SKA precursors, MeerKAT and ASKAP, astronomers cannot rely on downloading the data on a local machine to explore them as this would require too expensive storage units and powerful CPU /GPU. Thus, a shift in the way astronomers access, analyse and visualise the data is necessary, which, in some fashion, also entitles a shift in the astronomy culture. The data reduction and

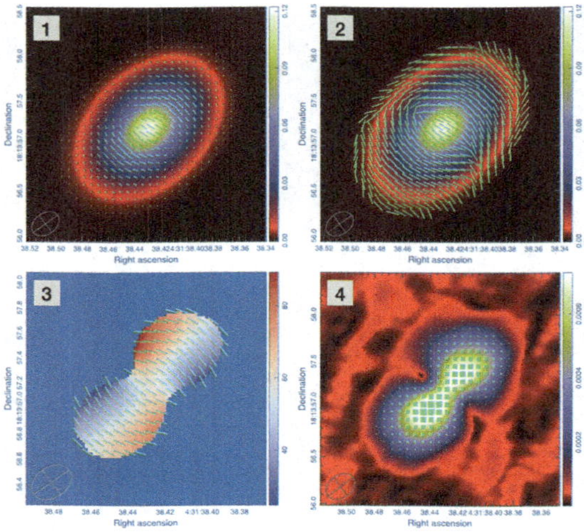

Fig. 11.4 *Vector Field Rendering*. Together with the raster and the contour rendering modes, a new GPU-accelerated rendering mode "Vector overlay" is added in CARTA. The vector overlay can be computed from a Stokes cube directly or be generated by interpreting pixel values directly as the angles of the vector line segments or as the mapped sizes of square markers

processing of Radio data have been moved to the Cloud and it is thus important that astronomers have also the opportunity to explore the data directly in the cloud. With this in mind, the Cube Analysis and Rendering Tool for Astronomy (CARTA) has been developed.

CARTA (Cube Analysis and Rendering Tool for Astronomy), is a next-generation image visualisation and analysis tool (Figs. 11.4 and 11.5) designed for very large instruments: ALMA, VLA, the SKA pathfinders MeerKAT and ASKAP, and coming up, the SKA and the ngVLA. It is specially architected to deal with big data, as it becomes no longer practical to handle extremely large data-sets on a personal computer or laptop. The primary mission of CARTA is to provide usability and scalability for the future by utilising modern web technologies and computing parallelisation.

Four partners make up the CARTA teams: IDIA, which sets direction under Russ Taylor, NRAO, ASIAA, and the University of Alberta (CIRADA). Russ was one of the original proposers and architects of CARTA, recognising the need for big data tools for astronomy visualisation. NRAO has an interest in CARTA in order to retire the CASA viewer, and as a next-generation visualisation tool to address the needs of the ngVLA. ASIAA is involved in the ALMA project and contributes developers including the Project Scientist. Finally, the University of Alberta has been associated with CARTA since the early days, and also contributes to development and direction.

History Many years ago, the CyberSKA initiative arose to develop a scalable and distributed platform to meet the evolving science needs of data intensive radio astronomy en route to the SKA. This was led by the University of Calgary with

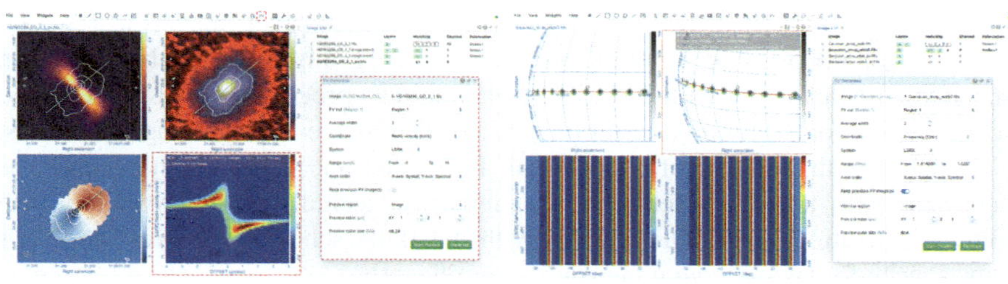

Fig. 11.5 *PV image generator*. A new re-engineered PV image generator, focusing on scalability and usability, is implemented in CARTA. It is possible to make a line region to generate a PV image. The image itself can be wide-field. In such cases, position offsets in fixed angular distance are sampled along the PV cut. Performance-wise, the PV image generator outperformed CASA-impv significantly for large cubes (FITS and CASA image formats). If the input image cube is in the HDF5-IDIA format, the performance is boosted further compared to the FITS and CASA format

several partner institutions from North America. The original CyberSKA project has now been overtaken by SKA endeavours, and one of the offshoots was CARTA.

The first CARTA release was in late 2018, and there have been four stable releases since, with several intermediate beta releases. It is currently among the foremost products used for analysis and visualisation by astronomers using large volumes of data.

Feature Portfolio Features include spectral line cube analysis tools, flexible multi-profile plotting capability, analysis tools such as profile fitting, profile smoothing, moment map generator, and spectral line query and labelling. Publication quality output is supported, as are widgets to provide polarimetric analysis and a scripting interface. The scripting interface allows users to create and share their own workflows for repeated operations, as well as customising CARTA for their specific needs. A full and current list can be found on the website.[3]

NRAO being one of the CARTA partners has meant that some of the requirements have been specified to enable CARTA to take over from the CASA viewer, and this goal is now fulfilled. Looking to the future, as well as looking back to the early aspirations of CyberSKA, collaboration tools are an important focus, enabling scientists to work in a distributed fashion on common data-sets. A thorough treatment of collaboration, including sharing and saving states, is released in version 4.0, with more enhancements on the way. Virtual Reality and Machine Learning integration are reasonably near term objectives.

Performance Performance is critical where very large data-sets are concerned, and several features and capabilities have been architected into CARTA to enhance speed when handling big data. These include memory-efficiency when loading

[3] https://cartavis.org/.

images: given ≈1 GB of ram, a 16,000 pixel × 16,000 pixel image, or a 16,000 pixel × 16,000 pixel × 1000 channel cube (≈1 TB in size), can be loaded in seconds. Parallelisation with CPUs is invoked in data processing. GPU-accelerated rendering is adopted for tiled raster images, contour images, and catalogue overlays. Catalogues with millions of sources can be loaded and rendered in seconds, as catalogue files are processed with parallelisation techniques and rendered with GPU acceleration. The HDF5 format with the IDIA schema improves the user's experience of viewing large image cubes significantly.

User Experience Excellent user experience was a key design goal from the outset, with the aim of providing a user interface such that the astronomer can focus on astronomical science rather than grappling with computer science. The CARTA Science Team and astronomer advisors help to draw up requirements for new features, setting priorities and guiding user interface design. Formal periods of usability testing have been staged to ensure compliance with de facto standards and expectations.

CARTA is equipped with a highly customisable and reusable graphical user interface for various use cases. Widgets can be moved around the screen, and coming up soon will be a feature to save the state for the next session.

Access to CARTA CARTA is available from the website https://cartavis.org/ for download, or with instructions for administrators installing on a server. Archive integration is a strong usability feature, with current access from the ALMA Archive. Delivering science from the SKA will be carried out via SKA Regional Centres, though given the size of SKA data it is likely that analysis will be carried out in situ, rather than the current common model of transferring data to a local site. CARTA was designed with federated cloud architectures in mind, such as ilifu (Sect. 6.8), the research compute infrastructure developed by IDIA.

11.2.2 Immersive Visualisation Tools: Virtual Reality and Digital Planetaria

Although astronomers may have access to increased GPU power and cloud-based architecture for data storage and processing, they cannot rely solely on automation and machine learning algorithms to analyse data of ever-increasing volume and complexity. In order to explore and disentangle complex behaviours hidden in the data, it is still necessary to efficiently inspect the data "visually", making use of all means available to enable the discoveries of the future.

The arrival of mainstream virtual reality (VR) headsets and the availability of versatile development tools for video games and digital planetaria are enabling scientists to finally deploy such technologies for their own needs and innovation. These represent an exciting novel avenue for astronomical data visualisation software development. In this section, we are going to explore some recent developments in this context.

11.2.2.1 Multi-Dimensional Analysis: Virtual Reality Tools

Since Virtual Reality (VR) headsets have become more easily accessible on the consumer market, the astronomy community has started to include VR in the list of instruments used to make astronomy more accessible to the public. Virtual Reality creates the unique opportunity to immerse users in artificially constructed spaces such as a beautiful starry night or a tour of the solar system and so on. The illusion of an immersive space created by VR headsets creates an experience for users that can lead to learning more about the research that is behind that beautiful environment in which he or she has just been so gently transported. In this context, many tools have been created for this outreach-driven end. Examples of such efforts are e.g., VIRUP,[4] developed by the EPFL or GAIA VR,[5] developed by ESA to support the dissemination of the GAIA mission. Most of these tools, developed primarily with the goal of serving at best the dissemination of science, usually are limited in interaction with the data as the attention is primarily focused on enabling the best visual experience.

Nevertheless, VR has great potential to also serve the professional astronomical community for scientific interrogation of multi-dimensional data-sets. Being immersed in the data provides the scientist with a new perspective on the data and at the same time, the 3-D environment provides a better-suited interactive platform for multi-parameter explorations with respect to the more conventional 2-D screen experience. In the context of Radio data visualisation and, in particular, HI data analysis, where the data are by definition 3-D data cubes, the introduction of VR seems a natural evolution for data exploration (Fig. 11.6). In the next session, we will describe one of the tools developed to support the analysis of astronomical data set using VR.

11.2.2.2 iDaVIE: Immersive Data Visualisation Interactive Explorer

iDaVIE or immersive Data Visualisation Interactive Explorer[6,7] (Sivitilli et al. 2020; Comrie et al. 2021; Jarrett et al. 2021) is a VR-based software suite developed by the Inter-University Institute for Data Intensive Astronomy[8] in collaboration with INAF-Osservatorio Astrofisico di Catania.[9] The goal of iDaVIE is to render

---

[4] https://www.epfl.ch/labs/lastro/public-outreach/virup/.

[5] https://sci.esa.int/web/gaia/-/60036-gaia-data-release-2-virtual-reality-resources.

[6] https://idavie.readthedocs.io/.

[7] https://zenodo.org/records/4614116.

[8] https://www.idia.ac.za/.

[9] https://www.oact.inaf.it/.

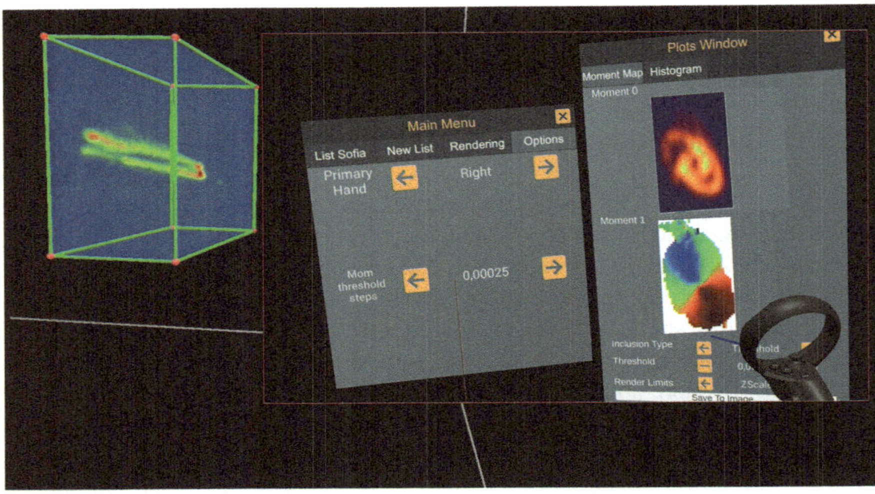

Fig. 11.6 Example of VR analytical tools available to the iDaVIE user. These include e.g., cropping to individual regions for higher resolutions along with the calculation and display of moment maps as shown here for an HI (neutral hydrogen) data cube of a nearby galaxy

data-sets in a virtual room-scale 3D space where users can intuitively view and uniquely interact with their data in ways unafforded by conventional flatscreen and 2D solutions.

iDaVIE consists of two primary flavours, iDaVIE-v and iDaVIE-p. The former is designed for 3D volumetric data-sets and renders them using a raymarching (gpu-programmable) shader while the latter is designed for rendering sparse multi-dimensional particle data-sets with billboard shaders. It is primarily with iDaVIE-v that radio astronomers have been able to implement scientific use-cases of VR, such as extra-galactic source-finding for survey-based observations.

As an immersive scientific visualisation tool, iDaVIE-v includes a wide range of interaction features, mostly invoked through the use of the VR controllers and voice commands. Because data cubes can be fairly large, on the order of hundreds of gigabytes, in the current software version it is necessary to downsample (temporarily reduce the volume size) cubes when rendering them in the VR space. This is mainly due to the minimum performance and hardware requirements. The user is then able to select sub-regions of the downsampled cube to crop and render to higher (i.e., native) resolutions. This is all done automatically by the software to provide both data volume flexibility and visual fidelity.

It is also possible to load optional catalogues and 3D masks with the input data cube. This allows the user to visually isolate areas of interest. Data masks can also be edited with a "painting" (edit) feature that enables the user to add or remove

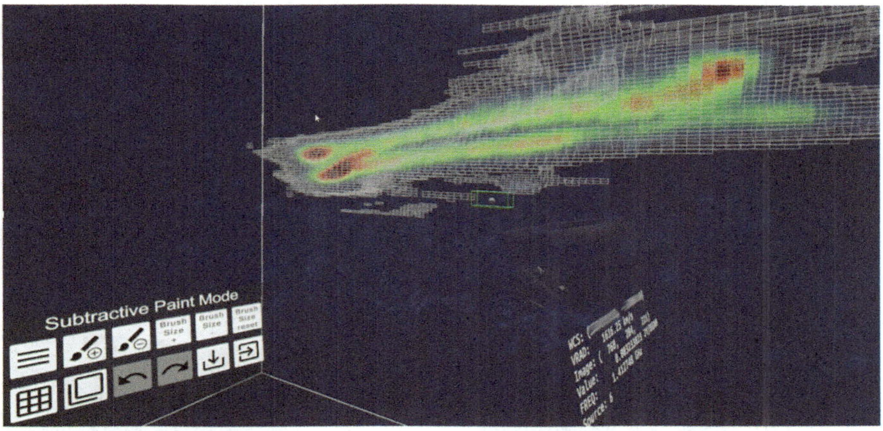

Fig. 11.7 iDaVIE mask editing feature. The user enters a "paint mode" for galaxy sources in which voxels can be individually added or removed from the cube's corresponding mask (e.g., with the "subtractive" mode shown in the picture). In this way the user is able to improve upon the source measurements as controlled by the mask. For convenience, the masked voxels appear as a grey mash on top of the HI data

regions from the mask (see Fig. 11.7). Rendering parameters can also be adjusted in realtime, including the colormap for example. Features are in constant development to respond to the evolving needs of the astronomical community.

11.2.2.3 Visualisation Using Digital Domes: Data to Dome

Along with VR, another immersive technology has shown promise in how radio astronomers visualise and interact with their 3D data. This is the fulldome digital planetarium (Fig. 11.8), the modern successor to the century-old analogue optomechanical planetarium (Kwasnitschka 2017; Marchetti et al. 2020; Jarrett et al. 2021). Introduced in the 1980s, the digital planetarium presents a paradigm shift for a facility traditionally limited by its specialised projection system of the night sky.

Digital planetaria have the capability of displaying a much broader range of visuals corresponding to whatever can be rendered on a computer. Using a network of such computers to form a cluster, multiple projectors are combined to display a coordinated fulldome image. Such systems provide high levels of visual rendering power, capable of presenting and real-time flying through astronomical data that comprise the cosmos we observe and understand today. Radio astronomers, e.g., are able to render their data in a 3D immersive environment, and unlike the VR experience, is designed to accommodate a potentially sizeable audience. Such a scenario can become a productive collaborative space while also amplifying the reach of research results through the established use of the planetarium as an education and outreach facility. The Data-to-Dome project exists is an organised

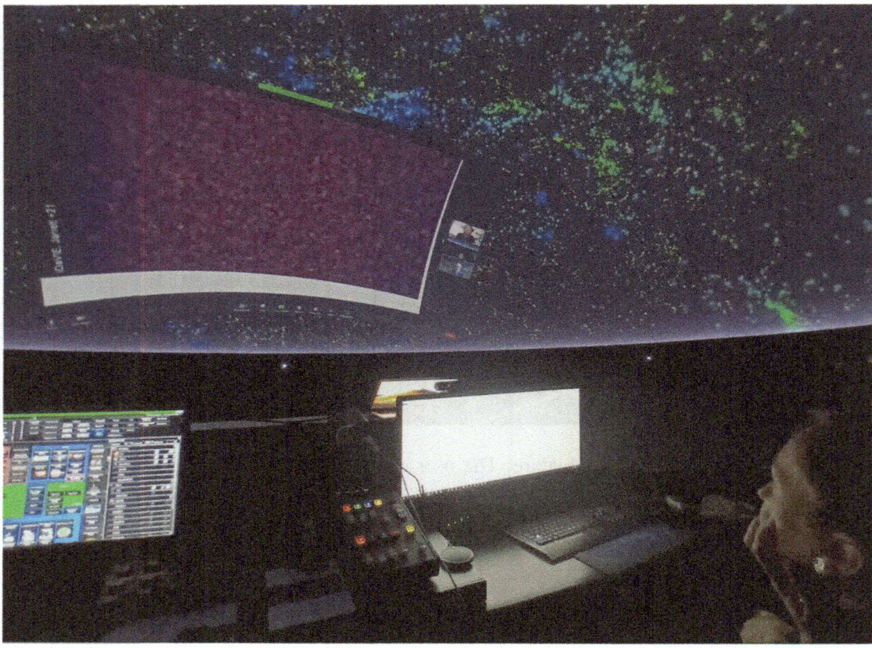

Fig. 11.8 Data-to-Dome research session at the Iziko Planetarium & Digital Dome in Cape Town, South Africa. Here the astronomer is rendering the "cosmic web of galaxies" to the planetarium dome in order to discern features that are accessible through 360-degree 'flying' immersion. This is done while also having a virtual meeting with colleagues abroad also projected on the dome

effort by the International Planetarium Society to use the digital planetarium as a visualisation space for scientific research.

There still exist numerous challenges with incorporating the digital planetarium medium into the workflow of radio astronomers. The digital planetarium market tends to be divided among several private companies that provide both installation services and proprietary software for interacting with the fulldome projection system. These software options are still generally controlled by a keyboard and mouse, which can be a challenge (as well as limiting) for users both new and experienced to navigate a virtual 3D space. Furthermore, the proprietary nature of the software means researchers do not have the freedom to incorporate essential features and capabilities on-the-fly that would be necessary to adapt the software to their specific needs.

A potential solution is the use of the iDaVIE VR software to render an interactive session for a VR-user that can then be observed by a virtual fisheye spectator camera and streamed to the digital dome in real time for the audience to view and experience. This provides the interaction advantages of VR for the presenter and audience, the dedicated volumetric rendering of iDaVIE, and the flexibility of opensource software that free up the digital planetarium space for radio astronomers. An example of how this works and looks is reported in Jarrett et al. (2021).

Other opensource software alternatives for the digital planetarium currently being explored include WorldWideTelescope[10] and OpenSpace.[11]

References

A. Comrie, K.-S. Wang, S.-C. Hsu, A. Moraghan, P. Harris, Q. Pang, A. Pińska, C.-C. Chiang, T.-H. Chang, Y.-H. Hwang, H. Jan, M.-Y. Lin, R. Simmonds, CARTA: The cube analysis and rendering tool for astronomy. Zenodo, June (2021)

E.M. Di Teodoro, F. Fraternali, 3D-Barolo: 3D fitting tool for the kinematics of galaxies. Astrophysics Source Code Library, record ascl:1507.001, July (2015a)

E.M. Di Teodoro, F. Fraternali, 3D BAROLO: a new 3D algorithm to derive rotation curves of galaxies. Mon. Not. R. Astron. Soc. **451**(3), 3021–3033 (2015b). https://doi.org/10.1093/mnras/stv1213

R. Gooch, Karma: a visualization test-bed, in *Astronomical Data Analysis Software and Systems V*, ed. by G.H. Jacoby, J. Barnes, volume 101 of Astronomical Society of the Pacific Conference Series (1996), p. 80

E.W. Greisen, The astronomical image processing system, in *Acquisition, Processing and Archiving of Astronomical Images* (1990), pp. 125–142

T.H. Jarrett, A. Comrie, L. Marchetti, A. Sivitilli, S. Macfarlane, F. Vitello, U. Becciani, A.R. Taylor, J.M. van der Hulst, P. Serra, N. Katz, M.E. Cluver, Exploring and interrogating astrophysical data in virtual reality. Astron. Comput. **37**, 100502 (2021). https://doi.org/10.1016/j.ascom.2021.100502

W.A. Joye, E. Mandel, New features of SAOImage DS9, in *Astronomical Data Analysis Software and Systems XII*, ed. by H.E. Payne, R.I. Jedrzejewski, R.N. Hook, volume 295 of Astronomical Society of the Pacific Conference Series (2003), p. 489

T. Kwasnitschka, Planetariums — not just for kids. Nature **544**(7651), 395 (2017). https://doi.org/10.1038/544395a

L. Marchetti, T.H. Jarrett, A. Comrie, A.K. Sivitilli, S. McFarland, A.R. Taylor, M. Clever, The DataToDome initiative at the Iziko Planetarium in Cape Town and the IDIA Visualization Lab, in *Astronomical Data Analysis Software and Systems XXIX*, ed. by R. Pizzo, E.R. Deul, J.D. Mol, J. de Plaa, H. Verkouter, volume 527 of Astronomical Society of the Pacific Conference Series (2020), p. 247

J.P. McMullin, B. Waters, D. Schiebel, W. Young, K. Golap, CASA architecture and applications, in *Astronomical Data Analysis Software and Systems XVI*, ed. by R.A. Shaw, F. Hill, D.J. Bell, volume 376 of Astronomical Society of the Pacific Conference Series (2007), p. 127

D. Punzo, J.M. van der Hulst, J.B.T.M. Roerdink, T.A. Oosterloo, M. Ramatsoku, M.A.W. Verheijen, The role of 3-D interactive visualization in blind surveys of H I in galaxies. Astron. Comput. **12**, 86–99 (2015). https://doi.org/10.1016/j.ascom.2015.05.004

D. Punzo, J.M. van der Hulst, J.B.T.M. Roerdink, Finding faint H I structure in and around galaxies: Scraping the barrel. Astron. Comput. **17**, 163–176 (2016). https://doi.org/10.1016/j.ascom.2016.09.002

[10] http://worldwidetelescope.org/.

[11] https://www.openspaceproject.com/.

D. Punzo, J.M. van der Hulst, J.B.T.M. Roerdink, J.C. Fillion-Robin, L. Yu, SlicerAstro: A 3-D interactive visual analytics tool for HI data. Astron. Comput. **19**, 45–59 (2017). https://doi.org/10.1016/j.ascom.2017.03.004.

A.K. Sivitilli, A. Comrie, L. Marchetti, T.H. Jarrett, Virtual reality and immersive collaborative environments: the new frontier for big data visualisation, in *Astronomical Data Analysis Software and Systems XXIX*, ed. by R. Pizzo, E.R. Deul, J.D. Mol, J. de Plaa, H. Verkouter, volume 527 of *Astronomical Society of the Pacific Conference Series* (2020), p. 221

R. Taylor, FRELLED: A realtime volumetric data viewer for astronomers. Astron. Comput. **13**, 67–79 (2015a). https://doi.org/10.1016/j.ascom.2015.10.002

R.P.W. Taylor, FRELLED: FITS realtime explorer of low latency in every dimension. Astrophysics Source Code Library, record ascl:1508.004, August (2015b)

J.M. van der Hulst, J.P. Terlouw, K.G. Begeman, W. Zwitser, P.R. Roelfsema, The Groningen Image Processing SYstem, GIPSY, in *Astronomical Data Analysis Software and Systems I*, ed. by D.M. Worrall, C. Biemesderfer, J. Barnes, volume 25 of Astronomical Society of the Pacific Conference Series (1992), p. 131

Part IV
Data Access and Reuse

Chapter 12
Exploitation Platforms and Virtual Observatory

François Bonnarel, Ian Bird, Mark Lacy, Marta Dembska, and Simon O'Toole

12.1 Virtual Observatory Framework and Science Platforms

The upcoming data avalanche that we have extensively described in the previous chapters is opening a huge challenge for both data providers and scientists. If we only take the example of SKA as was stated above (see Sect. 3.9) that's no less than 700 Pbytes/year of observatory data products which will be released every year. From these the SRC network will derive petabytes of advanced data products.

Full benefit of these data for the scientific communities will only be possible if they can be accessed, reused and processed in an open mode. To achieve that, data will have to come with standardised metadata and formats and data services will have to share standardised APIs and methods. Specifically, emphasis should be put on metadata describing the way the data have been obtained and produced in order to be able to navigate between the different levels of processing of the data. The

F. Bonnarel (✉)
CDS, Université de Strasbourg, CNRS, Observatoire astronomique de Strasbourg - UMR7550, Strasbourg, France
e-mail: francois.bonnarel@astro.unistra.fr

I. Bird
Laboratoire d'Annecy de Physique des Particules (LAPP) - CNRS, Annecy, France
e-mail: ian.bird@lapp.in2p3.fr

M. Lacy
National Radio Astronomy Observatory, Charlottesville, VA, USA
e-mail: mlacy@nrao.edu

M. Dembska
Institute of Data Science, German Aerospace Center (DLR), Jena, Germany
e-mail: marta.dembska@dlr.de

S. O'Toole
Australian Astronomical Optics, Macquarie University, Sydney, NSW, Australia

E. Vardoulaki et al. (eds.), *Data-Intensive Radio Astronomy*, Astrophysics and Space Science Library 472, https://doi.org/10.1007/978-3-031-58468-8_12

dramatic increase in the amount of available data requires at least two changes in data management. First of all, services and platforms have to implement a hierarchy of multi-resolution discovery and access capabilities from high level registry of data services towards highest resolution data retrieval going through various levels of coarse grain excerpts of data and metadata. In addition, all required data cannot be easily downloaded to local computer facilities, and processing and analysis software cannot always be run on these local facilities. The general concept of "code to the data" is at the heart of science or exploitation platforms.

This part of the book develops what data access and reuse actually means from now onward. We address the concrete examples of the Virtual Observatory framework and of science platforms, for the latter as experimented within the scope of the ESCAPE project. In addition, we relate that to the general emerging paradigm of data provenance and FAIRness of the data.

The current chapter summarises how radio astronomy data are exploited through a set of common standards for data description and data formats, and how those data can be delivered and exploited by scientists. In Sect. 12.2 the Virtual Observatory is briefly introduced, and the standards applicable to radio astronomy data are reviewed and described. In the following Sect. 12.3 the institutes and centres that deliver radio astronomy data according to the VO standards are surveyed, and described. Finally Sect. 12.4 introduces the European initiative for Open Science in general, and the projects relevant to radio astronomy, that are foreseen to act as a platform for delivery of astronomy data and services, not only to the astronomy community but also to a broader audience. The project relevant to astronomy, ESCAPE, has included work to bring the VO standards and services into the general Open Science framework.

12.2 VO Standards and Radio Astronomy

The Virtual Observatory (VO) is a set of data services, web services and tools following a set of standard specifications. The whole set defines an interoperable landscape for sharing astronomical data. VO standards are defined within the scope of the International Virtual Observatory Alliance (IVOA) since 2002. Radio astronomy data were already considered in the first decade of IVOA, within early VO data model and data access layer specifications. When multi-dimensional data science became a VO priority (2013), this drove the emergence of new protocols that included description discovery and access of polarisation and image cubes, both important for radio astronomy. Since that time, radio astronomy projects have further increased in size and variety, and many of them have already implemented VO services, or are actively working in this direction.

12.2.1 Transport Data Formats

In most of the cases, VO services are not redefining transport data formats and definitely not storage format which are exposed in Sect. 2.2. For astronomical datasets (images, cubes, visibility data) they are reusing de facto standards such as FITS and its dialects (FITS-IDI, PSRFITS, SDFITS, HDF5 with various schemata etc.) or MeasurementSets. Parquet may also be considered in the future.

There is an exception for tabular data, for which IVOA defined the VOTable standard XML format as early as 2002 (Ochsenbein et al. 2019). VOTable is largely used in astronomy as a transport format for catalogues of measurements of astronomical sources as well as for other types of datasets descriptions. The CDS VizieR service can deliver more than 20,000 astronomical catalogues this way and CADC and GAVO can also provide plenty of them.

Another specific recommendation is the Hierarchical Progressive Survey (HiPS) (Fernique et al. 2017) standard for pixel and catalogue data. But it's much more than a format, in such a way that we dedicate a specific subsection for it.

12.2.2 Data Models

Virtual Observatory makes use of data model mainly to describe metadata and emphasize the complex relationships embedded in the data. The ObsCore data model (Louys et al. 2017) including parts of the Characterisation data model (Louys et al. 2008) intends to describe datasets layout both on physical parameters axes and curation details. It is fully adapted for discovery of images and spectral cubes. Recently, description of radio data and interferometry visibility data has been also successfully achieved.[1]

The EPN-Core data model (Erard et al. 2021) has great similarities with ObsCore but tackle the specificities of solar system data. The IVOA Provenance data model (Servillat et al. 2020) may describe the history of the data: which rawer data, or calibration files it has been produced from, which software or workflow with, which parameters have been used, what were the observation configurations to obtain original data. More details will be given in Chap. 13. IVOA also developed a data model for spectra (McDowell et al. 2011). It's perfectly adapted for radio spectra (extracted from spectral cubes or directly obtained with single dish telescopes).

[1] Since we wrote this a more stable version may be found here: https://www.ivoa.net/documents/ObsCoreExtensionForRadioData/.

12.2.3 Data Access Layer

TAP (Table access protocol) is an universal access standard for astronomical relational databases (Dowler et al. 2019). TAP services registered in the IVOA registry, currently expose more than 22,000 tables. It makes use of an astronomy extended SQL-like query language called ADQL (Osuna et al. 2008). The database content is described in a standardised schema called the TAP_SCHEMA. Rich standardised metadata are included in the description. TAP services build up a query response in one single table. They extract data from the database following the structure given in the TAP_SCHEMA but provide only the selected columns in a response VOTable document. TAP services can distributes catalogues of radio sources as well as description of radio images, spectral cubes and datasets using the ObsCore or the EPN-Core standardised metadata.

SCS (Plante et al. 2008) is a very simple PARAMETER based interface delivering in VOTAble data or metadata in a given conic region of the sky. Simple image access (Dowler et al. 2016) defines a much richer PARAMETER based interface delivering responses following the ObsCore data model. Simple spectrum access (Tody et al. 2012) provides similar functionalities to discover and access spectra.

DataLink (Dowler et al. 2015) allows to associate different links to a single dataset or source record in a data access layer service response and to shortly describe this link in a standardised way.

All these standards are well adapted to create discovery and retrieval services for radio datasets and radio tabular data and are indeed widely used, as will be exposed in the next section. Interferometry visibility datasets may also be described and discovered using ObsCore according some specific precaution. In order to illustrate this latter aspect, Fig. 12.1 shows results of the JIVE ObsTAP service within AladinDesktop.

Current large projects paving the way to SKA such as LOFAR, ASKAP, MEERKAT are facing the "terabyte" retrieval issue. For this reason IVOA developed SODA, an interoperable standard interface for defining cutout services (Bonnarel et al. 2017). Figure 12.2 shows the result of a cutout query for an ASKAP dataset.

12.2.4 HiPS and MOC

HiPS (Fernique et al. 2017) is a data organization of pixel and catalogue data based on Healpix tesselation of the sky. For each data collection the HiPS is a specific directory which contains different branches for the different orders. Each tile in the tree contains a selection or combination of original values appropriate to the spatial resolution of the order.

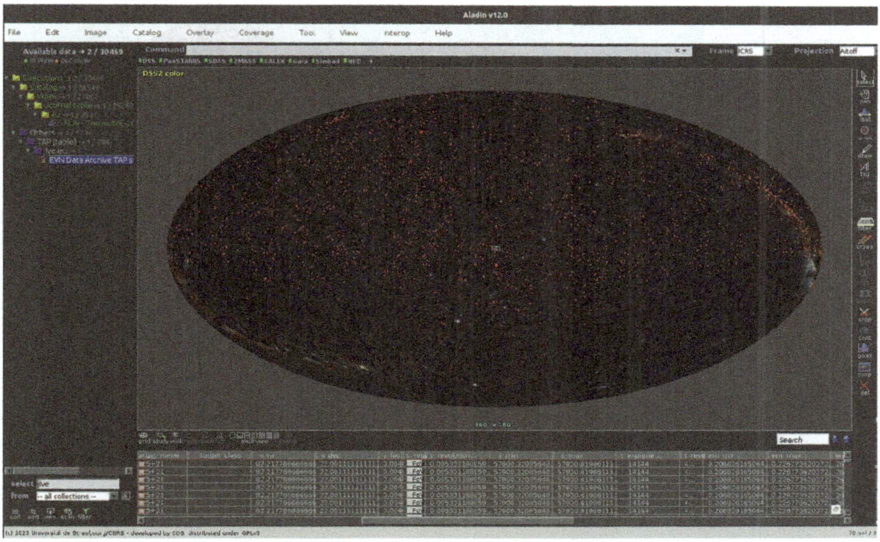

Fig. 12.1 JIVE ObsTAP service showing positions of visibility datasets on top of the sky with some datasets metadata around the M1 nebula

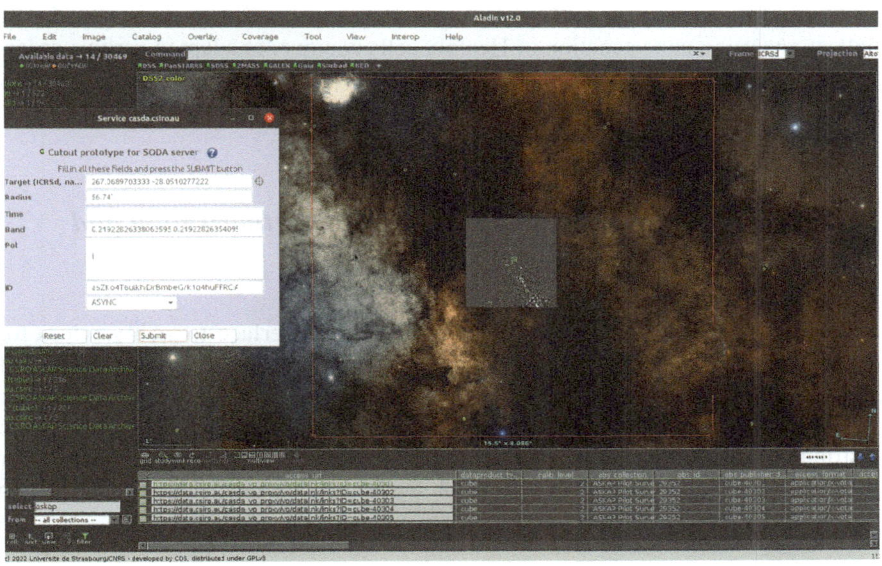

Fig. 12.2 Cutout in an ASKAP rapid survey dataset discovered thanks to SIA2 CASDA service. The full FOV of the dataset is shown in red and the SODA interface in Aladin are shown

Accessing an HiPS allows visualisation applications to zoom in and out in the data display by changing the retrieved order directory. Figure 12.3 shows a HiPS of

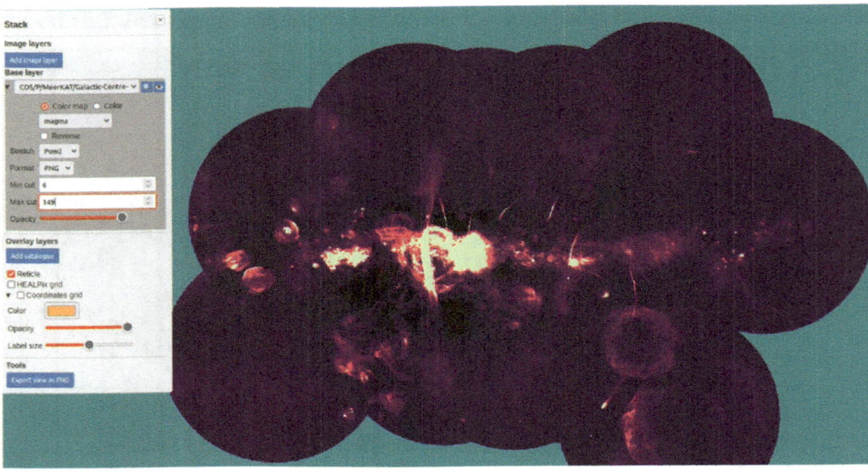

Fig. 12.3 HIPS of MeerKAT galactic center mosaic visualized with the help of AladinLite V3

the galactic center mosaic obtained by the MeerKAT project. HiPS has now been extended to process spectral cubes.

Multi-Order Coverage (MOC) is a standard for describing coverage of data collections using concatenation of Healpix cells contours at adapted various orders (Fernique et al. 2019). MOC has been recently extended in order to tackle the time coverage.

12.2.5 Web Services, Distributed Storage, Authorisation and Authentication

IVOA also defined several protocols to manage the interface to web services for data processing. UWS (Harrison and Rixon 2016) introduced standardisation of asynchronous modes. It can be used for discovery and access services as well as for more project defined processing. VOSpace (Graham et al. 2018) intends to make distributed storage resources interoperable. Eventually IVOA defined standards for description of authorization/authentication methods (Taffoni et al. 2017) and Group membership (Major et al. 2022).

12.2.6 Service and Resource Registration

The Virtual observatory services can be discovered via VO registries. The content and format of the registries has been standardized in a small set of specifications, such as ResourceMetadata (Hanisch et al. 2007), VOResource (Plante et al. 2018) and VO DataService (Demleitner et al. 2021)

12.3 VO Services Delivering Radio Data: An Implementation Review

The content of this section is a summary of the IVOA note "Radio astronomy in the VO: services implementation review".[2]

12.3.1 Short Overview of Radio Astronomy Services Implementing VO Standards

Table 12.1 summarizes how radio services implement the main VO standards. In addition to that, visibility data are discovered either through DataLink as progenitor information (ALMA, ASTRON, planned at INAF) or directly in ObsCore for JIVE, CADC, CASDA and MWA. ASTRON, JIVE and CDS collaborated to better map ObsCore to visibility observation fields and spectral windows and to propose an extension of ObsCore useful for visibility data. Other raw formats for single dish data (SDFITS) or pulsar data and FRBs (PSRFITS) are delivered through VO interfaces or will be soon by ASTRON, Nançay, ATOA, PSRDA, INAF.

Table 12.1 Projects implementing VO standards

Standard	Datatype	Projects	Comments
`SIA2/ObsTAP`	`spectral cubes`	ALMA, ASTRON (TGSSADR, Apertif, LoTSS), CADC, CASDA, ATOA	Planned for Nenufar, INAF, e-merlin
`DataLink`		ALMA, ASTRON, CADC CASDA	Planned at Nançay INAF, e-merlin
SCS, TAP	Measurements catalogs	CADC, CDS, MWA	
EPN-TAP services	Solar, Jupiter observations	MASER, Nançay	EPN-Core metadata served by TAP (Erard et al. 2021)
HiPS	Radio images cubes	ASTRON, CADC, CDS	

[2] https://www.ivoa.net/documents/Notes/RadioVOImp/20211119/RadioVOImplementation.pdf.

12.3.2 Data Centers Delivery

Data centers are gathering data from various projects including radio astronomy ones with some added value. The Canadian Astronomical Data Center (CADC) has always been involved in the IVOA and has taken on the effort to support VO protocols and models. CADC is delivering DRAO Observations through its TAP/ObsTAP interfaces also with Datalink and SODA cutout facility as well as CGPS and VGPS HiPSes. CADC services are also involved in distribution of the VLA Sky Survey data.

The Centre de Données astronomiques de Strasbourg (CDS) is a value added data centre (and not an observatory archive) whose goal is to distribute published data developed a solid expertise in VO standards. It was strongly involved in the definition and evolution of VO standards useful for radio data. See for example Louys et al. (2020). VizieR gives access to more than 2900 catalogues containing radio sources or radio measurements through TAP or SCS interfaces. Among those, the CDS SPECFIND catalog (Stein et al. 2020) cross-identifies 1,600,000 radio sources and provides reconstructed spectra for many of them of them. CDS is also storing dozens of HiPS belonging to the radio domain, from Effelsberg to SCUBA and from LOFAR to Apertif or MeerKAT galactic center mosaic (Fig. 12.3).

12.3.3 Single Dish and Pulsar Data Archives

The CSIRO (Commonwealth Scientific and Industrial Research Organisation)[3] delivers old spectral line Parkes 64 m radio telescope data as well as data from the latest Parkes receivers. All data are online and available to VO ObsTAP users for download through the Australia Telescope On line Archive, ATOA. CSIRO also operates the Pulsar Data Archive (PSRDA) which delivers indexes of observations via TAP.

The Istituto Nazionale di AstroFisica (INAF) manages three fully steerable reflector antennas for radio astronomy: the 32 m dishes at Medicina and Noto and the 64 m Sardinia Radio Telescope (SRT). The INAF radio archive stores continuum and spectropolarimetric raw data from single dish and pulsar observations. An ObsTAP service is under preparation for the holdings of the Radio Archive.

The Netherlands Institute for Radio Astronomy (ASTRON) is delivering the Apertif (see Sect. 3.4) FRB data (Apertif Radio Transient System Data Release 1, ARTS-DR1) through TAP and SCS interfaces.

[3] https://data.csiro.au.

12.3.4 VLBI Data Archives

The Joint Institute for VLBI in Europe (JIVE) hosts the archive of all radio-astronomical VLBI observations of the European VLBI Network (EVN). JIVE has been working towards making the raw visibilities and support information findable and downloadable through ObsTAP and DataLink services; the service has been released in late 2021.

INAF radio telescopes can also work together in VLBI mode or be joint to the EVN. The INAF radio archive ObsTap service also delivers them.

12.3.5 Long Term Radio Interferometry Projects and Surveys

The Atacama Large Millimeter Array (ALMA) archive (see Sect. 3.5) has implemented ObsTAP, SIAv2 and DataLink services in addition to its native website.

The Astron VO service is running an instance of DaCHS (Demleitner et al. 2014). Its VO services contain data from the Westerbork Synthesis Radio Telescope (WSRT) and the Apertif system DR1 and DR2 image surveys. In addition, the Alternative Data Release of the TIFR GMRT (Tata Institute of Fundamental Research Giant Metrewave Radio Telescope) Sky Survey (TGSSADR) is served through the Astron VO. Astron is also delivering HiPS for 2D survey images belonging to these projects.

ATOA serves spectral line observations from the Australia Telescope Compact Array (ATCA) in its ObsTAP service. Most ATCA data are in the form of visibilities.

The Murchison Wide-Field Array (MWA) ASVO (All-Sky Virtual Observatory)[4] provides TAP and SCS services to deliver catalog of measurements and ObsCore descriptions of observations. In addition to these services some MWA data are available through Data Central TAP[5] and SIA services.

The Nançay radio astronomy facility is hosting several radio telescopes allowing observation from 10 MHz to up to several GHz. The Nançay Data Centre (CDN) is publishing radio astronomy data products. For example data products from NDA (Nançay Decametric Array[6]) are accessible through the VO.

National Radio Astronomy Observatory (NRAO) operates the Very Large Array (VLA) of radio telescopes in Socorro (New-Mexico). One of the main programs was the VLA Sky Survey (VLASS). NRAO VLASS data are now available through an ObsTAP service recently released.

[4] https://asvo.mwatelescope.org/services#votap.

[5] https://datacentral.org.au/vo/tap.

[6] http://vogate.obs-nancay.fr/browse/nda/q/.

12.3.6 SKA, ngVLA and Pathfinders

Square Kilometer Array (SKA, see Sect. 3.9) is the "petabyte" project of the future for low and very low frequency radio astronomy. Advanced data processing and data distribution will be under the responsibility of the "SKA regional center network". This network is currently prototyping the storing, high performance computing, visualization and distribution choices. VO standards such as ObsCore, DataLink, SODA and HiPS are part of the sublayer on which data discovery and access, authorization and authentication and visualization will be built.

ngVLA will be the future project covering the mid frequency range in radio-astronomy filling the gap between ALMA and SKA. NRAO will distribute the data in its VO framework.

Several projects are considered as pathfinders to these huge projects and are already delivering data through VO services. The CSIRO ASKAP Science Data Archive (CASDA, see Sect. 3.3) aims to provide science ready data products from the Australian Square Kilometre Array Pathfinder (ASKAP) telescope. CASDA deliver visibility data (in MS format), images, cubes, spectra and moment maps, as well as Catalogue data. The TAP, SIAP and SSAP services query the PostgreSQL metadata database. Data access is implemented using the Datalink and SODA protocols. For the LOw Frequency ARray (LOFAR) (see Sect. 3.2), Astron is delivering the LoTSS (LOFAR Two-metre Sky Survey) DR1 and DR2 via its VO services (ObsTAP, DataLink and HiPS). The Nancay data center, on its side is developing prototypes for NenuFAR (New Extension in Nançay Upgrading LOFAR[7]).

12.3.7 Solar System Data

The MASER (Measurement, Analysis and Simulation of Emissions in the Radio range) project is providing the community with a science-ready toolbox for low frequency radio astronomy (below 100 MHz), which is the domain where Sun and planets are emitting radio waves through non-thermal radio emission processes and are dominant. Solar and planetary radio astronomy data products can be shared through EPN-TAP (combination of EPNcore and TAP). Radio astronomy Solar System observations are mostly dynamic spectra, but 2D images, event catalogues and timeseries are also available.

At Nancay prototypes of VO services are under development for the NRH (Nançay Radio Heliograph) and ORFEES (Observations Radio pour FEDOME et l'Etude des Eruptions Solaires).

[7] http://vogate.obs-nancay.fr/browse/nda/q.

12.4 European Open Science Cloud

The European Open Science Cloud (EOSC) is a framework for Open Science and FAIR (Findable, Accessible, Interoperable, Reusable) data in Europe. The vision is to create a federated and open environment for European researchers, ultimately also enabling access to industry, innovators and citizens. It is, by design, multi-disciplinary and cross-domain, providing facilities to enable researchers and users to "publish, find and reuse data, tools and services for research, innovation and educational purposes".[8] As such it is being developed through a number of individual projects funded by the European Commission under the Horizon 2020 Framework Programme, and in future under the Horizon Europe Programme. The most significant project at the moment is EOSC-Future, which builds on and integrates the results of previous projects, and importantly brings together the scientific communities and the e-infrastructure organisations that have been built up over many years. The scientific communities are represented through the five science clusters which themselves are important collaborative projects of domain-based science.

These cluster projects represent the domains of life sciences (EOSC-Life), environment (ENVRI-FAIR), social sciences (SSHOC), photon and neutron science (PANOSC), and Astronomy, Particle, and Nuclear Physics (ESCAPE). These clusters represent a huge proportion of European scientific research (some 50,000 researchers or more), and a strong representation of the scientific activities that will potentially benefit from EOSC. An important development is the realisation of the many synergies between these clusters and the desire of the clusters to work together to influence the direction of the EOSC and the relevant funding agencies. Although the EOSC is a European endeavour, it is important to recognise that all of the Research Infrastructures (RI) represented by the clusters are international or global in nature, and that the EOSC cannot exist in isolation, but must connect with other Open Science initiatives globally.

Although the EOSC is being implemented through these projects, the long term oversight and direction is foreseen through the EOSC-Association, which is a legal entity established to govern the EOSC, and is made up of representatives of research funders, research performing organisations, research infrastructures, and service providers. The vision for the long-term sustainability of the EOSC is vital for the full engagement of the scientific communities, and is a high level goal of the Association and the projects and organisations involved.

Thus, in addition to the high level aspirations of open science mentioned above, it is key that the EOSC provides value to the scientific community. This value can take the form of specific value-added services, such as a common Authentication and Authorisation Infrastructure (AAI) and similar fundamental services useful for all scientists; or resources provided free at the point of use to enable novel cross-

[8] https://ec.europa.eu/info/research-and-innovation/strategy/strategy-2020-2024/our-digital-future/open-science/european-open-science-cloud-eosc_en#documents.

discipline scientific activities that do not necessarily fit within the usual funding envelopes of individual RIs. In addition, it is important to recognise the building of synergies between disciplines and the capability of sharing not only experience and expertise, but also data, software and services developed by the individual science communities. Ultimately, having a structure such as EOSC should enable more cost-effectiveness through common service layers that can support and evolve with the needs of all of the scientific communities.

12.4.1 The ESCAPE Science Cluster

As one of the science clusters mentioned above, the European Science Cluster of Astronomy and Particle Physics ESFRI RIs (ESCAPE)[9] project brings together the ESFRI and other world-class research infrastructures in Astronomy, Astroparticle Physics, Gravitational Wave and Particle and Nuclear Physics in a single collaborative structure to create a domain-specific "cell" of the EOSC. The RIs are in Particle Physics: High Luminosity LHC (HL-LHC); in Nuclear Physics: Facility for Antiproton and Ion Research (FAIR); in Astrophysics: Cherenkov Telescope Array (CTA), European Gravitational Wave Observatory and the Virgo experiment (EGO-Virgo), European Solar Telescope (EST) , Extremely Large Telescope (ELT), Joint Institute for VLBI ERIC (JIVE), KM3Net neutrino observatory, Rubin Observatory Legacy Survey of Space and Time (LSST), and the Square Kilometre Array (SKA).

The RIs in ESCAPE have similar and aligned challenges for the future, in particular significant data volumes anticipated, at the scale from 10's of petabytes to several exabytes/year. These challenges, together with the scale of distributed computing infrastructures required, the requirement for cost efficiency and collaboration between infrastructures making use of common facilities, all required new approaches; while building on several years of expertise and experience.

ESCAPE has produced solutions to support the implementation of EOSC thanks to open data management, a cross-border and multi-disciplinary open environment, according to FAIR principles. ESCAPE has enabled interoperability between the facilities, minimising fragmentation, encouraging cross-fertilisation and has developed joint multi-wavelength/multi-messenger capabilities between astronomy, astrophysics and the particle astrophysics communities.

The output of ESCAPE is a domain-specific "EOSC cell" (illustrated in Fig. 12.4), being integrated in the developing EOSC infrastructure. The main results of ESCAPE are the following key points:

- Prototyped a full Exabyte-scale Data Management (Storage, Transfer and Access) Service (the "Data Lake") with a common AAI framework for

[9] https://projectescape.eu.

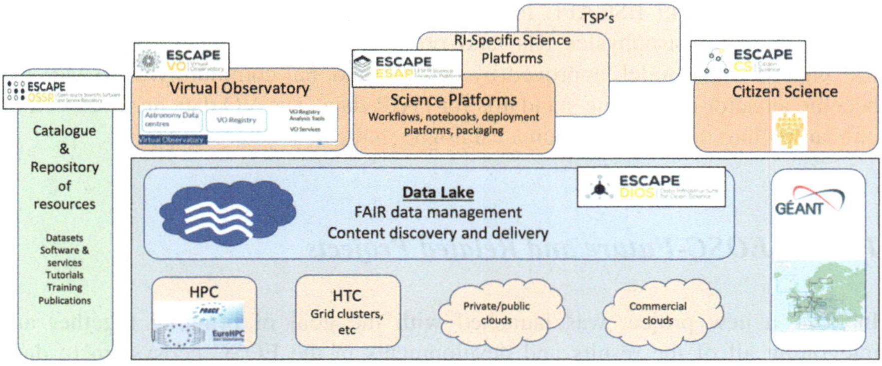

Fig. 12.4 The ESCAPE activities and relationships; user communities' access services and resources through the RI-specific platforms, Virtual Observatory, Science Projects (TSP), and as citizen scientists

distributed scientific computing. This has been validated by the RIs through a number of realistic Data Challenges;

- Developed a catalogue and software repository through which to publish digital scientific products of the ESCAPE communities. This is onboarded into the EOSC portal;
- Developed interoperability standards for astronomical data services, which enables a Virtual Research Environment of interoperable tools and services based on IVOA standards;
- Produced a reusable analysis toolkit for integrating diverse service offerings, and a Virtual Research Environment prototype that integrates ESCAPE services, already used by a number of the RIs in their analysis frameworks;
- Developed and produced training materials/schemes, and ran a number of highly successful schools and workshops;
- Several Citizen Science projects have been built upon ESCAPE developments, and attracted significant interest and involvement. Citizen Science is used as a research tool.

12.4.1.1 The Virtual Observatory in ESCAPE

The Virtual Observatory (VO) described previously (see Sects. 12.1, 12.2, and 12.3) is an active participant in ESCAPE where the goal has been to integrate VO data and services into the EOSC. It has implemented FAIR principles for astronomy data through the VO, and provided a number of community training events for both scientists and data producers or providers. It has also added value to the already trusted content in the astronomy archives through applying Deep Learning algorithms to the datasets, producing a prototype demonstrator.

The VO work in ESCAPE has successfully brought the common standards to many of the astrophysics RIs, and produced additional standards such as a Provenance Data model, approved by the international community. ESCAPE has become an actor in defining standards for FAIR data. The VO data archives will be onboarded into the EOSC resource catalogue in the near future.

12.4.2 EOSC-Future and Related Projects

In 2021 a new project was launched with the goal of bringing together and integrating all of the results and developments in the EOSC ecosystem to date, and building a first full implementation of the EOSC itself. That project is EOSC-Future,[10] and it brings together all of the significant stakeholders: the five science clusters, and the five e-infrastructures: EGI,[11] EUDAT,[12] OpenAire,[13] GEANT,[14] and the Research Data Alliance (RDA). [15]

The project engages with users in particular through ten innovative science projects that the science clusters are deploying into the prototype EOSC infrastructure. These are cross-domain, and in several cases cross-cluster, and are designed to demonstrate new capabilities that EOSC can bring, in particular being able to publish and reuse scientific data and services. In the case of ESCAPE, the two selected science projects are on Dark Matter, and the Extreme Universe, recognised in the roadmap of challenges in the field by the national consortia in the ESCAPE scientific domains.

- Dark Matter: aiming to produce new reproducible analyses through collecting all related data from complementary experiments, on a broad platform hosted in EOSC. The goal is to produce new constraints (or discovery) for dark matter properties.
- Extreme Universe: the idea is to develop a sustainable platform for multi-messenger astronomy, and to use it to study extreme energy astrophysical phenomena, as well as exploiting gravitational waves for fundamental physics.

The EOSC-Future project has designed an architecture for the EOSC, based on outputs from many EOSC working groups that had been in place previously. This architecture is shown in Fig. 12.5.

The core, the support services, and interoperability framework should be common to all communities, and represent the set of enabling services needed to operate

[10] https://eoscfuture.eu.

[11] https://egi.eu.

[12] https://eudat.eu.

[13] https://www.openaire.eu.

[14] https://geant.org.

[15] https://www.rd-alliance.org.

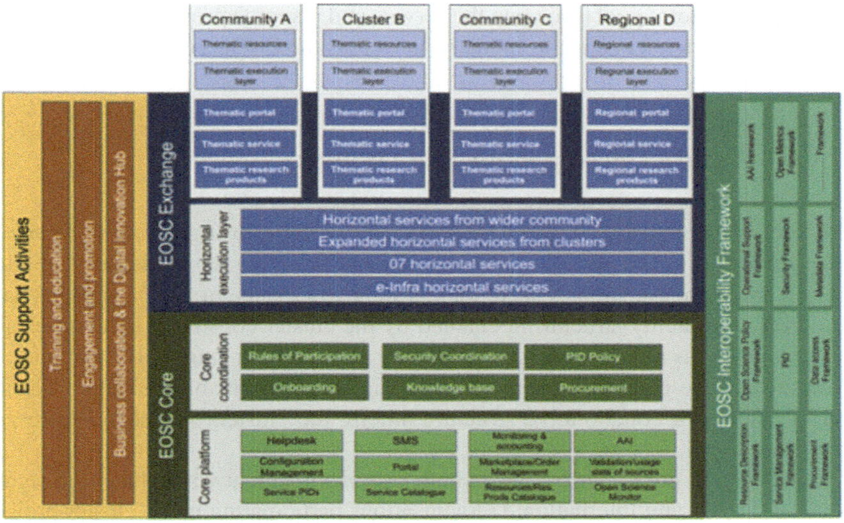

Fig. 12.5 The EOSC architecture: common services and facilities in the core; an Exchange layer of contributed services from the RIs or other projects; support structures; and an interoperability framework. The science communities and clusters are integrated, but also have thematic-specific resources and workflows

the EOSC. The Exchange layer is where science communities can onboard services they have developed which are potentially of broader use. In addition, the Exchange layer provides access to compute and storage resources provisioned through several related projects, specifically set up to provide resources. Additional details on the architecture can be found on the project web. It is recognised that the science communities have significant investment in their own infrastructures, so the goal is to abstract an underlying common infrastructure, and enable cross-fertilisation and sharing of expertise and research products of all kinds.

As an example of how a cluster may integrate with the EOSC architecture, we can consider ESCAPE. ESCAPE will publish infrastructure software such as the Data Lake (Exascale data management system), and other project-developed software into the Exchange layer. This will be done through the integration of the metadata of the ESCAPE software catalogue. This is the strategy of several of the clusters. Thus the Exchange layer can federate all of the research software and services produced in the clusters. As time goes on we anticipate additional common services to be provided, where such are interesting to a broad range of communities.

Importantly, the provision of resources (compute, storage) funded through the provisioning projects will enable new cross-discipline projects to be deployed. Such resources will of course be an attraction for users to the EOSC. Long term mechanisms for procurement of resources and services are an aspect of the EOSC that EOSC-Future together with the EC must develop during the lifetime of the project. A long term goal is to open the service and resource provision and operation to procurement tenders open to industry as well as the EOSC community.

12.4.3 Future Developments and Outlook

The five science clusters have worked together closely throughout the lifetime of these cluster projects, and in particular in guiding the setting up of the EOSC-Future project, where they are all key stakeholders and participants. In fact the clusters have found the community structuring enabled by the cluster constructs themselves and acting together as a broad science community to have been extremely beneficial. In all cases the five clusters have a plan for long term sustainability of their cluster, and in addition a collaborative structure among the clusters is also foreseen and welcomed by all.

ESCAPE has put in place a new collaboration agreement which has been signed by all of the RIs and International Organisations participating in the cluster project. The cluster will transition to the collaboration at the end of the ESCAPE project funding. The ESCAPE name will be kept. The new collaboration will be open to new RIs in the domain, as will the other clusters.

Many aspects of how the EOSC will develop and evolve are still to be understood. In particular it is important to understand how the sustainability of the EOSC in terms of funding will be managed. It is being implemented through project funding in Horizon 2020 and Horizon Europe framework programmes. However, that does not guarantee sustainability, and ultimately that will only come if the EOSC demonstrates to the science communities that it is really beneficial and attractive to users. Thus it is essential that the scientific communities drive the direction and implementation.

The potential strength of EOSC is its collaborative aspect, and the provision of a common set of core services required by all. This can be a huge cost benefit. In addition, the structuring of the scientific community and sharing of experience, expertise, as well as research products and software can be very important. The Exchange layer can capture that. Common strategies and policies for open and FAIR data, open access, and open science is a huge benefit to the science communities and potential data consumers.

References

F. Bonnarel, P. Dowler, M. Demleitner, D. Tody, J. Dempsey, IVOA Server-side Operations for Data Access Version 1.0. IVOA Recommendation 17 May 2017 (2017)

M. Demleitner, M.C. Neves, F. Rothmaier, J. Wambsganss. Virtual observatory publishing with DaCHS. Astron. Comput. **7**, 27–36 (2014). https://doi.org/10.1016/j.ascom.2014.08.003

M. Demleitner, R. Plante, A. Stébé, K. Benson, P. Dowler, M. Graham, G. Greene, P. Harrison, G. Lemson, T. Linde, G. Rixon, VODataService: A VOResource Schema Extension for Describing Collections, Services Version 1.2. IVOA Recommendation 02 November 2021 (2021)

P. Dowler, F. Bonnarel, L. Michel, M. Demleitner, IVOA DataLink Version 1.0. IVOA Recommendation 17 June 2015 (2015)

P. Dowler, D. Tody, F. Bonnarel, IVOA Simple Image Access. arXiv e-prints, arXiv:1601.00519, January (2016). https://ui.adsabs.harvard.edu/abs/2016arXiv160100519D

P. Dowler, G. Rixon, D. Tody, M. Demleitner, Table Access Protocol Version 1.1. IVOA Recommendation 27 September 2019 (2019)

S. Erard, B. Cecconi, P. Le Sidaner, M. Demleitner, EPN-TAP: Publishing solar system data to the virtual observatory, version 2.0. IVOA Working Draft (2021). https://ivoa.net/documents/EPNTAP/20201027/index.html

P. Fernique, M. Allen, T. Boch, T. Donaldson, D. Durand, K. Ebisawa, L. Michel, J. Salgado, F. Stoehr, HiPS - Hierarchical Progressive Survey Version 1.0. IVOA Recommendation 19 May 2017 (2017)

P. Fernique, T. Boch, T. Donaldson, D. Durand, W. O'Mullane, M. Reinecke, M. Taylor, MOC - HEALPix Multi-Order Coverage map Version 1.1. IVOA Recommendation 07 October 2019 (2019)

M. Graham, B. Major, D. Morris, G. Rixon, P. Dowler, A. Schaaff, D. Tody, VOSpace Version 2.1. IVOA Recommendation 21 June 2018 (2018)

R. Hanisch, IVOA Resource Registry Working Group, NVO Metadata Working Group, Resource Metadata for the Virtual Observatory Version 1.12. IVOA Recommendation 02 March 2007 (2007)

P.A. Harrison, G. Rixon, Universal Worker Service Pattern Version 1.1. IVOA Recommendation 24 October 2016 (2016)

M. Louys, A. Richards, F. Bonnarel, A. Micol, I. Chilingarian, J. McDowell, IVOA Data Model Working Group, Data Model for Astronomical DataSet Characterisation Version 1.13. IVOA Recommendation 25 March 2008 (2008)

M. Louys, D. Tody, P. Dowler, D. Durand, L. Michel, F. Bonnarel, A. Micol, IVOA Data Model Working Group, Observation Data Model Core Components, its Implementation in the Table Access Protocol Version 1.1. IVOA Recommendation 09 May 2017 (2017)

M. Louys, K. Lutz, Y. Stein, A. Egner, F. Bonnarel, Radio Astronomy visibility data discovery and access using IVOA standards. arXiv e-prints, arXiv:2012.09273, December (2020). https://ui.adsabs.harvard.edu/abs/2020arXiv201209273L

B. Major, P. Dowler, G. Taffoni, S. Zorba, A. Damian, S. Bertocco, M. Molinaro. Group Membership Service Version 1.0. IVOA Recommendation 22 February 2022 (2022)

J. McDowell, D. Tody, T. Budavari, M. Dolensky, I. Kamp, K. McCusker, P. Protopapas, A. Rots, R. Thompson, F. Valdes, P. Skoda, B. Rino, S. Derriere, J. Salgado, O. Laurino, IVOA Data Access Layer Group, Data Model Working Group, IVOA Spectrum Data Model Version 1.1. IVOA Recommendation 20 November 2011 (2011)

F. Ochsenbein, M. Taylor, T. Donaldson, R. Williams, C. Davenhall, M. Demleitner, D. Durand, P. Fernique, D. Giaretta, R. Hanisch, T. McGlynn, A. Szalay, A. Wicenec, VOTable Format Definition Version 1.4. IVOA Recommendation 21 October 2019 (2019)

P. Osuna, I. Ortiz, J. Lusted, P. Dowler, A. Szalay, Y. Shirasaki, M.A. Nieto-Santisteban, M. Ohishi, W. O'Mullane, VOQL-TEG Group, VOQL Working Group, IVOA Astronomical Data Query Language Version 2.00. IVOA Recommendation 30 October 2008 (2008)

R. Plante, R. Williams, R. Hanisch, A. Szalay, Simple Cone Search Version 1.03. IVOA Recommendation 22 February 2008 (2008)

R. Plante, M. Demleitner, K. Benson, M. Graham, G. Greene, P. Harrison, G. Lemson, T. Linde, G. Rixon, VOResource: an XML Encoding Schema for Resource Metadata Version 1.1. IVOA Recommendation 25 June 2018 (2018)

M. Servillat, K. Riebe, C. Boisson, F. Bonnarel, A. Galkin, M. Louys, M. Nullmeier, N. Renault-Tinacci, M. Sanguillon, O. Streicher, IVOA Provenance Data Model Version 1.0. IVOA Recommendation 11 April 2020 (2020)

Y. Stein, B. Vollmer, T. Boch, G. Landais, P. Vannier, M. Brouty, M.G. Allen, S. Derriere, P. Ocvirk, VizieR Online Data Catalog: SPECFIND V3.0 Catalog of radio continuum spectra (Stein+, 2020). *VizieR Online Data Catalog*, art. VIII/104, January (2020)
G. Taffoni, A. Schaaf, G. Rixon, B. Major, SSO - Single-Sign-On Profile: Authentication Mechanisms Version 2.0. IVOA Recommendation 24 May 2017 (2017)
D. Tody, M. Dolensky, J. McDowell, F. Bonnarel, T. Budavari, I. Busko, A. Micol, P. Osuna, J. Salgado, P. Skoda, R. Thompson, F. Valdes, Data Access Layer Working Group, Simple Spectral Access Protocol Version 1.1. IVOA Recommendation 10 February 2012 (2012)

Chapter 13
Historical Record of the Data and Its Origins: Data Provenance

Michael Johnson, Marta Dembska, Sirko Schindler, and François Bonnarel

13.1 Introduction to Data Provenance, Provenance Modelling and Provenance Standards

The term provenance originally gained prominence within the world of art, where it represented a record of ownership, authorship, and authenticity for any particular piece. The quality and content of this provenance record was therefore incredibly important for establishing the value of a work of art. Additionally, it could be used to ascertain whether the art had ever illegally changed hands, establishing the moral validity of it. More recently, provenance has also been adopted to describe the production of data, where qualities such as reliability and trustworthiness are equally as important.

Provenance for computational systems is a multi-faceted term which may describe different aspects of the data lifetime with different focuses depending on the domain. For example, the concepts of why-, how-, and where-provenance describe why the data was produced, how it was formed, and where did the data

M. Johnson (✉)
Max-Planck-Institut für Radioastronomie, Bonn, Germany

Institute of Data Science, German Aerospace Center (DLR), Jena, Germany

The University of Manchester, Manchester, UK
e-mail: michael.johnson-4@manchester.ac.uk

M. Dembska · S. Schindler
Institute of Data Science, German Aerospace Center (DLR), Jena, Germany
e-mail: marta.dembska@dlr.de; sirko.schindler@dlr.de

F. Bonnarel
CDS, Université de Strasbourg, CNRS, Observatoire astronomique de Strasbourg - UMR7550, Strasbourg, France
e-mail: francois.bonnarel@astro.unistra.fr

E. Vardoulaki et al. (eds.), *Data-Intensive Radio Astronomy*, Astrophysics and Space Science Library 472, https://doi.org/10.1007/978-3-031-58468-8_13

come from (more specifically, from which and what parts of the input data was the output data produced from) (Buneman et al. 2001; Cheney et al. 2009). Furthermore, computational provenance has also been separated into two distinct components—prospective and retrospective (Lim et al. 2010). Within this regime, prospective provenance captures a set of processes as a sort of data recipe, used to describe data yet to be created. Whereas, retrospective provenance captures information on data and data processing which has already occurred.

The data volumes of modern radio astronomy necessitate automated data processing. For example, the current largest radio telescope, LOFAR (Low Frequency Array) produced the LTA (Long Term Archive) dataset with a size of 50 PB. Whereas, the SKA will generate 600PB of scientific data products per year (Chrysostomou et al. 2020). The size of these data is greatly reduced from their raw size, for example SKALow will produce data at a rate of 2 PB/s, to be reduced to $\sim$7 TB/s and it will require $\sim$1 petaFLOPS to do so (An 2019). The size of the provenance data required to describe this processing scales with the number of operations. For the SKA, documenting each operation will amass a provenance dataset on the scale of the raw dataset size. This can however be reduced by abstracting the granularity with which the provenance is recorded. However, this abstracted view must also be a true and loyal account of the operations performed with sufficient detail such that the results may be reproduced exactly upon re-execution. Determining the granularity which satisfies this whilst also creating a provenance dataset of a reasonable size is an important challenge for the next generation of radio survey telescopes.

In addition to the different concepts within data provenance, the definition and focus of it may also vary depending on the domain of the data. The unification of these different concepts and definitions was a necessary step to make provenance data from different applications interoperable. This was the goal of PROV, which was designed to be a generalisable and extendable standard for recording provenance (Groth and Moreau 2013). The three major components within PROV are entities, activities, and agents. Entities represent things, whether digital or physical, activities can be thought of as actions or processes which generate or manipulate entities, and agents are what have the responsibility for running an activity, this may be things such as a person, organisation, or a piece of software. If we apply these terms to a simple astronomical example of a scientist using a telescope to observe an object, then the activity would correspond to the process of generating the data—the act of observing, the telescope and the data produced by the astronomer could be considered as entities, and the astronomer himself is the agent. Using these core concepts, the PROV definition for provenance is:

> Provenance is information about entities, activities, and people involved in producing a piece of data or thing, which can be used to form assessments about its quality, reliability or trustworthiness.[1]

[1] Groth and Moreau (2013).

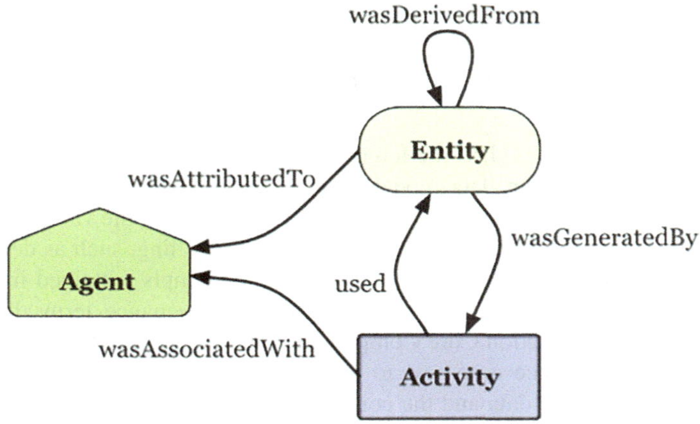

Fig. 13.1 PROV data model (from https://www.w3.org/TR/prov-primer/)

Another key component within PROV are the relations which exist between the three aforementioned kinds of objects and these may describe things such as: whether an entity was used or generated by an activity, which entities are derivations of others, or which entities or activities are attributed to or associated with an agent. Figure 13.1 shows an example diagram of the PROV data model containing all of the aforementioned components and relations.

PROV also specifies a number of mappings for serialisations from the PROV data model.[2] The native format is PROV-N, a text based file format for provenance documentation in a format which is the most closely related serialisation to the PROV Data Model.[3] Other mappings include PROV-XML[4] for serialisation into the Extensible Markup Language (XML) and PROV-O[5] for serialisation into linked data such as RDF, Turtle, or JSON-LD.

The PROV data model is intentionally very broad and welcomes extensions from specific fields to adapt PROV to their requirements. One such extension, aimed at the scientific community is ProvOne (Cao et al. 2016). Defined in 2015, the ProvOne extension was developed to provide a standard provenance representation for the complex computational pipelines which are becoming evermore prevalent within the scientific community.[6] Another example, targeted towards astronomers is the provenance model developed by the IVOA.

---

[2] https://www.w3.org/TR/prov-dm/.

[3] https://www.w3.org/TR/prov-n/.

[4] https://www.w3.org/TR/prov-xml/.

[5] https://www.w3.org/TR/prov-o/.

[6] http://jenkins-1.dataone.org/jenkins/view/Documentation%20Projects/job/ProvONE-Documentation-trunk/ws/provenance/ProvONE/v1/provone.html.

13.2 IVOA Provenance Data Model

13.2.1 History of IVOA Provenance Effort. Science Cases

The IVOA (see Sect. 12.1) had the long term goal to integrate a Provenance class in the overall Observation data model project. The intent was to trace back the history of science data back to observations of the sky across the various phases of processing. However due to other priorities in data modelling, such as defining the Characterisation Data Model of the data, these first attempts remained incomplete and were never implemented. Despite this, a few provenance terms have been integrated in the Observation Core Components Data Model (see Sect. 12.1).

One important science case was to use the provenance information to better tackle the quality of the data and the complexity of their production. Among other science cases, there was the possibility to go back to progenitors in order to allow better reprocessing. Around 2010, some astronomy projects had obvious pressure to make some progress in standardisation of provenance. That was peculiarly the case of the CTA (Cherenkov Telescope Array) project among others. CTA makes use of special data reduction processes including simulations and hypotheses. These details should be available as part of the data products metadata. Approximately at the same time the W3C PROV specification was published (see this chapter). It was a good opportunity to restart the IVOA provenance effort by importing from PROV the three basic concepts of entity, activity, and agent with their relationships.

13.2.2 IVOA Provenance as an Extension of W3C Provenance

IVOA provenance integrates the core of W3C PROV data model in the VO standards and specification landscape. First of all VO data models have to be designed along the lines of the VODML metamodel rules. This permits the model to be reused by other data models and to import some basic ones. VO data models also have to be serialised in some VO formats such as VOTable in order that metadata consistent with the model can be read and displayed by VO tools. In addition a data model mapping syntax has been recently developed and can be integrated in VOTable for a better integration of the model relationships in the documents. Moreover metadata consistent with the model have to be accessed by VO protocols consistent with the basic rules of the "Data Access Layer Interface". The adopted data model fulfil all these requirements. It is consistent with the VO-DML scheme, a VOTable serialisation has been developed and specific VO protocols such as ProvSAP and ProvTAP have been proposed. However PROV-JSON specific serialisation of this extension have also been experimented and visualisation using VOProv, an extension of the prov package is also used by CTA project.

In addition to these design constraints, The IVOA provenance has proposed several extensions to the core model.

First of all, in order to be able to tackle the prospective (see this chapter) we should be able to describe activities and entities in advance from actual processing.

Many entities can be grouped in families sharing the same specifications, such as the data product type, the level of reduction, the metadata content, etc. And many activities make use of the same software or the same instrument and differ only by the time of execution and the actual used and generated entities. In these conditions they will share metadata called "Entity Description" and "Activity Description". These descriptions are managed by specific classes.

In addition, activities not only make use of entities which exists independently from them but also of some configuration or input parameters which are fully part of the activity but are useful to be recorded in the provenance metadata. This is realised in the model via specific configuration and parameter classes.

The IVOA provenance data model is already used in the CTA pipeline to record both processing metadata and workflows. Provenance of a dedicated product can be retrieved using ProvSAP. It is also used in VizieR to record the origin and curation of the astronomical catalogs. Prototype implementation of ProvTAP exist at CDS, tracing the history of some HiPS.

13.2.3 IVOA Provenance for Radio Data

Various radio astronomy projects such as JIVE, and Astron plan to integrate provenance information in their services in order to trace back to original data and to store the processing in a standard way. Discussions are also going on within the scope of the SKA SRC network to integrate IVOA provenance metadata in order to manage workflow information back to the observation stage. The processing steps have to be traced and this information is to be kept, but the progenitors of the Observation data products will not be stored and some observations will have to be reproduced if the processing is to be modified. The size of SKA data (see Sect. 3.9) will be a challenge for the IVOA provenance data model.

13.3 Provenance Generation and Collection

When collecting provenance, a prime decision is which data should be collected. It needs to find the right balance between required information and the disc space needed to store it. Two main factors drive this decision. First, there are the use cases that have to be fulfilled. Each of them results in a certain set of requirements—data that needs to be collected in order to achieve the corresponding goal. If the use case is to document authorship, then all information needed are the contributors of datasets. If the use case is full (computational) reproducibility,[7] though, far more information needs to be gathered including data used, scripts applied,[8] and sometimes even the environment those scripts are run in. We will discuss more possible use cases later in this chapter (cf. Sect. 13.4), but these previous examples already hint to the wide range of provenance information.

The second factor is given by the granularity of provenance. Generally, a distinction is made between data provenance and workflow or process provenance (Tan 2009). Data provenance is concerned with the history of individual items or values. It is often studied for (relational) databases and follows the history of entries across a series of queries. The aforementioned concepts of why-, how-, and where- (Buneman et al. 2001; Cheney et al. 2009) have been developed and formalised in context of SQL-queries. Here, provenance information is often not materialised but computed on-the-fly by analysing the applied operations.

On the other hand, workflow or process provenance is concerned with the provenance of entire datasets. It documents the history of data products that are generated either though a dedicated workflow system such as Kepler[9] (Altintas et al. 2004), running individual scripts, or any other means. Contrary to data provenance, provenance records are almost always materialised and stored either as part of the resulting file (e.g., the processing history headers in PSRFITS (Hotan et al. 2004)) or in separate (sets of) files.

Scripting languages remain a popular choice in science both as standalone scripts or as part of Jupyter notebooks (Thomas et al. 2016). Unlike workflow systems, these environments do not provide a built-in capture of provenance information. To document the processing history nonetheless, different approaches have been proposed over time. They rely either on manual annotations within the source code or on monitoring the code execution similar to debugging or profiler tools.

One example for an annotation-based approach is YesWorkflow (McPhillips et al. 2015). As shown in Listing 13.1, the source code is structured into several nested blocks. Each block contains a computational step, e.g., a function, and is annotated with parameters, inputs, and outputs. YesWorkflow is now able to extract the relations among individual code blocks and can discover the corresponding data flow within the script. The extracted information can now be used to determine dependencies for each result of the script. It can also be visualised as shown in Fig. 13.2 to allow for a more human-accessible inspection.

[7] This use case is especially important when considering the current reproducibility crisis in many areas of science (Peng 2015; Hutson 2018; Hatton and van Genuchten 2019).

[8] Scripts by themselves are also subject to changes over time. So, it is not enough to point to a certain script from a broader collection but the used version has to be documented as well. This ties the provenance of datasets to the provenance of scripts and allows to answer question beyond the scope of individual datasets.

[9] https://kepler-project.org/.

Listing 13.1 YesWorkflow: Example script (from https://github.com/yesworkflow-org/yw-prototypes/). Annotations for YesWorkflow are highlighted in green and had to be added by the script authors. Individual blocks of code are defined using `@BEGIN` and `@END` and contain further information like parameters used in that block as well as inputs and outputs thereof. What constitutes a logical block is up to the respective authors but should at least include user-defined functions and similar structures

```
import netCDF4
import numpy as np
from netCDF4 import ma
import matplotlib.pyplot as plt
from matplotlib.backends.backend_pdf import PdfPages

# @BEGIN main
# @PARAM db_pth
# @PARAM fmodel
# @IN input_mask_file  @URI file:{db_pth}/land_water_mask/LandWaterMask_Global_CRUNCEP.nc
# @IN input_data_file  @URI file:{db_pth}/NEE_first_year.nc
# @OUT result_NEE_pdf  @URI file:result_NEE.pdf

def main(db_pth = '.', fmodel = 'clm'):

    # @BEGIN fetch_mask
    # @PARAM db_pth
    # @IN g  @AS input_mask_file  @URI file:{db_pth}/land_water_mask/LandWaterMask_Global_CRUNCEP.nc
    # @OUT mask  @AS land_water_mask
    g = netCDF4.Dataset(db_pth+'/land_water_mask/LandWaterMask_Global_CRUNCEP.nc', 'r')
    mask = g.variables['land_water_mask']
    mask = mask[:].swapaxes(0,1)
    # @END fetch_mask

    # @BEGIN load_data
    # @PARAM db_pth
    # @IN input_data_file  @URI file:{db_pth}/NEE_first_year.nc
    # @OUT data  @AS NEE_data
    f = netCDF4.Dataset(db_pth+'/NEE_first_year.nc', 'r')
    data = f.variables['NEE']
    data = data[:]
    data = data.swapaxes(0,2)
    adj = 60*60*24*(365/12)*1000
    data = data*adj
    # @END load_data

    # @BEGIN standardize_with_mask
    # @IN data @AS NEE_data
    # @IN mask @AS land_water_mask
    # @OUT data @AS standardized_NEE_data
    native = data.mean(2)
    latShape = mask.shape[0]
    logShape = mask.shape[1]
    for x in range(latShape):
        for y in range(logShape):
            if mask[x,y] == 1 and ma.getmask(native[x,y]) == 1:
                for index in range(data.shape[2]):
                    data[x,y,index] = 0
    # @END standardize_with_mask

    # @BEGIN simple_diagnose
    # @PARAM fmodel
    # @IN data @AS standardized_NEE_data
    # @OUT pp  @AS result_NEE_pdf  @URI file:result_NEE.pdf
    plt.imshow(np.mean(data,2))
    plt.xlabel("Mean␣1982-2010␣NEE␣[gC/m2/mon]")
    plt.title(fmodel + ":BGI")
    pp = PdfPages('result_NEE.pdf')
    pp.savefig()
    pp.close()
    # @END simple_diagnose

# @END main
```

Another approach, that does not require annotations, is noWorkflow (Murta et al. 2014; Pimentel et al. 2016).[10] Instead, it wraps the script during execution and is thus able to trace all operations within at runtime. By default, noWorkflow captures

[10] Not-only workflow.

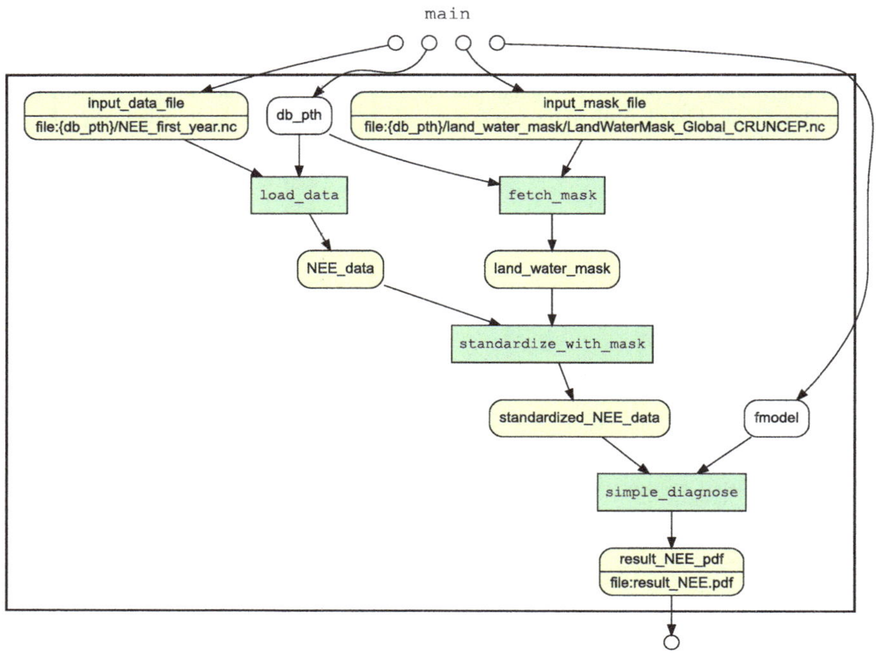

Fig. 13.2 YesWorkflow: Dataflow of the example script of Listing 13.1. Green rectangles represent the previously annotated code blocks. Parameters, inputs, and outputs are shown as white and yellow ovals. Arrows indicate the dataflow between the code blocks and thus their respective dependencies

all function calls including the current values for global variables, parameters, and the corresponding return values. This even extends to files that are accessed. Here, a copy of these files is kept alongside all other information gathered. Finally, noWorkflow collects information about the execution environment like Python version used, packages installed, or the operating system. The collected information can be visualised in two ways: A dataflow (cf. Fig. 13.3) highlights the dependencies of different functions and how they contributed to the final output. An activation graph (cf. Fig. 13.4), on the other hand, allows to profile the execution by showing the number of invocations and how much time is spent in each function.

By maintaining the provenance for different runs of a script, so called trials, noWorkflow can also follow the development of scripts similar to version control systems. Developers can switch between trials and continue working on their code. As noWorkflow keeps information about the dependencies here, it can also establish the relation across different trials as shown in Fig. 13.5.

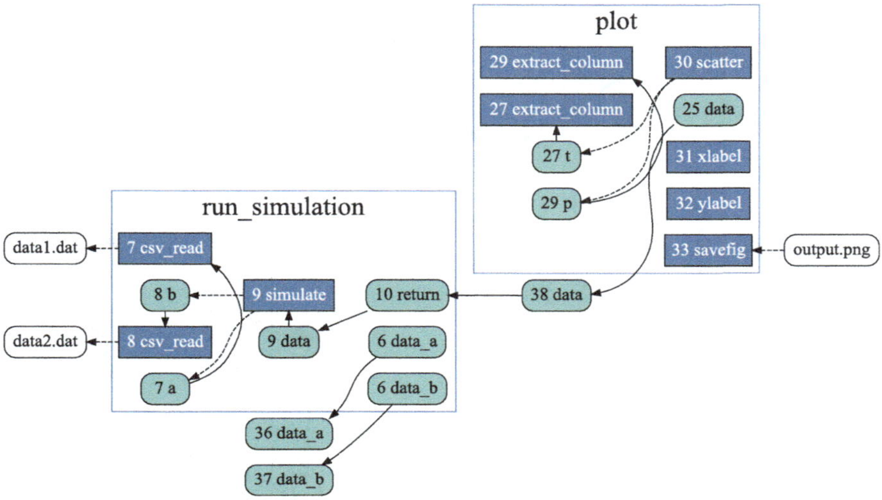

Fig. 13.3 noWorkflow: Dataflow (from noWorkflow demo project: https://github.com/gems-uff/noworkflow). Two functions (`run_simulation` and `plot`) are called directly from the main script and are shown as frames. Further function invocations from within those are displayed as rectangles with a number indication the line of code where the invocation happened. Variables passed between the functions are shown in ovals where the white ones are inputs and outputs to the entire script while the turquoise ones are internal to the script itself. Arrows point in the direction of data being created by an invocation or an invocation making use of a variable. Similar to Fig. 13.2, this visualises the dataflow and thus dependencies between different parts of the underlying script

13.4 Using Provenance in Radio Astronomy and Other Domains

In modern astronomy, it is not just data size that is big. The pipelines too are like complex automatically executed black boxes, and their level of complexity and automation will only increase in the future. This paradigm shift brings along a real need to collect information about the course of a pipeline run. As the whole process consists of many steps with different degrees of control by the user, data provenance, among others, can be used to ensure trust in scientific data products, reproducibility of a pipeline run or enable identifying outliers and their causes. In some other experimental fields, such as biology or chemistry, experiments are equivalent to pipeline runs in astronomy and records of these are a source of provenance data. It can well be that the process under consideration is a combination of both manual experiments and pipeline runs with the use of complex, heterogeneous data. Therefore, provenance data, if recorded, might serve as a tool to address a selection of scenarios that (radio) astronomy and other disciplines that are focused on data products resulting from pipelines and/or experiments have in common.

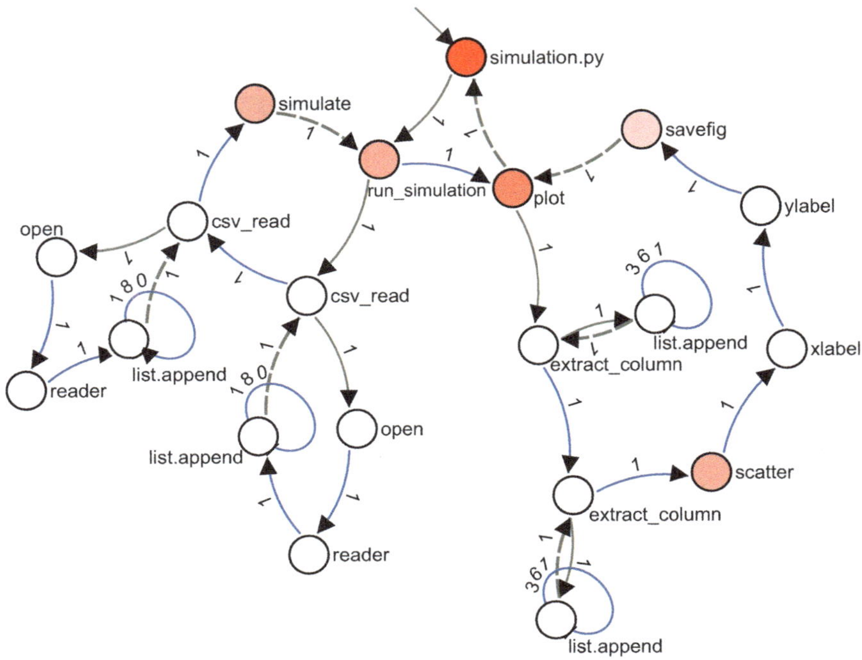

Fig. 13.4 noWorkflow: Details of trial execution (from noWorkflow demo project: https://github.com/gems-uff/noworkflow). Where Fig. 13.3 showed a high-level overview of a script dependencies, this figure visualises the details of a particular script run. It uses the same script (`simulate.py`) making the same direct and indirect function invocations as before (i.e., `run_simulation`, `plot`, `simulate`, etc.). However contrary to Fig. 13.3, this only visualises the respective function invocations and omits any dataflow. However, invocations also occur in loops as shown, e.g., with `list_append` being called 180 and 361-times by `reader` and `extract_column` respectively

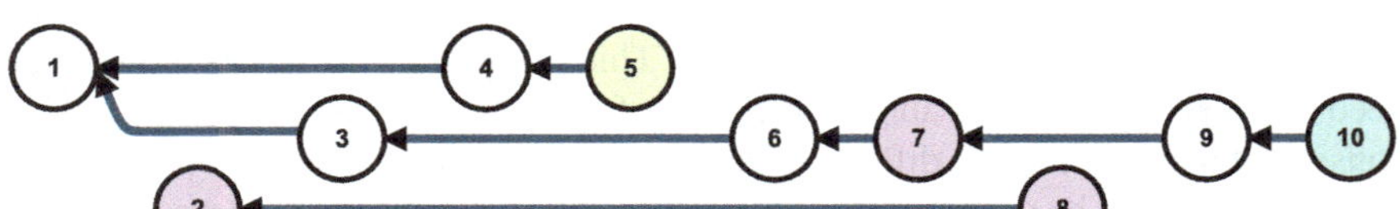

Fig. 13.5 noWorkflow: Trial history (from noWorkflow demo project: https://github.com/gems-uff/noworkflow). Similar to version control, noWorkflow allows to track the history of script development by connecting the provenance of different script runs. In the figure, each script run is given a sequential number. Script authors can switch between different version of their script and continue their work. Dependencies are shown in the figure by arrows pointing to the logical predecessor of a script run. This allows to derive the development history of the development. For example, from an initial version 1 two separate variants (3 and 4) were derived and independently developed in different branches

13.4.1 Reproducibility

One of the key aspects in scientific domains based on experiments and data analysis is establishing trust in scientific outcomes of these processes. Ideally, a re-run of a pipeline under the same conditions should bring the same result, ensuring reproducibility of data processing. It also applies to repeating an experiment. If that is not the case, data provenance can be used to compare these two occurrences and determine the source of the difference. Above all, however, the existence of records of these processes makes it possible to reproduce the conditions for repetition at all. Re-execution of a pipeline might be necessary to confirm repeatability of results. One of examples in radio astronomy showing practical use of data provenance in reproducibility and establishing trust in scientific results is to verify the cause of prior non-detection in case of phenomena that we know might be repetitive, such as fast radio bursts (FRBs, see Sect. 9.3.2) (Johnson et al. 2021). Such an approach enables the user to test their hypothesis with regard to the nature of a phenomenon under study. It also addresses the aspect of identifying and then handling technical failures on hardware or software side which previously went unnoticed but could have had impact on more than the one scientific product of the pipeline.

13.4.2 Impact of Errors

In experimental science, repeating an experiment and comparing the provenance records of the two occurrences in case of different results can be an invaluable source of information about the cause of the discrepancies. Again, provenance records facilitate better understanding fundamental natural phenomena. However, data provenance also makes it possible to estimate the impact of potential errors, if any, in the interpretation of scientific products and their further processing on the given domain. Miles et al. (2007) consider a scenario where an experiment is performed twice with the use of a sequence of data downloaded from a genetic database, with the process of data acquisition being repeated before the second experiment. Here the results are different and data provenance can be used to determine the cause, such as the experimental process itself, change of its configuration, change of the data or a combination of these. Similarly, by identifying an issue affecting one element of a pipeline one can draw a connection to all affected data products using provenance records. Further, this connection can be extended to indicate other pipelines making use of the element or the data products—now and in the past. In astronomy, detecting a faulty software version in a pipeline or another element, such as consumed data products (e.g. ephemeris files, calibration files) used as baseline in processing/analysis, would require identifying all results needing recomputation and affected output data products requiring inspection.

13.4.3 Outliers

Any unusual or unexpected result, whether it is desired and brings a novel scientific discovery or highly concerning and ends with a failure of a known process, requires a deeper insight by the scientist. Historical records of successful occurrences, such as past pipeline runs or experiment series, can be used to find the origins of the observed anomaly. An outlier is usually perceived as an outstanding data point in the sense of its value (see Sect. 10.6.2). Looking broader at the topic of anomalies, provenance records allow the user to go beyond value-based comparison. Instead, anomalies are considered at the level of process outcomes, their components and structure, and the relationship between these elements. In astronomy, a scientist might be interested in investigating the cause of different results of separate runs of the same pipeline, executed in other runtime environments or infrastructures. In chemistry, a laboratory technician might need to investigate obtaining unsatisfactory results of an experiment conducted according to a known procedure after receiving a new batch of one of the reagents. Unlike in case of reproducibility under identical conditions, in both scenarios only some elements are constant but it is still expected that the end product be the same. Any process failure that prevents its completion is also a potential use case for anomaly detection just as a completely surprising result, suggesting a possible scientific discovery. Interrupted events bring knowledge about the weaknesses of the process and what are the open points on the way to optimise it. Anomalies bearing the hallmarks of scientific discovery, on the other hand, require very careful inspection to distinguish them from interference or error. Provenance records can be especially useful to address these cases as they are beyond the scope of anomaly detection in a numerical sense. While the change or a known unusual factor can be the origin of the observed outlier, it is impossible to verify this hypothesis without a detailed comparison between elements of the processes. Moreover, if in doubt whether or not a given occurrence is unusual, the provenance records can be directly used to establish a baseline defining further constraints for what is and is not an anomaly in the first place.

13.4.4 Predictions

Repetitions are an integral part of the daily routine of scientific work, whether in radio astronomy or other disciplines. During multiple (re-)runs of a given procedure, whether we are dealing with the same pipeline or its modified version, new intermediate and final data products are created. Evaluating these outputs and comparing them with each other allows the users to gather experience and knowledge about a given process. It brings understanding on how a pipeline works and what impact the changes applied to it may have on its performance. Such a methodology requires (a) replicating the pipeline run repeatedly, even without its modification or after making changes whose impact on the results may be potentially negligible, (b) storing lots of similar or identical final data products, and collecting intermediate data products that may be otherwise completely redundant. These aspects are of great importance in the era of big data, when storing big data volumes is not a sustainable strategy, and repeated processing of these requires the use of resources whose impact is not indifferent to the environment and generates high costs. In the case of processes that generate physical artefacts, there is also the aspect of consumed materials, subsequent storage or disposal. Provenance records of pipeline runs can be used to predict the performance of a same, changed or similar pipeline run in the future. Data provenance can also be used to estimate only selected metrics of unknown pipeline processing. This includes domain-independent queries, such as expected run time or minimum resources required to successfully complete the process, along with more advanced metrics focusing on the needs of a particular scientific discipline. The latter group of queries requires supplementing provenance records with information on the data quality and defining the metrics framework focused to address issues that are relevant to the field. Each domain has their own metrics and while for some e.g. accuracy is of high value, for others it might have only a minor impact on the end result. Therefore, it is beneficial to identify the metrics before the provenance records are collected. One important aspect in (radio)astronomy and other e-Science domains is storing intermediate data products. Here provenance records can be used to compare storage versus recomputation analysis in terms of deciding which intermediate data products to keep for later (Johnson et al. 2021). This use case is similar in experimental science and intermediate products of experiments where making such a decision is even more crucial in the sense of reproducibility.

13.4.5 Recommendations

Each pipeline consists of a set of components and requires specific parameter values to execute. The choice of these elements determines the success of the pipeline execution and the quality of the results. It is highly recommended to learn from past runs to avoid failures and ensure optimal parameter values selection. It can be achieved by combining domain knowledge and experience enriched with provenance records of previous executions. Such recommendations are especially valuable for new in the domain or complete non-expert scientists to enable them to create a pipeline or its components, apply changes in an existing pipeline or decide on a sensible parameter configuration. In this case, due to inexperience or insufficient depth of knowledge, the challenge of a scientific cold start is particularly pronounced. As long as examples of execution or recommended values are present in provenance records, they can be used for parameter value recommendation without having to re-execute pipelines (Silva et al. 2021). An extensive collection of historical entries on multiple pipeline runs, their elements and versions thereof might serve as a foundation for recommendation of components of a pipeline, including data files (Huang et al. 2013). Finally, provenance records can be also used to identify pipelines similar or related to each other (Alawini et al. 2018; Johnson et al. 2021). It is also applicable for such cases as discovering the origin of a particular dataset, indicating a baseline pipeline for the one under consideration or evaluating the impact of the changes introduced in the pipeline on the state of the art.

13.4.6 Provenance Visualisation and Knowledge Graphs

As described above, usability of data provenance is potentially broad in scientific workflow analysis, which is a fundamental component of (radio) astronomy and scientific experimentation. However, in the era of big data the provenance records can be complex and of large-scale themselves. To achieve maximum understanding of them and enable interoperability across a variety of affiliations and domains, a user-friendly approach of presenting the provenance information is necessary. Provenance data can be easily modelled as a graph, visualising selected perspectives on a given pipeline execution. These might be relationships between pipeline components or data products, depicting data lineage, starting with the input data, through intermediate data products and ending with the final data output products. Producing such graphs for visual inspection might serve as a basis for comparison between multiple runs of the same pipeline, runs of pipelines that are expected to be similar (e.g. two different pulsar timing pipelines) or cross-domain structural comparison between applications having something in common (e.g. Groth et al. (2009) proposes a pipeline-centric for a class of deterministic applications, using astronomy as an example). It can also be useful for first inspection of a pipeline and an overview of its components for scientists who are working with a new pipeline without full understanding of its structure. Considering the complexity and automatisation of the pipelines, such an analysis might soon be a standard step for each domain expert starting at their new affiliation, not only users new to the field.

Provenance data can be visualised using scientific workflow management systems with built-in provenance visualisation, such as Kepler (Altintas et al. 2004), Apache Taverna (Hull et al. 2006) or VisTrails (Callahan et al. 2006), which have been used in astronomy, too. Even though they offer certain benefits in the context of provenance analysis, their usability for domain specialists might be limited and some of them have not been updated for a long time or completely not maintained anymore. As an alternative, there exist standalone provenance visualisation tools, often dedicated for specific purposes and compatible with the PROV, for example Provenance Viewer (Kohwalter et al. 2016) and PROV-O-Viz (Hoekstra and Groth 2014). Relationships between PROV types expressed in PROV-N can also be rendered as a diagram using Graphviz.[11]

[11] http://www.graphviz.org/.

Enriched provenance records with other (meta)data, stored in knowledge graphs (see Fig. 14.1 for an example of a knowledge graph in astronomy) represented as an ontology open doors to making (radio)astronomical and other domain knowledge machine-actionable, i.e. structured in such a way that they can be interpreted by machines unambiguously. A combination of a natural representation of provenance records as a graph and the possible (although time-consuming at the very beginning of the process) transformation of tabular (meta)data to knowledge graphs enable mapping between these two groups of data. But even provenance data alone can be used to extract knowledge performing graph queries assuming that the provenance records were previously stored in a database, such as a triple (RDF) store (using SPARQL[12]) or a Neo4j[13] database (using Cypher[14]). Bringing knowledge graphs into play allows to incorporate methods of data mining such as clustering which are otherwise not trivial to apply using other approaches. It also enables the users to make their provenance records artificial intelligence ready in the context data access.

13.5 Implementations of Provenance in the Context of Existing and Future Astronomical Observatories

Provenance is an emerging field within radio astronomy. Whilst there are examples of provenance models being purpose built for radio astronomical applications (see de Dios Santander-Vela et al., 2012), there are few examples of provenance recording implementations. Therefore, in order to learn from past implementations, we must look for those in other areas of the electromagnetic spectrum. As these other wavelengths are also entering the era of big data, they share the same issues for the future of provenance recording. Perhaps the most pertinent of which are scalability, understandability, and interoperability.

13.5.1 Scalability

Scalability is a challenge within many aspects of radio astronomy and provenance collection is no different. As radio observatories are capable of collecting some of the large datasets in astronomy, with current observatories like LOFAR creating final datasets of the order 50 PB large and future observatories such as the SKA expected to generate hundreds of petaBytes of reduced data each year.

[12] https://www.w3.org/TR/rdf-sparql-query/.

[13] https://github.com/neo4j/neo4j.

[14] https://neo4j.com/developer/cypher/.

A partial solution to the scalability problem comes in the form of virtual data, i.e. instead of storing the data, you store the recipe for the data and enable the generation of it on command. This concept was explored by Cohen et al. 2006 where they presented Chimera, a virtual data system that facilitated the tracing of provenance for data products and their on demand regeneration of them. In Chimera, pipelines are composed of: transformations, which refer to executable programs; derivations, that denote the execution of a transformation; and data objects, which are entities that may be consumed/produced by executions. Once the pipeline has been described using these building blocks, the Chimera virtual data language may be used to retrieve any data item which already exists. Chimera was used to describe the data processing of a pipeline which analysed data from the Sloan Digital Sky Survey (SDSS), an optical survey telescope which began operations in the year 2000 (York et al. 2000). The goal of the pipeline was to create a database of Galaxy clusters from all of the SDSS data. Chimera was successfully used to describe this system, retrieve target data products, and determine all data products that would need to be recomputed if an error was found at a specific point in the pipeline.

The concept of virtual data was further developed by Groth and Moreau (2013), where they presented a pipeline centric provenance model whereby the pipeline is modelled as a series of functions and only the information necessary to re-run the pipeline is stored—the initial input data, parameter settings and versions for each pipeline component, the execution environment, etc. Provenance modelled in this way also enables data products to be treated as virtual data. The combined concepts of pipeline centric provenance and virtual data can be seen within the provenance system designed for the Vera C. Rubin observatory (Juric and Tyson 2012) which is based upon the pipeline centric model and takes advantage of the possibility of virtual data for better allocation of computational resources.

13.5.2 Understandability

In addition to adding challenges for scalability, the large data volumes and complex computational pipelines present within radio astronomy can obfuscate the operation of the data analysis. Visualisation and exploration of the provenance of a system can enlighten the scientists on the data analysis, increasing trust in the results. VisTrails (Callahan et al. 2006) is an example infrastructure for the visualisation and dissemination of provenance. It has a focus on usability, demonstrating the flow of data, and was used as an effective tool for teaching scientific visualisation to students (Silva et al. 2011).

Though provenance visualisation can help with understanding computational processes, there is a limit to the amount of information that can be meaningfully displayed. Radio observatories often have complex, dedicated, data processing pipelines such as the e-Merlin Data Reduction Pipeline for the e-Merlin Correlator (Argo 2015) or the CoastGuard pipeline for reducing data from the Effelsberg Observatory (Lazarus et al. 2020). Data processing pipelines of future observatories are going to be more complex still, a prime example being the SKA Science Analysis Pipeline (Hollitt et al. 2016). Similar to the solution for scalability, abstraction of the provenance data can aid in scaling visualisation for datasets that describe radio astronomy pipelines. The summarisation of provenance data was explored by (Moreau 2015), in the paper the summarisation was created by aggregating provenance data by their provenance types.

13.5.3 Interoperability

The European Pulsar Timing Array (EPTA) is a combination of five 100 m class radio telescopes and as such, epitomises the importance of interoperable provenance data. The infrastructure and model for provenance generation within the EPTA is currently being developed under name of prov-PRAETOR (Pipeline pRovenance for Analysis Evaluation Trust Or Reproducibility) and is being designed to accommodate the use cases outlined in Sect. 13.4 (Johnson et al. 2021).

Interoperability is also essential for projects such as the Event Horizon Telescope (EHT) which was responsible for imaging Sagittarius A*. EHT was able to do this by combining observations from several very-long-baseline interferometry station around Earth, this array synthesised an aperture with sufficient angular resolution to observe the event horizon of the super massive black hole (Event Horizon Telescope Collaboration et al. 2019).

Within examples such as the EPTA and EHT where provenance from distinct observatories are required for a single purpose, these provenance datasets must be interoperable. Within this regime, a potential solution would be to use a single, multipurpose provenance system such as Chimera, to produce all provenance. However, the growing tendency is for bespoke provenance systems for each system, as is the case for the Vera C. Rubin observatory and the Cherenkov Telescope Array. Regardless of whether the implementation is bespoke or otherwise, one solution to ensure interoperability is for the provenance data to adhere to the FAIR principles.

References

A. Alawini, L. Chen, S. Davidson, S. Fisher, J. Kim, Discovering similar workflows via provenance clustering: A case study, in *Provenance and Annotation of Data and Processes - 7th International Provenance and Annotation Workshop, IPAW 2018, Proceedings*, Germany, ed. by K. Belhajjame, A. Gehani, P. Alper, Lecture Notes in Computer Science (including subseries Lecture Notes in Artificial Intelligence and Lecture Notes in Bioinformatics) (Springer, 2018), pp. 115–127. ISBN 9783319983783. https://doi.org/10.1007/978-3-319-98379-0_9. Funding Information: J. Kim—Evolutionary and Molecular Biology (Kim) Lab, University of Pennsylvania. Publisher Copyright: © Springer Nature Switzerland AG 2018.; 7th International Provenance and Annotation Workshop, IPAW 2018; Conference date: 09-07-2018 Through 10-07-2018

I. Altintas, C. Berkley, E. Jaeger, M. Jones, B. Ludascher, S. Mock, Kepler: an extensible system for design and execution of scientific workflows, in *16th International Conference on Scientific*

and Statistical Database Management, 2004. Proceedings, June (2004), pp. 423–424. https://doi.org/10.1109/SSDM.2004.1311241
T. An, Science opportunities and challenges associated with SKA big data. Sci. China Phys. Mech. Astron. **62**(8), 1–6 (2019)
M. Argo, The e-merlin data reduction pipeline. arXiv preprint arXiv:1502.04936 (2015)
P. Buneman, S. Khanna, W.C. Tan, Why and where: A characterization of data provenance, in *ICDT*, ed. by J.V. den Bussche, V. Vianu, volume 1973 of Lecture Notes in Computer Science (Springer, 2001), pp. 316–330. ISBN 3-540-41456-8
S.P. Callahan, J. Freire, E. Santos, C.E. Scheidegger, C.T. Silva, H.T. Vo, Vistrails: visualization meets data management, in *Proceedings of the 2006 ACM SIGMOD International Conference on Management of Data* (2006), pp. 745–747
Y. Cao, C. Jones, V. Cuevas-Vicenttin, M.B. Jones, B. Ludäscher, T. McPhillips, P. Missier, C. Schwalm, P. Slaughter, D. Vieglais, L. Walker, Y. Wei, Provone: extending prov to support the dataone scientific community, in *PROV: Three Years Later* (2016)
J. Cheney, L. Chiticariu, W.-C. Tan, *Provenance in Databases: Why, How, and Where* (Now Publishers, 2009)
A. Chrysostomou, C. Taljaard, R. Bolton, L. Ball, S. Breen, A. van Zyl, Operating the square kilometre array: the world's most data intensive telescope, in *Observatory Operations: Strategies, Processes, and Systems VIII*, volume 11449 of Society of Photo-Optical Instrumentation Engineers (SPIE) Conference Series (2020, December), p. 114490X. https://doi.org/10.1117/12.2562120
S. Cohen, S. Cohen-Boulakia, S. Davidson, Towards a model of provenance and user views in scientific workflows, in *International Workshop on Data Integration in the Life Sciences* (Springer, 2006), pp. 264–279
Event Horizon Telescope Collaboration et al., First m87 event horizon telescope results. IV. Imaging the central supermassive black hole. Preprint. arXiv:1906.11241 (2019)
J. de Dios Santander-Vela, E. García, S. Leon, V. Espigares, J.E. Ruiz, L. Verdes-Montenegro, E. Solano, Building a VO-compliant radio astronomical data model for single-dish radio telescopes (RADAMS). Exp. Astron. **34**(3), 623–652 (2012)
P. Groth, L. Moreau, PROV-overview. An overview of the PROV family of documents. World Wide Web Consortium (2013). http://www.w3.org/TR/2013/NOTE-prov-overview-20130430/
P. Groth, E. Deelman, G. Juve, G. Mehta, G.B. Berriman, Pipeline-centric provenance model, in *Proceedings of the 4th Workshop on Workflows in Support of Large-Scale Science, WORKS 2009, November 16, 2009, Portland, Oregon, USA*, ed. by E. Deelman, I.J. Taylor (ACM, 2009). https://doi.org/10.1145/1645164.1645168
L. Hatton, M. van Genuchten, Computational reproducibility: The elephant in the room. IEEE Softw. **36**(2), 137–144 (2019). https://doi.org/10.1109/ms.2018.2883805
R. Hoekstra, P. Groth, Prov-o-viz - understanding the role of activities in provenance, in *Provenance and Annotation of Data and Processes - 5th International Provenance and Annotation Workshop, IPAW 2014, Cologne, Germany, June 9–13, 2014. Revised Selected Papers*, ed. by B. Ludäscher, B. Plale, volume 8628 of Lecture Notes in Computer Science (Springer, 2014), pp. 215–220. https://doi.org/10.1007/978-3-319-16462-5_18
C. Hollitt, M. Johnston-Hollitt, S. Dehghan, M. Frean, T. Bulter-Yeoman, An overview of the SKA science analysis pipeline. Preprint arXiv:1601.04113 (2016)
A.W. Hotan, W. van Straten, R.N. Manchester, PSRCHIVE and PSRFITS: An open approach to radio pulsar data storage and analysis. Publ. Astron. Soc. Australia **21**(3), 302–309 (2004). https://doi.org/10.1071/as04022
X. Huang, T. Lu, X. Ding, N. Gu, Enabling data recommendation in scientific workflow based on provenance, in *2013 8th ChinaGrid Annual Conference* (2013), pp. 117–122
D. Hull, K. Wolstencroft, R. Stevens, C.A. Goble, M.R. Pocock, P. Li, T. Oinn, Taverna: a tool for building and running workflows of services. Nucleic Acids Res. **34**(Web-Server-Issue), 729–732 (2006). https://doi.org/10.1093/nar/gkl320
M. Hutson, Artificial intelligence faces reproducibility crisis. Science **359**(6377), 725–726 (2018). https://doi.org/10.1126/science.359.6377.725

M.A.C. Johnson, M. Paradies, M. Dembska, K. Lackeos, H.-R. Klöckner, D.J. Champion, S. Schindler, Astronomical pipeline provenance: A use case evaluation, in *13th International Workshop on Theory and Practice of Provenance (TaPP 2021)*, July (USENIX Association, 2021). https://www.usenix.org/conference/tapp2021/presentation/johnson
M. Juric, T. Tyson, LSST data management: Entering the era of petascale optical astronomy. Proc. Int. Astron. Union **10**(H16), 675–676 (2012)
T.C. Kohwalter, T.N. de Oliveira, J. Freire, E. Clua, L. Murta, PROV viewer: A graph-based visualization tool for interactive exploration of provenance data, in *Provenance and Annotation of Data and Processes - 6th International Provenance and Annotation Workshop, IPAW 2016, McLean, VA, USA, June 7-8, 2016, Proceedings*, ed. by M. Mattoso, B. Glavic, volume 9672 of Lecture Notes in Computer Science (Springer, 2016), pp. 71–82. https://doi.org/10.1007/978-3-319-40593-3_6
P. Lazarus, R. Karuppusamy, E. Graikou, R.N. Caballero, D.J. Champion, K.J. Lee, J.P.W. Verbiest, M. Kramer, Coastguard: Automated timing data reduction pipeline. Astrophysics Source Code Library, pages ascl–2003 (2020)
C. Lim, S. Lu, A. Chebotko, F. Fotouhi, Prospective and retrospective provenance collection in scientific workflow environments, in *2010 IEEE International Conference on Services Computing* (IEEE, 2010), pp. 449–456
T. McPhillips, T. Song, T. Kolisnik, S. Aulenbach, K. Belhajjame, K. Bocinsky, Y. Cao, F. Chirigati, S. Dey, J. Freire, D. Huntzinger, C. Jones, D. Koop, P. Missier, M. Schildhauer, C. Schwalm, Y. Wei, J. Cheney, M. Bieda, B. Ludäscher, YesWorkflow: a user-oriented, language-independent tool for recovering workflow information from scripts. Int. J. Digit. Curation **10**(1), 298–313 (2015). https://doi.org/10.2218/ijdc.v10i1.370
S. Miles, P. Groth, M. Branco, L. Moreau, The requirements of using provenance in e-science experiments. J. Grid Comput. **5**(1), 1–25 (2007). https://doi.org/10.1007/s10723-006-9055-3
L. Moreau, Aggregation by provenance types: A technique for summarising provenance graphs. Preprint. arXiv:1504.02616 (2015)
L. Murta, V. Braganholo, F. Chirigati, D. Koop, J. Freire, noworkflow: capturing and analyzing provenance of scripts, in *International Provenance and Annotation Workshop* (Springer, 2014), pp. 71–83. https://doi.org/10.1007/978-3-319-16462-5_6
R. Peng, The reproducibility crisis in science: A statistical counterattack. Significance **12**(3), 30–32 (2015). https://doi.org/10.1111/j.1740-9713.2015.00827.x
J.F. Pimentel, J. Freire, V. Braganholo, L. Murta, Tracking and analyzing the evolution of provenance from scripts, in *International Provenance and Annotation Workshop* (Springer, 2016), pp. 16–28. https://doi.org/10.1007/978-3-319-40593-3_2
C.T. Silva, E. Anderson, E. Santos, J. Freire, Using vistrails and provenance for teaching scientific visualization, in *Computer Graphics Forum*, vol. 30 (Wiley Online Library, 2011), pp. 75–84
D.J. Silva, E. Pacitti, A. Paes, D. de Oliveira, Provenance-and machine learning-based recommendation of parameter values in scientific workflows. PeerJ Comput. Sci. **7**, e606 (2021). https://doi.org/10.7717/peerj-cs.606. https://hal.archives-ouvertes.fr/hal-03418836
W.-C. Tan, Provenance, in *Encyclopedia of Database Systems* (Springer US, 2009), pp. 2202–2202. https://doi.org/10.1007/978-0-387-39940-9_283
K. Thomas, R.-K. Benjamin, P. Fernando, G. Brian, B. Matthias, F. Jonathan, K. Kyle, H. Jessica, G. Jason, C. Sylvain, I. Paul, A. Damián, A. Safia, W. Carol, Jupyter Development Team, Jupyter notebooks - a publishing format for reproducible computational workflows. Elpub **0**, 87–90 (2016). ISSN 0000-0000. https://doi.org/10.3233/978-1-61499-649-1-87
D.G. York, J. Adelman, J.E. Anderson Jr., S.F. Anderson, J. Annis, N.A. Bahcall, J.A. Bakken, R. Barkhouser, S. Bastian, E. Berman, W.N. Boroski, S. Bracker, C. Briegel, J.W. Briggs, J. Brinkmann, R. Brunner, S. Burles, L. Carey, M.A. Carr, F.J. Castander, B. Chen, P.L. Colestock, A.J. Connolly, J.H. Crocker, I. Csabai, P.C. Czarapata, J.E. Davis, M. Doi, T. Dombeck, D. Eisenstein, N. Ellman, B.R. Elms, M.L. Evans, X. Fan, G.R. Federwitz, L. Fiscelli, S. Friedman, J.A. Frieman, M. Fukugita, B. Gillespie, J.E. Gunn, V.K. Gurbani, E. de Haas, M. Haldeman, F.H. Harris, J. Hayes, T.M. Heckman, G.S. Hennessy, R.B. Hindsley, S. Holm, D.J. Holmgren, C.-H. Huang, C. Hull, D. Husby, S.-I. Ichikawa, T. Ichikawa, Ž.

Ivezić, S. Kent, R.S.J. Kim, E. Kinney, M. Klaene, A.N. Kleinman, S. Kleinman, G.R. Knapp, J. Korienek, R.G. Kron, P.Z. Kunszt, D.Q. Lamb, B. Lee, R.F. Leger, S. Limmongkol, C. Lindenmeyer, D.C. Long, C. Loomis, J. Loveday, R. Lucinio, R.H. Lupton, B. MacKinnon, E.J. Mannery, P.M. Mantsch, B. Margon, P. McGehee, T.A. McKay, A. Meiksin, A. Merelli, D.G. Monet, J.A. Munn, V.K. Narayanan, T. Nash, E. Neilsen, R. Neswold, H.J. Newberg, R.C. Nichol, T. Nicinski, M. Nonino, N. Okada, S. Okamura, J.P. Ostriker, R. Owen, A.G. Pauls, J. Peoples, R.L. Peterson, D. Petravick, J.R. Pier, A. Pope, R. Pordes, A. Prosapio, R. Rechenmacher, T.R. Quinn, G.T. Richards, M.W. Richmond, C.H. Rivetta, C.M. Rockosi, K. Ruthmansdorfer, D. Sandford, D.J. Schlegel, D.P. Schneider, M. Sekiguchi, G. Sergey, K. Shimasaku, W.A. Siegmund, S. Smee, J.A. Smith, S. Snedden, R. Stone, C. Stoughton, M.A. Strauss, C. Stubbs, M. SubbaRao, A.S. Szalay, I. Szapudi, G.P. Szokoly, A.R. Thakar, C. Tremonti, D.L. Tucker, A. Uomoto, D. Vanden Berk, M.S. Vogeley, P. Waddell, S.-I. Wang, M. Watanabe, D.H. Weinberg, B. Yanny, N. Yasuda, SDSS Collaboration, The Sloan digital sky survey: technical summary. Astron. J. **120**(3), 1579–1587 (2000). https://doi.org/10.1086/301513

Chapter 14
Data Documentation Beyond Provenance: Metadata, Research Data Management, FAIR Principles

Sirko Schindler, Marta Dembska, François Bonnarel, Mark Lacy, Michael Johnson, and Vandana Desai

Modern science is barely done in isolation. Instead, any advancement is built upon a foundation laid by innumerable other scientists in the various branches of science. In the current digital age, this collaboration among scholars relies heavily on exchanging not only ideas but also data, software, and digital services. The rather recent recognition of these artefacts as research outputs in their own right drew more attention to their lifecycle and in particular their creation, archiving, and sharing. Over time, best practises evolved and nowadays RDM is established as an integral part of the scientific process.

S. Schindler (✉) · M. Dembska
Institute of Data Science, German Aerospace Center (DLR), Jena, Germany
e-mail: sirko.schindler@dlr.de; marta.dembska@dlr.de

F. Bonnarel
CDS, Université de Strasbourg, CNRS, Observatoire astronomique de Strasbourg - UMR7550, Strasbourg, France
e-mail: francois.bonnarel@astro.unistra.fr

M. Lacy
National Radio Astronomy Observatory, Charlottesville, VA, USA
e-mail: mlacy@nrao.edu

M. Johnson
Max-Planck-Institut für Radioastronomie, Bonn, Germany

Institute of Data Science, German Aerospace Center (DLR), Jena, Germany
The University of Manchester, Manchester, UK
e-mail: michael.johnson-4@manchester.ac.uk

V. Desai
Infrared Processing & Analysis Center, California Institute of Technology, Pasadena, CA, USA
e-mail: desai@ipac.caltech.edu

E. Vardoulaki et al. (eds.), *Data-Intensive Radio Astronomy*, Astrophysics and Space Science Library 472, https://doi.org/10.1007/978-3-031-58468-8_14

Metadata, i.e. data about data, is a cornerstone of data exchange and summarises all relevant information about an artefact.[1] Its importance can be attributed to at least two main factors. First, it allows to attach additional information that is not strictly part oft the data itself but can be used to contextualise it. This can include, e.g., the means of data acquisition and processing, any measurement uncertainties, or describe the conditions of the underlying measurements. These aspects allow subsequent users to determine whether a dataset fits their current assumptions and preconditions. The second factor relates to metadata size. While the size of datasets can easily reach gigabytes and beyond, metadata usually remains rather small in comparison in the kilobyte to lower megabyte range. This impacts not only storage requirements but is especially useful for providing systems that need to access a large amount of datasets. Modern day search engines, e.g., would be virtually impossible without the reduction in size due to searching only over metadata and not the datasets themselves. Even when primary data search is done, it is usually only a second step after the set of candidate datasets has been substantially reduced by a metadata search.

As a pre-digital-era example, think of library catalogues. Instead having to read through all the books of a library to find the fact your are looking for, you first use the summaries in the catalogue to narrow down your search to a few promising books and then check only those in detail.

What exactly is considered relevant can vary across domains and use cases, though. This ambiguity has led to the development of various, sometimes competing metadata standards that emphasise different aspects relevant in a particular environment. Generally, elements of metadata standards for digital artefacts can be distinguished into the following categories although other more fine grained classifications also exist: *Descriptive* metadata is the core of all metadata and includes elements such as title, author, and a set of keywords describing the artefact. It is mainly used for identification and discovery. *Structural* metadata describe the inner structure of an artefact, e.g., the table of contents in a book or the individual columns in a table. Besides providing a more detailed description, this also allows for automated access to individual parts of a given artefact. *Administrative* metadata holds information concerning accessing the artefact. This ranges from location and format of the artefact to its licensing and possible embargo periods.[2] *Provenance* metadata, as discussed in the previous chapter, documents the history of an artefact. By examining the sources, contributors, and any modifications applied, this allows to establish the authenticity of an artefact and decide whether it is applicable for a new different purpose.

---

[1] Within this chapter "artefact" primarily refers to any scientific output. This includes papers, datasets, or code snippets. The same principles can be extended to everything that holds any importance within the scientific workflows: instruments for collecting data, authors of papers, projects, and the like.

[2] The embargo period of a digital artefact is the time after publication during which its creators have exclusive access to actual data. This allows them to publish their initial analysis before opening the artefact to the general public for further reuse.

14.1 FAIR Principles

Even the use of a common schema has historically not prevented a fragmentation of the metadata landscape. The intention of individual elements can still be interpreted quite differently and the lack of constraints leads to all kinds of heterogeneity including typographical errors. There have been different approaches to address these issues: For example, curation by librarians and other gatekeepers manually removes errors in metadata descriptions and increases the overall quality. Similarly, controlled vocabularies are introduced to constrain possible values in an effort to harmonise different descriptions. Yet despite these and other efforts, metadata descriptions varied quite substantially and automated processing especially across institutional borders remained next to impossible.

The latest attempt to organise metadata is based on the techniques of the Semantic Web (Berners-Lee et al. 2001) and is summarised in the FAIR-principles (Wilkinson et al. 2016). The Semantic Web transitions from textual representations of knowledge towards concepts that attempt to capture the very essence or meaning of things. Consequently, a single concept might have a multitude of labels to express the idea including synonyms and possibly abbreviations but also spanning across languages. Individual concepts are further linked to one another to express relationships like hierarchies (e.g., "radio waves" and "visible light" are both kinds of "electromagnetic radiation") or any other relation (e.g., "the Moon is gravitationally bound to the Earth"). From a technical point of view, the Semantic Web is founded on two main building blocks: RDF (Cyganiak et al. 2014) as a data model and serialisation as well as IRIs to uniquely identify concepts.

In general, knowledge is modelled as graphs whose nodes represent concepts and (labelled) edges denote relationships among them as illustrated in Fig. 14.1. The structure of these knowledge graphs is formalised in ontologies that define the schemata including classes to categorise concepts and the types of relations to be used. Knowledge graphs are generally domain specific, i.e. their content captures knowledge from a particular field of study. There are notable exceptions, though, including Wikidata (Vrandečić and Krötzsch 2014) and DBpedia (Lehmann et al. 2015) . They play an important role in linking knowledge graphs and provide a central nucleus to what is now called the Linked Data Cloud (Bizer et al. 2009).

The FAIR-principles harness knowledge graphs for (meta)data descriptions. In particular, they emphasise on making (meta)data not only accessible to human users but also enabling large scale automated processing. This is especially needed in areas like data discovery, integration, and processing as the increasing volume of (research) data has grown far beyond what humans can manage without support by automated systems. The FAIR-principles comprise four areas with the goal of making (meta)data more Findable, Accessible, Interoperable, and Reusable. In the

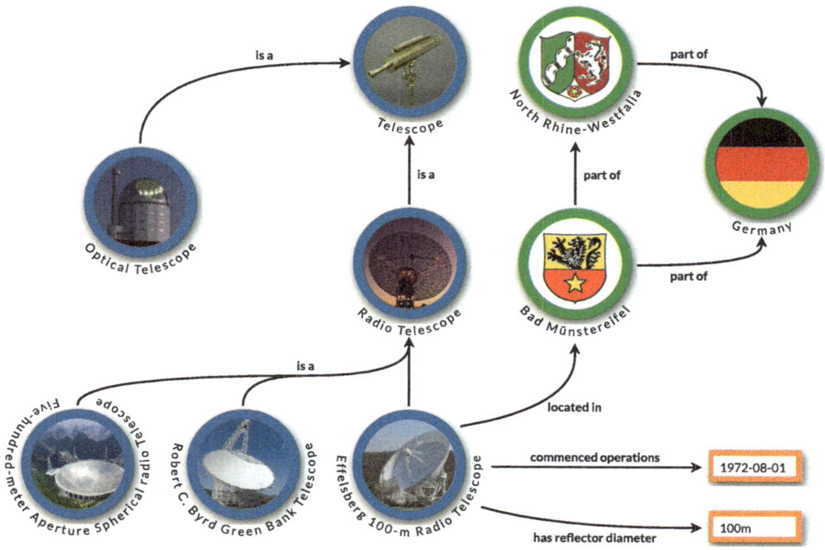

Fig. 14.1 Example knowledge graph

following, each of these aspects will be discussed in accordance with the GO-FAIR Initiative.[3]

14.1.1 Findability

There are myriads of datasets being created and published every day. So for any analysis that goes beyond ones own data, first, suitable datasets have to be discovered. As mentioned before, this is not limited to human users but extends also to automated processes. Consequently, data has to be described not only by free text meant for human consumption but has to include machine-readable and -interpretable metadata as well.

A core requirement are globally unique and persistent identifiers. The uniqueness allows to unambiguously refer to datasets, individual concepts, or any other object of interest. Persistence is in particular needed to maintain links between different objects. An obvious application of those identifiers has been discussed in Chap. 13. Here, datasets that contributed to a particular result need to be identified even way after the actual analysis.

[3] https://www.go-fair.org/.

The technological foundation to most identifiers are the previously discussed IRIs. They are a generalised version of more commonly known Uniform Resource Locators (URLs) and allow to retrieve documents via the World Wide Web. Based on IRIs, several services have been established to provide identifiers for different purposes. Listed below are some of the most important services but many more exist and are actively used.

- Knowledge graphs like Wikidata and DBpedia for individual concepts.
- The Open Researcher & Contributor ID (ORCID)[4] for individual researchers (Haak et al. 2012).
- Research Organization Registry (ROR)[5] for organisations.
- Digital Object Identifier (DOIs)[6] for scientific outputs like publications, datasets, or software (Paskin 2009).

A second requirement calls for extensive metadata descriptions. Metadata provides the means to discover data without having an identifier to start with. The more comprehensive metadata are the easier it is for both human and computers to quickly assess data and decide whether it fits, e.g., the current search. While most data are generated with a specific goal in mind, it might also be useful in a completely different context. So when authoring metadata one should not presume anything about potential users but provide as much context as possible.

Scope and extent of metadata varies across communities. However, the Dublin Core Metadata Initiative (DCMI) Metadata Terms (Kunze and Baker 2007) have emerged as the minimum requirement to virtually all metadata schemata. They include elements to annotate data with, e.g., its title, authors, or a human-readable description. An example of a domain specific metadata model has been discussed in Sect. 2.2.2: LOFAR's use of MeasurementSets. As shown in Fig. 2.4, its focus is on a dataset's provenance. So, the metadata includes information about the LOFAR station that collected the data, the respective antennae, and the configuration of data collection. But it also contains general information like the name of the observer (cf. author) of the dataset and a description of the corresponding project.

Data and metadata might be maintained in different places. In particular, metadata might also be stored in several different places while data, depending on its size, is usually stored at a single or only a hand full of locations. So for metadata to be useful in its task of easing access to the data, it is as important as it may be obvious to actually include a link to the data within the metadata description. Using persistent identifiers allows to distribute metadata across many repositories while still maintaining the connection described data.

[4] https://orcid.org/.

[5] https://ror.org/.

[6] https://doi.org/.

No data can be found and thus be used as long as it is hidden away. Consequently as the final requirement, metadata should be published. One way is to publish a website for the data that can subsequently be picked up by common search engines. A better approach is to register the data with a possibly public data repository. Depending on the repository, this may also involve an additional curation step and will advertise the data to those that might have an interest. A good resource for identifying repositories both for archiving and discovering data is re3data[7] which provides a search interface for data repositories (Pampel et al. 2013).

14.1.2 Accessibility

Besides being discoverable, both data and metadata should be retrievable using a commonly accepted protocol. Given an identifier, users should be able to access the underlying (meta)data without the need to any specialised tools or proprietary software.

Extending the previous notion, the access protocol should also be open and be free to implement. Common protocols fulfilling these requirements include HTTP and FTP. Note that even a well defined point of contact via email or phone can serve as FAIR access to the data.

FAIR data does not necessarily imply open data. So for data with restricted access, the protocol has to provide the means for authentication and authorisation. Further, an automated process should be able to recognise this requirement and act accordingly. This can, e.g., mean to provide an access token or to notify the user and ask for proper credentials.

The availability of data tends to decrease over time. In particular, the costs for keeping larger datasets that are rarely used are sometimes hard to justify. Being much smaller in size metadata are much easier and cheaper to keep. Depending on the level of detail in the metadata description, it has value in its own right and can provide, e.g., the contact details of the initial authors or describe the procedure to recreate the data itself. Consequently, metadata should be maintained even after the data being described has been lost for some reason.

14.1.3 Interoperability

Interoperability has different aspects. First and foremost, it refers to the ability to actually read and interpret the data. This is not limited to a human users but equally extends to automated access. The use of common and open file formats ensures that other users are able to import the data into their environment without

[7] https://www.re3data.org/.

major efforts. In particular, they won't need to resort to specialised or ad-hoc developed converter tools before starting their analysis. What constitutes a common format may depend on the type of data and the scientific domain like PSRFITS for pulsar astronomy (Hotan et al. 2004). If in doubt, human-readable formats like CSV (Shafranovich 2005) or JSON (JSON 2013), are generally a good place to start.

The use of common vocabularies to describe data and metadata has already been established as an essential pillar of an up-to-date RDM. But what about those vocabularies themselves? If a vocabulary can not be found or accessed, any annotation relying on it becomes far less useful. So, employed vocabularies by themselves have to adhere to the FAIR-principles. In particular, their findability and accessibility are important factors for the sustainable use of (meta)data annotations to its full potential.

Few datasets stand in complete isolation. More often, they are part of a network of data, algorithms, people, and other entities that contributed to their creation (cf. Chap. 13), can help to contextualise, or are otherwise related. All relevant links to other entries should be included within a dataset's metadata record. Further, the relation between them should be qualified to distinguish, e.g., between links that are essential to interpret a dataset like codelists and links that document the provenance of the current dataset.

Most astronomical analyses benefit from the integration of multiple datasets, which may have been collected from different observatories in different countries and made available through different data centers. Data integration is enabled by interoperability, which in turn requires standard metadata and data access protocols.

In astronomy, the "Virtual Observatory" is the vision of what can be accomplished with interoperability, namely that an astronomer can have easy access to astronomical observations collected from any observatory. In this case, "easy" means that the astronomer (or a machine used by the astronomer) doesn't have to learn new protocols or metadata vocabularies for each dataset. As the field matures and datasets proliferate, interoperability becomes ever more important for data integration. The International Virtual Observatory Alliance (IVOA) is the body that establishes the standards that support interoperability.

In recent years, the advent of cloud computing has popularized the idea of "data lakes": large, cloud-based (meta)data repositories with associated data services. In the extreme, a single data ocean containing all of the worldwide astronomical data would obviate the need for interoperability. More realistically, there are political and economic forces that make such a scenario difficult to impossible. Therefore, interoperability remains an important goal for any data center concerned with data reuse, which is the ultimate goal of FAIR.

14.1.4 Reusability

The ultimate goal of FAIR is to enable reuse of data. The previous principles focused on the technical aspects of this. This last principle addresses the decision whether specific data can actually be reused. So, metadata descriptions should not only contain information about the data's content but also describe its context to a sufficient level.

Besides technical and content-related aspects, there are also legal requirements to reusing data. Both machines and humans need to be able to decide whether the requested data can be used in a certain scenario. To this end, the metadata has to include a proper reference to a data license that defines usage policies. Among commonly used licences are the Creative Commons licences[8] or the MIT license.[9] Note again that this does not necessarily need to be an open license. The principles just call for a clear license that is unambiguous to both humans and machines.

As seen in Chap. 13, its provenance is a major factor when it comes to establishing trust in data. This provenance should also be included in the metadata. In particular, it should answer questions about the creator(s) of the data, the scope and intention of its creation, and whether this is raw or processed data. Similarly, any other data that has been used during the generation should be mentioned to establish a full trace for the data.

Finally, both data and metadata should adhere to whatever conventions are in place in the respective community. This can include using a commonly agreed upon metadata schema, established and sustainable file formats, and any other standard of organising the data. There may be exceptions that cause creators to deviate from such standards, but these need to be justified. Following common standards reduces the efforts needed by others to work with the data and thus increase its reusablility.

14.1.5 Measuring FAIRness

The FAIR-principles in their default form are general guidelines for how to manage research data. Answering the questions about how FAIR a particular dataset is or how its FAIRness can be improved is rather difficult. Consequently over time, many different approaches have been suggested to structure the assessment process and make results comparable.[10]

Many approaches rely heavily on manual efforts and self-assessment. Surveys like conducted by Dunning et al. (2017) examine a number of resources and grade the compliance with each principle. The next evolutionary step are templates or

[8] https://creativecommons.org/.

[9] https://opensource.org/licenses/MIT.

[10] https://fairassist.org.

checklists like the one provided by RDA's SHARC IG[11] (David et al. 2020). These provide evaluators with a predefined list of questions about a resource. Each question addresses a certain aspect of FAIRness and can usually be answered on a scale from unsupported to fully supported. The responses are then aggregated to summary metrics that quantify the degree of FAIRness.

Manual assessment are rather time consuming and as such can not be applied at scale. Recently, systems have been proposed to automate the evaluation and provide users with an immediate feedback and sometimes suggestions how to improve. One such system is F-UJI (Devaraju and Huber 2021). After providing a publicly available identifier for a resource (e.g., a DOI, or URL to a landing page), a series of metrics are evaluated each targeted at individual FAIR-principles. Take for example the requirement to use vocabularies which themselves are FAIR. Here, F-UJI first retrieves the metadata and, if possible, extracts all namespaces of semantic entities used within. These namespaces are subsequently compared to common registries like the Linked Open Vocabularies (LOV)[12] to verify that these namespaces are in common use. After all metrics are evaluated, users get a detailed feedback of the resource in question. This includes both a summary for a quick evaluation as well as a detailed report about the results of each metric. Systems like F-UJI allow to evaluate new resources rather quickly and may in the future become common practise for newly published data, software, and the like.

14.2 FAIR Data Share and RDM

Looking at modern data-driven science, one can identify several phases most projects have to go through: An initial idea is first cast into a *plan* to structure the corresponding research. Then, researchers have to *collect* data to support or disprove their hypothesis. The collected data also has to go through different levels of post-processing to *assure* its quality. During and after data collection, metadata has to be attached to *describe* the provenance and content of particular datasets. Subsequently, the prepared data should be stored at a sustainable location to *preserve* and possibly publish it. This way, other researchers may *discover* and use it beyond the scope of the initial project. Often a single dataset does not cover the entire information need, so researches need to *integrate* it with other sources of information. Finally, they can analyse this enriched dataset and answer the initial scientific question. Often enough, this leads to new ideas and questions restarting the lifecycle once again.

While the particular phases and their names may differ between authors, the general idea of a data lifecycle in science is generally adopted. The eight phases described above follow the structure used by DataONE (Allard 2012) and are

[11] Sharing Rewards and Credit (SHARC) IG: https://www.rd-alliance.org/groups/sharing-rewards-and-credit-sharc-ig.

[12] https://lov.linkeddata.es.

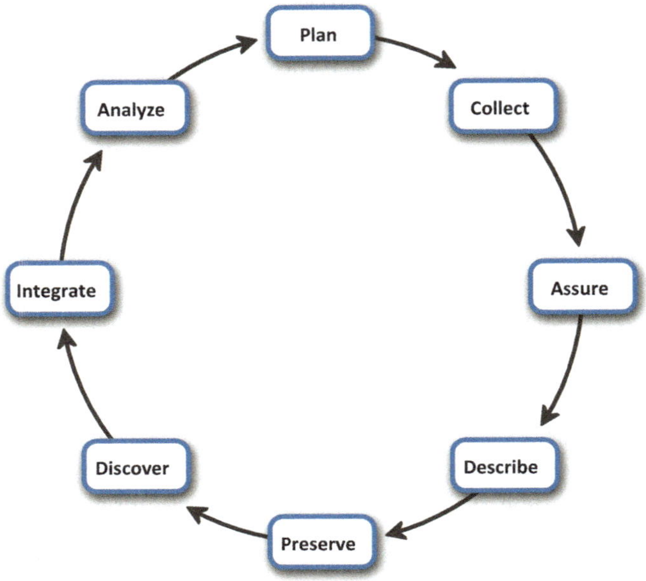

Fig. 14.2 Data lifecycle (after Allard (2012))

illustrated in Fig. 14.2. In practise, the order of phases is not fixed. Some phases might be repeated (e.g., when more data is needed) while others may be skipped entirely (e.g., when the data necessary for the analysis is already available). Nevertheless, this and similar lifecycles provide the mental model for structuring the tasks necessary in a data-driven project.

The goal of Research Data Management (RDM) is to support researchers throughout this data lifecycle and to provide tools or guidelines encouraging best practises. The FAIR principles, as discussed previously in Sect. 14.1, are a cornerstone of these efforts. But while they encourage sharing of data, this is by no means mandatory. Their recommendations apply both to published as well as partly or (for now) unpublished data. In certain environments like health care or industry in general, data can not easily be made available to a general audience. But even in such cases, proper data documentation and structured (albeit restricted) access will make sure that data remains usable even the initial creators become unavailable.

One aspect increasingly also picked up by funding agencies are data management plans (DMPs). In their contemporary incarnation, they require researchers to start considering data management requirements and how they are addressed before the start of a project. However, a DMP is a dynamic document that can be changed and adjusted should the circumstances change and render the previous planning inadequate. Topics included in a DMP may vary between projects and a agreed-upon set of requirements by funding agencies is yet to emerge. Nevertheless, some aspects have become commonplace. This includes a discussion about the type and volume of data expected to be generated or collected within a project. Further, the long term management has to be discussed: Where to store the data (also after the end of the project)? Who has access and which licenses or other restrictions do apply? Considering these aspects from the get go allows to plan accordingly and, e.g., allocate proper funds already in the initial proposal.

A core theme in RDM are extensive descriptions of data and other relevant artefacts. Emphasised by the FAIR principles as well, such metadata lays the ground for a large share of services provided. Besides the rather intuitive use case of discovering relevant information, it is also crucial for tasks like data integration. A lot of value is often generated when data from previously unrelated sources can be combined. The interoperability among such data is a usually a problem, though. Challenges can range from the heterogeneity in terms or abbreviations used for the same things to differences in methodology while creating the data. In such cases the metadata records of the involved datasets will provide researchers with the information needed to properly integrate all sources. With this in mind, also the importance of metadata standards becomes clear. They are the formalised consensus of the respective community and define the properties that are deemed essential. Further, they define the structure of properties: Some may be left as free text mostly targeted at human consumption while others are restricted to a certain set of possible values to allow also for their automated use. This automated processing of metadata is also one of the main motivations behind some of the FAIR principles. Formalising the relations between concepts makes them machine-actionable and allows to build knowledge-aware systems. These systems can emulate human understanding of a domain and draw conclusions based on the knowledge they have access to. A well-known example of such knowledge-aware systems are modern search engines that heavily rely on knowledge graphs to improve their search results.

A final example for a commonly used RDM practise are so called embargo periods for data publications. While metadata is usually published right away, access to the primary data sometimes remains restricted for a certain period of time ranging from weeks or months up to a few years depending on domain and content. During this period, the initials creators remain in full control of their data and can decide with whom to share the data. They can use this time to take full advantage of their efforts and perform all the analysis this data was meant for without the fear of someone else beating them to it. But contrary to storing the data just locally, the data is already stored with an established data provider and will automatically get publicly available once the embargo period runs out. A well defined process like this greatly reduces the risk of losing data due to unforeseen events or mere negligence by the creators. Similarly due to the fact that metadata is often already available, other researchers can already take note and collaborations can be established right away—still under full control of the data owners.

14.3 Making Radio Data FAIR

14.3.1 Open Radio Data

Many recent radio astronomy projects developed Science archives. Let's quote for example ALMA,[13] NRAO VLA,[14] MWA,[15] MeerKat.[16] Astronomy has a long tradition of data sharing and radio astronomy is not an exception. For all of these archives we generally find a web interface with some proprietary description of the data allowing to find and access those in standard (FITS and its dialects, MS, etc.) formats. As these science archive interfaces can be easily discovered using some generic web search engine, we can consider them as examples of "findable" and "accessible" data. But interoperability is another story because this obviously requires some standardization. Interoperability is almost a condition for reusability by analysis software developed by projects different from the initial data producers and providers. And indeed, interoperability and standardization has consequences on better findability and accessibility. As standardization of metadata, formats and access methods is the heart of Virtual Observatory paradigm, it's worth while looking now to the FAIRisation of data that VO services provide.

14.3.2 VO Services and Standards for Radio Astronomy and FAIRisation of These Data

Implementation of Radioastronomy VO services using VO standards is described in Sect. 12.2. These standards have been continuously developed for more than 20 years, starting well before the formal definition of FAIR principles. Nevertheless, VO services implement most of these principles in practice. S.O'Toole and J.Tocknell recently gave a review of the matching of VO standards and FAIR principles (O'Toole and Tocknell 2022). We summarize their main conclusions below and show schematically the mapping of the FAIR principles onto the VO architecture in Fig. 14.3.

[13] https://almascience.nrao.edu/aq/?result_view=observations.

[14] http://www.vla.nrao.edu/astro/nvas/ArchIndex.shtml.

[15] https://asvo.mwatelescope.org/services.

[16] https://skaafrica.atlassian.net/servicedesk/customer/portal/1.

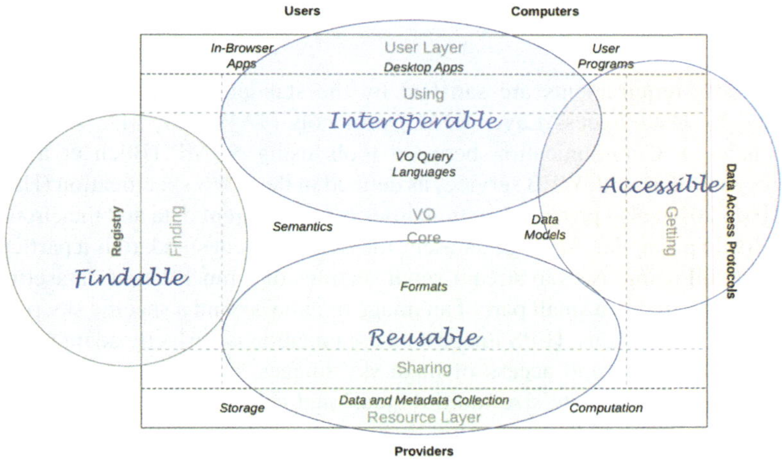

Fig. 14.3 A schematic mapping of the FAIR principles onto the VO architecture. The registry component ensures data are *Findable* in the VO, data access protocols ensure *Accessibility*, the core VO Query Languages and Vocabularies ensure *Interoperability* and the data provenance model and other metadata standards ensure *Reusability*

14.3.3 Findability

Data services and datasets served by these services can be discovered using the VO registries and some of VO DAL services (SIA, ObsTAP, etc.; see Sect. 12.1). Discovery is mainly done through selection made on rich metadata. When the VO started, nearly all searches were based on position and wavelength/frequency. Recently, time domain astronomy has become much more important, both in the optical and in the radio (with time domain surveys such as VLASS, VAST and ThunderKAT). New ways to represent survey coverage as a function of time as well as spatial dimension are being developed by the VO, in particular the expansion of the multi-order coverage (MOC) standard to include time information in version 2.0 of the standard.[17]

The IVOA provides ways to define unique identifiers for resources, which are perfectly sufficient inside the VO landscape and do not require setting of DOIs, although those may also be defined in addition to the IVOA identifiers. DOIs have the advantage of being persistent and citable, and the IVOA is looking into ways to include them.

[17] https://www.ivoa.net/documents/MOC/.

14.3.4 Accessibility

Accessibility requirements are satisfied by the standardisation of services' API provided by Data Access Layer (DAL) protocols (TAP, SCS, SIA, SSA, SLAP, Datalink etc.). Communication between tools using SAMP (Boch et al. 2012) features as well as with WEB services as defined in the UWS specification (Harrison and Rixon 2016) also provide functionalities for transparent data and the circulation of results. In particular, for large datasets, the SODA cutout standard is a particularly important, allowing users to stream small sections of a much larger dataset if they are only interested in a small part of an image or cube around a specific sky position, for example. Beside this, HiPS image data organization is perfectly adapted to multi resolution discovery and access of huge sky images. To go any further we also have to consider tera-byte size multi-dimensional cubes. Multi scale organization methods similar to HiPS are currently experimented for visualization purposes within the scope of the SKA SRC network.

The IVOA is also providing some standardisation of authorisation and authentication.

A challenge we anticipate in the near future is how data stored in the Cloud can be accessed through the VO. The simplest way is to use the Datalink standard to provide a link from a VO ObsCore table (for example) to a Cloud resource. Doing this in a manner transparent to the user can be challenging, however.

14.3.5 Interoperability

Interoperability is the main achievement of standardisation of vocabularies, data models, data formats, and data services via VO specifications. The VO was created mainly to achieve this goal. In particular, the IVOA has a semantics working group that maintains a set of vocabularies based on W3C RDF, a standard for Unified Content Descriptors (UCDs) to describe astrophysical quantities, and a standard way of representing units of physical quantities.

14.3.6 Reusability

The reusability requirements are satisfied through the existence of rich standardised metadata describing the data and their provenance (in the latter case following the IVOA provenance data model specification (see Sect. 13.2). The IVOA was developed before many of the current data licensing mechanisms came into common use, and thus one exception to the FAIR principles is that the IVOA does not define a common licensing policy. Currently, it is left up to the data providers to determine the appropriate license for their products.

14.3.7 Conclusion

The review by S.O'Toole and J.Tocknell cited above clearly shows that radio astronomy data made available in the VO landscape satisfies the vast majority of the FAIR requirements. Although not included in the VO standards, DOIs and licensing can be easily added to data and services by each data provider without introducing any incompatibilities with VO standards or services. Radio astronomy data available through VO services provide a very high level of FAIRness. On the other side it would be very difficult for accessible and findable data to be made interoperable without using common standards. And the definition of such standards is precisely the work of the IVOA.

14.4 Bidirectional Blueprint: Radio Astronomy and Other Domains

Promoting open science and FAIR data practices is part of a general movement towards sharing and disseminating research and its results, including software and data, across the scientific community. This paradigm is fostering the emergence of initiatives that promote multi-disciplinary collaboration. One example of such endeavours is the European Open Science Cloud (EOSC) aiming at proving the environment for science in Europe with the ultimate goal of creating a "Web of FAIR Data and services". EOSC consists of five thematic initiatives, gathering projects from the with the European Strategy Forum on Research Infrastructure (ESFRI), with ESCAPE Science Cluster bringing together astronomy, astroparticle and particle physics community along with their infrastructure (see Sect. 12.4 for a more detailed description of EOSC and ESCAPE).

Following a similar path, Germany created the Nationale Forschungsdateninfrastruktur (NFDI) e.V.[18]—a German national research data infrastructure which consists of a network of consortia focused around various scientific fields and areas of knowledge, brought together in a coordinated structure. Conceptually it is based on a scientific community and other bodies making up an infrastructure responsible for setting standards in data management to meet the FAIR principles. It also aims to address such scientific and research data management subjects as long-term storage, backup, and data availability and integration, at the national and international levels. Up to 30 consortia bring together domain specialists and facilities for specific research communities, e.g. FAIRmat (Consortium for Condensed Matter Physics and Chemical Physics of Solids, Division of Chemistry and Physics),[19] NFDI4Ing (National Research Data Infrastructure for Engineering Sciences, Division of

[18] https://www.nfdi.de.

[19] https://www.fair-di.eu/fairmat/.

Engineering and Mathematics)[20] or NFDI4Biodiversity (Biodiversity, Ecology, and Environmental Data, Division of Life Sciences).[21] They were selected to cover a broad span of scientific disciplines. The NDFI also facilitates collaboration in sections addressing cross-cutting topics. Consortia may differ significantly in their internal structure but their work programme should follow the same objectives referenced in the federal-state agreement. They focus on establishing and handling data management measures, procedures and standards (with a special emphasis on cross-disciplinary metadata standards), reusability of data as well as networking and collaboration, including cross-consortia involvement. Another example of an European association but at international level, involving 14 institutes from Netherlands, Belgium and Germany, is the FAIR Data Infrastructure for Physics, Chemistry, Materials Science, and Astronomy (FAIR-DI) e.V.,[22] working closely with the FAIRmat and the NOMAD Laboratory,[23] which is a platform for sharing and using computational materials science data.

Bringing together specialists from a number of related disciplines that share common points in relation to the goals and challenges in the data management approach promotes innovative insights into existing problems and proposed solutions. Above all, however, such cross-domain and international collaboration allows for a broadened perspective with regard to the formulation of key agenda items themselves. Using the example of the aforementioned FAIR-DI, FAIRmat and NOMAD Laboratory, the sharing aspect of scientific raw data that is one important element in making radio astronomy data FAIR, is also common with materials science or materials physics. It covers the aspect of big data accessibility to other institutions and making them ready for artificial intelligence. Additionally, it includes such services as hosting raw data for long enough, uniquely identified by a persistent identifier and sufficiently described by metadata. Scientists are encouraged to share their data that is currently not considered useful for their research. These enable future discoveries in historical data or re-run of a pipeline using the data if necessary, especially when data reproducibility is no longer possible. Interdisciplinary approach promotes ontologies and metadata schema development, building foundation of interoperability.

Data are essential for science within a wide range of different fields. Each scientific discipline has a different importance on the different aspects of RDM due to the requirements of their data. Therefore, though each field may encounter the same RDM problems (scalability, interoperability, etc.), they devote varying amounts of resources into each problem. Consequently, no singular field is most advanced in all aspects of RDM and all can benefit from the other's research. To exemplify this, the following section describes RDM from the perspective of: optical

---

[20] https://nfdi4ing.de/.

[21] https://www.nfdi4biodiversity.org/en/.

[22] https://www.fair-di.eu/fair-di.

[23] https://nomad-lab.eu/.

astronomy, Earth observation, ecology and environment sciences, and biomedical sciences.

14.4.1 Optical Astronomy

Similar challenges arise in optical astronomy as for radio astronomy when it comes to making data FAIR. In some cases, these challenges have been addressed in the same way for both disciplines. For instance, the use of DOIs as a findable and unique identifier for datasets and papers.

The data collection within optical astronomy is also increasing at an exponential rate. Making these quantities of data accessible to astronomers is a problem that has been tackled in various ways. Data access for the Sloan Digital Sky Survey (SDSS) is tailored to the type of data, with images, spectra, and catalogues all having bespoke solutions for data retrieval. These solutions employ the use of GUIs, SQL queries, and online science platforms for computationally intensive queries.

For the Vera C. Rubin Observatory, it is vital that the data are made accessible within a timely manner to enable follow-up observations on transient objects. As the expected data volumes are TB per night, the required real-time processing is inaccessible for the majority of end users. The Rubin Observatory therefore has a number of community brokers which will ingest, process, and serve data from the telescope. This includes operations such as catalogue matching, photometric classification, and prioritisation of objects for followup.

The FITS standard transcends any particular region of the electromagnetic spectrum and it has been a staple within optical astronomy for decades. Use of this common data format helps the interoperability between datasets from different telescopes. The FITS header also contains much of the image metadata important for both interoperability and reuse i.e position, date of observation, telescope etc. However, the naming format for FITS headers is somewhat free format and though they are often consistent from a natural language standpoint, there are frequent differences in the names of field (e.g. MJDObs, obsMJD, etc.) which can be the source of computational complexity when analysing data from multiple sources.

14.4.2 Earth Observation

Copernicus is an environmental satellite program which through Earth Observation, monitors the health of the planet by measuring a comprehensive set of planetary statistics. The Copernicus dataset includes data of the sea, land, atmosphere, and temperature of the Earth. Accessibility to the data is improved via mapping the data to an explorable GUI of the Earth. In addition, they offer users a variety of visualisation tools, each of which are aimed at users with different levels of experience.[24]

There is a wealth of open Earth observations data available from projects such as Copernicus and these data are typically accompanied by web based platforms for data discovery and navigation. In addition, the platforms are often accompanied by cloud computing resources which allow the users to analyse the data close to it. The Earth Observation Exploitation Platform Common Architecture (EOEPCA) is an initiative which is establishing a consensus of best practices for these Earth Observation platforms to improve their interoperability.[25]

Another consortium tackling standardisation for Earth Observation is the Open Geospatial Consortium (OGC) which is implementing standards for geospatial content and services. An important contribution is the Open GIS Web Map Service (WMS), an HTTP interface for requesting geospatial data from distributed datasets (de La Beaujardiere 2006).

14.4.3 Ecology and Environmental Sciences

The life cycle of astronomical radio data is to a high degree consistent with the model for empirical domains and shows numerous points in common with ecology and related environmental science disciplines. These domains can be considered as precursors for radio astronomy as they have been actively collecting and publishing data in digital information systems for a few decades. One of these, PANGAEA,[26] funded in 1994 by the German Ministry of Research and Technology, evolved from a data management service on the national level to a European and international multidisciplinary georeferenced (in space and time) data publisher for Earth and environmental sciences (Diepenbroek et al. 2002). PANGAEA is open to any institution or individual researcher to archive and publish their data and it is operated as an open access library with a guarantee of long-term preservation. From its early days, the service has focused on careful data curation and consistent archiving. It includes rich metadata description, (meta)data interoperability and extensive harmonisation of data inventory on the structural and semantic level. Efficient identification and citation are fostered with annotation of datasets provenance and using DOI since 2005 (published through DataCite[27]). The publisher allows also for dissemination of (meta)data to other data portals, e.g. World Data System (WDS)[28] or DataONE (Data Observation Network for Earth)[29] and infrastructures,

---

[24] https://marine.copernicus.eu/access-data/ocean-visualisation-tools.

[25] https://eoepca.org/.

[26] https://www.pangaea.de/.

[27] https://datacite.org/.

[28] https://worlddatasystem.org/.

[29] https://www.dataone.org/.

such as ORCID[30] and Google Dataset Search.[31] Similarly to radio astronomy data archives, PANGAEA offers an open data sharing policy with the exception of a small selection of datasets with an initial embargo period for ongoing projects.

Environmental science strongly relies on combining data from different sources, often collected over a long time span, similarly to radio astronomy. Hence, major challenges here are data integration and long-term access to heterogeneous information. To address these aspects Ecological Metadata Language (ELM)[32] standard was developed which formalises and standardises terminology required to describe ecological data. A widely used community accepted metadata specification for a whole domain, supporting rich documentation of environmental, ecological and Earth science data enables data reuse and fosters interoperability across the sub-disciplines. The unified structure makes it easier to develop software solutions and its modularity allows for adjusting the components to specific needs, including the creation of new modules if necessary, omitting the redundant ones. The modules included in ELM enable the user to not only describe their data content and structure but also research methods and annotate data with semantic vocabularies.

14.4.4 Biomedical Sciences

The challenges of biomedical sciences differ from those in radio astronomy, which makes the latter function without so much external formalisation at the level of metadata. However, unification and centralisation (e.g. of access to metadata resources) in radio astronomy seems inevitable with such a large and ever-growing amount of data, especially taking into account the differences in access standards for data from various sources. Data sharing and re-use are particularly prominent topics of interest for RDM in the biomedical and health sciences due to the large amounts of sensitive data, only part of the data available in a digitally structured form in health system and stronger unwillingness of health science researchers to share data than in other domains. Despite these aspects, biomedical ontologies and controlled vocabularies became significantly popular, especially in genomic research with the Gene Ontology (GO),[33] a community-based initiative that unifies the representation of gene product functions using controlled vocabularies to annotate gene products and enable analysis of genomic data. The success of GO led to forming the Open Biological and Biomedical Ontologies (OBO; formerly Open Biomedical Ontologies) Foundry[34] in 2001, a consortium of developers and users dedicated to collaboratively work on and share ontologies and controlled vocabularies for

[30] https://orcid.org/.

[31] https://datasetsearch.research.google.com/.

[32] https://eml.ecoinformatics.org/.

[33] http://geneontology.org/.

[34] https://obofoundry.org/.

the biological and medical domains. To foster interoperability, OBO Foundry provides a set of normative principles used for evaluation of ontologies submitted to them. They concern such aspect, as openness, format, versioning, documentation and more, and are based on good practise that every ontology should follow. The National Center for Biomedical Ontology (NCBO) Bioportal[35] is an online repository that can be used to browse, search for and share ontologies and terminologies actively used in the biomedical domains, including, but not limited to OBO ontologies. It also allows the online community of ontology developers and users to collaboratively work on the content. Two examples of widely used ontology-based terminology standard in biomedical sciences that appear in the Bioportal are Systematized Nomenclature of Medicine Clinical Terms (SNOMED CT)[36] and Medical Subject Headings (MeSH).[37] As the world's most comprehensive health terminology, SNOMED CT is a standardised, international and multilingual complex ontology for use by health care providers. It is recommended as the core general terminology for electronic health records. MeSH is a controlled vocabulary used to index, catalogue and search biomedical and health-related information in any kind of database, for example a collection of articles, books or documents. It is used, among others, for indexing citations in MEDLINE[38] and PubMed,[39] databases of references to journal articles in biomedicine.

14.4.5 Summary

From our examples, the field which has the most similar RDM challenges in common with radio astronomy is (rather unsurprisingly) optical astronomy. Scalability is a pertinent issue for both fields as the next generation of survey telescopes for each, the Vera C. Rubin Observatory (LSST) and the Square Kilometer Array (SKA), are both expected to produce unprecedentedly large datasets for their respective fields. Both projects have invested a large amount of resources into distributed data centres to process and disseminate their data to the scientific community. The LSST is expected to begin scientific operations approximately five years prior to SKA, therefore radio astronomy will likely be able to benefit from the lessons learnt by LSST during their operation with regards to data dissemination.

Earth observation does not yet create the same quantities of data as radio astronomy, however this is likely to change in the future and Earth observation will then benefit from the experience of scalability within radio astronomy. Interoperability and accessibility were both shown to be critical within Earth observation as it

---

35 https://bioportal.bioontology.org/.

36 https://www.snomed.org/.

37 https://www.nlm.nih.gov/mesh/meshhome.html.

38 https://www.nlm.nih.gov/medline/.

39 https://pubmed.ncbi.nlm.nih.gov.

requires simultaneous observations of the land, sea, and air. As a result, there are a plethora of web based platforms such as Copernicus and importantly, the EOEPCA which was tasked to set standards and best practices for these platforms. Establishing a similar initiative within radio astronomy would go a long way for improving the interoperability and accessibility for radio data.

Ecology and environmental sciences is another field which does not yet create the same quantities of data as radio astronomy but will likely do so in the future and have much to gain from the lessons of handling large datasets learnt from radio astronomy. The strengths of Ecology with regards to RDM lie in their standardisation of metadata and data preservation, showcased by their central data hub—PANGAEA. Though radio astronomy has some standardisation for metadata (e.g. within FITS headers or PSRFITS history tables) as well as some central data stores, radio astronomy could still strive for the level of standardisation and unification demonstrated within ecological data management.

The biomedical sciences have the restraint of personal information which makes open data a difficult task. In spite of this, their initiatives which outline the standards for annotation for shared data should be a source of inspiration for radio astronomical data annotation in the future.

As we have shown from analysing research data management in different scientific fields, the data life cycle is an integral component for scientific research, regardless of discipline. The importance of the FAIR principles is equally ubiquitous throughout data driven science, however the relative importance of each of the principles may vary from field to field. Consequently, each individual field invests resources in those aspects most important to them and each field can share in the benefits of this resource investment.

References

S. Allard, DataONE: Facilitating eScience through collaboration. J. eScience Librarianship, 4–17 (2012). https://doi.org/10.7191/jeslib.2012.1004

T. Berners-Lee, J. Hendler, O. Lassila, The semantic web. Sci. Am. **284**(5), 34–43 (2001). https://doi.org/10.1038/scientificamerican0501-34

C. Bizer, T. Heath, T. Berners-Lee, Linked data - the story so far. Int. J. Semantic Web Inf. Syst. **5**(3), 1–22 (2009). https://doi.org/10.4018/jswis.2009081901

T. Boch, M. Fitzpatrick, M. Taylor, A. Allan, J. Fay, L. Paioro, J. Taylor, D. Tody, Simple application messaging protocol version 1.3. IVOA Recommendation 11 April 2012 (2012)

R. Cyganiak, D. Wood, M. Lanthaler, RDF 1.1 Concepts and Abstract Syntax (2014). https://www.w3.org/TR/rdf11-concepts/

R. David, L. Mabile, A. Specht, S. Stryeck, M. Thomsen, M. Yahia, C. Jonquet, L. Dollé, D. Jacob, D. Bailo, H. Bravo, S. Gachet, H. Gunderman, J.-E. Hollebecq, V. Ioannidis, Y. Le Bras, E. Lerigoleur, A. Cambon-Thomsen, SHARC Community, Templates for fairness evaluation criteria - RDA-SHARC IG (2020). https://doi.org/10.5281/zenodo.3922069

J. de La Beaujardiere, Opengis® web map server implementation specification. version 1.3.0 (2006). https://repository.oceanbestpractices.org/handle/11329/1135

A. Devaraju, R. Huber, An automated solution for measuring the progress toward FAIR research data. Patterns **2**(11), 100370 (2021). https://doi.org/10.1016/j.patter.2021.100370

M. Diepenbroek, H. Grobe, M. Reinke, U. Schindler, R. Schlitzer, R. Sieger, G. Wefer, Pangaea - an information system for environmental sciences. Comput. Geosci. **28**(10), 1201–1210 (2002). https://doi.org/10.1016/s0098-3004(02)00039-0

A. Dunning, M. De Smaele, J. Böhmer, Are the FAIR data principles fair? Int. J. Digit. Curation **12**(2), 177–195 (2017). https://doi.org/10.2218/ijdc.v12i2.567

L.L. Haak, M. Fenner, L. Paglione, E. Pentz, H. Ratner, Orcid: a system to uniquely identify researchers. Learn. Publish. **25**(4), 259–264 (2012). https://doi.org/10.1087/20120404

P. A. Harrison, G. Rixon, Universal Worker Service Pattern Version 1.1. IVOA Recommendation 24 October 2016 (2016)

A.W. Hotan, W. van Straten, R.N. Manchester, psrchive and psrfits: An open approach to radio pulsar data storage and analysis. Publ. Astron. Soc. Australia **21**(3), 302–309 (2004). https://doi.org/10.1071/as04022

JSON, The JSON Data Interchange Format. Technical report, ECMA, 10 (2013). https://www.ecma-international.org/wp-content/uploads/ECMA-404_1st_edition_october_2013.pdf

J.A. Kunze, T. Baker, The dublin core metadata element set. 8 (2007). https://doi.org/10.17487/rfc5013

J. Lehmann, R. Isele, M. Jakob, A. Jentzsch, D. Kontokostas, P. Mendes, S. Hellmann, M. Morsey, P. van Kleef, S. Auer, C. Bizer, DBpedia - a large-scale, multilingual knowledge base extracted from wikipedia. Semant. Web J. **6**, 167–195 (2015). ISSN 15700844. https://doi.org/10.3233/SW-140134

S. O'Toole, J. Tocknell, FAIR standards for astronomical data. arXiv e-prints, art. arXiv:2203.10710, March (2022). https://doi.org/10.48550/arXiv.2203.10710

H. Pampel, P. Vierkant, F. Scholze, R. Bertelmann, M. Kindling, J. Klump, H.-J. Goebelbecker, J. Gundlach, P. Schirmbacher, U. Dierolf, Making research data repositories visible: The re3data.org registry. PLoS One **8**(11), e78080 (2013). https://doi.org/10.1371/journal.pone.0078080

N. Paskin, Digital object identifier (DOI®) system, in *Encyclopedia of Library and Information Sciences*, 3rd edn. (CRC Press 2009), pp. 1586–1592. https://doi.org/10.1081/e-elis3-120044418

Y. Shafranovich, Common format and MIME type for comma-separated values (CSV) files (2005). https://doi.org/10.17487/rfc4180. https://tools.ietf.org/html/rfc4180

D. Vrandečić, M. Krötzsch, Wikidata: A free collaborative knowledgebase. Commun. ACM **57**(10), 78–85 (2014). ISSN 0001-0782. https://doi.org/10.1145/2629489

M.D. Wilkinson, M. Dumontier, I.J. Aalbersberg, G. Appleton, M. Axton, A. Baak, N. Blomberg, J.-W. Boiten, L.B. da Silva Santos, P.E. Bourne, J. Bouwman, A.J. Brookes, T. Clark, M. Crosas, I. Dillo, O. Dumon, S. Edmunds, C.T. Evelo, R. Finkers, A. Gonzalez-Beltran, A.J.G. Gray, P. Groth, C. Goble, J.S. Grethe, J. Heringa, P.A.C. 't Hoen, R. Hooft, T. Kuhn, R. Kok, J. Kok, S.J. Lusher, M.E. Martone, A. Mons, A.L. Packer, B. Persson, P. Rocca-Serra, M. Roos, R. van Schaik, S.-A. Sansone, E. Schultes, T. Sengstag, T. Slater, G. Strawn, M.A. Swertz, M. Thompson, J. van der Lei, E. van Mulligen, J. Velterop, A. Waagmeester, P. Wittenburg, K. Wolstencroft, J. Zhao, B. Mons, The FAIR guiding principles for scientific data management and stewardship. Sci. Data **3**(1) (2016). https://doi.org/10.1038/sdata.2016.18

Chapter 15
Epilogue and Outlook

Beatriz Mingo, Hans-Rainer Klöckner, Hermann Heßling, Michael Kramer, and Stefan Wagner

A major task in radio astronomy is to find Extraterrestrial signals that are deeply buried in the data streams of radio telescopes. New methods in digitalisation and data analysis improved our ability to find even some of the weakest signals in the Universe, but with the caveat that the data operations and sizes are increasing since then. By the end of the next decade radio astronomy will enter the Exabyte regime (10^{18} bytes) with individual datasets of the order of a few petabytes (10^{15} bytes) and will face unforeseen challenges and opportunities to be mastered by then.

Currently radio astronomy is in a transition phase where the journey began with the science operation of radio facilities like ALMA, LOFAR, MWA around 2012. Since then these facilities provided an exceptional "radio" view of our Universe, but things didn't go as smoothly as just switching on a new "radio", instead one had to understand how to work with such datasets. New hardware and signal processing concepts plus the digitalisation provided new ways of data handling, quality assessment, and calibration implied that a fresh view on almost all aspects of the signal chain, the software and even the data structure in use needed to be taken. It also initiated a paradigm change to move the user to the data, that has not

B. Mingo (✉)
School of Physical Sciences, The Open University, Milton Keynes, UK
e-mail: bmingo@extragalactic.info

H.-R. Klöckner · M. Kramer
Max-Planck-Institut für Radioastronomie, Bonn, Germany
e-mail: hrk@mpifr-bonn.mpg.de; mkramer@mpifr-bonn.mpg.de

H. Heßling
Hochschule für Technik und Wirtschaft (HTW) Berlin, Berlin, Germany
e-mail: hessling@htw-berlin.de

S. Wagner
Zentrum für Astronomie der Universität Heidelberg, Heidelberg, Germany
e-mail: swagner@lsw.uni-heidelberg.de

E. Vardoulaki et al. (eds.), *Data-Intensive Radio Astronomy*, Astrophysics and Space Science Library 472, https://doi.org/10.1007/978-3-031-58468-8_15

ended yet, and we are slowly starting to appreciate new developments in fabricating automatic workflows to handle and analyse datasets in radio astronomy.

With ASKAP (early pilot projects in 2019) and MeerKAT (opening 2018) we have just entered a stopover to investigate the scalability and functionality of our new tool kits and software suits on our way to the "golden age of radio astronomy" with facilities such as the Square Kilometer Array (SKA) and the next-generation VLA (ngVLA).

This book provides an inventory of the facilities, the tools, and software suits that are currently in use enabling us to do the science we would like to pursue. However, are we prepared to cope with the avalanche of data that we will face in the future? Certainly not. Scalability in all aspects of the signal and data processing chain is a major concern and developments may need to evolve in incremental steps alongside with such demands, but we can address two basic questions to guide us. How can we analyse all the data, if we can only analyse a small fraction of it in full detail, and can we ever store all the data in a sustainable way.

The first question might be addressed by using a machine learning approach either with un- or supervised methods. But how do we fine tune the process? Generally, signal processing in astronomy is not a linear process, there are various things that can have an impact on the signal itself, the electronics in the receiver- and backend-system, the environmental changes of the observatory or telescopes, and the ageing of all mechanical and electronic components. A model that describes such a dynamical system would enable us to find the needle in the haystack, but it requires monitoring and modelling the receiving system in great detail. Such information would feed into a digital twin model of the data signal and observatory in order to generate provenance and data quality. This would be a first step in evaluating the data, however the necessary strong reduction of data volume needs to be implemented once a model has been determined and a real time system at the telescope or observatory has been installed to evaluate the signals. Such a real time system results in fundamental challenges how we may deal with data in the future and which compromise we have to accept.

Data Irreversibility

Traditionally, data taken by an experimental device are analysed and, finally, "derived datasets" are stored in long-term archives. A decisive element of this method is to minimise loss of information. The real time constraint, however, limits the effectiveness of traditional workflows as decisions have to be taken on the basis of missing information simply because there is not enough time to process workflows in full detail. The missing information cannot be recovered later on whereby an "arrow of time" is introduced, an essential characteristic of irreversibility.

Dynamic Archives

To reduce irreversibility effects as much as possible, feedback loops should be provided by analysing the quality of archived data. Firstly, the outcome of a "fast analysis" of online data could be used to optimise the control of sensors. Archived datasets have to be characterised by quality measures that are adjusted regularly. The outcome of this "slow analysis" could be used for steering the sensors via a second feedback. In other words, archives will no longer be static but dynamic entities. Accordingly, metadata schemes should be extendable dynamically to keep up with increasing knowledge.

In order to deal with upcoming data volumes of large infrastructure in radio astronomy, the "Traditional" data compression is not sufficient, but it is a matter of extracting "relevant information" already during the measuring process in near-realtime. In order to minimise the resulting "data irreversibility", a "Dynamic Life Cycle" (DLC) is needed that extends existing Big Data life cycle models by introducing two feedback loops: one between the sensors and a nearby computing centre, and one between world-wide distributed data archives and the sensors—both of which constantly optimise the receiver and backend control systems.

In everyday life data volumes and energy seem to have little in common, but already today the digital world has a large impact on global energy consumption. With increasing data volumes sustainability and Green Computing are needed to reduce energy and material consumption in information and communication technologies. Although energy-saving measures are being developed and put in place (e.g. by the Green Cube[1] at GSI[2]), they are not efficient enough to cope with the accelerated advances of digitisation.

15.1 Sustainability in Astronomy in the Data-Intensive Era

As we collectively work to attempt to stop global warming, and meet the Paris Agreement 1.5° target, the carbon footprint of research institutes and infrastructures has become a subject of increasing consideration. Astronomy is no exception to this scrutiny, as the carbon footprint of an astronomer is estimated to be far larger than that of the average person (Marshall et al. 2009; Burtscher et al. 2020; Flagey et al. 2020; Jahnke et al. 2020; Martin et al. 2022), though the actual numbers vary greatly from country to country, and depend on many factors. Astronomy infrastructures are also very vulnerable to climate change (Cantalloube et al. 2020; Haslebacher et al. 2022). The global astronomy community holds a collective responsibility to minimise its ecological impact, and there is growing concern and commitment to slashing emissions and meeting climate goals, as evidenced by the proliferation of dedicated publications[3] and meetings[4] in the past few years. Sustainability in astronomy is a fairly new, but very fast-evolving topic.

Many recent studies have aimed to provide measurements and estimates of our carbon footprint, from institute (Flagey et al. 2020; Jahnke et al. 2020; Martin et al.

[1] https://www.gsi.de/en/researchaccelerators/research_an_overview/green-it-cube.

[2] GSI Helmholtzzentrum für Schwerionenforschung in Darmstadt.

[3] https://www.nature.com/collections/fhfcdebecc.

[4] https://indico.icc.ub.edu/event/167/.

2022) to national level (e.g. Stevens et al. 2020; van der Tak et al. 2021, and the Swedish Climate Framework for Higher Education institutions.[5]) However, tools do not yet exist to collate all these reports into a single set of standard measurements, for direct comparison. The French 'GES 1point5' tool[6] developed by Labos 1point5 has provided the most exhaustive picture so far of the carbon footprint of the research community at a national level (Mariette et al. 2022), and it shows promise in terms of establishing some of these standards internationally. There are plans to port GES 1point5 to other countries, though this will require extensive analysis of the specifics of energy and transport infrastructure in each country.

Although there are many factors that contribute to the carbon footprint of astronomy, three large areas generally account for the bulk of it: infrastructures, travel, and computing.

The carbon footprint of astronomy infrastructures is huge, as shown by the recent study by Knödlseder et al. (2022), and might in fact be the largest contribution to the emissions of the average astronomer. While space-based observatories have large associated footprints, their intensity (measured as tons of CO_2 per paper or per author) is comparable to that of ground-based observatories, of which there is a larger number. Among ground-based facilities, radio observatories, and particularly radio interferometry arrays, feature quite prominently. The total carbon footprint of large radio arrays is of the same order of magnitude as that of very large (>3 m) optical facilities. Both building and operational costs need to be carefully assessed. The longer a facility is active the less intensive the initial carbon cost will be, but efforts should still be made to minimise it, by sourcing local, low-carbon materials whenever possible. Radio observatories tend to have very long lifespans, but so do many ground-based observatories operating at other wavelengths. From the data currently available, which are not comprehensive, it is difficult to quantify how instrument replacement, maintenance, and other factors contribute in the case of each facility, but it is clear that all institutes and consortia should start comprehensively measuring and reporting their carbon footprints, as some already are doing (e.g. ESO,[7] NOIRLab[8]), and that plans should be made to decarbonise existing infrastructures, with radio facilities being no exception.

The SKA and its precursors are all documenting their carbon footprints and planning or implementing sustainability plans. LOFAR has an expected lifetime carbon footprint of roughly one-third that of the VLA (Knödlseder et al. 2022). Its yearly carbon footprint is equivalent to that of 200–300 average EU citizens. In terms of electricity alone the usage is ~2000 times that of the average European. The electricity needed for storage and central processing accounts for roughly 10% of the total energy usage and total carbon footprint, though this does not account for further processing needed for science products, such as generating source catalogues

[5] https://www.kth.se/en/om/miljo-hallbar-utveckling/klimatramverket-1.903489.

[6] https://apps.labos1point5.org/ges-1point5.

[7] https://www.eso.org/public/about-eso/sustainability/green/.

[8] https://noirlab.edu/public/about/sustainability/.

(Kruithof et al. 2023). The carbon footprint is expected to reduce in the future, as the Netherlands is moving more of its electricity production to renewable sources.

In the case of the SKA, the headquarters and part of the facilities are expected to run on renewable energy.[9] The ngVLA is also working on power optimisation, and considering options for renewable energy.[10] Power-hungry infrastructures, particularly those in remote locations, such as radio telescopes, can put severe strain on existing energy networks, potentially exacerbating fossil fuel usage. Working locally to develop new green infrastructures to sustain observing and computing facilities, as well as the local population, can be a great way to make observatories more sustainable both environmentally and societally.

Travel, and in particular long-distance air travel, is often highlighted as one of the key contributors to our community's carbon footprint (Blanchard et al. 2022; Burtscher et al. 2022). However, this footprint is not equally distributed across the globe, nor in terms of career seniority (Berné et al. 2022). Travel is also one of the easiest carbon contributions to identify and tackle, particularly in view of the surge of remote collaboration and conferencing software, which has also proven to redress some of these accessibility barriers (Skiles et al. 2022).

As highlighted throughout this book, modern radio astronomy is very computationally intensive, more so than many other disciplines, requiring vast processing and storage capacity. As an example, the rates of data transfer to the processing facility in Cape Town from the SKA-Mid are expected to exceed the global average home broadband speed in 2022 by five orders of magnitude. The volume of data archived each year is expected to be $\sim$300 PB, roughly equivalent to the capacity of half a million average laptops.[11] Historically, even recently for facilities like the Event Horizon Telescope,[12] many radio observatories have physically transported their data, via road or air, as the fastest and most efficient way to get them to a processing facility. These processing, storage, and data transfer solutions have a substantial associated carbon footprint, in terms of construction, sourcing, and running.

For perspective, a 2018 article[13] estimated that as, a whole, data centres contribute 0.3% to global carbon emissions, and the information and communications technology (ICT) compendium, which encompasses televisions, mobile phones and their networks, among others, might be on par with the aviation industry, contributing >2% of the global carbon footprint. Data centres and ICTs associated with radio astronomy facilities contribute only a fraction of these emissions, but in absolute terms this is still substantial. It is undeniable that radio astronomers are more intensive users of ICTs than the average person, due to the data-intensive

---

[9] https://www.skao.int/en/explore/sustainability-skao.

[10] https://ngvla.nrao.edu/page/faq.

[11] https://www.skao.int/en/explore/big-data.

[12] https://eventhorizontelescope.org/faq/how-much-data-recorded-during-observation-and-how-it-transferred-central-processing.

[13] https://www.nature.com/articles/d41586-018-06610-y.

nature of their work. At the Max Planck Institute for Astronomy (MPIA), in Germany, it is estimated that between 75% and 90% of all electricity consumption is linked to computing resources, in large part due to their supercomputing facilities (Jahnke et al. 2020).

The carbon footprint linked to computing energy usage is greatly dependent on the country where the computing facilities are located, as it generally relies on the local energy infrastructure. As such, the carbon footprint of Australian astronomers, and thus their radio facilities, is currently very large, as only a small fraction of electricity in Australia stems from renewable energy sources (Stevens et al. 2020). For European facilities, which are generally more reliant on renewable energy, there are still some variations, with the carbon intensity of the electricity used at the Institute for Research in Astrophysics and Planetology (IRAP), in France, being a factor of $\sim$4 lower than that used by MPIA (Martin et al. 2022).

One possible way to reduce carbon emissions associated with computing, thus, is to remotely use hosting servers and processing facilities located in countries with less carbon-intensive energy production. However, 'green' credentials must be examined carefully for transparency, and if intensive use of computing adds too large a load to the underlying energy infrastructure, the risk that backup power generation might be needed should be considered, as this often stems from more carbon-intensive facilities. This solution could also be impractical when very large volumes of data need to be transferred back and forth. As such, it could be employed for less intensive computing and storage, while working locally to minimise energy usage and developing greener energy infrastructures should be the long-term aims.

Using more energy-efficient hardware could also help reduce computing carbon emissions, as long as some effort is invested in sourcing this hardware as sustainably as possible. For example, the SKA, like some existing facilities, will use graphics units (GPUs) for some of its data processing. Supercomputers are more carbon efficient than regular workstations when in single-core mode, but adding multiple cores increases their carbon footprint (while reducing processing time). The optimal balance of performance and carbon emissions is achieved with parallelisation and $\sim$1000 cores in a recent study (Portegies Zwart 2020). Use of supercomputers or GPU-equipped workstations should thus be encouraged for heavier processing tasks, but spreading the load to avoid increasing energy usage.

Another important consideration is what software and programming languages to use for data reduction and analysis. Recent studies show that Python, which is very commonly used in astronomy, has a much larger carbon footprint than other programming languages (Portegies Zwart 2020; Pereira et al. 2021). Carbon footprint is not only related to runtime, though it factors in heavily, but also to memory usage. As such, sometimes different algorithms perform better in terms of carbon footprint when implemented in different programming languages. C/C++, Pascal, and Fortran tend to be far less carbon intensive than (plain) Python or Java, but this lower carbon footprint can come at the cost of readability and ease of use. Coding the most carbon-intensive algorithms in (e.g.) C/C++ and wrapping them in Python could be a possible approach to reach a compromise. Alternatively, or concurrently, languages like Julia, or more optimal implementations of Python, such as Pythran (combined with OpenMP), could prove very carbon efficient (Augier et al. 2021).

It is also important to train programmers (and provide them with the time and tools) to optimise their code, as the recent work by Augier et al. (2021) found that developer skills and optimisation were significant factors in the carbon efficiency of several benchmark tests. Radio astronomy, being data intensive, is less dependent on 'amateur' coding than other astronomy disciplines, but it is important that carbon efficiency is considered at all levels, from providing training and clear guidelines to the core developers to optimising the datasets and software tools that the end users of the data-sets need to produce scientific results. Tools such as Green Algorithms (Lannelongue et al. 2021) can help provide standard carbon footprint measurements for software, and identify key areas where optimisation might be needed. New guidance is also emerging on how to estimate the carbon footprint of computational research (Lannelongue and Inouye 2023).

In conclusion, it is important to acknowledge that radio astronomy has a substantial carbon footprint. We must urgently work on coordinated solutions to optimise and decarbonise existing resources, and to minimise the carbon footprint of any ongoing and future developments, in order to make radio astronomy sustainable for future generations.

References

P. Augier, C.F. Bolz-Tereick, S. Guelton, A.V. Mohanan, Reducing the ecological impact of computing through education and Python compilers. Nat. Astron. **5**, 334–335 (2021). https://doi.org/10.1038/s41550-021-01342-y

O. Berné, L. Agier, A. Hardy, E. Lellouch, O. Aumont, J. Mariette, T. Ben-Ari, The carbon footprint of scientific visibility. Environ. Res. Lett. **17**(12), 124008 (2022). https://doi.org/10.1088/1748-9326/ac9b51

M. Blanchard, M. Bouchet-Valat, D. Cartron, J. Greffion, J. Gros, Concerned yet polluting: A survey on French research personnel and climate change. PLoS Clim. **1**(9), 1–20 (2022). https://doi.org/10.1371/journal.pclm.0000070

L. Burtscher, D. Barret, A.P. Borkar, V. Grinberg, K. Jahnke, S. Kendrew, G. Maffey, M.J. McCaughrean, The carbon footprint of large astronomy meetings. Nat. Astron. **4**, 823–825 (2020). https://doi.org/10.1038/s41550-020-1207-z

L. Burtscher, L. Balaguer-Núñez, V. D'Orazi, D. Barret, T. Beuchert, E. Dil, A. Janiuk, B. Mingo, E. Poggio, Astronomy organizations should lead in our battle against the climate crisis. Nat. Astron. **6**, 764–764 (2022). https://doi.org/10.1038/s41550-022-01722-y

F. Cantalloube, J. Milli, C. Böhm, S. Crewell, J. Navarrete, K. Rehfeld, M. Sarazin, A. Sommani, The impact of climate change on astronomical observations. Nat. Astron. **4**, 826–829 (2020). https://doi.org/10.1038/s41550-020-1203-3

N. Flagey, K. Thronas, A. Petric, K. Withington, M.J. Seidel, Measuring carbon emissions at the Canada-France-Hawaii Telescope. Nat. Astron. **4**, 816–818 (2020). https://doi.org/10.1038/s41550-020-1190-4

C. Haslebacher, M.E. Demory, B.O. Demory, M. Sarazin, P.L. Vidale, Impact of climate change on site characteristics of eight major astronomical observatories using high-resolution global climate projections until 2050. arXiv e-prints, art. arXiv:2208.04918, August (2022). https://doi.org/10.48550/arXiv.2208.04918

K. Jahnke, C. Fendt, M. Fouesneau, I. Georgiev, T. Herbst, M. Kaasinen, D. Kossakowski, J. Rybizki, M. Schlecker, G. Seidel, T. Henning, L. Kreidberg, H.-W. Rix, An astronomical institute's perspective on meeting the challenges of the climate crisis. Nat. Astron. **4**, 812–815 (2020). https://doi.org/10.1038/s41550-020-1202-4

J. Knödlseder, S. Brau-Nogué, M. Coriat, P. Garnier, A. Hughes, P. Martin, L. Tibaldo, Estimate of the carbon footprint of astronomical research infrastructures. Nat. Astron. **6**, 503–513 (2022). https://doi.org/10.1038/s41550-022-01612-3

G. Kruithof, C. Bassa, I. Bonati, W. van Cappellen, A. Doek, N. Ebbendorf, M. Gerbers, M. van Haarlem, R. Halfwerk, H. Holties, S. Kajuiter, V. Kondratiev, H. Meulman, R. Pizzo, T. Shimwell, J. Swinbank, The energy consumption and carbon footprint of the LOFAR telescope. Exp. Astron. (2023). https://doi.org/10.1007/s10686-023-09901-z

L. Lannelongue, M. Inouye, Carbon footprint estimation for computational research. Nat. Rev. Methods Primers **3**(1), 9 (2023). ISSN 2662-8449. https://doi.org/10.1038/s43586-023-00202-5

L. Lannelongue, J. Grealey, M. Inouye, Green algorithms: Quantifying the carbon footprint of computation. Adv. Sci. **8**(12), 2100707 (2021). https://doi.org/10.1002/advs.202100707. https://onlinelibrary.wiley.com/doi/abs/10.1002/advs.202100707

J. Mariette, O. Blanchard, O. Berné, O. Aumont, J. Carrey, A. Ligozat, E. Lellouch, P.-E. Roche, G. Guennebaud, J. Thanwerdas, P. Bardou, G. Salin, E. Maigne, S. Servan, T. Ben-Ari, An open-source tool to assess the carbon footprint of research. Environ. Res. Infrastruct. Sustain. **2**(3), 035008 (2022). https://doi.org/10.1088/2634-4505/ac84a4

P.J. Marshall, N. Bennert, E.S. Rykoff, K.J. Shen, J.D.R. Steinfadt, J. Fregeau, R.R. Chary, K. Sheth, B. Weiner, K.B. Henisey, E.L. Quetin, R. Antonucci, D. Kaplan, P. Jonsson, M.W. Auger, C. Cardamone, T. Tao, D.E. Holz, M. Bradac, T.S. Metcalfe, S. McHugh, M. Elvis, B.J. Brewer, T. Urrutia, F. Guo, W. Hovest, R. Nakajima, B.Q. For, D. Erb, D. Paneque, Low-energy astrophysics: stimulating the reduction of energy consumption in the next decade, in *astro2010: The Astronomy and Astrophysics Decadal Survey*, vol. 2010 (2009), p. P35. https://doi.org/10.48550/arXiv.0903.3384

P. Martin, S. Brau-Nogué, M. Coriat, P. Garnier, A. Hughes, J. Knödlseder, L. Tibaldo, A comprehensive assessment of the carbon footprint of an astronomical institute. Nat. Astron. **6**, 1219–1222 (2022). https://doi.org/10.1038/s41550-022-01771-3

R. Pereira, M. Couto, F. Ribeiro, R. Rua, J. Cunha, J.P. Fernandes, J. Saraiva, Ranking programming languages by energy efficiency. Sci. Comput. Program. **205**, 102609 (2021). ISSN 0167-6423. https://doi.org/10.1016/j.scico.2021.102609. https://www.sciencedirect.com/science/article/pii/S0167642321000022

S. Portegies Zwart, The ecological impact of high-performance computing in astrophysics. Nat. Astron. **4**, 819–822 (2020). https://doi.org/10.1038/s41550-020-1208-y

M. Skiles, E. Yang, O. Reshef, D.R. Muñoz, D. Cintron, M.L. Lind, A. Rush, P.P. Calleja, R. Nerenberg, A. Armani, K.M. Faust, M. Kumar, Conference demographics and footprint changed by virtual platforms. Nat. Sustain. **5**(2), 149–156 (2022). ISSN 2398-9629. https://doi.org/10.1038/s41893-021-00823-2

A.R.H. Stevens, S. Bellstedt, P.J. Elahi, M.T. Murphy, The imperative to reduce carbon emissions in astronomy. Nat. Astron. **4**, 843–851 (2020). https://doi.org/10.1038/s41550-020-1169-1

F. van der Tak, L. Burtscher, S.P. Zwart, B. Tabone, G. Nelemans, S. Bloemen, A. Young, R. Wijnands, A. Janssen, A. Schoenmakers, The carbon footprint of astronomy research in the Netherlands. Nat. Astron. **5**, 1195–1198 (2021). https://doi.org/10.1038/s41550-021-01552-4

Glossary

A3COSMOS Automated Mining of the ALMA Archive in the COSMOS Field
ACA Atacama Compact Array
Acceleration derivative Commonly called 'jerk' in time-domain astronomy
Accessibility The ability of users to access and retrieve data even if restricted, with open metadata associated with it
Adaboost Adaptive Boosting
Adaptation When we transfer between distributions, but for the same task
ADIOS Adaptable I/O System
AE Auto-Encoder, an artificial neural network that encodes a signal to learn features, which then are decoded to refine and validate the encoding
AGN Active galactic nucleus/nuclei
AI Artificial intelligence
AIPS Astronomical Image Processing System
Aladin Aladin Sky Atlas, Virtual Observatory Image Application developed by CDS
AladinDeskTop Aladin standalone version, developed in Java
AladinLite Aladin javascript version, embedded in Web browsers
ALMA Atacama Large Millimeter/submillimeter Array
ALTA Apertif Long-Term Archive
Amplitude Highest elongation of a periodic signal
Anomaly Outlier or irregularity compared to usual data
Argabrightening events Instrumental noise in Kepler space telescope observations
ASDM Archival Science Data Model (ALMA)
ASKAP Australian Square Kilometre Array Pathfinder
ASKAPsoft Software package for ASKAP
ASIAA Institute of Astronomy & Astrophysics, Academia Sinica
Astropy Python package for astronomical applications
ATCA Australia Telescope Compact Array

E. Vardoulaki et al. (eds.), *Data-Intensive Radio Astronomy*, Astrophysics and Space Science Library 472, https://doi.org/10.1007/978-3-031-58468-8

APEX Atacama Pathfinder Experiment telescope
API Application Programming Interface
Apertif Aperture Tile In Focus (Instrument)
ARC ALMA Regional Centre
ASPECS ALMA Spectroscopic Survey in the Hubble Ultra Deep Field
ASTRON The Netherlands Institute for Radio Astronomy
ASVO All-Sky Virtual Observatory
ATOA Australia Telescope Online Archive
AU Astronomical unit
AusSRC Australian SKA Regional Center
Barycentred Reference frame brought to the solar system barycentre
Bandpass Filter that allows to pass only a certain frequency range
Baseline Line connecting two separate antennas of an interferometer or array
Baseband Frequency range in which a signal was not modulated to higher frequencies
BAF Biparametric Adaptive Filter
Beam-forming Technique to optimise the response pattern of an array of telescopes into a certain direction
BMU Best matching unit
BNN Binarized Neural Network
Birefringence Phenomenon exhibited by certain materials in which an incident ray of light is split into two rays, linear or circular polarised
Bit It measures the amount of data that is transferred in a second between two telecommunication points
Blender An open-source 3D graphics package
BNN Bayesian Neural Networks learn the posterior distribution over weights, and are thus able to quantify the network-model uncertainty
Bremsstrahlung Bremsstrahlung radiation, or braking radiation, is the radiation from deaccelerated charged particle
Byte A unit of data, measuring memory or the volume of information in a computer system, and it is eight binary digits long
C Completeness; the fraction of real sources detected as a function of signal-to-noise or flux density
CADC Canadian Astronomical Data Center
Calibration Process in data reduction to minimize the difference between model and observed visibilities
Calibrator Source which is used as a reference for the calibration of radio-interferometric data
Caesar Compact And Extend Source Automated Recognition
CARMA Combined Array for Research in Millimeter Astronomy
CARTA Cube Analysis and Rendering Tool for Astronomy
CASA Common Astronomy Software Applications
CASDA CSIRO ASKAP Science Data Archive
CD Cosmic dawn
CDFS The Chandra Deep Field-South Survey

CDN Nançay Data Centre
CDS Centre de Données astronomiques de Strasbourg
CHIME Canadian telescope
CHORD The Canadian Hydrogen Observatory and Radio-transient Detector
CIRADA Canadian Initiative for Radio Astronomy Data Analysis
Classification The task of mapping data samples to one (or more) pre-defined classes
CLEAN Algorithm to perform deconvolution of radio astronomical images
CNN Convolutional Neural Networks learn representations by assembling features in a hierarchical manner via learned filters, or kernels
COALAS CO ATCA Legacy Archive of Star-forming galaxies
COLDz VLA CO Luminosity Density at High Redshift survey
Collecting area Total area of a telescope from which it collects the received signal, e.g., the dish area
Component Radio component or blob; a flux density peak within an island typically characterised by an elliptical Gaussian
Continuum Electromagnetic spectrum characterised by bands of astrophysical emission/measurement, e.g. radio, infrared, optical, UV, X-ray, γ-ray
Controlled vocabulary A structured set of terms used to describe the concepts relevant to a particular field or domain of knowledge
Correlation Process of combining the signal of two separate antennas of an interferometer or array usually through multiplication and integration of both signals
COSMOS Cosmic Evolution Survey
Covariate shift In machine learning, this is a specific type of data-set shift, i.e. when the distribution of input data shifts between the training environment and live environment
CPU Computer's central processing unit
CRAFTS Commensal Radio Astronomy FAST Survey
CSIRO Commonwealth Scientific and Industrial Research Organisation
CTA Cherenkov Telescope Array
CTDS Casacore Table Data System
CUDA Compute Unified Device Architecture
Curvature Refers to the radio spectral index, and the shape of the source under the effects of spectral ageing, adiabatic expansion, etc.
CuTEx Curvature Threshold Extractor
CyberSKA Is a Research Software platform developed by the University of Calgary to share data and visualisations and support collaborative research around big data
CV Computer vision
CWL Common Workflow Language
DAL Data access layer
DALiuGE The Data Activated Liu Graph Engine

Data interrogation Selecting, displaying and characterising data (or a subset) from a multidimensional dataset to explore structures and relationships between data parameters/properties

DataLink DataLink protocol, IVOA standard

DCC Digital Curation Centre

DCMI Dublin Core Metadata Initiative

DDF Direction-dependent Facet

Deblending Techniques for resolving radio image islands into components. In machine learning, it is a technique to best combine model predictions from multiple contributing models

Deconvolution Process in radio-interferometric imaging to extract a model image from the dirty image

Delay Frequency-dependent shift of the correlated phase signal

Denoising In machine learning, a technique to reduce the noise in an image while preserving the information and quality. The cleaning of (high levels of) multi-component noise, foreground removal and removal of radio beam artefacts

Depolarisation Process of losing polarised signal

DESI Dark Energy Spectroscopic Instrument

DETR DEtection TRansformer

Diffeomorphism An invertible function that maps one differentiable manifold to another

Dimensionality reduction The process of reducing the dimension of a dataset while preserving the important characteristics of the data

Dirty beam Beam achieved from the Fourier transform of the uv-coverage (sampling function)

Dirty image Image of the observed sky brightness distribution convolved with the dirty beam

Dispersion A delay proportional to the observing wavelength squared, which arises from the group velocity of radio waves in a cold plasma

DL Deep learning; it designs a group of machine learning methods that rely upon artificial neural networks

DLC Dynamic Life Cycle

DM Dispersion measure, the integrated column density of the free electrons along our line-of-sight

DS9 An astronomical imaging and data visualisation application; also known as SAOImageDS9

DMP Data management plan

Downsample Or downsampling: to temporarily reduce the volume size

DOI Digital Object Identifier

DRAO Dominion Radio Astrophysical Observatory

DRSP ALMA Design Reference Science Plan

DSA-100 The Deep Synoptic Array 10-dish prototype

Dynamic range The ratio between the largest and lowest signal power, in radio images typically between the brightest and the faintest pixel value

Dynamic spectrum The intensity of the pulsed emission as a function of frequency and time

e-Science Electronic science, a term used to describe the use of advanced computing, data management, and communication technologies to enable scientific research that is data-intensive, computationally intensive, or both

EB Exabyte, 10^{18} bytes (decimal) or 1,152,921,504,606,846,976 bytes (binary)

Effelsberg Effelsberg telescope

EGO-Virgo European Gravitational Wave Observatory and the Virgo experiment

EHT Event Horizon Telescope

Einstein@Home A volunteer-distributed computing project that uses spare CPU and GPU cycles from volunteer machines to search not only for gravitational waves but also binary radio pulsars in circular orbits

Ellipticity Measure of flattening with respect to a circular or spherical geometry

ELT Extremely Large Telescope

ELM Ecological Metadata Language

EMU Evolutionary Map of the Universe

EMU-PS EMU Pilot Survey

Emulation The reproduction of the function or action of a different computer, software system; the method to make computers 'think' like humans

EoR Epoch of reionisation

EOSC European Open Science Cloud

EOEPCA Earth Observation Exploitation Platform Common Architecture

EPNCore EuroPlanet Core metadata dictionary.

EPNTAP TAP service delivering EPNCore table.

EPTA European Pulsar Timing Array

ESCAPE European Science Cluster of Astronomy & Particle physics ESFRI research infrastructures

ESA European space agency

ESFRI European Strategy Forum on Research Infrastructures

ESO European Southern Observatory

ESOReflex ESO pipeline package for reducing VLT/VLTI data

EST European Solar Telescope

EVN European VLBI Network

Facetting Process of dividing the image plane in facets and image them separately

FAIR Nuclear Physics: Facility for Antiproton and Ion Research

FAIR-DI FAIR Data Infrastructure for Physics, Chemistry, Materials Science, and Astronomy

FAIR principles A set of guidelines for making data findable, accessible, interoperable, and reusable

FAIRmat Consortium for Condensed Matter Physics and Chemical Physics of Solids, Division of Chemistry and Physics

FDF Faraday depth function

Feed Receiving unit of a radio telescope

FFA Fourier Fringe Analysis

FFT Fast Fourier Transform

FDR False detection rate. The probability that any measured source is not real
Flagging Process of marking data being bad
Flux density In astronomy usually meaning radiative flux, i.e. the amount of power radiating through a surface per unit area
FIRST Faint Images of the Radio Sky at Twenty cm survey
Findability Degree to which data can be identified and located through a standard identification mechanism, enhanced by the use of metadata
First null Closest area to the main lobe of the primary beam where the sensitivity of the telescope is zero
FITS Is a file format designed to store, transmit, and manipulate scientific images and associated data
FPGA Field Programmable Gate Array
FRB Fast radio burst
Frelled 3D rendering package
Fringes Resulting wave signal of the interference of the signal of two antennas
FUDS FAST Ultra-Deep Surveys
GAIA Global Astrometric Interferometer for Astrophysics
Gain (in the context here) visibility value in amplitude and phase
Galah GALactic Archaeology with Hermes survey
Galaxy Zoo Is a citizen science astronomy project which invites people to assist in the morphological classification of large numbers of galaxies
GALFACTS The G-ALFA Continuum Transit Survey
Galfit Galfit is a widely used galaxy parametric photometry tool
GANs Generative Adversarial networks
GB Gigabytes, 10^9 bytes (decimal) or 1,073,741,824 bytes (binary)
Gbps Gigabits per second; Billions of bits per second
GBT Green Bank Telescope
GCC Gnu Compiler Collection
GH-VFDT Gaussian Hellinger Very Fast Decision Tree
GIF Graphics Interchange Format
Gigabit One million bits
GILDAS Grenoble Image and Line Data Analysis Software
GIPSY Software package
GLT Greenland Telescope
GMRT Giant Metrewave Radio Telescope
GPU Graphics processing unit
GPS Global Positioning System
GRG Giant radio galaxy typically above $\sim$1Mpc in linear projected size
Grid computing Grid computing is a distributed architecture of multiple computers connected by networks to accomplish a joint compute-intensive task
Gridding Process of calculating the grid in which the radio image will be reconstructed
GRU Gated recurrent units
GUI Graphic User Interface
Halogas survey Hydrogen Accretion in LOcal GAlaxieS (HALOGAS) Survey

Harmonic summing Is a technique used to improve the sensitivity of Fourier domain search methods in time-domain astronomy
HBA High Band Antennas
HDF5 Hierarchical Data Format version 5
HERA Hydrogen Epoch of Reionization Array, an SKA precursor radio interferometric instrument
Hessian Hessian matrix or Hessian is a square matrix of second-order partial derivatives of a scalar-valued function, or scalar field
HiPS Hierarchical Progressive Survey
HIRAX The Hydrogen Intensity and Real-time Analysis eXperiment
HL-LHC High Luminosity LHC
Holography Technique to measure and reconstruct an electromagnetic wave
HPC High-Performance computing
HST Hubble Space Telescope
HTC High-Throughput Computing
HTRU High Time Resolution Universe
HTTP Hypertext Transfer Protocol. HTTP is an application layer protocol in the Internet protocol suite model for distributed, collaborative, hypermedia information systems
HUDF Hubble Ultra Deep Field
I/O Input/Output
ID Identification; refers to source identification
iDaVIE immersive Data Visualisation Interactive Explorer
IDIA Inter-University Institute for Data Intensive Astronomy
IDL Interactive Data Language; a programming language
IFCA International Federation of Automation Control
IGM Intergalactic medium, the physical space between galaxies
IISM Ionised interstellar medium
ILT International LOFAR Telescope
IR Infrared
IRAF Image Reduction and Analysis Facility
IRAM Institut de radioastronomie millimétrique
IRI Internationalized Resource Identifier
iRODS Integrated Rule-Oriented Data System. An open software stack which provides a rule-based interface for data management
IRQ Interquartile Range
IPTA International Pulsar Timing Array
Imaging Process of reconstructing the sky brightness distribution from the input visibilities
Interoperability Ability to exchange and reuse data by researchers, institutions, organisations, or countries because its structured in a way that can be analysed and understood both in terms of its syntax and its meaning
IMSAD Image Search and Destroy
INAF Istituto Nazionale di AstroFisica

Inference Is the process of running data points into a machine learning model to calculate an output
Interference Combination of two wave signals at every single point in space and time
Interferometer Instrument that uses interference of superimposed signals
Ionosphere Layer of the Earth's atmosphere which is mainly ionized
Island Island of radio emission; an image region determined by its flux density distribution above a given threshold
ISM Interstellar medium
Isolation forests A machine-learning technique suitable for, e.g., anomaly detection
IVOA International Virtual Observatory Alliance
JCMT James Clerk Maxwell Telescope
JIVE Joint Institute for VLBI in Europe
JSON JavaScript Object Notation
JSON-LD JSON for Linking Data
JWST James Webb Space Telescope
JWT JSON Web Tokens
Karma Is a toolkit for interprocess communications, authentication, encryption, graphics display, user interface and manipulating the Karma network data structure
Kepler Kepler space telescope
Kepler Scientific workflow system
LASSO Least Absolute Shrinkage and Selection Operator
Latency The delay before a transfer of data begins
LBA Low Band Antennas
LDAP Lightweight Directory Access Protocol. A common use of LDAP is to provide a central place to store usernames and passwords
LHC Large Hadron Collider
LINC LOFAR Initial Calibration Pipeline
LiLF Library for Low Frequencies
LMT Large Millimeter Telescope
LOF Local outlier factor; it measures how isolated the object is with respect to the surrounding neighbourhood
LOFAR Low-Frequency Array
LOFAR-EoR Low Frequency Array Epoch of Reionisation key project
LoLSS LOFAR LBA Sky Survey
LOTAAS LOFAR Tied-Array All-Sky Survey
LoTSS LOFAR two-metre sky survey, that currently maps the northern sky at 144 MHz central frequency
LSDCat Line Source Detection and Cataloguing
LSDMA Helmholtz Association project "Large Scale Data Management and Analysis"
LSTM Long short-term memories
LSST Rubin Observatory Legacy Survey of Space and Time

LTA Long Term Archive
Macaroons Macaroons are authorisation credentials that support decentralised delegation between principals
MADFM Mean Absolute Deviation From the Median (sometimes shortened to MAD)
Magnetar A highly-magnetised neutron stars
MASER Measurement, Analysis and Simulation of Emissions in the Radio range
MaxFinder A blind line detection code to automate the source finding and multi-wavelength counterpart cross-matching
Main lobe Area of the primary beam with the highest sensitivity
Maximum likelihood estimation A method for estimating parameters of a probabilistic distribution
MOC Multi-Order Coverage
Monte Carlo Dropout A realisation that the use of the regular dropout can be interpreted as a Bayesian approximation of a well-known probabilistic model
MCMC Markov chain Monte Carlo is a class of algorithm to sample from a probability distribution
Measurement equation Equation which refers the visibilites of the true sky brightness distribution to the visibilities of the observed sky brightness distribution
Medicina Medicina Radio Observatory, an astronomical observatory located 30 km from Bologna, operated by INAF
MeerKAT MeerKAT International GHz Tiered Extragalactic Exploration, also known as Meerkat Karoo Array Telescope, the predecessor for SKA-Mid situated in the Karoo desert, South Africa
MeSH Medical Subject Headings
MF Matched Filter
MF3D A 3D blind line emitter finding code with Matched Filtering technique
MIDAS Munich Image Data Analysis System
MIGHTEE MeerKAT International GHz Tiered Extragalactic Exploration, an extragalactic radio survey conducted by MeerKAT radio telescope
ML Machine learning
MLP Multi-layer perceptron
mm Millimetre
MMGPS-L MPIfR MeerKAT-L band Galactic Plane survey
MPI Message Passing Interface
MS MeasurementSet
MSA Modular Supercomputing Architecture
MTC Many Task Computing
MTMFS Multi-Term Multi Frequency Synthesis
Multi-frequency synthesis Technique to use measurements of more than one frequency to fill the uv-plane
MUSE Multi Unit Spectroscopic Explorer
MWA Murchison Widefield Array
NCBO National Center for Biomedical Ontology

NenuFAR New Extension in Nançay Upgrading LOFAR

Neural network A method in artificial intelligence that teaches computers to process data in a way that is inspired by the human brain

Neuron In artificial intelligence, neurons are interconnected nodes or neurons in a layered structure that resembles the human brain

Nextflow Workflow language

NFDI Nationale Forschungsdateninfrastruktur

ngVLA Next generation Very Large Array

NOEMA NOrthern Extended Millimeter Array, IRAM

NOMAD Laboratory The Novel Materials Discovery Laboratory

Noto Noto Radio Observatory, VLBI Station located on Sicily, operated by the Istituto di Radioastronomia di Bologna.

Novelty detection A special case of anomaly detection where the observed and detected behaviour never happened before

NRAO National Radio Astronomy Observatory

NRH Nançay Radio Heliograph

NRRD Nearly raw raster data is a library and file format for the representation and processing of n-dimensional raster data

NxGen Next Generation

Nyquist limit Nyquist frequency of the highest frequency to be recovered

OBO Open Biological and Biomedical Ontologies Foundry

OGC Open Geospatial Consortium

Ontology Is a model or representation of the entities and relationships within a particular domain of knowledge

OpenACC Open Accelerators

OpenCL Open Computing Language

OpenMP Open Multi-Processing

Optical depth Refers to a specific level of transparency, given by the observed to the emitted ratio of the radiation

ORCs Odd radio circles

ORCID The Open Researcher & Contributor ID

ORFEES Observations Radio pour FEDOME et l'Etude des Eruptions Solaires

OT ALMA Observing Tool

OTFM On-the-fly mapping

Oversampling Over sampling is used when the amount of data collected is insufficient, e.g. with the technique SMOTE

PALFA Pulsar Arecibo L-band Feed Array

PANGAEA Data Publisher for Earth & Environmental Science

Parkes Parkes telescope

PAWS Plateau de Bure Arcsecond Whirlpool Survey

Pegasus Scientific workflow system

PB Petabyte, 10^{15} bytes (decimal) or 1,125,899,906,842,624 bytes (binary)

Pb Petabits, 8 Petabits = 1 Petabyte

PCA Principal Component Analysis

PCs Principal components

PHANGS The Physics at High Angular resolution in Nearby GalaxieS survey
Phased array Array of antennas that can use beam-forming to point to a certain direction in the sky without physically moving them
PI Principle Investigator
PICS Pulsar Image Classification System
PINK Parallelized Rotation/Flipping INvariant Kohonen Maps, an unsupervised learning ML method
Pipeline An organised collection of modules designed to analyse data
Pipeline component An individual module used within pipeline to process data
Pipeline run An individual instance of a pipeline execution
Pixel A basic unit of a computer image
PMPS Parkes Multibeam Pulsar Survey
Polarisation Spatial orientation of the signal wave
Polarisation leakage Leakage of unpolarised signal into the polarisation channels
POSSUM Polarisation Sky Survey of the Universe's Magnetism
POSIX Portable Operating System Interface
PROV Standard for representing and sharing metadata that describes the origin, context, and history of digital objects, including data, documents, and software (short for provenance)
PSF Point spread function
P-Threads POSIX threads
Pulsar Pulsating star; neutron star
PyBDSF Python Blob Detector and Source Finder, for example used for source detection for LOFAR currently
PySE Python Source Extractor
Python Is a computer programming language
PV Position-Velocity diagram
PV telescope Pico Veleta IRAM 30 m telescope
Primary beam Sensitivity of the (single) telescope as a function of direction
PSRDA Pulsar Data Archive
QFitsView Is a FITS file viewer
QSO Quasi-stellar object; type of active galactic nucleus identified using optical data
Quasar Type of active galactic nucleus; nowadays similar to QSO
R Reliability; the fraction of detected sources that are real as a function of signal-to-noise or flux density
RAM Random-Access Memory, often referred to as main memory
RDM Research data management
RDF Resource Description Framework
Regression The task of finding a functional relationship within data, i.e. mapping data samples to a continuous value
ReLu Rectified linear unit
ResNet Residual neural network where skip connections between blocks of layers are used

Reproducibility Ability to replicate research findings, computational analyses, and experimental results using the same data, code, and methodology

Reusability Degree of sufficient detail of data description and sharing it with licenses that impose minimal restrictions, enabling maximum reuse and seamless integration with other data sources

RF Random Forest; a machine learning technique that's used to solve regression and classification problems

RFI Radio frequency interference; caused by transmitters at frequencies close or at the frequency bands assigned to radio astronomy

RGZ Radio Galaxy Zoo, a citizen science project

RI Research infrastructure

ROCm Radeon Open Compute platform

ROR Research Organization Registry

RM Rotation measure

RMS Root Mean Square

RNN Recurrent neural networks

RPNs Region proposal networks

RSR Redshift Search Receiver

RSRO Resident Shared Risk Observing

Rule-based anomaly Anomaly detection based on pre-defined rules, for example, given by domain experts

SAD Search and Destroy

SAGE Space Alternating Generalized Expectation Maximization

SANReN South African National Research Network

SARAO South African Radio Astronomy Observatory

SCS Simple Cone Search protocol, IVOA standard

SCUBA Submillimetre Common-User Bolometer Array, instrument used for detecting submillimetre radiation on the James Clerk Maxwell Telescope in Hawaii.

Segmentation Image segmentation is a computer vision technique used to understand what is in a given image at a pixel level

SDC1 Square-Kilometre Array Science Data Challenge 1

SDSS Sloan Digital Sky Survey

SED Spectral energy distribution is a graph of the energy emitted from an astrophysical object versus wavelength

SEFD System Equivalent Flux Density

Self-calibration Calibration routine that uses a model extracted from an image created from previously calibrated data

SExtractor Source Extractor

SF Star formation

SFG Star-forming galaxy

SFR Star-formation rate

SFIND Source Find

SIA Simple Image Access protocol, IVOA standard

Sideband In particle physics, the background dominated invariant mass spectrum

Sidelobe Area of the primary beam beyond the first null
Spectral index Change in brightness of an astronomical source as a function of frequency
SMA Submillimeter Array
SMT Submillimeter Telescope
Snakemake Workflow language
Standard model Standard Model of Cosmology (SMC) or "Concordance Cosmological Model" or the "ΛCDM Model"; it assumes the universe was created in the "Big Bang" from pure energy, and is now composed of about $\sim$5% ordinary or baryonic matter, $\sim$27% dark matter, and $\sim$68% dark energy
SKA Square-Kilometre Array
SKAO Square-Kilometre Array Observatory
SKOS Simple Knowledge Organization System
SN Signal-to-noise; also S/N
SMOTE Synthetic Minority Oversampling Technique
SNOMED CT Systematized Nomenclature of Medicine Clinical Terms
SODA Server Side Operations for Data Access protocol, IVOA standard
SoFiA Source Finding Application
Softmax output A probability of class membership for each class label
SOM Self-organising map
Source An astronomical object characterised by continuum data
SPARK Workflow management system
Spaxel A single spatial pixel
SPT South Pole Telescope
SQL Structured Query Language
Sr Steradians
SRC SKA Regional Center
Stokes Stokes parameters are a set of values that describe the polarisation state of electromagnetic radiation.
SRT Sardinia Radio Telescope
SSA Simple Spectrum Access protocol, IVOA standard
SSD Solid-state drive
SUPERB SUrvey for Pulsars and Extragalactic Radio Bursts
SVM Support vector machine
Synchrotron Electromagnetic radiation emitted when charged particles travel in curved paths; relativistic particle's bremsstrahlung radiation
System temperature Total noise power along the telescope's signal chain
TAP VO-compliant Table Access Protocol
Taverna Scientific workflow system
TaQL Table Query Language
TB Terabyte, 10^{12} bytes (decimal) or 1,099,511,627,776 bytes (binary)
TCP Transmission Control Protocol
TEC Total electron content
Terabit One trillion bits
TESS Transiting Exoplanet Survey Satellite

TFLOPS Floating Point Operations Per Second
TIFR Tata Institute of Fundamental Research, National Research Centre of the Government of India
TGSS TIFR GMRT Sky Survey
TGSSADR TGSS Alternative Data Release
Thermal noise Noise occurring in electronic devices caused by its non-zero temperature
Thresholding Selecting a range of data based on a certain threshold, mostly done during flagging
TOPCAT Tool for OPerations on Catalogues And Tables
t-SNE t-distributed Stochastic Neighbour Embedding
Uncertainty When a measurement is uncertain
U-Net Convolutional neural network with (symmetric) up- and down-sampling developed for image segmentation
Upsample Or upsampling: inserting zero-valued samples between original samples to increase the sampling rate
uv**-coverage** Points in the uv-plane covered by a particular interferometric observation
uv**-plane** Plane defined for interferometric observations, for which the axes are conventionally called u and v. It is a fixed coordinate system whose values change with pointing direction, relative location of the antennas to each other, and observational frequency
UWS Universal Worker Service protocol, IVOA standard
VAE Variational Auto-Encoder, an artificial neural network where an encoder maps to parameters of a variational distribution, from which the decoder generates data
VAST The ASKAP Variables and Slow Transients Pilot Survey
VESPA Virtual European Solar and Planetary Access
VIPERS VIMOS Public Extragalactic Redshift Survey
Visibility A measured value of an interferometer or array, typically a result of correlation of signals of two separate antennas
VLA Very Large Array
VLASPECS VLA-ALMA Spectroscopic Survey in the Hubble Ultra Deep Field
VLASS VLA Sky Survey
VUDS VIMOS Ultra-Deep Survey
VVDS VIMOS VLT Deep Survey
VLBA Very large baseline array
VLBI Very large baseline interferometry
VLT Very Large Telescope
Voxel A single data point, a single sample, in a 3-dimensional data-set/grid
VO Virtual Observatory
VR Virtual reality
WEAVE William Herschel Telescope Enhanced Area Velocity Explorer instrument

WebDAV Web Distributed Authoring and Versioning. WebDAV is a set of extensions to the Hypertext Transfer Protocol (HTTP), which provides a framework for users to create, change and move documents on a server

WIDAR Wideband Interferometric Digital ARchitecture

WISE Wide-field Infrared Survey Explorer telescope

Workflow A sequence of tasks or activities that processes a set of data which are required to achieve a specific goal or outcome

w-**projection** Converting all visibilites of an observation to the same plane of projection

WST Wavelet scattering transforms

WSRT Westerbork Synthesis Radio Telescope

XAI eXplainable Artificial Intelligence

XGBoost eXtreme Gradient Boosting

xCOLD GASS Extended CO Legacy Database for GASS

XML Extensible Markup Language

YOLO You-only-look-once methods based on artificial neural networks that are efficient for object detection

Zettabyte 10^{21} bytes (decimal) or 1,180,591,620,717,411,303,424 bytes (binary)

Zooniverse A citizen science platform at zooniverse.org

Index

E. Vardoulaki et al. (eds.), *Data-Intensive Radio Astronomy*, Astrophysics and Space Science Library 472, https://doi.org/10.1007/978-3-031-58468-8

K

L

The manufacturer's authorised representative in the EU is Springer Nature Customer Service Centre GmbH, Europaplatz 3, 69115 Heidelberg, Germany. If you have any concerns regarding our products, please contact ProductSafety@springernature.com

Printed and bound by CPI Group (UK) Ltd, Croydon, CR0 4YY
28/09/2026
02224056-0011